Modern Abstract

Algebra

Note to the Student

Dear Student,

If you winced when you learned the price of this textbook, you are experiencing what is known as "sticker shock" in today's economy. Yes, textbooks are expensive, and we don't like it any more than you do. Many of us here at PWS have sons and daughters of our own attending college, or we are attending school part time ourselves. However, the prices of our books are dictated by the cost factors involved in producing them. The costs of paper, designing the book, setting it in type, printing it, and binding it have risen significantly each year along with everything else in our economy. You might find the following table to be of some interest.

Item	1967 Price	1986 Price	The Price Increase
Monthly Housing Expense	$114.31	$686.46	6.0 times
Monthly Automobile Expense	82.69	339.42	4.1 times
Loaf of Bread	.22	1.00	4.6 times
Pound of Hamburger	.39	1.48	3.8 times
Pound of Coffee	.59	2.45	4.2 times
Candy Bar	.10	.35	3.5 times
Man's Dress Shirt	5.00	25.00	5.0 times
Postage	.05	.22	4.4 times
Resident College Tuition	294.00	1,581.00	5.4 times

Today's prices of college textbooks have increased only about 2.8 times 1967 prices. Compare your texts sometime to a general trade book, i.e., a novel or nonfiction book, and you will easily see significant differences in the internal design, quality of paper, and binding. These features of college textbooks cost money.

Textbooks should not be looked on only as an expense. Other than your professors, your textbooks are your most important source of what you hope to learn in college. Additionally, the textbooks you keep can be valuable resources in your future career and life. They are the foundation of your professional library. Like your education, your textbooks are one of your most important investments.

We are concerned and we care. Please write to us at the address below with your comments. We want to be responsive to your suggestions, to give you quality textbooks, and to do everything in our power to keep their prices under control.

Wayne Barcomb

Wayne A. Barcomb
President

PWS Publishers
20 Park Plaza
Boston, MA 02116

Modern Abstract

Algebra

David C. Buchthal
Douglas E. Cameron

The University of Akron

Prindle, Weber & Schmidt

Boston

PWS PUBLISHERS

Prindle, Weber & Schmidt • ✲ • Duxbury Press • ♠ • PWS Engineering • △ • Breton Publishers • ⊛
20 Park Plaza • Boston, Massachusetts 02116

PWS Publishers is a division of Wadsworth, Inc.

Library of Congress Cataloging-in-Publication Data

Buchthal, David C.
    Modern abstract algebra.

    Bibliography: p.
    Includes index.
    1. Algebra, Abstract. I. Cameron, Douglas E.
II. Title.
QA162.B83   1987   512'.02   86–22504

ISBN 0–87150–057–4

Printed in the United States of America

87 88 89 90 91 — 10 9 8 7 6 5 4 3 2 1

Sponsoring Editor: David Pallai
Production Coordinator: Ellie Connolly
Production: Eileen Katin
Manuscript Editor: Eileen Katin
Interior and Cover Design: Ellie Connolly
Cover Photo: Harold M. Lambert Studios, Inc./E. P. Jones Co.
Typesetting: H. Charlesworth & Co., Ltd., Huddersfield, England.
Cover Printing: New England Book Components
Printing and Binding: The Maple-Vail Book Manufacturing Group

Preface

to Instructors

This text is an outgrowth of over fourteen years of teaching courses in abstract algebra at a state university. Our objective is to present the subject of abstract algebra in a clear, straightforward manner so that students will come to learn and appreciate the major structures of algebra, such as groups, rings, and fields. A second goal is to develop students' abilities to prove results in algebra. To convey to the students the importance of algebra, we have incorporated historical perspective in the development of algebra through numerous footnotes and discussions. A number of applications are included, which present the usefulness of abstract algebra. These range from classical questions concerning euclidean constructions and polynomial solvability via radicals to more modern considerations, such as group and polynomial coding. This blending of past with present will facilitate the students' appreciation of the aesthetic qualities of abstract algebra.

This text can be best utilized in a one- or two-semester course for upper-class students. While a course in calculus could be a useful prerequisite, this is only to ensure mathematical maturity. The best preparation for this text is a variety of mathematics courses that lay the groundwork for studying the different structures we present here. However, we do not presume that the student has any previous exposure to abstract algebra.

Since we do not require any extensive experience with logic and the theory of proofs*, the text includes a quick introduction to this material in Chapter 1. The wisdom of the instructor will determine the depth in which this material is covered or assumed. On the other hand, we have made a conscious effort to get on with the main purpose of this text, the presentation

---

*We indicate the end of proofs by a box: □.

of abstract algebra. Some abstract algebra texts include so much introductory material in the first few chapters that the student sees little abstract algebra until well into the course. We have deliberately delayed the presentation of some introductory material, such as types of mappings, equivalence relations, and the proofs of certain preliminary results, until later in the text where it can be applied immediately.

A number of different tracks are available to those using this text. Instructors teaching a two-course sequence could cover almost all the material in the twenty-three chapters with the possible exceptions of the discussions of algebraic coding theory contained in Chapter 12 and in Sections 19.4 and 21.2. Some instructors prefer to discuss linear transformations in greater depth, while others want to present lattices and Boolean algebras. We have included material to permit either approach. If instructors wish to take a more contemporary approach and use modern applications of abstract algebra to computer science, they will find that Section 1.5 on switching circuits, Chapter 12 and Sections 19.4 and 21.2 on algebraic coding theory, and Section 15.4 on the construction of fast adders are especially appropriate. While we have included enough material for a two-semester sequence, we have successfully used parts of this text, namely, Chapters 1–8, 13–14, and parts of Chapters 15, 16, and 17, to teach a one-semester course in abstract algebra.

Because of the varied interests and backgrounds of those who might use this text, we have included as many different types of problems as possible. Consequently some problems stress the theory and some the applications, but all of them emphasize the knowledge and understanding of the material. For this reason, it is important that the student work as many problems as possible. None of the problems included in this text is beyond the abilities of the students for which they were written, but admittedly some are more difficult than others. We have made no effort to grade the exercises according to difficulty, and the ordering is somewhat random in each exercise set. This is due to our observation that students can be intimidated by problems labelled as difficult and consequently spend little time on these problems before quitting. We believe students should attempt each problem with the same intensity since the difficult problems often provide the deepest insight. Because many students now enter college with their own microcomputers, each chapter includes a number of problems that involve some elementary programming experience. We indicate these problems with an asterisk (∗) so instructors can assign them when appropriate.

The answer section includes hints and answers to selected exercises as well as some completely worked-out problems. While we recommend that students make a serious effort to solve the problems before turning to the answer section, students can use the hints to start on some exercises and can use the partial proofs to develop facility in writing convincing arguments.

We wish to thank a number of people who aided us in the preparation of this text. Our editor, David Pallai, provided much guidance and encouragement. We appreciate the efforts of our copy editor, Eileen Katin, who polished our prose and served capably in the role of devil's advocate, and Jerry Lyons for his contributions to the project. Numerous typists, including Pat Matthews, Rebecca DiAntonio, and Cecelia Bearer, contributed to various versions of this text. Our wives, Janice and Nancy, helped with the typing of the final manuscript and often surprised us with their prowess in algebra. To them and to our children, who endured the long process of developing this text, we offer a special thanks. We are grateful to the developer of Gutenberg, a multi-purpose programmable word processor, which we used to produce the final manuscript. We must also express our appreciation to the many reviewers whose comments and suggestions were most welcome: Joseph Adney, Michigan State University; Ezra Brown, Virginia Polytechnic Institute and State University; Carl Droms, James Madison University; Todd Feil, Denison University; Nikolas Heerema, Florida State University; John Higgins, Brigham Young University; Richard Laatsch, Miami University; Anne Ludington, Hamilton College; William H. Meeks, III, Rice University; Larry J. Morley, Western Illinois University; Edwin P. Oxford, Baylor University; and Daniel Shapiro, Ohio State University. Finally we must cite the contributions of our abstract algebra students, especially Cindy Barb, who proofread and tested earlier versions of this manuscript.

Preface

to Students

Why study abstract algebra? This is a question that we have heard since we first began teaching. An easy response dismisses the question with the reply that abstract algebra makes students think. A better answer is that there is a great deal of beauty to be encountered in the study of abstract algebra—beauty in the simplicity of the axiomatic definitions, beauty in the construction of powerful theories from simple propositions, beauty in the form of ingenious proofs to conjectures. But beauty is in the eye of the beholder, and if you never study abstract algebra, you will never have the opportunity to decide if the subject matter indeed contains elements of beauty. From a more practical standpoint, most of abstract algebra, although a field of study in itself, serves as a foundation for other areas of mathematics. Algebra is as much a tool for the mathematician as mathematics is for the engineer or physicist.

How should you use this text, and what should you expect to find as you proceed? First, no matter how simple the introductory material may seem to you, start at the beginning. We have found an increasing number of students who enroll in our algebra course have little or no experience in mathematical logic or proofs.* For this reason, the text begins with a brief look at logic and set theory so that you will understand the nature of the arguments presented in verifying results. If you already possess a strong background in these subjects, you can omit a detailed study of this material. Nevertheless, you should quickly read over the Sections 1.1–1.3 covering sets and proofs and spend some time on Section 1.4, which deals with basic results that will be used in subsequent chapters. If you have had little experience with

---

*We indicate the end of proofs by a box: □.

proofs, the time spent in carefully reading Chapter 1 will be repaid many times over when you work in later chapters. Chapter 1 contains an optional section in which we discuss the application of logic to switching circuits and give you a feel for how this material may be applied in an area other than mathematics. Chapter 2 discusses mappings and operations and analyzes sets of symmetries. This important material should not be passed over too quickly since it "sets the stage" for the discussion of groups in the next chapter. The theory of groups is developed in Chapters 3–11. You will find that quite interesting results can be derived from a few basic axioms. Chapters 13–17 deal with rings, structures that have a few more axiomatic properties than groups have. In Chapters 18–19, you will deal with vectors in a more abstract way than you have handled them previously in physics or mathematics courses. In Chapters 20–22, you will see how these various algebraic structures come together to form a beautiful construction known as Galois Theory. Finally, in Chapter 23, you will see how lattices are used in algebra and how they can be applied to illuminate some questions on switching circuits.

The introduction to each chapter will alert you to the highlights of the material to come. Pay attention to the closing paragraphs of each section, where we have provided you with insight into the connection between the material just finished and that which you will encounter next. Before you read a section in depth, look over the material. When you see how things fit together, learning new material is much easier. List the definitions and major results. You really need to have a firm grasp of the basic definitions and ideas if you hope to understand the material. Beside each definition on your list, write down at least one example. Having a sufficient supply of examples is one "almost sure" technique for mastering the subject matter. Finally, carefully read the material one more time. These suggestions have worked for our students; they should work for you!

How should you approach the exercises your instructor assigns? To start, make sure you have read the appropriate material carefully. Then review the definitions and theorems that apply to the problem, and try to determine some connection between what is given and what is required. If, after making a concentrated effort, you still find yourself hopelessly lost, check the answers and hints at the end of the text. Too many students make the mistake of taking this last step first and trying to force a proof in the direction indicated by the answer. Do not be one of them. You will only gain proficiency in proofs by trying to do them. Insight might come slowly, but it will come if you give yourself a chance.

Contents

1 | Preliminaries

The study of mathematics is basically a study of what is true (what can be done) within a framework governed by established laws or axioms and what is false (what cannot be done). In order to understand just what some person or some text is saying, you must have a basic understanding of the laws of logic. To read an argument or proof without knowing the logical requirements or the type of proof used would be wasted effort. You would be just spinning your wheels!

Then how does one acquire competency in logic and proof? Some students gain this ability by a process akin to osmosis. They take algebra and geometry in high school, read the books, listen to the instructor, and before long, find they can write a proof that will not be rejected by the teacher. Most first attempts at proofs are altered copies of similar results. Better efforts usually follow. Nevertheless, some students never overcome the hurdle in proofs. They try but somehow get lost in the midst of the effort.

There is some question of whether or not the ability to create a proof can be taught. Not everyone can be an expert architect or a polished computer programmer, and the mental processes involved in designing a complex structure or complicated computer program are much like those necessary to manufacture a proof. This is especially true if it is desired that the proof make as much sense to the person reading it as to the person who wrote it. There are, however, certain facts about logic and proof that can aid you not only in following someone else's proofs but also in designing effective arguments of your own. We will discover many of them in this chapter.

1.1

Mathematical Logic

In this first section, we will list the basic rules of logic. In the next, we will provide simple arguments to illustrate various types of proofs. Neither is intended to be completely authoritative but will be an introductory survey of the concepts that you will need for an in-depth investigation of these two topics.

Definition 1.1

A **TF statement** (true-false statement) is a statement that is either true (always true in the time and locative sense provided) or false (at least not true once in the time or locative sense given).

Definition 1.2

If p and q are TF statements, then the TF statement "p and q" is called a **conjunction**. We will denote such a statement symbolically by $p \wedge q$.

Conjunction

p	q	$p \wedge q$
T	T	T
T	F	F
F	T	F
F	F	F

Figure 1.1

The possible cases for a conjunction are represented in Figure 1.1, which is called a **truth table**. Note that since either p or q can be true or false, there are $2 \cdot 2 = 4$ possibilities for truth values. If a TF statement involves three variables, say p, q, and r, then there are $2 \cdot 2 \cdot 2 = 8$ possibilities for truth values.

Definition 1.3

If p and q are TF statements, then the TF statement "p and/or q" is called the **disjunction** of p and q and is denoted symbolically by $p \vee q$. Figure 1.2 displays the truth table for disjunction.

Definition 1.4

If p is a TF statement, then the statement "not p" is called the **negation** of p and is denoted by $\sim p$. Figure 1.3 displays the truth table for negation.

Disjunction

p	q	$p \vee q$
T	T	T
T	F	T
F	T	T
F	F	F

Figure 1.2

Negation

p	$\sim p$
T	F
F	T

Figure 1.3

Definition 1.5

If p and q are TF statements, then the TF statement "if p then q" is called a **conditional statement** and is denoted by $p \to q$.

Conditional

p	q	$p \to q$
T	T	T
T	F	F
F	T	T
F	F	T

Figure 1.4

The truth table for conditional, depicted in Figure 1.4, shows us that a conditional statement is false only when the "if" statement (called the **hypothesis**) is true and the "then" statement (called the **conclusion**) is false. This tends to be confusing, but let us clarify this point using an example.

Suppose your teacher tells you "If you get an 85 on the next test, then you will pass this course." When is this statement false; that is, under what conditions has the teacher "lied"? If you get an 85 on the test but fail the course, the statement is false. If you do not get an 85 on the test but still pass the course, the statement is true. If you do not get an 85 on the test and fail the course, the statement is true, nevertheless. If you get an 85 and pass the course, the statement is certainly true. Thus this conditional statement is false only when the hypothesis is true and the conclusion is false.

The statement "if $\cdots$" in mathematics is often confusing. Remember that "if $\cdots$" means "suppose $\cdots$" even though you might know that what you are to assume is never true. "If the moon is made of green cheese $\cdots$" is obviously false, but it should be accepted in the stated context.

The last type of TF statement we will consider is the biconditional statement.

Definition 1.6

If p and q are TF statements, then the TF statement "p if and only if q" is called a **biconditional statement** and is denoted by $p \leftrightarrow q$.

Biconditional

p	q	$p \leftrightarrow q$
T	T	T
T	F	F
F	T	F
F	F	T

Figure 1.5

The truth table for the biconditional is shown in Figure 1.5. A biconditional statement "p if and only if q" may be read "(p if q) and (p only if q)." "p if q" is the conditional statement "if q, then p," and we say that q is **sufficient** for p; that is, in order to know that p occurs, it is sufficient to know that q occurs. The statement "p only if q" is the **converse** of the conditional statement "if q then p"; that is, "p only if q" means "if p then q." In this case we say that q is **necessary** for p; that is, in order for p to occur, it is necessary for q to occur. For example, the condition "$x = 2$" is sufficient to know that "$x^2 = 4$," but it is not necessary since "$x = -2$" also guarantees that "$x^2 = 4$." In calculus we learn that continuity is necessary for differentiability but that continuity is not sufficient for differentiability. We also learn that continuity is sufficient for integrability but is not necessary for integrability.

Exercises 1.1

Determine the truth of each of the TF statements 1–13:

1. If oranges are purple, then cows are green.
2. If the sun is not shining, then the earth revolves.

3. If rain is falling water, then Babe Ruth is the President of the United States of America.

4. One plus one is not equal to two if and only if two plus two is not equal to four.

5. Cows are green if and only if horses have four legs.

6. If x is an even integer, then $3x$ is an even integer.

7. If a right triangle has two sides of equal integral length, then the third side cannot have integral length.

8. If a polygon has five sides, then a polygon is a pentagon.

9. If r is a real number and r^2 is irrational, then r is a rational number.

10. If Napoleon was French, then Winston Churchill was English.

11. If a, b, and c are real numbers and $ac > bc$, then $b > c$.

12. Whenever r is a real number such that $r^3 > 0$, then $|r| = r$.

13. If two plus five equals eight, then three plus six equals ten.

14. Find two conditional statements in the exposition of this section. Comment on the truth of these statements.

15. The negation of a statement of the form "there exists an x such that p is true" is the statement "p is false for all x." Similarly, the negation of a statement of the form "p is true for all x" is "there exists at least one x such that p is false." Find the negation of each of the following:
    a. There is a person who hates maths.
    b. $x^2 \geqslant 0$ for every integer x.
    c. Hard work is appreciated by all.
    d. Nobody cuts a mathematics class.

1.2

Tautologies and Methods of Proof

We stated earlier that the statement "p only if q" is the converse of the conditional "if q then p." There are three different statements that can be related to the conditional "if p then q."

Definition 1.7

If p and q are TF statements, then

a. the conditional $q \to p$ is the **converse** of the conditional $p \to q$;

b. the conditional $\sim p \to \sim q$ is the **inverse** of the conditional $p \to q$; and

c. the conditional $\sim q \to \sim p$ is the **contrapositive** of the conditional $p \to q$.

For example, the converse of the statement "If wishes were horses, then beggars would ride" is "If beggars would ride, then wishes would be horses." The contrapositive of the original statement is "If beggars would not ride, then wishes would not be horses" while the inverse is "If wishes were not horses, then beggars would not ride."

At this point, let us digress briefly to discuss the meaning of the terms *axiom*, *postulate*, *theorem*, and *corollary*, which we use in our study. The terms **axiom** and **postulate** refer to facts that are to be accepted without proof, and in fact, for which there may be no proof. **Theorem** is a term for a fact that has been proved or verified. A **lemma** is a theorem whose primary purpose is to aid in the proof of another result. A **corollary** is a theorem which follows immediately from the proof of another theorem or which may have been proven during the proof of that result.

Definition 1.8

> A TF statement that has truth value T in every possible instance is called a **tautology**.

To determine whether or not a TF statement is a tautology, one usually needs to construct a truth table.

Theorem 1.1

The following conditional and biconditional statements are tautologies:

a. $[p \leftrightarrow q] \leftrightarrow [(p \rightarrow q) \wedge (q \rightarrow p)]$
b. $(p \rightarrow q) \leftrightarrow [(\sim q) \rightarrow (\sim p)]$ (Law of Contraposition)
c. $\sim(\sim p) \leftrightarrow p$ (Law of Double Negation)
d. $(p \rightarrow q) \leftrightarrow [(\sim p) \vee q]$
e. $[p \rightarrow (q \wedge r)] \leftrightarrow [(p \rightarrow r) \wedge (p \rightarrow q)]$
f. $[(p \wedge q) \rightarrow r] \leftrightarrow [p \rightarrow (q \rightarrow r)]$
g. $[\sim(p \wedge q)] \leftrightarrow [(\sim p) \vee (\sim q)]$
h. $[\sim(p \vee q)] \leftrightarrow [(\sim p) \wedge (\sim q)]$
i. $[\sim(p \rightarrow q)] \leftrightarrow [p \wedge (\sim q)]$
j. $[\sim(p \leftrightarrow q)] \leftrightarrow [(p \wedge (\sim q)) \vee ((\sim p) \wedge q)] \leftrightarrow [p \leftrightarrow (\sim q)]$
k. $[p \wedge (p \rightarrow q)] \rightarrow q$ (Law of Detachment)
l. $[(p \rightarrow q) \wedge (q \rightarrow r)] \rightarrow [p \rightarrow r]$ (Law of Syllogism)

Proof

We will demonstrate the proofs for a and f and leave the verification of the others to the exercises (Exercise 3). The truth tables in Figures 1.6 and 1.7 show that a and f, respectively, are tautologies. Since each entry in the last column in each table has value "true," the two statements are tautologies. ☐

p	q	$p \rightarrow q$	$q \rightarrow p$	$(p \rightarrow q) \wedge (q \rightarrow p)$	$p \leftrightarrow q$	$[p \leftrightarrow q] \leftrightarrow [(p \rightarrow q) \wedge (q \rightarrow p)]$
T	T	T	T	T	T	T
T	F	F	T	F	F	T
F	T	T	F	F	F	T
F	F	T	T	T	T	T

Figure 1.6

p	q	r	$p \wedge q$	$(p \wedge q) \rightarrow r$	$q \rightarrow r$	$p \rightarrow (q \rightarrow r)$	$[(p \wedge q) \rightarrow r] \leftrightarrow [p \rightarrow (q \rightarrow r)]$
T	T	T	T	T	T	T	T
T	T	F	T	F	F	F	T
T	F	T	F	T	T	T	T
T	F	F	F	T	T	T	T
F	T	T	F	T	T	T	T
F	T	F	F	T	F	T	T
F	F	T	F	T	T	T	T
F	F	F	F	T	T	T	T

Figure 1.7

We are now ready to discuss some of the methods of proof that you will encounter during the course of this book.

1. **Direct Proof:** To prove $p \rightarrow q$ is true (that is, $p \rightarrow q$ is a tautology), the straightforward method is to assume that p is true and then show that q must be true.

Suppose you wished to prove the statement "If x is an even integer, then x^2 is an even integer." Since an even integer is one that is divisible by 2 or, equivalently, is a multiple of 2, the statement that x is even says that $x = 2n$ for some integer n. But then $x^2 = 4n^2$ is also a multiple of 2, so that x^2 is even. Thus the proof is complete.

But now suppose you wanted to prove "If x is an integer such that x^2 is odd, then x is an odd integer." In an attempt to prove this statement, we recall that an odd integer is one expressible in the form $x = 2n + 1$ for some integer n. To complete the proof, you might try to take the square root of $2n + 1$, but that seems impossible.

To get out of this dilemma, let us consider another method of proof:

2. **Proof by Contraposition:** In proving $p \rightarrow q$ is true, it is sometimes easier to use the Law of Contraposition of Theorem 1.1 and prove that $(\sim q) \rightarrow (\sim p)$ is true.

In trying to prove "If x is an integer such that x^2 is odd, then x is an odd integer" we have already seen that the direct method of proof appears impossible. Let us therefore consider proof by contraposition. If an integer is not odd, then it must be even. Thus the contrapositive of our statement is "If x is an even integer, then x^2 is an even integer." But this statement was already proved true by the direct method. Thus the result follows.

   Warning: Sometimes people mistakenly assume that if the statement $p \rightarrow q$ is true, then the converse $q \rightarrow p$ is also true. For example, in calculus it is shown that "differentiability implies continuity." Many students then assume that "continuity implies differentiability," but you know that this is not true. (For example, the function $f(x) = |x|$ is continuous for all values of x but is not differentiable at $x = 0$.) In general, if a conditional statement is true, one must not assume the converse is true—it must be verified.

   There is a third type of proof:

   3. **Proof by Contradiction** (*reductio ad absurdum*): To prove the implication $p \rightarrow q$, one can show that $[p \wedge (\sim q)] \rightarrow (\sim p)$.

   This latter statement $\sim p$ is obviously wrong since p cannot be both true and false at the same time. This method of proof follows from the tautology (Exercise 5)

$$[p \rightarrow q] \leftrightarrow [(p \wedge (\sim q)) \rightarrow (\sim p)]$$

   Let us use proof by contradiction to prove that "If x is an integer such that x^2 is even, then x is even." We assume that x^2 is even and the statement "x is even" is false, so that x is odd. But then $x = 2n + 1$ for some integer n and

$$x^2 = (2n + 1)^2 = 4n^2 + 4n + 1 = 2(2n^2 + 2n) + 1$$

which is an odd integer. This contradicts the assumption that x^2 is even. Therefore the statement "x is even" cannot be false and, hence, x must be even. The original statement is thus true.

   We close this section with a few comments. Often a TF statement p is a conjunction. If this is the case, then "p is false" means that at least one part of the conjunction is not true. This follows from the tautology $[\sim(p \wedge q)] \leftrightarrow [(\sim p) \vee (\sim q)]$. Also, to prove a result, it is not sufficient to show that a statement is true in a specific case. Specifics do not guarantee generalities; an example is not a proof. However, if you desire to show that a result is not (always) true, then you can do this by producing a single counterexample, that is, one instance in which the hypothesis is true and the conclusion false. For example, suppose you were asked to prove "if a function is continuous, then it is differentiable." You point out that $f(x) = |x|$ is continuous at $x = 0$ but not differentiable there. Therefore you have found a counterexample and have shown the statement to be wrong.

Exercises 1.2

1. State the converse, inverse, and contrapositive for the following statements:
   a. If cows are green, then apples are not fruit.
   b. If $1 + 1 = 2$, then mathematics is fun.
   c. If the sun shines, then green grass grows and the sky is blue.

d. If a, b, and c are the lengths of the sides of a right triangle and c is the length of the hypotenuse, then $c^2 = a^2 + b^2$.

e. If the cost of gasoline rises, then fewer people will drive or the number of high-mileage cars sold will increase.

2. Find the truth values of the variables that make the following statements true:

a. $(p \wedge q) \to \sim q$

b. $[(\sim p) \vee (\sim q)] \leftrightarrow \sim (p \vee q)$

c. $[p \wedge q \wedge r] \to [(p \vee r) \wedge q]$

d. $[p \wedge (\sim r)] \to [q \vee r]$

3. Prove that the following conditionals and biconditionals are tautologies:

a. $(p \to q) \leftrightarrow [(\sim q) \to (\sim p)]$

b. $[p \wedge (p \to q)] \to q$ (Law of Detachment)

c. $[(p \to q) \wedge (q \to r)] \to [p \to r]$ (Law of Syllogism)

d. $[\sim (\sim p)] \leftrightarrow p$ (Law of Double Negation)

e. $(p \to q) \leftrightarrow [(\sim p) \vee q]$

f. $[p \to (q \wedge r)] \leftrightarrow [(p \to q) \wedge (p \to r)]$

g. $[\sim (p \wedge q)] \leftrightarrow [(\sim p) \vee (\sim q)]$

h. $[\sim (p \vee q)] \leftrightarrow [(\sim p) \wedge (\sim q)]$

i. $[\sim (p \to q)] \leftrightarrow [p \wedge (\sim q)]$

j. $[\sim (p \leftrightarrow q)] \leftrightarrow [(p \wedge (\sim q)) \vee ((\sim p) \wedge q)] \leftrightarrow [p \leftrightarrow (\sim q)]$

4. Give an example other than the one in the text to show that a conditional and its converse do not necessarily have the same truth value.

5. Prove the tautology for *reductio ad absurdum*; that is, show

$$[p \to q] \leftrightarrow [(p \wedge (\sim q)) \to (\sim p)]$$

is a tautology.

1.3

Set Theory

The study of algebra is an investigation of different mathematical systems, their properties, and their applications. These systems, such as the integers, real numbers, and various collections of geometrical symmetries, have basic underlying structural (algebraic) similarities. It is these similarities that we will study in this text.

Any mathematical system is composed of a collection of objects. Therefore, it is important to review the basic ideas of set theory. Developed by Georg Cantor[1], set theory is the study of collections of objects (elements) and their properties; together these form the building blocks of mathematics.

---

[1]Georg Cantor (1845–1918) was the Russian-born son of Danish parents who was reared in Germany. Among his many accomplishments was the proof of the countability of the rational numbers and the proof of the non-countability of the reals.

This section is intended as a brief review of set theory and as an introduction to the notation that we will use throughout this text. For this reason, we will state our results without proof and present no examples. If you are familiar with the use of sets, it is likely that you will need only to skim this section and proceed. If you have not studied set theory previously, read this section carefully and do the exercises.

A **set** is a mathematician's term for a well-defined collection of distinct objects, which are called **elements**. The terminology *well-defined* means that the contents (elements) of the set are such that we can easily inspect an object and determine whether or not that object is an element of the set. Some examples are the set of real numbers, the set of rational numbers, or the set of students enrolled in your class.

The standard mathematical symbol used to represent a set is a capital letter:

A, P, X

The elements or members of a set are denoted by small letters:

a, p, x

If the object b is an element of the set A, we will write $b \in A$. If the element b is not an element of A, we write $b \notin A$.

There are several sets that we will be using throughout this book. We will use the following special symbols to denote these sets:

$\mathbb{R}$: the set of all real numbers;
$\mathbb{Q}$: the set of all rational numbers;
$\mathbb{Z}$: the set of all integers;
$\mathbb{N}$: the set of all natural numbers, also known as $\mathbb{Z}^+$, the positive integers;
$\mathbb{C}$: the set of all complex numbers.

In standard notation, we enclose the elements of a set by using curly braces, $\{$ and $\}$. For example, $A = \{a, b, c\}$ is the set containing the elements $a, b,$ and c. A set may also be described by the use of a representative element and a defining condition. To represent the above set A in this form, we would write

$A = \{x \mid x$ is one of the first three letters of the alphabet$\}$

The vertical bar $\mid$ is read "such that." Thus the set A is the set of all x such that x is one of the first three letters of the alphabet.

$\emptyset = \{\ \}$ is the **empty set** (alternately known as the **null set** or the **void set**), the set which contains no elements.

Two sets A and B are **equal**, written $A = B$, if they have precisely the same elements. Thus, two sets are equal if and only if $x \in A$ implies $x \in B$ and $x \in B$ implies $x \in A$.

In a particular discussion, if all the elements being considered are of the same set X, then X is called the **universal set**.

If A is a set containing only a finite number of distinct elements, we denote the number of elements in A by $|A|$. The number of distinct elements in a set A is known as the **cardinality of** A.

A set B is a **subset** of a set A if every element of B is an element of A; that is, if $x \in B$ then $x \in A$. The notation used is $B \subseteq A$ (B is a subset of A or B is contained in A or A contains B). If B is not a subset of A, we will write $B \nsubseteq A$.

For a given set X, $\mathscr{P}(X)$ is the set of all subsets of X and is called the **power set of** X. In mathematical notation, we write $\mathscr{P}(X) = \{B | B \subseteq X\}$.

A set B is a **proper subset** of a set A if B is a subset of A, but $A \neq B$ (A contains at least one element that is not in B). The notation used is $B \subset A$. A subset B of A is termed **improper** if $B = A$, although some mathematicians say that the empty set is also an improper subset of any set.

The following proofs can be easily derived from the definitions and can be done in Exercise 4.

Theorem 1.2

The empty set is a subset of every set.

Corollary 1.1

The empty set is a proper subset of every set except itself.

It is important to be able to construct new subsets from those already given or obtained. The next three definitions describe methods for doing this.

Definition 1.9

The **union** of two sets A and B is the set

$$A \cup B = \{x | x \in A \text{ or } x \in B\}$$

In other words, the union of two sets A and B is the set of all elements that are either in A or in B or in both.

Definition 1.10

The **intersection** of two sets A and B is the set

$$A \cap B = \{x | x \in A \text{ and } x \in B\}$$

That is, the intersection of two sets is the set of all elements that the sets have in common.

Definition 1.11

A and B are **disjoint** if their intersection is the empty set; that is, $A \cap B = \varnothing$.

Definition 1.12

The **difference**, $A - B$, **of two sets** A **and** B is the set

$$A - B = \{x \mid x \in A \text{ and } x \notin B\}$$

Definition 1.13

If X is the universal set and $A \subseteq X$, then the set difference $X - A$ is called the **complement of** A **(in** X**)**.

Properties of union, intersection, and set difference are listed below.

Theorem 1.3

If A, B, and C are arbitrary sets, then the following hold:

a. $A \subseteq B$ if and only if $A \cup B = B$
b. $A \subseteq B$ if and only if $A \cap B = A$
c. $A \cap (B \cup C) = (A \cap B) \cup (A \cap C)$
d. $A \cup (B \cap C) = (A \cup B) \cap (A \cup C)$
e. $A \cup B = B \cup A$
f. $A \cap B = B \cap A$
g. $A \cap (B \cap C) = (A \cap B) \cap C$
h. $A \cup (B \cup C) = (A \cup B) \cup C$
i. $A \cap (A \cup B) = A$
j. $A \cup (A \cap B) = A$

Parts i and j of this theorem are referred to as **absorption laws**.

Theorem 1.4

For any two sets A and B, the following properties of set difference (complementation) are true:

a. $A \cap (B - A) = \emptyset$
b. $A \cup (B - A) = B \cup A$
c. $A - B = A - (A \cap B)$

A useful concept in set theory is the idea of the partition of a set.

Definition 1.14

A **partition** of a nonempty set X is a collection $\{A_1, A_2, A_3, ...\}$ of nonempty subsets of X such that

a. $X = A_1 \cup A_2 \cup A_3 \cup ...$, and
b. for $i \neq j$, either $A_i = A_j$ or $A_i \cap A_j = \emptyset$.

In other words, a partition of X is a collection of nonempty subsets of X whose union is X and such that unequal subsets of this collection are disjoint. If $\varnothing \subset A \subset X$, then A and $X - A$ form a partition of X.

In calculus, graphing a function as a collection of points in the euclidean plane involves a special case of the next and last definition of the section.

Definition 1.15

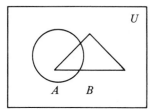

Figure 1.8

The **Cartesian product of the sets A and B** is the set

$$A \times B = \{(a, b) \mid a \in A \text{ and } b \in B\}$$

Thus $A \times B$ consists of all ordered pairs of elements the first of which is from A and the second of which is from B. The **Cartesian product of the sets A_1, A_2, ..., A_n** is the set

$$\prod_{i=1}^{n} A_i = \{(a_1, a_2, \ldots, a_n) \mid a_i \in A_i, i = 1, 2, \ldots, n\}$$

If $A = A_i$ for all i, we will write $A^n = \prod_{i=1}^{n} A_i$.

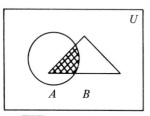

(a) ▨ $= A \cap B$ and $B \cap A$

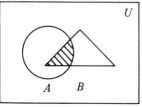

(b) ▧ $= A \cap B$

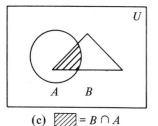

(c) ▨ $= B \cap A$

Figure 1.9

As you attempt to prove your first theorem, you may not know quite how to begin or you may not even believe the result that you are setting out to verify. A handy creation that helps you visualize the set relations under investigation in this chapter is a **Venn diagram**[2]. This is a pictorial representation of the relationships between sets.

For example, if you want to verify that set intersection is commutative or that $A \cap B = B \cap A$ for any two sets A and B (Theorem 1.3f), you could draw the Venn diagram shown in Figure 1.8. The rectangle (labeled U) represents the universal set, the circle represents the set A, and the triangle represents the set B. To see whether or not the result is true, we proceed by shading the part of the diagram that represents $A \cap B$ and the portion that represents $B \cap A$. You may wish to do this on the same diagram (Figure 1.9a) or in two separate diagrams (Figures 1.9b and 1.9c). In either case, the result may be considered true if the shaded areas coincide. *Warning:* This does not mean that the theorem is true but only that it holds in your particular case.

The problem is not in the representation of A as a circle and B as a triangle but simply that there are many different relationships which the sets A and B might have. For example, maybe $A \subseteq B$ (Figure 1.10a) or $A \cap B = \varnothing$ (Figure 1.10b). To actually have a valid proof by Venn diagrams, you would have to represent all possible cases. The real value of Venn diagrams is in seeing that a result may be true (Exercise 11) or to provide a counterexample

[2] Venn Diagrams were developed by John Venn in 1880.

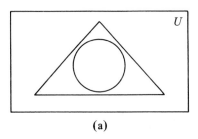
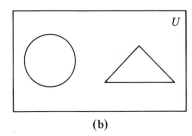

Figure 1.10   (a)   (b)

if your diagram shows that the result is not true. It is important to remember that a proof of a result by itemization (or listing of all possible cases) is valid only if you have listed all possible cases and verified in each the result you are trying to obtain.

Exercises 1.3

1. If $A = \{1, 3, 5\}$, $B = \{2, 6, 7, 8\}$, $C = \{1, 3, 5, 6\}$, $D = \{2, 4, 7, 9, 10\}$, and $X = \{1, 2, 3, 4, 5, 6, 7, 8, 9, 10\}$, find the following:
   a. $A \cap B$ b. $(D \cup C) \cup A$
   c. $A \cup C$ d. $B \cap (C \cup D)$
   e. $A \cap C$ f. $(B \cap C) \cup (B \cap D)$
   g. $A - C$ h. $B - C$
   i. $X - A$ j. $X - (A \cup B)$

2. For A, B, and C as given in Exercise 1, show that
   a. $A \cup (B \cap C) = (A \cup B) \cap (A \cup C)$ b. $A \cap (B \cup C) = (A \cap B) \cup (A \cap C)$
   c. $A \cup B = B \cup A$ d. $A \cap C = C \cap A$
   e. $A \cap (B \cap C) = (A \cap B) \cap C$ f. $A \cup (B \cup C) = (A \cup B) \cup C$

3. Let $X = \{a, b, c\}$, $Y = \{1, 2\}$, $Z = \{I, II\}$. Find
   a. $X \times Y$ b. $Y \times Z$ c. $Y \times X$
   d. $X \times Y \times Z$ e. $Z \times X \times Y$

4. Prove Theorem 1.2 and Corollary 1.1.

5. Use a proof by contradiction to verify that there is only one empty set.

6. Use the definition of subset to prove the statements in Theorem 1.3.

7. For any sets A and B, show
   a. $A \subseteq (A \cup B)$ and $B \subseteq (A \cup B)$ b. $(A \cap B) \subseteq A$ and $(A \cap B) \subseteq B$

8. Prove De Morgan's Laws[3]: If A and B are subsets of X, then
   a. $X - (A \cap B) = (X - A) \cup (X - B)$
   b. $X - (A \cup B) = (X - A) \cap (X - B)$

---

[3]Augustus De Morgan (1886–1871), who helped found the British Association for the Advancement of Science in 1831, refused to submit to the required religious test at Oxford and Cambridge and could not obtain a teaching position. Instead he found a position at London University, later known as University College of the University of London. He was one of the first to use letters and symbols for numbers to aid in the abstraction of mathematics.

9. How many partitions of the set A are there if (a) $|A| = 3$, or (b) $|A| = 4$?

10. Let X be the set of all triangles. For each triangle $x \in X$, let $T(x)$ be the set of all triangles similar to x. Is $\{T(x) | x \in X\}$ a partition of X?

11. The operation to be defined in this exercise and denoted by $\triangle$ is called the **symmetric difference of A and B**. If A and B are subsets of a set X, define

$$A \triangle B = (A - B) \cup (B - A)$$

(It might help to draw a Venn diagram.) Prove:
a. $A \triangle B = B \triangle A$
b. $A \triangle B = (A \cup B) - (A \cap B)$
c. If $C \in \mathscr{P}(X)$, $A \triangle (B \triangle C) = (A \triangle B) \triangle C$.

1.4

Fundamental Results

To be able to prove some of our early results as well as provide examples illustrating them, we will state a number of theorems in this section. At a later point in this text, we will provide you with proofs for them. Rest assured that our later proofs will not rely upon the results that we present now; that is, we will not use circular reasoning. Our motive for listing these theorems now is pedagogical. These results are needed, but we doubt that the long involved proofs would be appreciated at this time.

Most of our basic results are statements of "facts" that have been utilized throughout your study of mathematics. Recall that $p \in \mathbb{N}$ is **prime** if $p > 1$, and the only divisors of p in $\mathbb{N}$ are 1 and p.

The Fundamental Theorem of Arithmetic

For any natural number $n > 1$, there exist primes $p_1, p_2, \ldots, p_m$ and natural numbers $e_1, e_2, \ldots, e_m$ such that

$$n = p_1^{e_1} p_2^{e_2} \cdots p_m^{e_m}$$

where the factorization is unique up to a rearrangement of the prime powers.

For example, $40 = 2^3 \cdot 5$, which we may also write as $5 \cdot 2^3$ (a rearrangement of the prime powers). Of course this theorem is an important one in the study of algebra and arithmetic as it provides the opportunity for simplification of fractions leading to a unique result. Imagine the difficulties that would be encountered if a fraction such as 24/14 had two different simplified forms!

Another important result involves the ability to divide one integer by another in a unique manner.

The Division Algorithm for $\mathbb{Z}$

If $a, b \in \mathbb{Z}$, $b \neq 0$, then there exist unique $q, r \in \mathbb{Z}$ such that $a = bq + r$, where $0 \leqslant r < |b|$.

For example, if $a = 22$ and $b = 6$, then $22 = 6 \cdot 3 + 4$; so $q = 3$ and $r = 4$. If $a = 14$ and $b = -4$, then $14 = (-4) \cdot (-3) + 2$; so $q = -3$ and $r = 2$. When $a = -27$ and $b = 4$, we have $-27 = 4(-7) + 1$.

A third basic result concerns the concept of greatest common divisor. Recall that a positive integer d is the **greatest common divisor** of $a, b \in \mathbb{Z}$, where both are not equal to 0, if:

    a. d divides both a and b, and
    b. c divides both a and b, then c divides d.

The greatest common divisor of a and b is usually denoted by (a, b). Although this notation is the same as that for the point (a, b) in the Cartesian product, the context of the discussion will usually indicate which concept is involved.

Using this notation, we can list the following greatest common divisors:

$(4, 6) = 2$

$(39, -63) = 3$

$(15, 8) = 1$

When two integers have greatest common divisor 1, we say that the integers are **relatively prime**. An interesting connection between two integers and their greatest common divisor is given by the next result.

Theorem 1.5

If $a, b \in \mathbb{Z}$, where both are not equal to 0, then there exist integers s and t such that $(a, b) = sa + tb$.

As examples of this theorem, note that

$(4, 6) = 2 = (-1)(4) + (1)(6)$

$(39, 63) = 3 = (-8)(39) + (-5)(-63)$

$(15, 8) = 1 = (-1)(15) + (2)(8)$

It is not always easy to find the values of s and t that enable you to write the greatest common divisor in terms of a and b. Happily there is a result, known as the **Euclidean Algorithm**, that gives a systematic way to calculate s and t. We will consider it at a later time. For our purposes now, it is enough to know that such numbers do exist.

The proofs of many results involve one of the basic concepts in mathematics—the Principle of Mathematical Induction. For many students, mathematical induction is one of the great mysteries of mathematics. They can learn to use it but sometimes still doubt its validity. In actuality, there are things in mathematics, such as the natural numbers, whose existence simply must be assumed. In mathematics, the natural numbers are defined in terms of axioms (one example being the Peano Postulates[4]), and these axioms are used to derive the basic properties of the natural numbers (such as commutativity of addition). One of these basic axioms is equivalent to the following:

The Principle of Mathematical Induction

Let $P(n)$ be a statement involving a natural number n. If

   a. $P(1)$ is true (the statement is valid for $n = 1$), and
   b. for each $k \in \mathbb{N}$, $P(k + 1)$ is true whenever $P(k)$ is true,

then $P(n)$ is true for all $n \in \mathbb{N}$.

We want to emphasize that this statement is an axiom of the natural numbers and, as such, does not require a proof. Later we will state a result that is logically equivalent to the Principle of Mathematical Induction and is easier for the average student to accept. For now, concentrate on the following example.

| Example 1.1 |

The sum of the first n natural numbers is $[n(n + 1)]/2$; that is,

$$1 + 2 + 3 + \cdots + n = \frac{n(n + 1)}{2}$$

To prove that this statement is true for all natural numbers, first rewrite it to permit the use of mathematical induction. Let $P(n)$ be the statement

$$1 + 2 + \cdots + n = \frac{n(n + 1)}{2}$$

The statement $P(1)$ states that $1 = [1(1 + 1)]/2$, which is certainly true.

Now we need to verify part b. To do this, assume that $P(k)$ is true for

---

[4]Giuseppe Peano (1858–1932) was an Italian mathematician who was one of the foremost contributors to the arithmetization of mathematics. He reduced the theory of natural numbers to the smallest set of suppositions and undefined terms from which the theory could be defined. His work appeared in 1889. Among his students was Bertrand Russell.

some $k \in \mathbb{N}$ so that

$$1 + 2 + \cdots + k = \frac{k(k+1)}{2}$$

We want to prove that $P(k+1)$ is true; that is,

$$1 + 2 + \cdots + k + (k+1) = \frac{(k+1)[(k+1)+1]}{2} = \frac{(k+1)(k+2)}{2}$$

Now since $1 + 2 + \cdots + k = [k(k+1)]/2$ by hypothesis, we know that

$$1 + 2 + \cdots + k + (k+1) = (1 + 2 + \cdots + k) + (k+1)$$

$$= \frac{k(k+1)}{2} + (k+1)$$

$$= (k+1)\left(\frac{k}{2} + 1\right)$$

$$= (k+1)\frac{k+2}{2}$$

$$= (k+1)\frac{(k+1)+1}{2}$$

But this is what we wanted to prove. The result then follows for all n by the Principle of Mathematical Induction.

There is a story about the renowned nineteenth century mathematician Carl Friedrich Gauss involving the equality of this example. Supposedly, he and his young classmates were too raucous for the teacher who, in order to gain a respite, ordered his charges to find the sum of the first 100 integers. The students had only slates and chalk with which to calculate. You can imagine the teacher's surprise when, within a few minutes, Gauss laid his slate on the teacher's desk with the answer "5050." Gauss had discovered the formula we proved by induction by observing that if

$$S = 1 + 2 + \cdots + 99 + 100$$

then

$$S = 100 + 99 + \cdots + 2 + 1$$

so that

$$2S = 101 + 101 + \cdots + 101 + 101 \ \ (100 \text{ times})$$

Therefore

$$2S = (100)(101) \quad \text{and} \quad S = \frac{(100)(101)}{2} = 5050$$

Such are the stories in the folklore of mathematics.

Returning to our discussion of induction, consider another example.

Example 1.2

Prove that if n is a natural number, then $n^3 - n$ is a multiple of 3.

To prove this result by induction, let $P(n)$ be the statement that $n^3 - n$ is a multiple of 3. Clearly $P(1)$ is true, since $1^3 - 1 = 0 = 0 \cdot 3$. Suppose that $P(k)$ is true; that is, that $k^3 - k = 3m$ for some $m \in \mathbb{Z}$. Then

$$(k+1)^3 - (k+1) = k^3 + 3k^2 + 3k + 1 - k - 1$$
$$= (k^3 - k) + (3k^2 + 3k)$$
$$= 3m + 3(k^2 + k)$$
$$= 3(m + k^2 + k)$$

which is a multiple of 3, and thus $P(k+1)$ is true. By the Principle of Mathematical Induction, we conclude that $P(n)$ is true for all $n \in \mathbb{N}$.

We stated earlier that the Principle of Mathematical Induction was equivalent to an axiom of the natural numbers. As such, it cannot be proved; more precisely, it cannot be proved without accepting another premise, one which, in fact, is itself mathematically equivalent to the Principle of Mathematical Induction. Such is the following:

Well-Ordering Property for $\mathbb{N}$

Every nonempty subset of the natural numbers $\mathbb{N}$ has a least element; that is, if $U \subseteq \mathbb{N}$ and $U \neq \varnothing$, then there exists $m \in U$ such that $m \leqslant u$ for each $u \in U$.

Most students have very little trouble accepting the Well-Ordering Property. It seems to be a matter of common sense. Nevertheless it is equivalent to the Principle of Mathematical Induction. By that we mean that you can use the Well-Ordering Property to prove the Principle of Mathematical Induction, and conversely. Before we demonstrate this, however, let us consider some examples of how the Well-Ordering Property is used in proofs.

Recall that we proved by induction

$$1 + 2 + \cdots + n = \frac{n(n+1)}{2}$$

for all $n \in \mathbb{N}$. Let us establish this same result by the Well-Ordering Property.

Set $U = \left\{ n \in \mathbb{N} \mid 1 + 2 + \cdots + n \neq \frac{n(n+1)}{2} \right\}$. We want to show that $U = \varnothing$. Let's suppose not. Then U contains a least element m such that

$$1 + 2 + \cdots + m \neq \frac{m(m+1)}{2}$$

Since m is the least element of U, $(m - 1) \notin U$. Clearly $m > 1$ since $1 \notin U$, so that

$$1 + 2 + \cdots + (m - 1) = (m - 1)\frac{(m - 1) + 1}{2}$$

$$= \frac{(m - 1)m}{2}$$

Then

$$1 + 2 + \cdots + m = 1 + 2 + \cdots + (m - 1) + m$$

$$= \frac{(m - 1)m}{2} + m$$

$$= m\left(\frac{m - 1}{2} + 1\right)$$

$$= m\frac{(m - 1) + 2}{2}$$

$$= \frac{m(m + 1)}{2}$$

contradicting the assumption that $m \notin U$. Therefore $U = \emptyset$, and the result follows.

As a second example, let us show that every natural number greater than 1 is either prime or a product of primes. Let us set U equal to the set of natural numbers greater than 1 that are neither prime nor a product of primes. We claim that $U = \emptyset$. If $U \neq \emptyset$, then the Well-Ordering Property states that U contains a least element m. By the definition of U, $m \in U$ implies that m is not a prime. Therefore $m = m_1 m_2$ where $m_1, m_2 \in \mathbb{N}$, $1 < m_1 < m$, $1 < m_2 < m$. Since m is the least element of U, $m_1 \notin U$ and $m_2 \notin U$. Therefore both m_1 and m_2 are either primes or products of primes. This forces $m = m_1 m_2$ to be a product of primes, contrary to the fact that $m \in U$. This contradiction implies that U must be the empty set, and the result follows.

What we have just done is to use the Well-Ordering Property to prove *part* of the Fundamental Theorem of Arithmetic. The uniqueness of the factorization will be shown later.

Our first example, following the statement of the Well-Ordering Property, was a result that can be proved using either induction or the Well-Ordering Property. This is not surprising since, as indicated, they are equivalent.

Theorem 1.6

The Well-Ordering Property implies the Principle of Mathematical Induction.

Proof

Let $P(n)$ be a sequence of statements such that $P(1)$ is true, and whenever k is natural number such that $P(k)$ is true, then $P(k + 1)$ is true. Let

$$U = \{n \in \mathbb{N} \,|\, P(n) \text{ is false}\}$$

We must show that $U = \emptyset$. Suppose not; then U has a least element m, and since $P(1)$ is true, $m > 1$.

Now $(m - 1) \notin U$ by the fact that m is the least number in U. Thus $P(m - 1)$ is true. But then $P((m - 1) + 1) = P(m)$ must be true by our induction assumption, so $m \notin U$. This contradiction forces $U = \emptyset$, and the result follows. $\qquad \square$

We leave the proof of the converse of Theorem 1.6, as well as some variations on the Principle of Mathematical Induction, to the exercises. In the chapters ahead, we will often use the results obtained in this chapter to establish properties of the various algebraic structures we investigate. However mathematical induction has many applications far removed from the domain of abstract algebra. In computer science, the "DO" and "FOR" statements and program verification both rely on the concept of mathematical induction. For example, if you write a program that works for $n = 1$, and you can show that whenever it works for $n = k$, it also works for $n = k + 1$, then you can conclude that it works for all $n \in \mathbb{N}$. In your future studies, whether in mathematics, statistics, natural science, or computer science, look for instances where induction is or could be used. Your knowledge of the Principle of Mathematical Induction should aid in your understanding of that material.

Exercises 1.4

1. Find the prime factorization of each of the following:
    a. 84   b. 269   c. 861   d. 1028   e. 4122

2. Find the greatest common divisor of each pair of integers:
    a. (84, 16)   b. (−27, 36)   c. (8, 48)   d. (125, 63)

3. Use the definition of greatest common divisor to show that the greatest common divisor of two integers is unique.

4. Use the Principle of Mathematical Induction to prove each of the following for all $n \in \mathbb{N}$:

    a. $1^2 + 2^2 + 3^2 + \cdots + n^2 = \dfrac{n(n + 1)(2n + 1)}{6}$

    b. $1^3 + 2^3 + 3^3 + \cdots + n^3 = \left[\dfrac{n(n + 1)}{2} \right]^2$

    c. $1 + 3 + 5 + \cdots + (2n - 1) = n^2$
    d. $4^n - 1$ is divisible by 3
    e. $n < 2^n$
    f. $2^{n-1} \leqslant n!$ (Recall $n! = n(n - 1)(n - 2) \cdots 2 \cdot 1$.)

5. Use the Well-Ordering Property to prove each of the results in Exercise 4.

6. There is another version of the Principle of Mathematical Induction wherein the statement "$P(1)$ is true" is replaced by "$P(m)$ is true," where m is some natural number. Then the conclusion reads "$P(n)$ is true for all $n \geqslant m$." Use this modified induction to prove
   a. $(2n + 1) < 2^n$ if $n \geqslant 3$ b. $n^2 < 2^n$ if $n \geqslant 5$

7. Prove that the Principle of Mathematical Induction implies the Well-Ordering Property.

8. Does every nonempty set of integers have a least element? Does every nonempty set of negative integers have a greatest element?

9. Is the statement "$n^2 + n + 1$ is a prime integer for all natural numbers n" true for $n = 1$? for $n = 2$? for $n = 3$? Can you prove that the statement is true for all n? Why? If not, what can be proved?

10. There is another version of induction called the Second Principle of Mathematical Induction. The statement is: "Let $P(n)$ be a set of statements, one for each $n \in \mathbb{N}$. Suppose $P(1)$ is true and, for each $k \in \mathbb{N}$, the fact that $P(m)$ is true for all $m < k$ implies that $P(k)$ is true. Then $P(n)$ is true for all $n \in \mathbb{N}$."
    a. Prove that the Second Principle follows from the Well-Ordering Property.
    b. Use the Second Principle to prove that if $n \in \mathbb{N}$, $n > 1$, then n is a prime or a product of primes.

11. The binomial symbol $\binom{n}{k}$ is defined by the equation
    $$\binom{n}{k} = \frac{n!}{k!(n-k)!}$$
    where $n! = n(n-1)(n-2) \cdots 2 \cdot 1$ and $0! = 1$.
    a. Use induction on k to prove that if p is a prime integer, then $\binom{p}{k}$ is divisible by p for $k = 1, 2, \ldots, p - 1$.
    b. Verify $\binom{n}{k} + \binom{n}{k+1} = \binom{n+1}{k+1}$.
    c. Use induction on n to prove the Binomial Theorem:
       If $a, b \in \mathbb{R}$ and $n \in \mathbb{N}$, then
       $$(a+b)^n = a^n + \binom{n}{1}a^{n-1}b + \binom{n}{2}a^{n-2}b^2 + \cdots + \binom{n}{n-1}ab^{n-1} + b^n$$

12. Joey came home from school one day and announced to his parents that he had just learned that all people have the same sex! His concerned parents demanded to know where he had learned this. Joey responded that the source was his math teacher who that day had taught his class the Principle of Mathematical Induction. Joey argued in the following way:
       Let $P(n)$ be the statement "all sets of n people have the same sex." $P(1)$ is clearly true. Suppose for some $k \in \mathbb{N}$ that $P(k)$ is true; that is, all sets of k people have the same sex. Then consider any set of $k + 1$ people. Line them up. The first k people have the same sex by our induction hypothesis, as do the last k people. Since these two sets of people overlap in a set of $k - 1$ people, all $k + 1$ must have the same sex and $P(k + 1)$ is true. By the Principle of Mathematical Induction, $P(n)$ is true for all $n \in \mathbb{N}$. Thus any set of n people have the same sex.
       Please help Joey's parents show him the error in his logic.

1.5

Switching Circuits

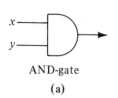

AND-gate

(a)

x	y	Output
1	1	1
1	0	0
0	1	0
0	0	0

(b)

Figure 1.11

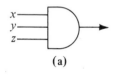

(a)

x	y	z	Output
1	1	1	1
1	1	0	0
1	0	1	0
1	0	0	0
0	1	1	0
0	1	0	0
0	0	1	0
0	0	0	0

(b)

Figure 1.12

Binary switching or logic circuits are at the basis of our present day digital computers. These circuits are combined to form switching networks that allow the compilers in these computers to perform numerous intricate computations. In this section, we will mathematically model elementary switching circuits and show how our work with truth tables and tautologies in Section 1.1 can aid us in understanding and simplifying switching networks.

There are a number of basic assumptions we will make about our logic circuits. We will use block diagrams or *gates* to represent these elementary circuits. Each gate will have one or more input wires and one output wire. The wires transmit electricity, and they have two states—either "on" or "off." (In practice the wires may carry high or low voltages.)

Figure 1.11a depicts an example of an AND-gate. We represent "on" by a "1" and "off" by a "0," and show a state diagram of the AND-gate in Figure 1.11b. Note that this state diagram is simply the truth table for the statement "x and y are 'on'" when 1 takes the place of T and 0 takes the place of F. For this reason, we denote the AND-gate by $x \wedge y$ or xy.

An AND-gate can have more than two inputs. For example, the gate of Figure 1.12a has the truth table or state diagram of Figure 1.12b. Note that the first three columns of this truth table contain the binary numbers from $2^3 - 1 = 7$ to 0 in descending order. This observation can help you construct truth tables for AND-gates of more than three variables.

In addition to the AND-gate, there are two other elementary logic circuits that we will combine to form our gating networks. An OR-gate is "on" if any one of its inputs is "on." An example of an OR-gate and its state diagram is given in Figure 1.13a and 1.13b. This gate is symbolized by $x \vee y$ or $x + y$ since it parallels our logical "or" variable statement.

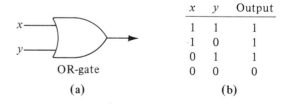

OR-gate

Figure 1.13   (a)

x	y	Output
1	1	1
1	0	1
0	1	1
0	0	0

(b)

Our last elementary circuit is the **inverter** or NOT-gate (given with its accompanying state diagram in Figure 1.14) and is symbolized by $\bar{x}$. This gate switches the voltage or state of x and corresponds to a logical negation.

We will examine switching circuits made up of these three elementary circuits. Although there are other types of switching devices such as delay flipflops and NOR-gates, we will limit our analysis to the three types already

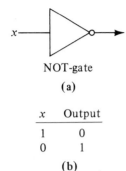

NOT-gate

(a)

x	Output
1	0
0	1

(b)

Figure 1.14

discussed. We will also make use of the material we developed concerning truth tables and logical identities. In practice, two switching circuits f_1 and f_2 are termed **equivalent** if they both realize the same output for identical inputs. In other words, f_1 and f_2 are equivalent if $f_1 \leftrightarrow f_2$ is a tautology.

Consider the two circuits pictured in Figure 1.15:

These two circuits are equivalent as shown in Figure 1.16.

Some simplification can be made in drawing circuits. In combining inverters with AND-gates and OR-gates, you can save space by omitting the triangle in the NOT-gate and write

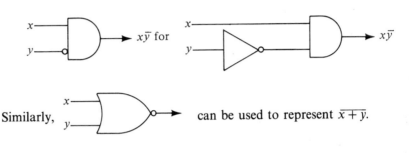

Similarly, can be used to represent $\bar{x} + \bar{y}$.

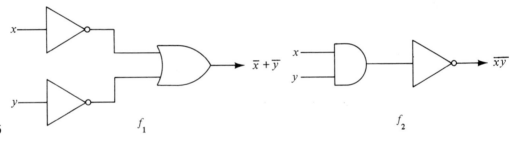

Figure 1.15

Figure 1.16

x	y	$\bar{x}$	$\bar{y}$	$\bar{x}+\bar{y}$	xy	$\overline{xy}$	$(\bar{x}+\bar{y})\leftrightarrow\overline{xy}$
1	1	0	0	0	1	0	1
1	0	0	1	1	0	1	1
0	1	1	0	1	0	1	1
0	0	1	1	1	0	1	1

The analysis of switching networks involved circuits obtained by combining sequences of AND-gates, OR-gates, and NOT-gates. As an example, suppose you wished to analyze the circuit g of Figure 1.17. The three circuits on the left represent xy, $\overline{y+z}$, and $\bar{x}z$, respectively. The AND-gate on the right then yields the output

$$g = xy + \overline{y+z} + \bar{x}z$$

Figure 1.18 tells us that the circuit g is "on" unless x and z are "on" and y is "off," or unless x and z are "off" and y is "on."

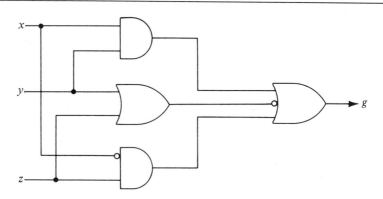

Figure 1.17

x	y	z	xy	$y+z$	$\overline{y+z}$	$\bar{x}$	$\bar{x}z$	$(xy)+(\overline{y+z})+(\bar{x}z)$
1	1	1	1	1	0	0	0	1
1	1	0	1	1	0	0	0	1
1	0	1	0	1	0	0	0	0
1	0	0	0	0	1	0	0	1
0	1	0	0	1	0	1	0	0
0	1	1	0	1	0	1	1	1
0	0	1	0	1	0	1	1	1
0	0	0	0	0	1	1	0	1

Figure 1.18

x	y	z	f
1	1	1	0
1	1	0	0
1	0	1	1
1	0	0	0
0	1	1	1
0	1	0	0
0	0	1	0
0	0	0	0

Figure 1.19   (a)   (b)

Our knowledge of truth tables and tautologies enables us to design circuits also. Suppose that we wish to construct a circuit f that accepts three inputs and produces the output of Figure 1.19a.

Since an AND-gate is 1 if and only if each input is 1, it follows that $x\bar{y}z = 1$ if and only if $x = z = 1$ and $y = 0$. Also $\bar{x}y\bar{z} = 1$ if and only if $x = z = 0$ and $y = 1$. Therefore the network $x\bar{y}z + \bar{x}y\bar{z}$ is one that has the desired output. This circuit is diagramed in Figure 1.19b. Note that the output of this circuit network f is the opposite of the output of the network g. We write $g = \bar{f}$. This observation allows us to design another circuit that is equivalent to g ($\bar{f}$ in Figure 1.20).

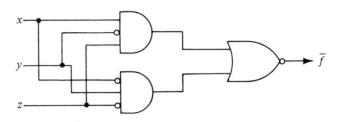

Figure 1.20

The question of whether circuit $\bar{f}$ is *simpler* than circuit g depends upon the definition of *simpler*. We will investigate this question in a later chapter. We are content if you now appreciate the importance of mathematical logic in the design of computer circuits and the applicability of some preliminary material in abstract algebra. We will endeavor to point out additional applications as we proceed.

Exercises 1.5

1. Determine the state diagrams or truth tables that correspond to each of the following circuits.

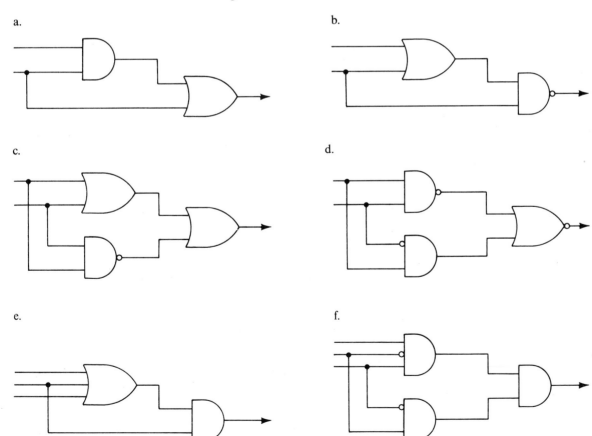

a.

b.

c.

d.

e.

f.

g.

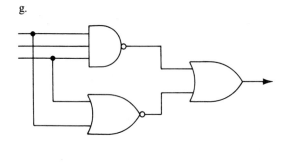

h.

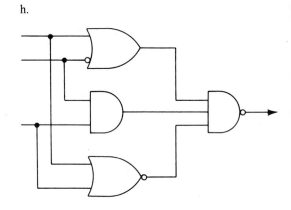

2. Use a truth table to show that the pairs of the following circuits are equivalent.

a.

b.

c.

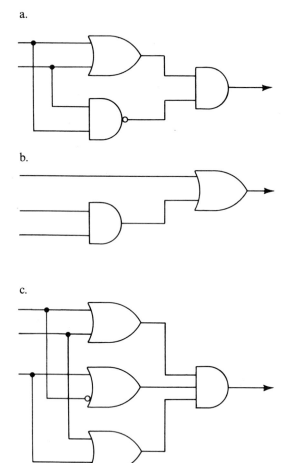

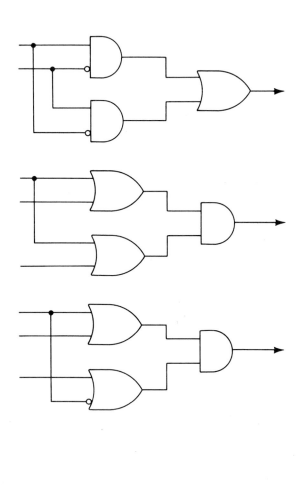

d.

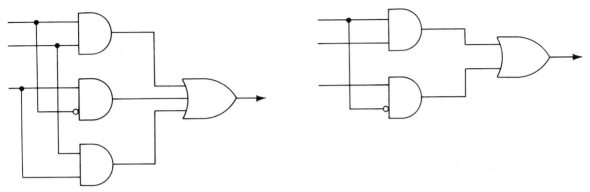

3. Use AND-gates, OR-gates, and NOT-gates to design circuits to achieve the following:

a. $x(\bar{x} + y)$ b. $\overline{(x + \bar{y})y}$

c. $x\bar{y} + \bar{x}z + xyz$ d. $(x + \bar{y})(x + \bar{z})(z + y)$

4. Construct circuit networks that represent the state diagram f, g, and h given in the table.

x_1	x_2	f	g	h
1	1	1	1	0
1	0	1	1	1
0	1	0	1	1
0	0	0	1	0

5. Construct circuit networks that achieve the state diagrams f, g, and h shown in the table.

x_1	x_2	x_3	f	g	h
1	1	1	0	0	0
1	1	0	1	1	0
1	0	1	0	1	0
1	0	0	0	1	1
0	1	1	0	1	1
0	1	0	0	1	1
0	0	1	0	1	0
0	0	0	1	1	1

*6. Write a program to take two input variables x and y, given in terms of their column arrays, and produce the output arrays corresponding to $x + y$, xy, and $\bar{x}$.

*7. Write a program that takes a given integer $n > 1$ and produces the column arrays in a truth table for $x_1, x_2, \ldots, x_n$.

2 Mappings, Operations, and Symmetry

Algebra is much more than the study of sets. The manner in which elements of a set interact with one another is the focal point of modern algebra. For this reason, the interesting sets in algebra are those for which there exist ways to combine elements. Methods of combining elements are investigated in the study of binary operations. In this chapter, we will define operations, study their properties, and discuss a number of important examples. Among these examples are modular arithmetic and symmetries of geometric figures.

2.1

Mappings

In your high school algebra, trigonometry, and calculus courses, you spent a great deal of time working with functions or, as we will refer to them in this text, mappings. Recall that a **function** or **mapping** or **map** $f: X \rightarrow Y$ is a rule of correspondence between two sets such that for each $x \in X$, there exists a unique $y \in Y$ associated with it. X is called the **domain** of the mapping f while Y is termed the **range** of f. If $y \in Y$ is the unique element associated with $x \in X$ by the mapping f, we say that y is the **image** of x, and x is the **preimage** of y; we write $y = f(x)$. The set of all elements of Y that are images of elements of f is called the **image** of f and denoted $f^{\rightarrow}(X)$. If $X = Y$, we say that f is a **mapping on** X. We diagram this definition as in Figure 2.1.

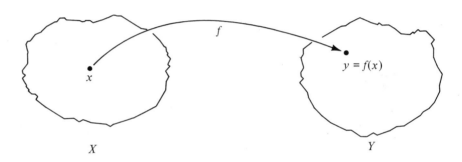

Figure 2.1

Example 2.1

Let $X = \{a, b, c, d\}$, $Y = \{1, 2, 3, 4\}$, and $f: X \rightarrow Y$ be given by

Then f is a mapping, and 4 is the image of both b and d.

Example 2.2

If $X = \{1, 2, 3\}$ and $Y = \{a, b, c\}$, then the correspondence g given by

is not a mapping since 2 does not have a unique image.

Example 2.3

If $X = Y + \mathbb{R}^+ = \{x \in \mathbb{R} | x > 0\}$, then the rule $y = f(x) = 1/x$ defines a mapping on $\mathbb{R}^+$.

The mapping $1_X : X \rightarrow X$ given by the formula $1_X(a) = a$ for all $a \in X$ arises in all areas of mathematics. It is known as the **identity mapping on** X.

If $f: X \rightarrow Y$ is a mapping and $Y \subseteq \mathbb{R}$, then f is termed a **real-valued mapping**. If k is some element of a set Z, then the mapping $g: X \rightarrow Z$ given by $g(x) = k$ for all $x \in X$ is a **constant mapping** since the image has a constant value.

The important thing to remember about mappings is that three distinct objects are required to define a mapping properly: two nonempty sets and a correspondence between them. The correspondence in many cases is not sufficient to determine the mapping. For example, if $f(x) = 2x$, do you know the mapping? The answer is no, since you do not know the domain. The mapping $f: \mathbb{N} \rightarrow \mathbb{N}$ given by the formula $f(x) = 2x$ and the mapping $g: \mathbb{R} \rightarrow \mathbb{R}$ given by $g(x) = 2x$ are different mappings even though they have the same rule of correspondence.

This poses an interesting question: When are two mappings equal? If f and g represent the same mapping, then both must have the same domain. Moreover, for each element x in their common domain, it must be true that $f(x) = g(x)$. Using this reasoning, $f(x) = x$ and $g(x) = x^2/x$ represent the same mapping on $\mathbb{R} - \{0\}$, while $f(x) = x$ and $g(x) = x^2/x$ do not represent the same mapping on $\mathbb{R}$, since $g(0)$ is undefined. In fact, g is *not* a mapping on $\mathbb{R}$.

If one has a mapping $f: X \rightarrow Y$, then information about sets X and Y can be derived by investigating properties of the mapping. For example, if X is a finite set and no two elements of X have the same image in Y, then it can be argued that "clearly" there must be at least as many elements in Y as there are in X. We will investigate these concepts in more detail later. For now, it is enough to understand that mappings can be used to make such comparisons.

Often mappings $f: X \rightarrow Y$ and $g: Y \rightarrow Z$ are combined to form a new mapping from X to Z. This permits comparison to be made between X and Z by utilizing information about f and g. To be more precise, if $f: X \rightarrow Y$ and $g: Y \rightarrow Z$ are mappings, then the **composition mapping of** f **and** g is the mapping $g \circ f: X \rightarrow Z$ defined by $(g \circ f)(x) = g(f(x))$. The mapping $g \circ f$ is frequently referred to as the **composite of** g **and** f.

We may visualize this concept in Figure 2.2.

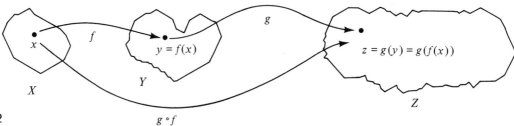

Figure 2.2

Example 2.4

Let $f: \mathbb{R} \to \mathbb{R}$ be defined by $f(x) = 2x - 1$ and $g: \mathbb{R} \to \mathbb{R}$ be defined by $g(x) = x^2$. Then the mapping $g \circ f: \mathbb{R} \to \mathbb{R}$ has the formula

$$(g \circ f)(x) = g(2x - 1)$$
$$= (2x - 1)^2$$
$$= 4x^2 - 4x + 1$$

The mapping $f \circ g: \mathbb{R} \to \mathbb{R}$ is given by

$$(f \circ g)(x) = f(x^2)$$
$$= 2(x^2) - 1$$
$$= 2x^2 - 1$$

which is different than $g \circ f$.

Example 2.5

If $f: \mathbb{Z} \to \mathbb{N}$ is defined by $f(x) = x^2 + 1$ and $g: \mathbb{N} \to \mathbb{Q}$ by $g(x) = (x - 1)/x$, then the composite $g \circ f: \mathbb{Z} \to \mathbb{Q}$ is given by

$$(g \circ f)(x) = g(x^2 + 1)$$
$$= \frac{(x^2 + 1) - 1}{x^2 + 1}$$
$$= \frac{x^2}{x^2 + 1}$$

Note that the composite $f \circ g$ is not defined on $\mathbb{N}$ in this case.

Comparison of three systems may be extended by the next result, which we will find useful.

Theorem 2.1

If $f: X \to Y$, $g: Y \to W$, and $h: W \to Z$, then $h \circ (g \circ f) = (h \circ g) \circ f$.

Proof

$h \circ (g \circ f)$ and $(h \circ g) \circ f$ are both mappings from X to Z. We need to show that they have the same rule of correspondence. Let $x \in X$. Then

$$[h \circ (g \circ f)](x) = h[(g \circ f)(x)]$$
$$= h[g(f(x))]$$

On the other hand,

$$[(h \circ g) \circ f](x) = (h \circ g)(f(x))$$
$$= h[g(f(x))]$$

The result follows since these two mappings act the same on all elements in their common domain. $\square$

In subsequent chapters we will have the need for two specialized mappings that are associated with a mapping $f: X \to Y$. These mappings are concerned with images and preimages of subsets of X and Y respectively.

If $f: X \to Y$ is a mapping and $B \subseteq Y$, then define

$$f^{\leftarrow}(B) = \{x \in X \mid f(x) \in B\}$$

This mapping is called the **preimage mapping associated with** f. Properties of this mapping are discussed in Exercise 6.

If $f: X \to Y$ is a mapping and $A \subseteq X$, then define

$$f^{\rightarrow}(A) = \{y \in Y \mid y = f(a) \text{ for some } a \in A\}.$$

This mapping is called the **image mapping associated with** f. Properties of this mapping are discussed in Exercise 7.

Exercises 2.1

1. For the following pairs of mappings, compute the composites $f \circ g$ and $g \circ f$:

   a. $f(x) = x^2 + 1, g(x) = 2x + 5$ b. $f(x) = \dfrac{1}{1-x}, g(x) = x^2$

   c. $f(x) = x^{1/3}, g(x) = 2^x$ d. $f(x) = \sqrt{3-x}, g(x) = x^3 - 2x + 3$

2. The domain of the composite $g \circ f$ is $\{x \in \text{domain } f \mid f(x) \in \text{domain } g\}$. For the following pairs, find the largest subset of $\mathbb{R}$ that can serve as the domain of $g \circ f$:

   a. $f(x) = \sqrt{x}, g(x) = 2x + 5$ b. $f(x) = \sqrt{x}, g(x) = x^2$

   c. $f(x) = 2x - 3, g(x) = \sqrt{x}$

3. Verify that $h \circ (g \circ f) = (h \circ g) \circ f$ if $f(x) = x^2 - 1$, $g(x) = 3x + 5$, $h(x) = \sqrt{x}$.

4. Recall that the mapping $1_X: X \to X$ defined by $1_X(x) = x$ for any $x \in X$ is called the **identity mapping on** X. Verify that $g \circ f = 1_X$ for each of the following pairs of mappings; that is, show $(g \circ f)(x) = x$ for all $x \in X$:

   a. $f(x) = 2x - 1$; $g(x) = \frac{1}{2}(x + 1)$, $X = \mathbb{R}$

   b. $f(x) = x^2$; $g(x) = \sqrt{x}$, $X = \mathbb{R}^+$

   c. $f(x) = (x + 1)^{1/3}$; $g(x) = x^3 - 1$, $X = \mathbb{R}$

5. a. Let a correspondence $f: \mathbb{Q} \to \mathbb{Z}$ be given by $f(a/b) = a + b$. Why is f **not** a mapping?

   b. Suppose a correspondence $f: \mathbb{Z} \times \mathbb{Z} \to \mathbb{Z}$ is given by $f((a, b)) = a + b$. Show that f is a mapping.

   c. Discuss the difference between the correspondences of parts a and b of this exercise.

6. If $f: X \to Y$ is a mapping and $B \subseteq Y$, then define $f^{\leftarrow}(B) = \{x \in X \mid f(x) \in B\}$. For $B_1, B_2 \in \mathscr{P}(Y)$, prove that

   a. $f^{\leftarrow}(B_1 \cup B_2) = f^{\leftarrow}(B_1) \cup f^{\leftarrow}(B_2)$ b. $f^{\leftarrow}(B_1 \cap B_2) = f^{\leftarrow}(B_1) \cap f^{\leftarrow}(B_2)$

   c. $f^{\leftarrow}(Y - B) = X - f^{\leftarrow}(B)$ d. $f^{\leftarrow}(\varnothing) = \varnothing$

   e. $f^{\rightarrow}(f^{\leftarrow}(B)) \subseteq B$ f. $A \subseteq f^{\leftarrow}(f^{\rightarrow}(A))$

7. If $f: X \to Y$ is a mapping and $A \subseteq X$, then define

   $$f^{\rightarrow}(A) = \{y \in Y \mid y = f(a) \text{ for some } a \in A\}$$

   For $A_1, A_2 \in \mathscr{P}(X)$, verify the following:

   a. $f^{\rightarrow}(A_1 \cup A_2) = f^{\rightarrow}(A_1) \cup f^{\rightarrow}(A_2)$

b. $f^{\rightarrow}(A_1 \cap A_2) \subseteq f^{\rightarrow}(A_1) \cap f^{\rightarrow}(A_2)$. Construct an example to show why equality does not always hold.

c. $f^{\rightarrow}(X - A) \supseteq Y - f^{\rightarrow}(A)$. Construct an example to show why equality does not always hold.

d. Determine necessary and sufficient conditions for equality to hold in parts b and c of this exercise.

2.2

Binary Operations

When studying a set, it is often useful to know if the set elements interact with one another. If every two elements can be combined to form a third, then there is some structure connected with the set. Binary operations control the manner in which two elements in a set produce a third. The goal of this section is to investigate these operations.

We now use our knowledge of mappings to investigate binary operations. Let A be a nonempty set and $X = A \times A$. A **binary operation on a set** A is a mapping from $X = A \times A$ to A. If $*$ is a binary operation on A, we will denote the image of (x, y) under $*$ by $x * y$.

A familiar example of a binary operation is addition of integers. If $m, n \in \mathbb{Z}$, then $m + n$ is another integer. Subtraction of integers is another binary operation on $\mathbb{Z}$, but subtraction is not a binary operation on $\mathbb{N}$. To see this, note that $(2, 5) \in \mathbb{N} \times \mathbb{N}$, but $2 - 5 = -3 \notin \mathbb{N}$. We say that $\mathbb{N}$ is *not closed* with respect to subtraction. On the other hand, $\mathbb{N}$ is *closed* with respect to integer addition. In a similar manner, multiplication is an operation on $\mathbb{Z}$ but division is not, since $3 \div 2 \notin \mathbb{Z}$.

The power set $\mathscr{P}(X)$ of a set X provides other illustrations of operations. If $A, B \in \mathscr{P}(X)$, then $A \cup B$, $A \cap B$, and $A - B$ are all subsets of X, so that union, intersection, and difference are binary operations on $\mathscr{P}(X)$. The Cartesian product $A \times B$ is not a binary operation on $\mathscr{P}(X)$, since $A \times B \notin \mathscr{P}(X)$.

Matrices can be used to introduce other examples of binary operations. Let

$$M_2(\mathbb{Z}) = \left\{ \begin{bmatrix} a & b \\ c & d \end{bmatrix} \middle| a, b, c, d \in \mathbb{Z} \right\}$$

Each element of $M_2(\mathbb{Z})$ is a display or an array of two rows and two columns of integers and is termed a 2×2 (read "two-by-two") matrix. Define addition of matrices by

$$\begin{bmatrix} a & b \\ c & d \end{bmatrix} + \begin{bmatrix} r & s \\ t & u \end{bmatrix} = \begin{bmatrix} a+r & b+s \\ c+t & d+u \end{bmatrix}$$

and multiplication of matrices by

$$\begin{bmatrix} a & b \\ c & d \end{bmatrix} \times \begin{bmatrix} r & s \\ t & u \end{bmatrix} = \begin{bmatrix} ar+bt & as+bu \\ cr+dt & cs+du \end{bmatrix}$$

For example,

$$\begin{bmatrix} 1 & -1 \\ 2 & 3 \end{bmatrix} + \begin{bmatrix} 4 & 5 \\ 1 & 6 \end{bmatrix} = \begin{bmatrix} 1+4 & -1+5 \\ 2+1 & 3+6 \end{bmatrix} = \begin{bmatrix} 5 & 4 \\ 3 & 9 \end{bmatrix}$$

while

$$\begin{bmatrix} 1 & -1 \\ 2 & 3 \end{bmatrix} \times \begin{bmatrix} 4 & 5 \\ 1 & 6 \end{bmatrix} = \begin{bmatrix} (1)(4)+(-1)(1) & (1)(5)+(-1)(6) \\ (2)(4)+(3)(1) & (2)(5)+(3)(6) \end{bmatrix}$$

$$= \begin{bmatrix} 3 & -1 \\ 11 & 28 \end{bmatrix}$$

Since addition and multiplication are binary operations on $\mathbb{Z}$, the sum and product of two elements in $M_2(\mathbb{Z})$ is another element in $M_2(\mathbb{Z})$. We conclude that matrix addition and matrix multiplication are binary operations. Later we will see that matrices provide a rich source of examples and counterexamples.

Note that we used the same symbols to denote matrix operations as we did to denote their integer counterparts, even though they are clearly distinct operations. This should not cause any great concern, since the nature of the problems at hand should indicate whether you are working with integers or matrices of integers. Since most of the operations encountered in this text are binary operations, we sometimes omit the word *binary*.

Binary operations on a finite set are often defined by means of tabular displays, which are called **Cayley tables**[1]. Let $A = \{a, b, c\}$, and consider the example of Figure 2.3.

*	a	b	c
a	b	c	a
b	c	b	a
c	a	a	b

Figure 2.3

Figure 2.3 defines $*$ in the following way: to determine $x * y$ locate the term in the row with x on the far left and in the column with y on the top. For example, $a * b = c$ and $b * c = a$. Since every element in the table belongs to A, we see that $*$ is a binary operation on A.

Operations are classified and analyzed according to their properties.

Definition 2.1

A binary operation $*$ on a set X is

i. **commutative** if $a * b = b * a$ for all $a, b \in X$
ii. **associative** if $(a * b) * c = a * (b * c)$ for all $a, b, c \in X$.

Clearly addition and multiplication of integers are both commutative

---

[1] Arthur Cayley (1821–1895) was an Englishman who also taught in the United States. Primarily an algebraist, he published extensively in other areas as well; for example, he used determinants to study geometry. He was one of the first mathematicians to utilize matrices. His work made him one of the founders of modern abstract algebra.

and associative operations. Subtraction of integers is neither commutative nor associative since, in general, $m - n \neq n - m$ and $m - (n - t) \neq (m - n) - t$. Similarly, set union and intersection are both commutative and associative on $\mathcal{P}(X)$. These results are contained in Theorem 1.3. If $X \neq \varnothing$, then set difference is not commutative; that is, $A - B$ is not always equal to $B - A$. The operation of set difference is not associative either.

You might have noticed that the given examples that were not commutative were not associative either. Is this always the case? The answer is no. Let X be a set, and let $A = \{f : X \to X \,|\, f \text{ is a mapping}\}$. Then the composition of two mappings in A is another mapping in A so that composition is a binary operation on A. In Theorem 2.1, if $X = Y = Z = W$, then we have shown that composition of mappings is an associative operation on A. Composition is not commutative, however, as the following examples show.

Let

$$A = \{0, 1\} \qquad f \colon \begin{cases} 0 \to 0 \\ 1 \to 0 \end{cases} \qquad g \colon \begin{cases} 0 \to 1 \\ 1 \to 0 \end{cases}$$

Then

$$g \circ f \colon \begin{cases} 0 \to 1 \\ 1 \to 1 \end{cases} \qquad \text{while} \qquad f \circ g \colon \begin{cases} 0 \to 0 \\ 1 \to 0 \end{cases}$$

As another example, consider matrix multiplication. It is a straightforward (though tedious) exercise to show that matrix multiplication is an associative operation on $M_2(\mathbb{Z})$, but the operation is not commutative. On the other hand, in Exercise 5 we give an example of an operation that is commutative but not associative. Thus, the properties of associativity and commutativity are indeed independent of one another.

In considering operations given by means of a table, commutativity is quite easy to check. If the elements of the table are $a_1, a_2, \ldots, a_n$, then the elements in the ith row and jth column of the table is $a_i * a_j$. To check that $a_i * a_j = a_j * a_i$ for each a_i and a_j, simply observe whether or not the element in the ith row and jth column agrees with the element in the jth row and ith column. In geometric terms, if you draw a line from the term in the upper left corner to the term in the lower right corner, then you have constructed the **main diagonal** of the table. If the table defines a commutative operation, then the corresponding terms on either side of this diagonal must agree; that is, the table must be symmetric. Consider our previous example of Figure 2.4. Since the terms connected by arrows agree, the operation $*$ is commutative.

Establishing associativity for operations defined by tables is another matter. If the set A contains n elements, then a term-by-term check of associativity involves the computation of n^3 pairs of elements of the form $(a * b) * c$ and $a * (b * c)$. Figure 2.4 would require that $3^3 = 27$ pairs be checked to prove associativity. Of course, you need only find one unequal pair to show that the operation is not associative. There are table tests for

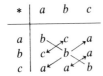

Figure 2.4

associativity, but these tests are more involved than the test we presented for commutativity.[2]

There are two more operations that are extremely important. They involve the concept of modular arithmetic. Let $n > 1$ be a positive integer, and consider the set

$$\mathbb{Z}_n = \{0, 1, 2, \ldots, n-1\}$$

$\mathbb{Z}_n$ is called the set of **integers modulo** n or the set of **residues modulo** n. Note that the elements of $\mathbb{Z}_n$ are precisely the set of remainders obtained by dividing each of the integers by n. (This follows from the Division Algorithm for $\mathbb{Z}$.) This observation is the key to defining an operation on $\mathbb{Z}$ called **modular addition** and symbolized by $\oplus$.

Definition 2.2

$\oplus$	0	1	2	3	4
0	0	1	2	3	4
1	1	2	3	4	0
2	2	3	4	0	1
3	3	4	0	1	2
4	4	0	1	2	3

Figure 2.5

Let $a, b \in \mathbb{Z}_n$. Since the integer $a + b$ can be written as $qn + r$ where $q, r \in \mathbb{Z}$ and $0 \leqslant r < n$, we define $a \oplus b = r$.

Since the Division Algorithm assures us that r (and also q) is unique, we have that $\oplus$ is indeed an operation on $\mathbb{Z}_n$. Moreover, since integer addition is commutative, both $b + a$ and $a + b$ have the same remainder upon division by n. Therefore, $a \oplus b = b \oplus a$ for all $a, b \in \mathbb{Z}_n$. A similar argument can be used to conclude that $a \oplus (b \oplus c) = (a \oplus b) \oplus c$ for all $a, b, c \in \mathbb{Z}_n$.

Example 2.6

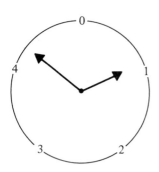

Figure 2.6

In Figure 2.5, we have constructed the Cayley table for modular addition in $\mathbb{Z}_5$.

The term modular addition does not refer to a single operation; in fact, for each $n > 1$ there exists a different operation. For example, $3 \oplus 4 = 2$ in $\mathbb{Z}_5$, but $3 \oplus 4 = 1$ in $\mathbb{Z}_6$. In order to specify which operation is being considered, we must tell you the value of the **modulus** n.

Modular addition is also called **clock addition**. The reason for this is clear if we consider $\mathbb{Z}_5$ and a clock with the numbers 0, 1, 2, 3, and 4. (See Figure 2.6.) If you add four hours to three o'clock, you obtain two o'clock; that is, $3 \oplus 4 = 2$. Similarly, two hours past four o'clock is one o'clock; that is, $4 \oplus 2 = 1$.

We can do the same thing with multiplication as we did with addition to obtain the following definition.

---

[2] Burn, R. P. "Cayley tables and associativity," *The Mathematical Gazette* Vol. 62 (1978), pp. 278–281.

Definition 2.3

Let $a, b \in \mathbb{Z}_n$. Inasmuch as the integer $ab = qn + r$, where $q, r \in \mathbb{Z}$ and $0 \leqslant r < n$, we define $a \odot b = r$.

Modular multiplication, denoted by $\odot$, is another binary operation on $\mathbb{Z}_n$. Arguments similar to those used with modular addition can be utilized to show that $\odot$ is associative and commutative on $\mathbb{Z}_n$. (See Exercise 14.)

Example 2.7

The Cayley tables for modular multiplication on $\mathbb{Z}_4$ and $\mathbb{Z}_5$ are given in Figure 2.7.

$\odot$	0	1	2	3
0	0	0	0	0
1	0	1	2	3
2	0	2	0	2
3	0	3	2	1

$\odot$	0	1	2	3	4
0	0	0	0	0	0
1	0	1	2	3	4
2	0	2	4	1	3
3	0	3	1	4	2
4	0	4	3	2	1

Figure 2.7

Exercises 2.2

1. Which of the following are operations on $\mathbb{N}$? on $\mathbb{Z}$? on $\mathbb{Q}$?

   a. $a * b = a + b - ab$ b. $a \$ b = \dfrac{a^2 - b^2}{a - b}$ c. $a \& b = \dfrac{ab - 2}{a}$

2. Show that $a * b = a + b + ab$ defines a commutative and associative operation on $\mathbb{N}$.

3. Verify that the operation $\#$ on $\mathbb{Z}$ defined by $a \# b = ab - a$ is neither commutative nor associative.

4. Compute the sum and product of each of the following pairs of elements in $M_2(\mathbb{Z})$.

   a. $\begin{bmatrix} 2 & 3 \\ 1 & 1 \end{bmatrix}, \begin{bmatrix} 2 & -3 \\ -1 & 2 \end{bmatrix}$

   b. $\begin{bmatrix} 2 & 3 \\ 3 & 4 \end{bmatrix}, \begin{bmatrix} 1 & 2 \\ 3 & -4 \end{bmatrix}$

   c. $\begin{bmatrix} 1 & 4 \\ 1 & 3 \end{bmatrix}, \begin{bmatrix} -1 & 4 \\ 1 & -3 \end{bmatrix}$

5. Define the operation $*$ on $M_2(\mathbb{Z})$ by
   $$\begin{bmatrix} a & b \\ c & d \end{bmatrix} * \begin{bmatrix} u & v \\ w & x \end{bmatrix} = \begin{bmatrix} b+v & 0 \\ 0 & c+w \end{bmatrix}$$
   Prove that $*$ is commutative but not associative.

6. a. Verify that matrix addition + and matrix multiplication × are associative operations on $M_2(\mathbb{Z})$.
   b. Prove that + is commutative on $M_2(\mathbb{Z})$, but find a counterexample to show that × is not.

7. Determine if the following tables define commutative and/or associative operations.

*	a	b	c	d
a	a	d	c	b
b	d	d	d	d
c	c	d	a	b
d	b	d	b	c

#	a	b	c	d
a	b	c	d	a
b	c	d	a	b
c	d	a	b	c
d	a	b	c	d

△	a	b	c	d
a	a	b	c	d
b	b	c	d	a
c	c	d	d	d
d	d	b	d	a

8. If * is an operation on a set A, then an element $a \in A$ is termed an **identity** for * if $a * b = b * a = b$ for all $b \in A$. For the operations of Exercise 7, determine all identities.

9. Define * on $\mathbb{Z}$ by $a * b = b$ for any $a, b \in \mathbb{Z}$. Prove that * is associative but not commutative on $\mathbb{Z}$.

10. Construct the Cayley tables for modular addition on $\mathbb{Z}_n$ for
    a. $n = 2$ b. $n = 3$ c. $n = 4$ d. $n = 6$

11. Construct the Cayley tables for modular multiplication on $\mathbb{Z}_n$ for
    a. $n = 2$ b. $n = 3$ c. $n = 4$ d. $n = 6$

12. Verify that modular multiplication is an operation on

$$\mathbb{Z}_5^* = \mathbb{Z}_5 - \{0\} = \{1, 2, 3, 4\}$$

but not an operation on

$$\mathbb{Z}_4^* = \mathbb{Z}_4 - \{0\} = \{1, 2, 3\}$$

13. Suppose * is a commutative and associative binary operation on X and A is a subset of X closed under *; that is, * is also a binary operation on A. Prove that * is commutative and associative on A.

14. Verify that modular multiplication is associative and commutative on $\mathbb{Z}_n$.

15. Show that the operation * defined by the following Cayley table is a commutative but nonassociative operation.

*	a	b	c
a	b	a	a
b	a	c	a
c	a	a	c

2.3

Symmetry

Symmetry is one of the unifying concepts of nature. The idea of symmetry pervades art, music, and dance. In science and mathematics, symmetry is invoked to aid in exploring phenomena, making conjectures, and extending

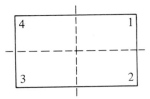

Figure 2.8

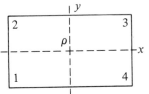

Figure 2.9

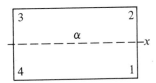

Figure 2.10

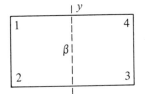

Figure 2.11

proofs. In geometry, symmetry transformations form a major area of interest. In this section we wish to analyze how regular geometric constructions, polygons and polyhedra, are affected by symmetry transformations. Some properties possessed by the symmetry transformations of a geometric object are independent of the shape of the object. We wish to isolate these properties since they will provide us with axioms for a basic structure of algebra—the group. We will study groups in detail throughout the next ten chapters.

In this section we will consider length-preserving motions of the plane that preserve the shape of the rectangle and the equilateral triangle. These length-preserving motions are also known as **symmetries** (or **rigid motions** in the case of two-dimensional shapes). They must preserve not only the shape but also the outline of these geometric objects. Not every rotation is a symmetry; the rotation must preserve the outline as well. Similar considerations can be made of three-dimensional objects, but our examples using two-dimensional objects will be enough to illustrate the important properties.

Let us first consider a nonsquare rectangle with vertical and horizontal sides whose vertices are labeled as in Figure 2.8. What rigid motions maintain or preserve its shape? In considering this question, we will be moving the rectangle by rotations of the rectangle in the plane as well as in space. One of the simplest changes we could effect is to leave the rectangle alone; this preserves its outline. Let us symbolize this motion by e. What else can we do? If we rotate the rectangle $180°$, we keep the original shape, but now the vertices have new labels (see Figure 2.9). We label this rotation ρ. If we pick up the original rectangle and flip it about the dashed x-axis, we obtain the orientation in Figure 2.10. Let us label this symmetry motion α. Similarly, we can flip the original rectangle about the dashed y-axis to obtain the configuration in Figure 2.11, which we label β.

Are there any other symmetry motions of the rectangle? What will happen if we rotate by ρ and then flip the resulting rectangle by α? The sequence illustrated in Figure 2.12 shows that combining ρ and α results in flip β. We symbolize this by writing $\rho\alpha = \beta$.[3] Similarly if we first flip by β and then flip by α, we obtain the configuration for ρ, so $\beta\alpha = \rho$.

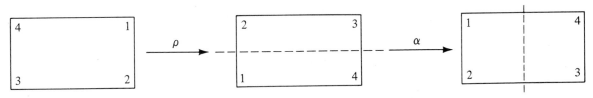

Figure 2.12

---

[3] In this development we prefer to consider the motions as we have described them and not as mappings on the set of vertices. If we did view them as mappings, then $\rho\alpha$ would correspond to the mapping $\alpha \circ \rho$. At this introductory level, we prefer to combine two motions left-to-right.

	e	ρ	α	β
e	e	ρ	α	β
ρ	ρ	e	β	α
α	α	β	e	ρ
β	β	α	ρ	e

Figure 2.13

Let us construct a Cayley table for these motions, where the "product" of β with α is located in the row corresponding to β and the column below α. (See Figure 2.13.) Note that each element is its own inverse; that is, if we perform a rotation or a flip twice, we return to the original configuration e. Also e is an identity in that no matter what element we combine with e, we obtain that element again. Notice that no new elements arise in our table so that the set $\{e, \rho, \alpha, \beta\}$ is closed under "product" of symmetry motions. Now this in itself does not mean that we have discovered all the symmetries of the rectangle. It only means that the symmetries we have discovered produce no new symmetries. In fact we *have* listed all the possible symmetries of the rectangle. We can show this by arguing that any symmetry must keep 4 across from 1 and 3 across from 2, and vertices 1 and 2 must both be either on the right or on the left. Thus there is but one motion with a 1 in any given corner—that means four motions in all.

As we proceed with our discussion of symmetries, you might find it useful to construct a copy of the shape under consideration. For example, for the preceding discussion, you would make a rectangle and label the vertices. (Be certain to label both sides of the piece of paper.)

The "product" operation defined in our table corresponds to composition of motions. Each motion can be thought of as a special type of mapping on the labels of the vertices $\{1, 2, 3, 4\}$. Since composition of mappings is associative, we can conclude that our "product" table is associative. (See Exercise 11.) In fact, this same argument can be used to prove that the Cayley table of any set of symmetries is associative. We note that our table is symmetric, so that our operation is commutative on the set $\{e, \rho, \alpha, \beta\}$. This, however, is not always the case as our next example illustrates.

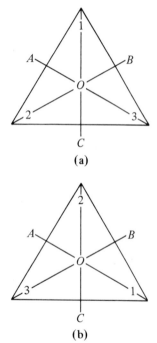

(a)

(b)

Figure 2.14

Let us now consider an equilateral triangle with one side horizontal, vertices labeled 1, 2, 3, and the perpendicular bisectors by A, B, C. (See Figure 2.14a.) Our problem is to describe the symmetries of the triangle, that is, to find what sort of movements or motions can be applied to this triangle so that its outline remains the same. At this point we want to emphasize that in the process we are using, although the vertices change position, the axes do not. That is, the vertical axis C (through the vertex labeled 1 in Figure 2.14a) will always remain unchanged although vertex 2 might replace vertex 1. (See Figure 2.14b.)

Again the outline of the shape remains the same if we leave the triangle alone. Let us denote this change, which we will describe as a rotation about the circumcenter O through an angle of $0°$ by the symbol e.[4] Let us rotate the triangle clockwise through an angle of $120°$ so that 2 is at the top. We label the rotation ρ. (See Figure 2.14b, and note, as we have just mentioned, that the vertical axis C is still vertical and passes through vertex 2.) If we rotate

---

[4]The circumcenter of a triangle is the point at which the perpendicular bisectors of the sides intersect.

Figure 2.15 e ρ ρ^2 τ_A τ_B τ_C

again through an angle of 120°, we obtain a triangle with 3 at the top. This amounts to a rotation of 240°. Symbolize this motion by ρ^2. If we apply ρ again, that is, rotate 360°, we see that the triangle has returned to its original position. Therefore ρ^3 has the same effect on the triangle as the motion e. Now suppose that we pick up the triangle and turn it over. Let τ_A denote what happens when we reflect ("flip") the triangle 180° about its A-axis. Then we obtain a motion leaving the outline of the triangle fixed. If we apply τ_A again, then we return to our original triangle; that is, τ_A^2 is the same as e. Similarly, we can define τ_B and τ_C as reflections about axes B and C, respectively. We list the motions so far defined in Figure 2.15.

We might now ask ourselves if there are any other motions that leave the outline of the triangle fixed. What if we apply ρ and follow this by the motion τ_A? You should check that we then obtain τ_B. What if we first apply τ_A and then apply ρ? Then we obtain τ_C. In fact, as we see below, the motions described are the only motions leaving the outline of the triangle fixed.

Let us digress a bit from this discussion to describe an easy way for you to determine the result of two motions. Construct a triangle, label the vertices, and on another piece of paper, draw the triangle and the axes. Since the triangle must always have a horizontal base and there are three bases and two sides, we can argue that there are only $6 = 3 \times 2$ combinations of bases and sides that can be at the bottom and thus only six motions. To determine the result of each motion, place the cutout triangle on top of the triangle with the axes drawn. Label the lower base of the triangle with an e, which is the motion that brings that side and base to that position. (Label it just above the midpoint or where axis C cuts the base). Now perform motion ρ (120° clockwise rotation) and label the new base side ρ. Performing ρ once again (or the composite ρ^2), label the new base side (still blank) appropriately. Note that performing another rotation brings the vertices back to their initial position and the e to the base. Next, perform motion τ_A and label the new base side (blank, isn't it?) τ_A. Again perform motion τ_A and we are back to base e. Repeat this last procedure for motions τ_B and τ_C. You may now check that all bases are now labeled. Thus there can be no other bases.

Return the triangle to its initial position and perform two consecutive motions: for example, τ_A followed by ρ. You will find that the new base side is labeled τ_C; thus $\tau_A\rho = \tau_C$. This procedure serves two useful purposes. It provides a simple method to determine the number of motions, and it also determines the motion that is the product of any two motions.

	e	ρ	ρ^2	τ_A	τ_B	τ_C
e	e	ρ	ρ^2	τ_A	τ_B	τ_C
ρ	ρ	ρ^2	e	τ_B	τ_C	τ_A
ρ^2	ρ^2	e	ρ	τ_C	τ_A	τ_B
τ_A	τ_A	τ_C	τ_B	e	ρ^2	ρ
τ_B	τ_B	τ_A	τ_C	ρ	e	ρ^2
τ_C	τ_C	τ_B	τ_A	ρ^2	ρ	e

Figure 2.16

Let us use a table to describe these motions of our triangle. The table in Figure 2.16 reads from left to right; so to find $\rho\tau_A$ (that is, the motion ρ followed by the motion τ_A) we look for the entry in the ρ-row and the τ_A-column: τ_B. You should verify all entries in the table that have not been previously established. (See Exercise 1.)

Another interesting question is whether the axes need to remain fixed? That is, what happens if axis C always passes through vertex 1, axis B through 2, and axis A through 3? The answer is that if you do not fix the axes, the resulting operation is not a group operation. (See Exercise 13.)

If we use our method to combine our motions, we have produced a table with an operation that is associative and with the set closed under the operation. We also have an identity element e. Note, as well, that for each motion, there exists another motion that, combined with the first, returns the triangle to its original shape. For example, $\rho\rho^2 = e$ and $\tau_A\tau_A = e$. We say that these motions are **inverses** of each other. Unlike the situation with the rectangle, however, our "product" operation is not commutative for these symmetries. We might ask whether the rotation needs to be clockwise. The answer is no; a counterclockwise rotation would work just as well with the only change being that the new motion ρ is the old ρ^2 and the new ρ^2 is the old ρ. (See Exercise 12.)

In the next chapter we will see that the motions we have described form an algebraic structure called a **group**. The groups we have studied here are known as the group of symmetries of the rectangle and the group of symmetries of the equilateral triangle. Similar groups can be studied for other geometric figures; for example, squares, cubes, or any shape or solid that possesses some form of rotation, reflection, or other type of symmetry.[5] Results from these studies have important applications in crystallography, chemical bonding, X-ray spectroscopy, lasers, and many other areas of

---

[5] The symmetry motions of a Rubik's cube form a group. If you take a cube and twist a face or the middle level of a face, you obtain a symmetry. The total number of such motions or combinations of such motions is about 4×10^{19}. For an excellent discussion of these symmetries and the application of the theory of this chapter to the solution to the Rubik's cube problem read Douglas R. Hofstadter, "Metamagical Themas," *Scientific American* (March 1981).

physics and chemistry. Students are often amazed to find that abstract algebra has any applications at all! In this book, we restrict ourselves mainly to applications within mathematics and computer science, with our emphasis on algebraic structures themselves, their definitions, and properties.

Exercises 2.3

1. Verify the entries in the table of symmetries of an equilateral triangle (Figure 2.16).

2. What is the table of motions of an isosceles triangle that is not equilateral? Describe the motions of a scalene (that is, nonisosceles) triangle.

3. Suppose that in describing the motions of an isosceles triangle, we were not allowed to lift the triangle off the paper. Construct the Cayley table of the motions that are still allowed. Do the same for an equilateral triangle.

4. Consider a square. Construct the Cayley table of all symmetry motions that leave the outline of the square fixed but do not require the lifting of the square from the paper.

5. To the motions of Exercise 4, add all other symmetries of the square, and construct the resulting Cayley table.

6. Describe all the symmetries of a regular tetrahedron, a four-sided prism with each face being an equilateral triangle.

7. List all the symmetries of a regular five-pointed star.

8. Find all the symmetries of a regular pentagon.

9. Considering the number of symmetries of the equilateral triangle, the square, and the regular pentagon, can you conjecture how many symmetries there are for a regular n-sided polygon for $n > 5$?

10. How many symmetries are there for the cube?

11. Suppose we define the product of two motions of a rectangle $\alpha\beta$ as the composition of mappings $\beta \circ \alpha$. Use the fact that composition of mappings is associative to prove that our product operation is associative.

12. Determine the Cayley table for the symmetries of each of the following if the motions of rotation are counterclockwise rather than clockwise:
    a. Rectangle
    b. Equilateral triangle
    c. Square
    d. Regular pentagon
    e. Regular five-pointed star

13. Determine the Cayley table for the symmetries of an equilateral triangle if the axes rotate with the vertices rather than remaining fixed. Also show that the operation thus determined is not a group operation.

3 | Groups

Abstract algebra is the study of sets and operations. In the first two chapters, we have investigated both. In the remainder of this text, we will see how focusing on set operations with certain properties can provide rich and interesting areas of investigation. It is thus reasonable to commence our studies by considering a set together with a single binary operation. We want to consider a simple type of algebraic structure, but at the same time, the operation must have enough properties to make our investigation interesting. A basic structure that satisfies these requirements is known as a "group."

The theory of groups is one of the most unifying concepts in all mathematics. It has been utilized in almost all branches—geometry, topology, functional analysis, and computer design, to name a few. In areas of applied mathematics such as quantum mechanics, X-ray spectroscopy, and algebraic coding, group theory has become an indispensible tool, not only in leading to new discoveries, but also in minimizing calculation and enabling the statement of deep results in a condensed form.

The ideas of group theory were used as early as 1770 by Lagrange[1]. Around 1830, Galois[2] extended Lagrange's work in the

---

[1] J. L. Lagrange (1736–1813), generally regarded as the keenest mathematician of the eighteenth century, investigated numerous topics including the theory of equations. One of his methods of attack on the problem of finding roots of polynomials was to study permutations of these roots.

[2] Evariste Galois (1811–1832) was a brilliant French mathematician whose unorthodoxy contributed twice to his failing the entrance exams for the Ecole Polytechnique. At the age of 20, he was jailed twice as a radical. A year later, he died as a result of a duel, years before his most significant work was published and the true worth of his mathematical discoveries was recognized.

investigation of roots of equations and first introduced the term *group*. At that time, mathematicians worked with *groups of transformations*. These were sets of mappings that, under composition, possessed certain properties. Mathematicians such as Felix Klein[3] adopted the idea to unify areas of geometry. While Kronecker[4] gave a set of postulates for a group in 1870, it was not until the first decade of the twentieth century that the modern axioms of an abstract group were first published[5].

---

[3]Felix Klein (1849–1925) used group theory as his tool in his efforts to unify the mathematical sciences. In an address known as the "Erlangen Program," he declared that every geometry is actually the theory of invariants of a particular transformation group.

[4]Leopold Kronecker (1823–1891) was a prosperous businessman who accepted a professorship at the University of Berlin in 1883. He made his most significant contribution in algebra, although he worked in many areas.

[5]You may find it interesting to consult the original papers on this axiomatization; for example, L. E. Dickson's "Definition of a group and a field by independent postulates," *Trans. of Amer. Math. Soc.*, Vol. 6 (1905), pp. 198–204; and E. V. Huntington's "Note on the definition of abstract groups and fields by sets of independent postulates," *Trans. of Amer. Math. Soc.*, Vol. 6 (1905), pp. 181–197.

3.1

The Definition

Considering the symmetries investigated in the last chapter, we can make the following general observations:

1. If we perform one rigid motion and then another, the result is again one of our motions.
2. If we perform any three rigid motions, r, s, t, we get the same symmetry whether we find the result of r and s and then follow it by t or if we follow r by the result of s and t; that is, $(rs)t = r(st)$.
3. There exists a motion e such that, for all motions m, we have $me = em = m$.
4. For each motion m, there exists a particular motion m' such that $mm' = m'm = e$.

It is these properties that we wish to abstract and study in our first algebraic structure. Since they have a natural basis, it seems logical to include them in our basic definition.

Definition 3.1

A *group* $(G, *)$ is a set G (sometimes called the **ground set**) together with a binary operation $*$ on G such that

GA I: If $a, b, c \in G$, then $(a * b) * c = a * (b * c)$.
GA II: There exists $e \in G$ such that $e * a = a * e = a$ for all $a \in G$.
GA III: For each $a \in G$, there exists $a' \in G$ such that
        $a' * a = e = a * a'$.

Note: Inherent in the statement that $*$ is a binary operation on G is

GA 0:   If $a, b \in G$, then $a * b \in G$.

GA 0 is known as the **closure axiom**, and it is usually the axiom most overlooked by students in checking that the group properties hold for a given set and operation. GA II implies that $G \neq \varnothing$. The element $e \in G$ is called an **identity** for G and the element a' is termed an **inverse** of a. The symbol a' is read "a inverse." Note that e is an identity for all $a \in G$, while the element a' is related to a particular element of G. The **order** of a group $(G, *)$, denoted $|G|$, is the cardinality of its ground set G. A group $(G, *)$ is termed **finite** or said to have **finite order** if $|G|$ is finite. If $(G, *)$ is not finite, we say that $(G, *)$ is an **infinite group**.

The motions of an equilateral triangle provided us with an example of a group in which the group operation was not commutative. Therefore commutativity is not a property that can be derived from the other group axioms. A group in which the operation is commutative is given a special name.

Definition 3.2

A group $(G, *)$ is said to be **abelian**[6] or **commutative** if it has the additional property

GA IV: For all $a, b \in G$, $a * b = b * a$.

In the remainder of this section, we will present some examples of groups. We will not bother to justify the details of statements concerning

---

[6] The term *abelian* is derived from the name of Neils Henrik Abel (1802–1829), a Norwegian mathematician who, at the age of 22, showed that there exists no general formula for finding the roots of an arbitrary polynomial of degree five. In the standards of European scientific notation, the use of the lower case "a" in abelian indicates that the term is generally accepted; that is, the name has "made it."

examples with which we feel you should be familiar; instead we leave the verification of the axioms to you as an exercise.

| Example 3.1 |

The set of symmetries of a regular n-sided polygon, or a regular polyhedron such as a tetrahedron or cube, constitutes a group. For a particular example, recall the set of motions of the rectangle or the equilateral triangle we discussed in the last chapter.

| Example 3.2 |

Consider the set of integers $\mathbb{Z}$ under addition. Clearly addition is a commutative and associative operation on $\mathbb{Z}$, $e = 0$ is an additive identity, and $a' = -a$ is the additive inverse of a. Thus the integers under addition, denoted $(\mathbb{Z}, +)$, form an abelian group. The integers under the operation of multiplication, denoted $(\mathbb{Z}, \cdot)$, fail to form a group, for although $e = 1$ serves as an identity for the operation of multiplication, no integers other than 1 and -1 possess multiplicative inverses.

| Example 3.3 |

Let $G = 2\mathbb{Z}$, the even integers, and let $*$ be $+$. Then it is a straightforward exercise to show that $(2\mathbb{Z}, +)$ is an abelian group. Likewise, it is not difficult to show that $(\mathbb{Q}, +), (\mathbb{Q} - \{0\}, \cdot), (\mathbb{R}, +), (\mathbb{R} - \{0\}, \cdot), (\mathbb{C}, +)$, and $(\mathbb{C} - \{0\}, \cdot)$ are abelian groups under the appropriately defined operations of $+$ and $\cdot$.

| Example 3.4 |

Let X be a set, and let $\mathscr{P}(X)$ be the power set of X. Define

$$\triangle: \mathscr{P}(X) \times \mathscr{P}(X) \to \mathscr{P}(X)$$

by $\quad A \triangle B = (A - B) \cup (B - A)$

The fact that $\triangle$, known as the **symmetric difference**, is commutative and associative on $\mathscr{P}(X)$ is a consequence of Exercise 11 of Section 1.3. GA II is satisfied with $e = \varnothing$ since

$$A \triangle \varnothing = (A - \varnothing) \cup (\varnothing - A) = (A - \varnothing) = A$$

GA III is satisfied because every subset is its own inverse:

$$A \triangle A = (A - A) \cup (A - A) = \varnothing \cup \varnothing = \varnothing$$

Thus $(\mathscr{P}(X), \triangle)$ is an abelian group.

| Example 3.5 |

For each $n \in \mathbb{N}$, $\mathbb{Z}_n$ forms an abelian group under modular addition $\oplus$. To verify this, note first that we have already shown that modular addition is associative and commutative on $\mathbb{Z}_n$. Clearly 0 serves as an identity for $\oplus$ on $\mathbb{Z}_n$ since the remainder of $k + 0$ upon division by n is the same as the remainder of k. Now that the existence of an identity has been established, we can proceed to verify the existence of inverses. To do this, note that if $k \in \mathbb{Z}_n$, then $n - k \in \mathbb{Z}_n$, also. We have

$$k + (n - k) = n$$

and the remainder upon dividing n by n is 0. Thus

$$k \oplus (n - k) = 0 \qquad \text{and} \qquad k' = n - k$$

$\odot$	1	2	3	4
1	1	2	3	4
2	2	4	1	3
3	3	1	4	2
4	4	3	2	1

Figure 3.1

$\odot$	1	2	3	4	5
1	1	2	3	4	5
2	2	4	0	2	4
3	3	0	3	0	3
4	4	2	0	4	2
5	5	4	3	2	1

Figure 3.2

What about modular multiplication? Does $(\mathbb{Z}_n, \odot)$ constitute a group? The answer is no, since the element 1 is clearly the only possibility for a multiplicative identity and yet 0 then could never have a multiplicative inverse; that is, if $k \in \mathbb{Z}_n$, then $k \odot 0 = 0 \neq 1$. You might ask whether or not we can get a group by ignoring the element 0. In other words, if $\mathbb{Z}_n^* = \mathbb{Z}_n - \{0\}$, then is $(\mathbb{Z}_n^*, \odot)$ a group? The answer is that it depends upon the value of n. For $n = 5$, the Cayley table in Figure 3.1 shows that $\odot$ is a binary operation on $\mathbb{Z}_5^*$. We know that $\odot$ is associative and commutative on $\mathbb{Z}_5$. The element 1 is an identity, and $1' = 1$, $2' = 3$, $3' = 2$, $4' = 4$. Thus each element has an inverse, and we have shown that $(\mathbb{Z}_5^*, \odot)$ is an abelian group.

Now consider $\mathbb{Z}_6^*$, whose Cayley table under $\odot$ is given in Figure 3.2. Clearly $\odot$ is not a binary operation on $\mathbb{Z}_6^*$, since $2 \odot 3 = 0$ and $0 \notin \mathbb{Z}_6^*$. Thus $(\mathbb{Z}_6^*, \odot)$ is *not* a group. What distinguishes $n = 5$ from $n = 6$? The answer is that 5 is a prime and 6 is not. If $n = st$, where $1 < s < n$ and $1 < t < n$, then, in $\mathbb{Z}_n^*$, we have $s \odot t = 0$. This proves that if n is not a prime, then $(\mathbb{Z}_n^*, \odot)$ is not a group. If $n = p$, a prime, then $(\mathbb{Z}_p^*, \odot)$ *is a group*. We leave the verification of this as Exercise 11.

Example 3.6

On $M_2(\mathbb{Z})$, the set of all 2×2 matrices with integer entries, the operation of addition defined in the last chapter is a commutative and associative operation. The matrix

$$\begin{bmatrix} 0 & 0 \\ 0 & 0 \end{bmatrix}$$

serves as an identity and, since

$$\begin{bmatrix} a & b \\ c & d \end{bmatrix} + \begin{bmatrix} -a & -b \\ -c & -d \end{bmatrix} = \begin{bmatrix} 0 & 0 \\ 0 & 0 \end{bmatrix}$$

every matrix has an additive inverse. Thus $(M_2(\mathbb{Z}), +)$ is an abelian group.

Example 3.7

The set $M_2(\mathbb{Z})$ under matrix multiplication fails to form a group. Although the matrix

$$\begin{bmatrix} 1 & 0 \\ 0 & 1 \end{bmatrix}$$

can be shown to be an identity for multiplication, not every matrix has a multiplicative inverse. Details of this discussion are contained in the exercises.

Example 3.8

This example involves the idea of rigid motions or symmetries developed in the last chapter. Let P_n denote a regular polygon in the plane with n sides, $n \geqslant 3$. The set of symmetries of P_n forms a group D_n. The elements of this group are

    a. the rotations about the center of the polygon through angles of measure $0, 2\pi/n, 4\pi/n, \ldots, 2(n-1)\pi/n$; and

    b. the n reflections about the lines joining vertices of P_n to midpoints of opposite sides (if n is odd), or the n reflections about the lines joining vertices of P_n opposite to one another and sides of P_n opposite to one another (if n is even).

The group D_n is termed the **dihedral group of order** $2n$. If ρ denotes the rotation corresponding to the angle of measure $2\pi/n$ and δ denotes any of the reflections, then it can be shown that $\rho^n = \delta^2 = 1$ and $\rho\delta\rho = \delta$. By ρ^n we mean the motion resulting from performing ρ n times. The elements of D_n are therefore

$$1, \rho, \rho^2, \rho^3, \ldots, \rho^{n-1}, \delta, \rho\delta, \rho^2\delta, \ldots, \rho^{n-1}\delta$$

One way to describe this group is to write

$$D_n = \langle \rho, \delta \mid \rho^n = \delta^2 = 1, \delta\rho\delta = \rho^{n-1} \rangle$$

This procedure is known as defining a group in terms of **generators** and **relations**.

Example 3.9

Consider the set C of all real-valued continuous functions on $\mathbb{R}$. If $f, g \in C$, define $f + g$ by $(f + g)(x) = f(x) + g(x)$. In calculus you learned that the sum of two continuous functions is also continuous. Therefore, function addition is a binary operation on C. The function $0(x) = 0$ is an identity for this operation and if $f \in C$, then the function $-f$ defined by $(-f)(x) = -(f(x))$ is the inverse of f. We conclude $(C, +)$ is a group.

Exercises 3.1

1. Why do the following sets together with the designated operations not constitute a group?

    a. $(\mathbb{N}, +)$ b. $(\mathbb{Z}, -)$ c. $(\mathbb{Z}, \div)$

    d. The set of all integer-valued constant mappings on the integers with the operation being composition of mappings.

    e. The odd integers with the operation being addition of integers.

2. Let $G = \mathbb{Z} \times \mathbb{Z}$. Define $(a, b) \# (c, d) = (ac - bd, ad + bc)$. Show that G is a commutative operation and that $(G, \#)$ is not a group.

3. Let $*$ be defined on $\mathbb{Z}$ by $a * b = a + b - 1$ for $a, b \in \mathbb{Z}$, and $\circ$ be defined on $\mathbb{Z}$ by $a \circ b = a + b + ab$ for $a, b \in \mathbb{Z}$. Find identities for $*$ and $\circ$. Is $(\mathbb{Z}, *)$ or $(\mathbb{Z}, \circ)$ a group? Why or why not?

4. Let $G = \mathbb{Z} \times \mathbb{Z}$, and define $*$ on G by $(a, b) * (c, d) = (a + c + 1, b + d)$. Prove that $(G, *)$ is an abelian group.

5. Set $G = \mathbb{Z} \times \mathbb{Z}$ and define $*$ on G by $(a, b) * (c, d) = (a - d, b - c)$. Show that $(G, *)$ is not a group. Is $(0, 0)$ an identity for $*$?

6. Construct the Cayley table for symmetric difference $\triangle$ on $\mathscr{P}(X)$ if $X = \{a, b, c\}$.

7. Verify that if X is a nonempty set, then $(\mathscr{P}(X), \cup), (\mathscr{P}(X), \cap),$ and $(\mathscr{P}(X), -)$ are not groups. In each case, explain what goes wrong.

8. Verify that the set of symmetries of the square forms a group of order 8. List the inverse of each element. Is the group abelian?

9. Construct the Cayley table for the set of mappings $\{e(x) = x, f(x) = 1/x\}$ on the set $\mathbb{R}^+$ under composition of functions. Is this set a group? Do the same for

    $$\{e(x), g(x) = 1 - x\} \qquad \text{and} \qquad \{e(x) = x, f(x) = 1/x, g(x) = 1 - x\}$$

10. If $G = \{u\}$ and $\circ: G \times G \rightarrow G$ is defined by $u \circ u = u$, verify that $(G, \circ)$ is a group.

11. Recall that if p and k are relatively prime integers, then there exist $s, t \in \mathbb{Z}$ such that $1 = sp + tk$. Use this result to prove that if p is a prime and $k \in \mathbb{Z}_p^*$, then k has a multiplicative inverse in $\mathbb{Z}_p^*$. Use this result to conclude that if p is a prime, then $(\mathbb{Z}_p^*, \odot)$ is a group.

12. Construct the Cayley tables for $(\mathbb{Z}_2^*, \odot), (\mathbb{Z}_3^*, \odot),$ and $(\mathbb{Z}_7^*, \odot)$.

13. Verify that the elements $1, 3, 5, 7 \in \mathbb{Z}_8$ form a group under $\odot$.

14. Which elements in $\mathbb{Z}_9$ have multiplicative inverses? Show that they form a group under $\odot$.

15. If

    $$M_2(\mathbb{Z}) = \left\{ \begin{bmatrix} a & b \\ c & d \end{bmatrix} \middle| a, b, c, d \in \mathbb{Z} \right\} \qquad \text{and}$$

    $$\begin{bmatrix} a & b \\ c & d \end{bmatrix} \times \begin{bmatrix} r & s \\ t & u \end{bmatrix} = \begin{bmatrix} ar + bt & as + bu \\ cr + dt & cs + du \end{bmatrix}$$

    a. Verify that $\begin{bmatrix} 1 & 0 \\ 0 & 1 \end{bmatrix}$ is an identity for $\times$.

    b. Prove that $\begin{bmatrix} 2 & 3 \\ 2 & 3 \end{bmatrix}$ has no inverse.

    c. Set $G = \left\{ \begin{bmatrix} a & b \\ c & d \end{bmatrix} \middle| ad - bc = 1 \right\}$

    Prove that $(G, \times)$ is a group with inverse given by

    $$\begin{bmatrix} a & b \\ c & d \end{bmatrix}^{-1} = \begin{bmatrix} d & -b \\ -c & a \end{bmatrix}$$

    Is this group abelian?

16. Let $(G, *)$ be a group and $M_2(G) = \left\{ \begin{bmatrix} a & b \\ c & d \end{bmatrix} \middle| a, b, c, d \in G \right\}$

Define

$$\begin{bmatrix} a & b \\ c & d \end{bmatrix} \oplus \begin{bmatrix} e & f \\ g & h \end{bmatrix} = \begin{bmatrix} a*e & b*f \\ c*g & d*h \end{bmatrix}$$

Is $(M_2(G), \oplus)$ a group? Under what conditions? Under what conditions is $(M_2(G), \oplus)$ an abelian group?

17. Is $(\mathscr{P}(X), \triangle)$ a group if $X = \varnothing$? Why?

18. Let p be a prime. In the group $(\mathbb{Z}_p^*, \odot)$, prove that $p - 1$ is its own inverse and that $(p - 1)! = p - 1$ in this group.

19. Let D_n be the dihedral group of order $2n$ described in Example 3.9. If ρ denotes the rotation corresponding to the angle of measure $2\pi/n$ and δ denotes any of the reflections, then verify that $\rho^n = \delta^2 = 1$, $\rho\delta\rho = \delta$, and that the elements of D_n are

$$1, \rho, \rho^2, \rho^3, \ldots, \rho^{n-1}, \delta, \rho\delta, \rho^2\delta, \ldots, \rho^{n-1}\delta$$

20. Construct the Cayley tables for the dihedral groups D_3 and D_4.

3.2

Basic Properties and Alternate Definitions

While the definition that we gave for a group is the most standard one, there are other forms of the definition that could have been used. Before we consider some of these alternate versions, let us look to see what properties groups have in general. The most useful of all properties is the cancellation property.

Theorem 3.1

Cancellation Property

Let $(G, *)$ be a group and $a, b, c \in G$. If $c * a = c * b$, then $a = b$, and if $a * c = b * c$, then $a = b$.

Proof

Since $c \in G$, there exists $c' \in G$ such that $c' * c = e$ (GA III). If $c * a = c * b$, then the fact that $*$ is a binary operation implies $c' * (c * a) = c' * (c * b)$ and

$$a = e * a$$
$$= (c' * c) * a$$
$$= c' * (c * a)$$
$$= c' * (c * b)$$
$$= (c' * c) * b$$
$$= e * b$$
$$= b$$

by repeated use of GA I and GA II. The proof of the second part of the theorem is similar.   □

Corollary 3.1

Let $(G, *)$ be a group and $a, b \in G$. Then either

$$a = b * a \quad \text{or} \quad a = a * b \qquad \text{implies } b = e$$

One of the problems that one is faced with when attempting to do mathematics is that of notation. We will usually use "·" or simply juxtaposition[7] to denote the operation of a group. Sometimes the symbol "+" will be used to indicate the operation in an **abelian** group. You should be careful and realize that "+" and "·" will denote many different operations and not just addition and multiplication in $\mathbb{N}$. If we use the additive symbol "+" in our discussion of a group, we regularly will write 0 for the identity of G and $-a$ for the inverse of the element $a \in G$. If we use the symbol "·" or juxtaposition, we will usually write 1 for the identity of G and indicate a' by a^{-1}. Often we will abuse notation and say "let G be a group" when of course we mean "let $(G, \cdot)$ be a group." If any danger to understanding might arise because of this notation or lack of it, we will try to alert you.

If we use juxtaposition to denote our operation, we can restate our last corollary as "In a group G, $a = ba$ or $a = ab$ implies $b = 1$." This certainly is more concise and there should be no confusion as to the meaning.

Because of the Cancellation Property of groups, an element of a finite group appears exactly once in each row and each column of the group's Cayley table. To see this, suppose the element b appears in the row labeled by a and in the columns labeled by both c and d. Then $b = ac = ad$. (See Figure 3.3.) But then $c = d$ by the Cancellation Property, contradicting the fact that c and d are distinct.

	a	b	c	d
a			b	b
b				
c				
d				

Figure 3.3

Often these ideas can be used to complete a Cayley table for a group operation given some partial information. For example, suppose we know that Figure 3.4a is part of the Cayley table of a group whose elements are a, b, and c. By the fragment, we see that $bc = a$. Therefore, neither b nor c can be the identity of G. Thus, a must be the identity, and we obtain Figure 3.4b. Using the fact that an element of a group appears exactly once in any row or column, we must conclude that $bb = c$, $cc = b$, and $cb = a$. The completed table appears in Figure 3.4c.

The next theorem contains important results about groups.

Figure 3.4

	a	b	c
a			
b			a
c			

(a)

	a	b	c
a	a	b	c
b	b		a
c	c		

(b)

$\cdot$	a	b	c
a	a	b	c
b	b	c	a
c	c	a	b

(c)

---

[7]Juxtaposition is the placing of two elements side-by-side; that is, writing ab for $a \cdot b$.

Theorem 3.2

Let G be a group. Then

    i. the identity is unique;
    ii. a^{-1} is unique for all $a \in G$;
    iii. $(ab)^{-1} = b^{-1}a^{-1}$ for all $a, b \in G$;
    iv. $(a^{-1})^{-1} = a$ for all $a \in G$.

Proof

    i. Suppose that e and f are both identities of G. Then $e = ef = fe = f$. Thus G possesses exactly one identity.
    ii. If b is another element of G such that $ab = 1$, then $ab = 1 = aa^{-1}$ and $b = a^{-1}$ follows by the Cancellation Property.
    iii. By definition of the inverse of an element, $(ab)^{-1}(ab) = 1$. But

$$(b^{-1}a^{-1})(ab) = b^{-1}(a^{-1}a)b$$

$$= b^{-1}1b$$

$$= b^{-1}b$$

$$= 1$$

    Since $(ab)^{-1}(ab) = 1$, the Cancellation Property implies $(ab)^{-1} = b^{-1}a^{-1}$.
    iv. By GA III, $(a^{-1})^{-1}a^{-1} = 1$, but $aa^{-1} = 1$. Since we have shown that inverses of elements are unique, we must have that $a = (a^{-1})^{-1}$.

                                                    □

Part iii of Theorem 3.2 causes a good deal of difficulty among students. Most students incorrectly assume that "the inverse of a product is the product of the inverses"; that is $(ab)^{-1} = a^{-1}b^{-1}$. This statement, however, is true only for abelian groups. To see this, recall the group of symmetry motions of an equilateral triangle given in Figure 2.14. Here $\rho\tau_A = \tau_B$ so that

$$(\rho\tau_A)^{-1} = \tau_B^{-1} = \tau_B$$

Sure enough,

$$\tau_A^{-1}\rho^{-1} = \tau_A\rho^2 = \tau_B$$

but

$$\rho^{-1}\tau_A^{-1} = \rho^2\tau_A = \tau_C \neq \tau_B$$

Therefore

$$(\rho\tau_A)^{-1} = \tau_A^{-1}\rho^{-1}$$

but

$$(\rho\tau_A)^{-1} \neq \rho^{-1}\tau_A^{-1}$$

We have mentioned that there are alternate versions for the definition of a group. We will state three such forms, one applying only to finite groups. The first shows that our original definition contained more than was necessary.

Theorem 3.3

Let G be a nonempty set with a binary operation $*$ such that

GA I′: $a * (b * c) = (a * b) * c$ for all a, b, $c \in G$.

GA II′:   There exists $e \in G$ such that $e * a = a$ for all $a \in G$ (e is a **left identity**).

GA III′:   For each $a \in G$, there exists $a' \in G$ such that $a' * a = e$ (a' is a **left inverse**).

Then $(G, *)$ is a group.

Proof

Since GA I′ agrees with GA I, we need only show that a left identity is also a right identity and that a left inverse is also a right inverse. First observe that our proof of the Left Cancellation Property ($c * a = c * b$ implies $a = b$) involved only associativity, left inverses, and a left identity. Thus the left cancellation property holds for this set of axioms also.

Suppose $a \in G$ and $a' * a = e$. Then

$$a' * a = e = e * e = (a' * a) * e = a' * (a * e)$$

By the Left Cancellation Property, we have $a = a * e$. Thus a left identity is a right identity.

To prove that a left inverse is also a right inverse, consider $a' = a' * e$. Now

$$a' = e * a' = (a' * a) * a' = a' * (a * a')$$

by repeated use of the new axioms. By Corollary 3.1, $e = a * a'$. Since a was an arbitrary element of G, the result follows. □

It should be clear to you that we also obtain a group if simultaneously we replace GA II by the statement

GA II″:   There exists $e \in G$ such that $a * e = a$ for all $a \in G$.

and GA III by the condition

GA III″:   For each $a \in G$ there exists $a' \in G$ such that $a * a' = e$.

In view of these results, it is natural to inquire whether or not a set closed under an associative binary operation with **right** inverses and a **left** identity is a group. In Exercise 11, we give an example to show that the answer is no.

The next result shows that, for a finite group, the identity and inverse axioms can be replaced by the Cancellation Property.

Theorem 3.4

Let G be a **finite** nonempty set together with an associative binary operation $*$ on G. Suppose, moreover, that for all a, b, $c \in G$,

$$a * c = b * c \quad \text{implies} \quad a = b$$

and

$$c * a = c * b \qquad \text{implies} \qquad a = b$$

Then G is a group under the operation $*$.

Proof

Let $a \in G$ and $G = \{a = a_1, a_2, \ldots, a_n\}$. Consider the set

$$H = \{a * a_1, a * a_2, \ldots, a * a_n\}$$

If $a * a_i = a * a_j$, then $a_i = a_j$. (Why?) Therefore, all the elements of H are distinct. Since H is a subset of G (Why?) containing the same number of elements as G and G is a finite set, then $H = G$. Since $a \in G$, there must exist some k such that $1 \leqslant k \leqslant n$ and $a * a_k = a$.

We first show that a_k is not only a right identity for a, but also is a left identity for a. Since $a * a_k = a$,

$$a * a = (a * a_k) * a = a * (a_k * a)$$

By the Left Cancellation Property in the hypothesis, $a = a_k * a$. We now claim that a_k is a left identity for all other elements of G. To do this, let $b \in G$. Since $G = H$, $b = a * a_j$ for some j such that $1 \leqslant j \leqslant n$. Then

$$a_k * b = a_k * (a * a_j) = (a_k * a) * a_j = a * a_j = b$$

and a_k is a left identity.

It remains to show that every element in G possesses a left inverse in G. Let $c \in G$. Consider the set $K = \{a_1 * c, a_2 * c, \ldots, a_n * c\}$. Using an argument similar to that in the first paragraph but with the other cancellation property of the hypothesis, we have $G = K$. Since our identity a_k is an element of G, there exists some t such that $1 \leqslant t \leqslant n$ and $a_t * c = a_k$; that is, a_t is a left inverse of c. Since c was an arbitrary element of G, every element of G possesses a left inverse. Therefore G is a group. $\square$

We leave the proof of our last result to the exercises. (See Exercise 13.)

Theorem 3.5

Let $*$ be an associative binary operation on a set G with properties that

    i. for every $a \in G$, there exists $a' \in G$ such that $a * a' * a = a$ and $a' * a * a' = a'$; and
    ii. there exists a unique element $e \in G$ such that $e * a = a$ for all $a \in G$.

Then $(G, *)$ is a group.

Exercises 3.2

1. Give an example of a group G where $ab = bc$ for some $a, b, c \in G$ does not imply $a = c$.

2. Let G be a group. Use only the definition of group to prove that if $x \in G$ and $xx = x$, then $x = 1$.

3. If G is a group in which $(ab)^{-1} = a^{-1}b^{-1}$ for all $a, b \in G$, show that G must be abelian.

4. Let G be a group. Prove that G is abelian if and only if $(ab)^{-1} = a^{-1}b^{-1}$ for all $a, b \in G$.

5. Let G be a group. Prove that G is abelian if and only if $(ab)^{-1} = a^{-1}b^{-1}$ for all $a, b \in G$.

6. Prove: G is an **abelian** group if and only if G is a set together with a binary operation (denoted by juxtaposition) such that GA II and GA III hold and that $a(bc) = (ba)c$ for all $a, b, c \in G$.

7. Use only the definition of group to prove that if $a, b, c \in G$ and $ac = bc$, then $a = b$.

8. Use induction to demonstrate that if $a_1, a_2, \ldots, a_n \in G$, a group, then

$$(a_1 a_2 \cdots a_n)^{-1} = a_n^{-1} a_{n-1}^{-1} \cdots a_2^{-1} a_1^{-1}$$

9. Using Cayley tables, show that any group with fewer than five elements must be abelian.

10. Beachcomber Bill found this note on his deserted isle:
    "A chest of gold for him who is able
    To determine x in this group's Cayley table.
    a. under giant rock   b. in lagoon
    c. in cave   d. near large palm tree."
    Help Bill find the fortune. (See table at left.)

	a	b	c	d
a	x			
b				
c			b	
d		b		

11. Let $*$ be defined on $\mathbb{N}$ by $a * b = b$ for all $a, b \in \mathbb{N}$. Show that $*$ is an associative binary operation on $\mathbb{N}$ with left identity 1. Show that every $a \in \mathbb{N}$ possesses a right inverse. Is $(\mathbb{N}, *)$ a group? Why?

12. Consider the example of the group described in Exercise 15 of the preceding section. Show that

$$\begin{bmatrix} 1 & 3 \\ 1 & 4 \end{bmatrix}^{-1} = \begin{bmatrix} 4 & -3 \\ -1 & 1 \end{bmatrix} \quad \text{and} \quad \begin{bmatrix} 2 & 3 \\ 3 & 5 \end{bmatrix}^{-1} = \begin{bmatrix} 5 & -3 \\ -3 & 2 \end{bmatrix}$$

Calculate

$$\left(\begin{bmatrix} 1 & 3 \\ 1 & 4 \end{bmatrix} \begin{bmatrix} 2 & 3 \\ 3 & 5 \end{bmatrix} \right)^{-1}$$

13. Prove Theorem 3.5.

14. In a group G, the associative property is given by $(ab)c = a(bc)$ for all $a, b, c \in G$. If $a_i \in G$ for $i = 1, 2, \ldots, n$, then define $a_1 a_2 \cdots a_n = (a_1 a_2 \cdots a_{n-1})a_n$. Prove that $a_1 a_2 \cdots a_n$ is the same no matter where the parentheses are located. This is called **generalized associativity**.

15. If $(G_1, \cdot)$ and $(G_2, \circ)$ are groups, define $*$ on $G_1 \times G_2$ by

$$(a_1, b_1) * (a_2, b_2) = (a_1 \cdot a_2, b_1 \circ b_2)$$

Prove that $(G_1 \times G_2, *)$ is a group.

16. In the proof of Theorem 3.4, show that every element in G possesses a right inverse.

4 | Subgroups

Most interesting results arise through the investigation of the inner structure of groups—the way things fit together. The axioms of a group permit exceedingly complex structure. One way of analyzing a group of large order is to investigate certain subsets that it contains. In Section 3.2, we saw that there can exist groups contained in other groups. For example, the group of even integers lies in the group of all integers. By studying subsets of a group that are themselves groups, called *subgroups*, one can often determine vital information about the larger group.

In this chapter, we will first consider a class of groups that are very simple in structure. The groups in this class are called *cyclic groups*. An interesting fact about cyclic groups is that they can be found as subgroups in any group. In the second section, we will see that there also can be subgroups that are not cyclic. We will investigate the nature of subgroups in general, and consider how properties of a group affect its subgroups. In later chapters, questions concerning the relationship between the set of subgroups of a group and the set of mappings on the group that preserve group properties will be answered.

Cyclic Groups

We noticed in Chapter 3 that one of the basic group concepts is closure; that is, if G is a group and $a, b \in G$, then $ab \in G$. Applying this property to a single element a, we see that $a^2 = aa$, $a^3 = a^2 a$, and $a^4 = a^3 a$ are all elements of G. In fact, each successive composite is also in G. An interesting question is, "When does this procedure produce *all* the elements of G?" In other words, given a group G, when does an element $a \in G$ exist such that every other element in G is obtained by successively composing a with itself?

As an example, consider the group $\mathbb{Z}_5^*$ of nonzero integers modulo 5 under the operation of modular multiplication. Here in $\mathbb{Z}_5^*$, $2^2 = 4$, $2^3 = 3$, and $2^4 = 1$. Therefore every element of $\mathbb{Z}_5^*$ is a power of 2.

As another example, recall the group of rigid motions of a rectangle discussed in Section 2.3. An examination of the Cayley table in Figure 2.13 reveals that $x^2 = e$ for all x in this group. Therefore, if you take successive powers of any element in this group, you never will obtain any new elements other than the identity e.

In order to discuss the differences between these two examples, we must adopt some new notation. If G is a group and $a \in G$, define

$$a^1 = a$$
$$a^2 = aa$$

and

$$a^{n+1} = (a^n)a \qquad \text{for } n \in \mathbb{N}$$

This is called a **recursive definition** in that it allows us to determine a given power by using the preceding element in a sequence of powers.

If we are using additive notation, this definition takes the form

$$1a = a$$
$$2a = a + a$$

and

$$(n+1)a = na + a \qquad \text{for } n \in \mathbb{N}$$

Be careful: The plus sign on the left side of the last equation is addition of natural numbers while the plus on the right side refers to the group operation.

Consider the group of integers $\mathbb{Z}$ under addition. If $a = 1$, then the set of successive composites of 1 with itself is the set of all natural numbers. This set is not a group, however, because the identity of the group and inverses of the elements are lacking. Let us extend our notational definition as follows:

Multiplicative Notation	Additive Notation
$a^0 = 1$	$0a = 0$ (the group identity)
$a^{-n} = (a^{-1})^n$	$(-n)a = n(-a)$

There are advantages to this notation. For one, many of the traditional laws of exponents continue to hold.

Theorem 4.1

If G is a group, $a \in G$, and $m, n \in \mathbb{Z}$, then

    i. $a^m a^n = a^{m+n}$
    ii. $(a^n)^{-1} = a^{-n}$
    iii. $(a^m)^n = a^{mn}$

Proof

We divide the proof of part i into a number of cases. If $m = 0$ or $n = 0$, then the result is clear. Suppose then that $n > 1$. We proceed by induction. For $n = 1$, we have

$$a^m a^1 = a^m a$$
$$= a^{m+1}$$

Now suppose that the result is true for $n = k$. Then

$$a^m a^{k+1} = a^m (a^k a)$$
$$= (a^m a^k) a$$
$$= (a^{m+k}) a$$
$$= a^{(m+k)+1}$$
$$= a^{m+(k+1)}$$

By induction, part i holds for all $n \in \mathbb{N}$.

If $m < 0$ and $n < 0$, then $-m > 0$ and $-n > 0$ so that

$$a^m a^n = (a^{-1})^{-m} (a^{-1})^{-n}$$
$$= (a^{-1})^{(-m)+(-n)}$$
$$= (a^{-1})^{-(m+n)}$$
$$= a^{m+n}$$

So all that remains in the proof of our first identity is the case where one exponent is negative and the other is positive. We leave this for the exercises. (See Exercise 11.)

To prove part ii, note that our first identity implies

$$a^{-n} a^n = a^{-n+n}$$
$$= a^0$$
$$= 1$$

Thus $a^{-n} = (a^n)^{-1}$.

The proof of our third identity (part iii), in the case where $n = 0$, is straightforward. For $n \geqslant 1$, we again use induction. By definition,

$$(a^m)^1 = a^{m1}$$

$$= a^m$$

If the result holds for $n = k$, then

$$(a^m)^{k+1} = (a^m)^k a^m$$

$$= a^{mk} a^m \qquad \text{by induction}$$

$$= a^{mk+m}$$

$$= a^{m(k+1)}$$

If $n < 0$, set $t = -n$. Then

$$(a^m)^n = (a^m)^{-t}$$

$$= [(a^m)^{-1}]^t \qquad \text{by definition}$$

$$= (a^{-m})^t \qquad \text{by part ii}$$

$$= a^{(-m)t} \qquad \text{since } t > 0$$

$$= a^{m(-t)} = a^{mn} \qquad \qquad \square$$

This theorem provides us with all the tools necessary to prove the following result:

Theorem 4.2

Let G be a group and $a \in G$. If $H = \{a^k | k \in \mathbb{Z}\}$, then H is an abelian group.

Proof

If a^m, $a^n \in H$, then $a^m a^n = a^{m+n} \in H$ by part i of Theorem 4.1. Therefore composition is an operation on H. Since $(a^m a^n)a^t$ and $a^m(a^n a^t)$ both equal a^{m+n+t}, the associativity of composition on H follows from part i of Theorem 4.1 as well as the associativity of addition in $\mathbb{Z}$. Clearly $1 = a^0 \in H$, and if $a^n \in H$, then $(a^n)^{-1} = a^{-n} \in H$ by part ii of Theorem 4.1.

To complete the proof and show that H is abelian, we need only point out that if a^m, $a^n \in H$, then $a^m a^n = a^{m+n} = a^{n+m} = a^n a^m$. $\square$

Motivated by the preceding theorem, we make the following definition:

Definition 4.1

If G is a group and there exists $a \in G$ such that $G = \{a^k | k \in \mathbb{Z}\}$ then G is called a **cyclic group** and a is called a **generator** of G. We denote this by writing $G = \langle a \rangle$.

For example, consider first $(\mathbb{Z}, +)$. Here we need to interpret Definition 4.1 in terms of additive notation. If $a = 1$, then $H = \langle 1 \rangle = \{n1 | n \in \mathbb{Z}\} = \{n | n \in \mathbb{Z}\} = \mathbb{Z}$. Therefore $\mathbb{Z}$ is a cyclic group. For $a = 2$, we have $H = \langle 2 \rangle = \{n2 | n \in \mathbb{Z}\} = 2\mathbb{Z}$, the group of even integers.

Let $G = (\mathbb{Z}_5^*, \odot)$. Then $\langle 1 \rangle = \{1^n | n \in \mathbb{Z}\} = \{1\}$, but $\langle 2 \rangle = \{2^n | n \in \mathbb{Z}\} = \{1, 2, 4, 3\} = \mathbb{Z}_5^*$; so $\mathbb{Z}_5^*$ is a cyclic group.

We have seen already that the motions of a rectangle do not form a cyclic group. We leave it to the exercises to show that neither the symmetries of a square nor those of an equilateral triangle constitute a cyclic group.

Two questions arise from the examples above. The first concerns whether a cyclic group can have more than one generator. In the example $(\mathbb{Z}_5^*, \odot)$, it is not hard to show that 3 is another generator. The integers $(\mathbb{Z}, +)$ are also generated by -1.

The second question involves the need to consider negative powers of a generator. Under what conditions is a cyclic group generated by just the positive powers of an element? The answer is contained in the next result.

Theorem 4.3

If $G = \langle a \rangle$ is cyclic of finite order, then $G = \{a, a^2, a^3, \ldots, a^m = 1\}$ with $|G| = m$.

Proof

Since G is a finite group, the infinite set of powers $\{a^n | n \in \mathbb{N}\}$ must contain some duplication. If $a^s = a^t$ with $s < t$, then

$$a^{t-s} = a^t a^{-s}$$

$$= a^s a^{-s}$$

$$= a^0$$

$$= 1$$

Thus the set $W = \{n \in \mathbb{N} | a^n = 1\}$ is nonempty. Let m' be the least element of W. Then $a^{m'} = 1$. Moreover if $k \in \mathbb{Z}$, the Division Algorithm for $\mathbb{Z}$ asserts that there exist $q, r \in \mathbb{Z}$ such that $k = qm' + r$ where $0 \leqslant r < m'$. Then

$$a^k = a^{qm' + r}$$

$$= (a^{m'})^q a^r$$

$$= 1^q a^r$$

$$= a^r$$

This shows that $G = \langle a \rangle = \{a^k | k \in \mathbb{Z}\} \subseteq \{a, a^2, \ldots, a^{m'} = 1\} \subseteq G$. The proof of the fact that these elements are distinct is left to you as Exercise 9. Then $m' = |G|$. $\square$

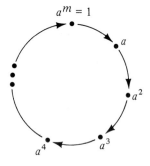

$a^m = 1$

a

a^2

a^4 a^3

Figure 4.1

This result shows that finite cyclic groups are generated by positive powers. It also allows you to visualize the group action in a cyclic group G generated by an element a. If the elements of G are located in order about a circle, then the effect of combining a with the other elements of G can be depicted by arrows. See Figure 4.1 for an example.

If $G = \langle a \rangle$ is a group of order m, then m is the smallest positive integer such that $a^m = 1$. We symbolize this by $o(a) = m$ and say that m is the **order of** a. By Theorem 4.3, the order of a is equal to the order of the cyclic group it generates. The concept of order is quite important in the study of groups. The order of a group is related to the order of the elements of the group in a number of ways. Often, the structure of a group is influenced by the orders of its elements. In Exercise 5 of Section 3.2, you were asked to prove that a group is abelian if every element in that group has order 1 or 2. A similar result is contained in Exercise 15 of this section, and many others will be encountered in future sections.

We will eventually show how cyclic groups form the building blocks of more involved groups. The structure of cyclic groups, however, is quite simple. In fact, we will see in a later chapter that there are essentially only two types of cyclic groups. Infinite cyclic groups are alternate versions of the integers $(\mathbb{Z}, +)$, while finite cyclic groups can be thought of as groups of the form $(\mathbb{Z}_n, \oplus)$.

Exercises 4.1

1. Which of the following groups are cyclic? If cyclic, find a generator.
   a. $3\mathbb{Z}$, the integer multiples of 3
   b. The rotations of an equilateral triangle
   c. $(\mathbb{Z}_4, \oplus)$
   d. $(\mathbb{R}, +)$
   e. The rotations of a regular pentagon
   f. $(\mathbb{Z}_7^*, \odot)$
   g. $(M_2(\mathbb{Z}_3), +)$, the set of all 2×2 matrices with entries in $(\mathbb{Z}_3, \oplus)$
   h. The symmetries of an equilateral triangle
   i. The symmetries of a square

2. Prove that 1 and -1 are the only generators of $(\mathbb{Z}, +)$.

3. Find all generators of the following cyclic groups.
   a. $(\mathbb{Z}_6, \oplus)$ b. $(\mathbb{Z}_5^*, \odot)$ c. $(2\mathbb{Z}, +)$ d. $(\mathbb{Z}_{11}^*, \odot)$

4. Prove that if $G = \langle a \rangle$ has order n, then $a^n = 1$.

5. Show that if $G = \langle a \rangle$ has order n and $b \in G$, then $b^n = 1$.

6. If $G = \langle a \rangle$ has order n and $k < n$, verify that the inverse of a^k is a^{n-k}. Use this result to show that every cyclic group of even order contains at least two elements that are their own inverses.

7. If $G = \langle a \rangle$ and there exists $k \in \mathbb{N}$ such that $a^k = 1$, prove that G is a finite group.

8. Show that $(\mathbb{Z}_n, \oplus)$ is cyclic for all $n \in \mathbb{N}$.

9. Let $G = \langle a \rangle$ have order n, and suppose $1 \leqslant s < n$ and $1 \leqslant t < n$. If $a^s = a^t$, prove $s = t$.

10. For each $n \in \mathbb{N}$, find a cyclic group of order n.

11. Complete the proof of part i of Theorem 4.1.

12. Suppose $G = \langle a \rangle$ has order n. Prove that a^k is another generator of G if and only if $(k, n) = 1$. (If n is a natural number, the **Euler**[1] **phi-function** of n, denoted $\phi(n)$, is defined by $\phi(1) = 1$ and, for $n > 1$, $\phi(n)$ equals the number of natural numbers both less than n and relatively prime to n. This exercise says that a cyclic group of order n has $\phi(n)$ generators.)

13. Find all generators of the cyclic groups $(\mathbb{Z}_9, \oplus), (\mathbb{Z}_{12}, \oplus), (\mathbb{Z}_{21}, \oplus)$, and $(\mathbb{Z}_{13}^*, \odot)$.

14. Determine the number of elements in the groups

$$G = \langle a, b | a^2 = b^2 = 1 \text{ and } ab = ba \rangle \text{ and } H = \langle a | a^4 = 1 \rangle$$

See Example 3.9 for the meaning of this notation. Are these groups the same or is there something fundamentally different about them?

15. Suppose G is a group such that $(ab)^k = a^k b^k$ for $k = 3, 4$, and 5. Prove that G must be abelian.

4.2

Definition of Subgroups

In Section 4.1, we developed and investigated the concept of cyclic group. A fundamental result was that if G is a group and $a \in G$, then $H = \{a^k | k \in \mathbb{Z}\}$ is a group. If G is cyclic and a is a generator, then $G = H$. But even if a is not a generator or G is not cyclic, nevertheless H is a group. What is the operation of the group H? Ostensibly, it is the same operation as that of G. But the domain of the group operation of G is $G \times G$, while that of H is $H \times H$. Thus the operations are not identical. The next definition should clear up the confusion.

Definition 4.2

If $*$ is an operation on a set G, $H \subseteq G$, and H is closed with respect to the operation $*$, then the **restriction** of $*$ to H (denoted $*_H$) is the mapping

$$*_H: H \times H \to H \qquad \text{defined by} \qquad h_1 *_H h_2 = h_1 * h_2.$$

While $*_H$ is formally different from $*$, we can simply compose elements of H as we compose them in G. We do this when we add two fractions; we get the same answer whether we add them as rational numbers or as their corresponding real counterparts. Since addition in $\mathbb{Q}$ is a restriction of

---

[1] Leonhard Euler (1707–1783) produced an enormous amount of mathematical research in areas as diverse as analytic number theory and hydrodynamics. Many of the problems he studied have inspired present-day theories.

addition in $\mathbb{R}$, we "identify" the two operations as one. With this "bookkeeping" detail out of the way, we can now state the basic definition of this chapter.

Definition 4.3

> If $(G, *)$ is a group and $H \subseteq G$, then $(H, *_H)$ is a **subgroup** of $(G, *)$ if $(H, *_H)$ is itself a group.

Theorem 4.2 says that if G is a group and $a \in G$, then $H = \langle a \rangle$ is a subgroup of G. Now these subgroups of G are always cyclic, but we will see that groups can contain, as well, subgroups that are not cyclic. In fact, our definition allows G to be a subgroup of itself. Since not every group is cyclic, subgroups do not have to be cyclic either.

Example 4.1

If G is a group, then $\{1\}$ and G are both subgroups of G. $\{1\}$ is termed the **trivial subgroup**, and G is called an **improper subgroup**. Subgroups of G other than G are termed **proper** subgroups.

Example 4.2

The even integers $2\mathbb{Z}$ under addition form a subgroup of $(\mathbb{Z}, +)$. This follows from the facts that the sum of two even integers is even, addition is associative on $\mathbb{Z}$, $0 = 2(0)$ is the identity, and the inverse of an even integer is even.

Example 4.3

The odd integers fail to form a subgroup of $(\mathbb{Z}, +)$ since the closure and identity axioms do not hold.

Example 4.4

The rotations of a square form a subgroup of the set of rigid motions of a square. The counterparts of the symmetries of a rectangle form a subgroup of the rigid motions of a square. This second subgroup is not cyclic.

Example 4.5

Let H consist of just the integer 1. Then $(H, \cdot)$ is a group and H is a subset of the group $G = (\mathbb{Z}, +)$. But H is **not** a subgroup of G because multiplication, the operation of H, is not the restriction of the operation in G.

Example 4.6

Let us consider the set $X = \{1, 2, 3, 4, 5, 6\}$ and the group $(\mathscr{P}(X), \triangle)$, where $\triangle$ is the symmetric difference operator described in Section 3.1. The singleton subsets generate cyclic subgroups of order 2. If we try to find a noncyclic subgroup of $\mathscr{P}(X)$, it must contain at least two different nonempty subsets, say $\{1\}$ and $\{2\}$. If we include the empty set $\varnothing$, then $\mathscr{P}(X)$ contains an identity and inverses of each element, since each element is its own inverse. If the closure axiom is to hold, however, we must include the element $\{1\} \triangle \{2\} = \{1, 2\}$. We leave it as Exercise 3f for you to show that $H = \{\varnothing, \{1\}, \{2\}, \{1, 2\}\}$ is indeed a subgroup and that it is not cyclic.

We cannot stress too much the importance of checking closure in attempting to verify that a subset of a group is indeed a subgroup. To illustrate this, we consider again the group $(M_2(\mathbb{Z}), +)$ of 2×2 matrices with integer entries under the operation of matrix addition. Let

$$H = \left\{ \begin{bmatrix} a & b \\ c & d \end{bmatrix} \middle| a, b, c, d \in \mathbb{Z} \text{ and either } b = 0 \text{ or } c = 0 \right\}$$

The identity matrix of $(M_2(\mathbb{Z}), +)$, the matrix with all entries equal to 0, is an element of H. Matrix addition is associative on H; in addition, the additive inverse of any element of H is another element of H. Nevertheless, H is **not** a subgroup of $(M_2(\mathbb{Z}), +)$ since H is not closed under matrix addition. The example below establishes this assertion.

$$\begin{bmatrix} 1 & 0 \\ 2 & 4 \end{bmatrix} + \begin{bmatrix} 5 & 6 \\ 0 & 3 \end{bmatrix} = \begin{bmatrix} 6 & 6 \\ 2 & 7 \end{bmatrix}$$

It is common practice in abstract algebra to suppress the operation in referring to a subgroup and simply say "H is a subgroup of G." This is symbolized by the notation "$H \leqslant G$". When you see this, you are expected to realize that H is a subset of the group G, H is a group, and the operation in H is the restriction of the operation on G.

In the last chapter, we introduced the concept of a direct product of two groups; that is, if H and K are groups, then $G = H \times K$ is another group whose operation is given by

$$(h_1, k_1) * (h_2, k_2) = (h_1 h_2, k_1 k_2)$$

In this situation, neither H nor K is a subgroup of G since the elements of G are ordered pairs. Then how are H and K related to G? If 1_H and 1_K denote the identities of H and K, then $H' = H \times \{1_K\}$ and $K' = \{1_H\} \times K$ **are** subgroups of G that correspond to H and K. We will explain this correspondence in detail in Chapter 8.

In checking subsets of a group G to see that they form subgroups of G, we went through the complete set of axioms. In Section 4.3, we will discuss some shortcuts to this process as well as investigate some properties of a subgroup that are *inherited* from the parent group G.

Exercises 4.2

1. If the real numbers $\mathbb{R}$ are identified with the complex numbers of the form $a + 0i = a$, then show that real number multiplication is the restriction of complex number multiplication.

2. Determine which of the following sets on the left are subgroups of the groups on the right.
   a. $\mathbb{N}$ $(\mathbb{Z}, +)$
   b. $\{1\}$ $(\mathbb{Q} - \{0\}, \cdot)$
   c. $\{1, 4\}$ $(\mathbb{Z}_5^*, \odot)$
   d. $3\mathbb{Z}$ $(\mathbb{Z}, +)$
   e. $\{1, 3\}$ $(\mathbb{Z}_5^*, \odot)$
   f. $\{0, 2, 4\}$ $(\mathbb{Z}_6, \oplus)$

3. Which of the following are subgroups of $(\mathscr{P}(X), \triangle)$, where $X = \{1, 2, 3, 4, 5, 6\}$? Which are cyclic subgroups?
   a. $\{\varnothing, \{5\}\}$
   b. $\{\varnothing, \{1\}, \{1, 3\}\}$
   c. $\{\{1\}, \{2\}, \{1, 2\}\}$
   d. $\{\varnothing, \{3\}, \{4\}, \{5\}, \{3, 4\}, \{3, 5\}, \{4, 5\}, \{3, 4, 5\}\}$
   e. $\{\varnothing\}$
   f. $\{\varnothing, \{1\}, \{2\}, \{1, 2\}\}$

4. Which of the following are subgroups of $(M_2(\mathbb{Z}), +)$, the group of all 2×2 matrices with integer entries?

   a. $M_2(\mathbb{N})$ b. $H = \left\{ \begin{bmatrix} 0 & 0 \\ 0 & 0 \end{bmatrix} \right\}$

   c. $K = \left\{ \begin{bmatrix} a & 0 \\ 0 & b \end{bmatrix} \middle| a, b \in \mathbb{Z} \right\}$ d. $L = \left\{ \begin{bmatrix} a & 0 \\ 0 & 1 \end{bmatrix} \middle| a \in \mathbb{Z} \right\}$

5. Verify that the rotations of a square form a cyclic subgroup of the group of symmetries of a square.

6. Prove that the symmetries of a rectangle form a noncyclic subgroup of the symmetries of a square.

7. Find $o(a)$ for each $a \in \mathbb{Z}_6$. Do the same for the elements of the group of symmetries of an equilateral triangle.

8. Suppose $H \leqslant G$ and $K \leqslant H$. Verify $K \leqslant G$.

9. a. Prove that

$$G = \left\{ \begin{bmatrix} a & b \\ c & d \end{bmatrix} \middle| a, b, c, d \in \mathbb{Q} \text{ and } ad - bc \neq 0 \right\}$$

   is a group under the operation of matrix multiplication $\times$ defined in Section 2.2.
   b. Show that

$$H = \left\{ \begin{bmatrix} a & b \\ c & d \end{bmatrix} \in M_2(\mathbb{Z}) \middle| ad - bc = 1 \text{ or } -1 \right\}$$

   is a subgroup of G.
   c. Verify that

$$K = \left\{ \begin{bmatrix} a & b \\ c & d \end{bmatrix} \in M_2(\mathbb{Z}) \middle| ad - bc = 1 \right\}$$

   is a subgroup of H.

10. In Section 3.1, we encountered the group C of all real-valued continuous functions on $\mathbb{R}$.

a. Does the set of all differentiable functions on $\mathbb{R}$ form a subgroup of C?

b. A function $f \in C$ is termed **bounded** if there exists some constant M (depending upon the function f such that $|f(x)| \leq M$ for all $x \in \mathbb{R}$. Both the sine and cosine functions are bounded. Prove that the set of all bounded functions forms a subgroup of C.

11. Prove that $H' = H \times \{1_K\}$ and $K' = \{1_H\} \times K$ are subgroups of $G = H \times K$.

*12. Write a computer program to determine the order of each element of $\mathbb{Z}_n$ for $n > 1$.

4.3
Properties

With some examples at our disposal and a good understanding of what a subgroup is, let us review how you would check a subset H of a group G to see if H is a subgroup. First, you would make sure that the subset was nonempty, since a group must have at least an identity element. Next you would see if H was closed with respect to the group operation on G, or, in other words, if the operation on G restricted to H is an operation on H. Once you determine that the closure axiom is satisfied, you turn to associativity. But is this necessary? The answer is no! The reason for this is subtle; if the associative law holds for elements of G, then it holds automatically for elements in any subset of G. More precisely, associativity is an *inherited* property.

Therefore, now you really have to check only that the identity and inverse axioms hold. Suppose the inverse axiom holds and $a \in H$. Then $a^{-1} \in H$ and, by closure, $e = aa^{-1} \in H$. If the identity of H is the same as the identity of G, then you have shown that the identity axiom holds in H. The fact that the identity of any subgroup is the same as the identity of the group follows immediately from the Cancellation Property for groups. (See Exercise 1.)

The point of the preceding discussion is that we have obtained an alternate definition of a subgroup—one that is easier to apply to an arbitrary **nonempty** subset.

Theorem 4.4

Let H be a nonempty subset of the group G. Then $H \leq G$ if and only if

i. $a, b \in H$ implies $ab \in H$, and

ii. $a \in H$ implies $a^{-1} \in H$.

Example 4.7

Let G be the group of symmetries of an equilateral triangle and $H = \langle \rho \rangle$, where ρ corresponds to a rotation of $120°$. While we already know that H is a cyclic subgroup of G, we want to point out that the Cayley table for H demonstrates both closure and existence of inverses. (See Figure 4.2.)

	e	ρ	ρ^2
e	e	ρ	ρ^2
ρ	ρ	ρ^2	e
ρ^2	ρ^2	e	ρ

Figure 4.2

Example 4.8

Let $G = \mathbb{Z}_5^*$, the group of nonzero integers modulo 5 under the operation of modular multiplication, $H = \{1, 4\}$ and $K = \{1, 3\}$. Figure 4.3a and Theorem 4.4 show that $H \leqslant G$, but Figure 4.3b demonstrates that K is not a subgroup of G.

$\odot$	1	4
1	1	4
4	4	1

(a)

$\odot$	1	3
1	1	3
3	3	4

(b)

Figure 4.3

Example 4.9

Let H and K be subgroups of an arbitrary group G. We claim that $H \cap K$ is a subgroup of G. We will use Theorem 4.4. Since $H \leqslant G$ and $K \leqslant G$, both subgroups contain the identity 1 of G. Thus $1 \in H \cap K$ and $H \cap K$ is a nonempty subset. Now we need only check closure and inverses. If $a, b \in H \cap K$, then $a, b \in H$ and $a, b \in K$. Since H and K are both subgroups, $ab \in H$ and $ab \in K$. We conclude $ab \in H \cap K$. Similarly if $a \in H \cap K$, then $a \in H$ and $a \in K$ so that $a^{-1} \in H$ and $a^{-1} \in K$. Again, $a^{-1} \in H \cap K$.

Example 4.10

Let G be an arbitrary group and set

$$Z(G) = \{z \in G \mid zg = gz \text{ for all } g \in G\}$$

This set is extremely important in group theory and is known as the **center of** G. The center of a group consists of all the elements in G that commute with every other element of G. An abelian group coincides with its center. The size of the center is a measure of how close G is to being abelian. The center is in fact a subgroup of G. To see this, note first that the identity $1 \in Z(G)$. If $a, b \in Z(G)$, then $ag = ga$ and $bg = gb$ for all $g \in G$. Then the equation

$$(ab)g = a(bg) = a(gb) = (ag)b = (ga)b = g(ab)$$

shows that $ab \in Z(G)$. Since $ag = ga$ for all $g \in G$, $a^{-1}(ag)a^{-1} = a^{-1}(ga)a^{-1}$ and hence $ga^{-1} = a^{-1}g$ for all $g \in G$ and $a^{-1} \in Z(G)$. By Theorem 4.4, $Z(G) \leqslant G$.

In Examples 4.7 and 4.8, we used a Cayley table to check for closure. Note that we were then immediately able to determine the existence of inverses. This demonstrates that, for finite groups, it is not necessary to check for the existence of each and every inverse.

Theorem 4.5

If H is a nonempty **finite** subset of the group G, then $H \leqslant G$ if and only if $a, b \in H$ implies $ab \in H$.

Proof

By Theorem 4.4 we need only show that inverses exist for each $a \in H$. Consider $Ha = \{ha | h \in H\}$. If $H = \{h_1, h_2, ..., h_m\}$, then

$$Ha = \{h_1 a, h_2 a, ..., h_m a\}$$

If H has m distinct elements, then the Cancellation Property can be used to show that Ha also has m distinct elements. By closure, $a \in H$ implies $Ha \subseteq H$. Since these sets have the same number of elements, $Ha = H$. This says that it is possible to write $a \in H$ in the form $h_j a$ for some $h_j \in H$. But then $a = h_j a$ implies $h_j = 1 \in H$. If $1 \in H = Ha$, then there exists $h_k \in H$ such that $1 = h_k a$. Therefore $a^{-1} = h_k \in H$, and the theorem is proved. □

If we remove the condition that H is finite, then Theorem 4.5 is not true. (See Exercise 8.) However, there is a single-step way to check whether or not a subset is a subgroup. Note that the single step of Theorem 4.6, whose proof we leave as Exercise 22, involves both the closure and inverse axioms.

Theorem 4.6

If G is a group and H a nonempty subset of G, then $H \leqslant G$ if and only if $a, b \in H$ implies $ab^{-1} \in H$.

Before we conclude this chapter, we should make some mention about the relationships between the structural nature of a group and that of its subgroups. Commutativity, like associativity, is an inherited property. Thus the subgroups of an abelian group are also abelian. The converse to this result is false, however. There exist nonabelian groups whose proper subgroups are all abelian. (See Exercise 9.) The property that a group is cyclic is also inherited by subgroups, but the demonstration of this fact is somewhat more involved than that of the other properties discussed to this point.

Theorem 4.7

Let $G = \langle a \rangle$ and $H \leqslant G$. Then H is cyclic.

Proof

Let $W = \{m \in \mathbb{N} | a^m \in H\}$. If $W = \varnothing$, then $H = \{1\}$ (Why?). In this case, $H = \langle 1 \rangle$ is cyclic. If $W \neq \varnothing$, let k be the least element of W. We claim $H = \langle a^k \rangle$.

If $a^t \in H$, write $t = kq + r$, where $0 \leqslant r < k$; then

$$a^t = a^{kq+r} = (a^k)^q a^r$$

and

$$a^r = a^t (a^{kq})^{-1} = a^{t-kq} \in H$$

Since $r < k$ and $a^r \in H$, the choice of k forces $a^r = 1$ and $a^t = (a^k)^q \in \langle a^k \rangle$. Then we have $H \subseteq \langle a^k \rangle$. But $\langle a^k \rangle \subseteq H$ so the result follows. □

Exercises 4.3

1. Prove that if $H \leqslant G$, then the identity of H is also the identity of G.

2. Use Theorem 4.4 to check if the following subsets are subgroups of the group of symmetries of an equilateral triangle:
   a. $\{e, \rho^2\}$ b. $\{e, \tau_A\}$ c. $\{e, \rho, \tau_B\}$

3. Show that $\{1, -1\}$ and $\{1, i, -1, -i\}$ form subgroups of $(\mathbb{C} - \{0\}, \cdot)$.

4. Is $H = \{bi | b \in \mathbb{R}\} \leqslant (\mathbb{C}, +)$?

5. Is $\{bi | b \in \mathbb{R}, b \neq 0\} \leqslant (\mathbb{C} - \{0\}, \cdot)$?

6. Verify that if $G = \{a_1, a_2, \ldots, a_n\}$ is a group and $b \in G$, then $Gb = \{a_1 b, a_2 b, \ldots, a_n b\} = G$.

7. If $H \leqslant G$ and $K \leqslant G$, is $H \cup K \leqslant G$? Why?

8. The set of nonzero integers is closed under multiplication. Show that this set is not a subgroup of the group $(\mathbb{R} - \{0\}, \cdot)$. Doesn't this show that the word *finite* is needed in Theorem 4.5?

9. Give an example of a nonabelian group all of whose proper subgroups are cyclic.

10. In the proof of Theorem 4.7, why does $W = \varnothing$ imply $H = \{1\}$?

11. Is Theorem 4.4 still true if the word *nonempty* is omitted?

12. Prove that if $G = \langle a \rangle$ has order n, then the proper subgroups of G are of the form $H = \langle a^k \rangle$ where k is a proper divisor of n.

13. Verify that if $G = \langle a \rangle$ has order n, then $a^m = 1$ for $m \in \mathbb{N}$ implies that n divides m.

14. Let $G = \langle a \rangle$ have order n. Prove that if k divides n, then there exists a subgroup H of G with $|H| = k$.

15. Let G be a group. Prove that $Z(G)$ is an abelian subgroup of G. Find $Z(G)$ if G is the group of symmetries of an equilateral triangle.

16. Determine $Z(G)$ if:
    a. G is the group of symmetries of a rectangle;
    b. G is the group of symmetries of a square.

17. Suppose $H \leqslant G$ and $K \leqslant G$. Prove $H \cap K \leqslant K$.

18. Suppose $H \leqslant G$. Verify that $Z(G) \leqslant Z(H)$.

19. Let G be a finite group and $H \leqslant G$. The **normalizer of H in G** is defined by

$$N_G(H) = \{x \in G | xhx^{-1} \in H \text{ for all } h \in H\}$$

Show $N_G(H) \leqslant G$.

20. Let G be the group of symmetries of an equilateral triangle. (See Figure 2.16.) In the notation of Exercise 19, determine $N_G(H)$ if H is
    a. $\langle \rho \rangle$ b. $\langle \tau_A \rangle$

21. Let $n \in \mathbb{N}$. Prove that the set of natural numbers less than n and relatively prime to n is closed with respect to modular multiplication in $\mathbb{Z}_n$. Does this set form a subgroup of $\mathbb{Z}_n$?

22. Prove Theorem 4.6: If G is a group and H is a nonempty subset of G, then $H \leqslant G$ if and only if $a, b \in H$ implies $ab^{-1} \in H$.

23. Let G be an abelian group and $n \in \mathbb{N}$. Define $nG = \{ng | g \in G\}$ and $G[n] = \{g \in G | ng = 0\}$. Prove that $nG \leqslant G$ and $G[n] \leqslant G$.

24. Let G be a group such that $x^2 = 1$ for all $x \in G$. If $a, b \in G$, find the smallest subgroup H of G containing $\{a, b\}$.

5 | Permutation Groups

Most of the early work in group theory involved the study of mappings that rearranged or permuted the roots of a polynomial. These *groups of transformations* were studied before the axioms of an abstract group were codified and provided some of the earliest applications of group theory. Groups of this type are called *permutation groups* and traditionally have provided a readily accessible supply of examples. It is appropriate, therefore, that we study these groups in some detail.

In this chapter we first review some basic properties of mappings in order to define permutations. Next we show how composition of mappings provides us with an operation, and we use that operation to establish a group structure for permutations. A shorthand notation for permutation composition will be developed, and, from that notation, the concept of odd and even permutations will be studied. Finally, we will look at an important class of permutation groups called the set of alternating groups.

5.1

Mappings and Permutations

We begin this section with a definition:

Definition 5.1

> A mapping $f: X \to Y$ is said to be **injective** if $f(x_1) = f(x_2)$ implies $x_1 = x_2$ for $x_1, x_2 \in X$. (An injective mapping is sometimes said to be **one-to-one**.)

This definition says that each element in the image of the mapping comes from just one element of the domain. At first glance, this is quite similar to the definition of mapping wherein each element of the domain produces a unique element of the range. In actuality, however, these two definitions are quite dissimilar as evidenced by the following examples.

Example 5.1

Consider the mapping on $\mathbb{R}$ defined by $f(x) = x^2$. Now f is a mapping since for each x there is a unique x^2 associated with it. But f is not injective since for each positive real number x, $f(x) = f(-x)$ but $x \neq -x$.

Example 5.2

Any linear mapping $f: \mathbb{R} \to \mathbb{R}$ whose graph is a nonhorizontal line is an injective mapping. You learned in calculus that the equation of any nonvertical line may be expressed as $y = f(x) = mx + b$, where m is the slope and b is the y-coordinate of the y-intercept. (Recall that a vertical line is not the graph of a mapping. It should be obvious that a horizontal line is not the graph of an injective mapping since the image of every real number is the same.) To verify that a mapping is injective, one usually proceeds in the following manner.

Suppose $f(x_1) = f(x_2)$. In our example, this means $mx_1 + b = mx_2 + b$ for $x_1, x_2 \in \mathbb{R}$. Then, using the additive cancellation law of $(\mathbb{R}, +)$, we obtain $mx_1 = mx_2$. Since $m \neq 0$, we can again use cancellation, this time in $(\mathbb{R} - \{0\}, \cdot)$, to conclude $x_1 = x_2$. This shows that if two real numbers have the same image under a linear mapping, then they must be the same element. Thus, by the definition of injective mapping, linear mappings whose graphs are nonhorizontal lines are injective.

Definition 5.2

> A mapping $f: X \to Y$ is said to be **surjective** if for each element $y \in Y$ there is an element $x \in X$ such that $f(x) = y$. (Another term for surjective is **onto**.)

In the definition of mapping, we used the term *into* meaning that each element of the domain had an image that was an element of the range, not necessarily that every element in Y had a preimage in the domain. That is the distinction between the terms *into* and *surjective*. Remember that a surjective mapping is, by definition, *into* but that an into mapping is not necessarily *surjective*.

Example 5.3

Every nonhorizontal and nonvertical line represents the graph of a surjective mapping.

Example 5.4

The mapping $g(x) = x^3$ is both injective and surjective, while the trigonometric sine mapping from $\mathbb{R}$ to $[-1, 1]$ is surjective but not injective. The mapping $h: \mathbb{R}^+ \to \mathbb{R}$ given by $h(x) = 1/x$ is injective but not surjective. The verifications of these assertions are left as Exercise 1.

Mappings that satisfy both of the foregoing definitions are of special importance in mathematics.

Definition 5.3

A mapping $f: X \to Y$ is **bijective** if it is both injective and surjective. (Another, and possibly more familiar, term for this type of mapping is **one-to-one correspondence**.)

In Examples 5.1–5.4, the linear mappings of the form $f(x) = mx + b$, $m \neq 0$, and the mapping given by $g(x) = x^3$ are bijections of $\mathbb{R}$ onto itself.

Definition 5.4

A bijection $f: X \to X$ is called a **permutation on** X.

The term permutation comes from the fact that on a finite set (for example, $\{1, 2, \ldots, n\}$ for some positive integer $n > 1$), a permutation is essentially a mapping that permutes or changes the order of the elements, that is, a rearrangement.

In order to consider the set of permutations on a given set as a group, we must have an operation. The natural operation to consider is that of composition of mappings. But in order for composition to be a binary operation on the set of permutations, we must show that if f and g are permutations (bijections or one-to-one correspondences) on a set X, then $g \circ f$

is also a permutation on the set X. In fact, this result follows as a corollary to the following general theorems.

Theorem 5.1

If $f: X \to Y$ and $g: Y \to Z$ are injections, then $g \circ f: X \to Z$ is an injection.

Proof

By definition of composition of mappings, $(g \circ f)(x) = g(f(x))$. Thus if $(g \circ f)(x_1) = (g \circ f)(x_2)$, then $g(f(x_1)) = g(f(x_2))$. However, since g is injective, $f(x_1)$ must equal $f(x_2)$; and since f is injective, x_1 must equal x_2. Therefore, $g \circ f$ is injective. ☐

It is worthwhile to note that the converse of this theorem is not true: that is, if $g \circ f$ is injective, it does not necessarily follow that both f and g are injective. Exactly what is true about the converse to Theorem 5.1 is given in Exercise 5 for you to work out.

Theorem 5.2

If $f: X \to Y$ and $g: Y \to Z$ are surjections, then $g \circ f: X \to Z$ is a surjection.

Proof

Since g is surjective, then, for each $z \in Z$, there is an least one $y \in Y$ such that $g(y) = z$. Since f is surjective, there is an $x \in X$ such that $f(x) = y$. Therefore, $g \circ f(x) = g(f(x)) = z$. We conclude that $g \circ f$ is surjective. ☐

As with the converse of Theorem 5.1, the converse of Theorem 5.2 is not true. For exactly what is correct, we refer you to Exercise 6.

Corollary 5.1

If $f: X \to Y$ and $g: Y \to Z$ are bijections, then $g \circ f: X \to Z$ is a bijection.

Corollary 5.2

If f and g are permutations on a set X, then $g \circ f$ is a permutation on X.

Therefore we have obtained the result that composition of permutations is a binary operation on the set of all permutations on a set X. Also by Theorem 2.1, composition of permutations is associative. Consequently, to verify that the set of all permutations on a set X is a group, we need only show that there is an identity and that each permutation has an inverse.

The fact that there is an identity comes quickly (and also shows that whenever a set X is nonempty, the set of all permutations of X is also nonempty). The mapping $1_X: X \to X$ defined by $1_X(x) = x$ is the identity for the operation of composition. The verification of this is straightforward and is left as an exercise. (See Exercise 7.)

To show that the set of permutations of X forms a group, we must show that for each permutation of X there exists an inverse permutation. We will see this from the following discussion of inverse mappings.

Definition 5.5

A mapping $g: Y \to X$ is the **inverse** of the mapping $f: X \to Y$ if $g \circ f$: $X \to X$ is the mapping 1_X and $f \circ g: Y \to Y$ is the mapping 1_Y.

The definition says that $g: Y \to X$ is the inverse of $f: X \to Y$ if $f(g(y)) = y$ for all $y \in Y$ and $g(f(x)) = x$ for all $x \in X$.

Now we consider the question: "When does a mapping have an inverse?" We will obtain our answer by investigating the properties that an inverse must possess and then seeing if we can reverse them.

We claim first that if $g: Y \to X$ is the inverse of $f: X \to Y$, then f must be surjective. To see this, let $y \in Y$ be given. Let $x = g(y)$. Then $f(x) = f(g(y)) = 1_Y(y) = y$, so f is surjective. Secondly, f must be injective. To see this, simply note that if $f(x_1) = f(x_2)$, then $x_1 = g(f(x_1)) = g(f(x_2)) = x_2$.

On the other hand, suppose $f: X \to Y$ is both injective and surjective. We claim f has an inverse $g: Y \to X$. Let $y \in Y$. The fact that f is surjective implies that there exists at least one $x \in X$ such that $f(x) = y$. Define $g: Y \to X$ by setting $g(y) = x$. To show that g is a mapping we must prove that $g(y)$ is unique. Suppose there exists another $x_1 \in X$, $x \neq x_1$, such that $f(x_1) = y$. Is $g(y) = x_1$? No, for then $f(x) = f(x_1)$, but $x \neq x_1$, contradicting the assumption that f is injective. Thus $g(y)$ is unique. It is a straightforward matter to show $g \circ f = 1_X$ and $f \circ g = 1_Y$.

We have therefore obtained the following result:

Theorem 5.3

A mapping $f: X \to Y$ has an inverse if and only if f is injective and surjective; that is, f is a bijection.

A mapping that has an inverse is termed **invertible**. The inverse of a mapping $f: X \to Y$ is usually denoted by f^{-1}. Our discussion about inverses now yields this important result.

Theorem 5.4

If S_X is the set of all permutations on a set X, then $(S_X, \circ)$ is a group, called the **symmetric group on** X.

One of our main objectives in studying groups of permutations is to obtain a new class of examples of finite groups. For this reason, we will consider the structure of groups of permutations on a finite set in Section 5.2.

Exercises 5.1

1. Verify the assertions in Examples 5.3 and 5.4.
2. State which of the following mappings are injective and which are surjective.
   a. $f: \mathbb{N} \to \mathbb{N}$ $f(n) = n + 1$
   b. $f: \mathbb{N} \to \mathbb{N} \cup \{0\}$ $f(n) = n - 1$
   c. $f: \mathbb{N} \to \mathbb{N}$ $f(n) = 3n$

    d. $f: \mathcal{P}(X) \to \mathcal{P}(X)$ $f(A) = A \cup B$, for $B \subseteq X$
    e. $f: \mathcal{P}(X) \to \mathcal{P}(X)$ $f(A) = X - A$
    f. $f: \mathbb{Z} \times \mathbb{Z} \to \mathbb{Z}$ $f(m, n) = m + n$
    g. $f: \mathbb{Z} \times \mathbb{Z}^+ \to \mathbb{Q}$ $f(m, n) = m/n$
    h. $f: \mathbb{Z} \times \mathbb{Z} \times \mathbb{Z} \to \mathbb{Z}$ $f(m, n, t) = m^2 + n^2 + t^2$

3. Which of the following mappings on $\mathbb{R}$ (or a subset of $\mathbb{R}$) are injective and/or surjective?

    a. $f: \mathbb{R} \to \mathbb{R}$ $f(x) = x^2 + x$
    b. $f: \mathbb{R} \to \mathbb{R}$ $f(x) = x^3 + x^2$
    c. $f: \mathbb{R}^+ \to \mathbb{R}^+$ $f(x) = x^2$
    d. $f: \mathbb{R}^+ \to \mathbb{R}$ $f(x) = x^2 + x$
    e. $f: \mathbb{R}^+ \to \mathbb{R}^+$ $f(x) = (1/x) + x$
    f. $f: \mathbb{R} \to \mathbb{R}$ $f(x) = x$
    g. $f: (-\pi/2, \pi/2) \to \mathbb{R}$ $f(x) = \tan x$
    h. $f: \mathbb{R} \to \mathbb{R}$ $f(x) = \sin x$

4. For each mapping in Exercise 2 and Exercise 3 that is a bijection, give a formula for the inverse.

5. For mappings $f: X \to Y$ and $g: Y \to Z$ such that $g \circ f: X \to Z$ is injective, show that f must be injective. Construct an example to show that g need not be injective.

6. For mappings $f: X \to Y$ and $g: Y \to Z$ such that $g \circ f: X \to Z$ is surjective, show that g must be surjective. Construct an example to show that f need not be surjective.

7. For any nonempty set X, show that the mapping $1_X: X \to X$ defined by $1_X(x) = x$ is the identity for the operation of composition of mappings.

8. Suppose both $f: X \to Y$ and $g: Y \to Z$ are invertible mappings. Prove that the mapping $g \circ f$ is invertible by showing that $(g \circ f)^{-1} = f^{-1} \circ g^{-1}$.

9. Let $f: X \to X$ be a mapping and suppose X is a finite set.
    a. Prove that if f is surjective, then f is injective.
    b. Prove that if f is injective, then f is surjective.

10. Let S be the group of all permutations on $\mathbb{R}$. A mapping $f \in S$ is termed **order-preserving** if $x_1 < x_2$ implies $f(x_1) < f(x_2)$.
    a. Prove that the set of all order-preserving permutations on $\mathbb{R}$ is a subgroup of S.
    b. Determine the set of all order-preserving permutations of $\mathbb{Z}$.

5.2

Cycle Notation

The symmetric group on a set X, denoted S_X, was the subject of the last section. For the rest of this chapter, we wish to consider this group when X is a finite set. If $X = \{1, 2, \ldots, n\}$, then S_X is commonly written as S_n and S_n is termed the **symmetric group on n letters** (or **numbers** or **symbols**). The elements of S_n are then bijections or one-to-one correspondences of $\{1, 2, \ldots, n\}$. How many such mappings are there?

To determine the answer to this question, we will proceed in the following manner. Let σ be an element of S_n, and consider the number of choices for $\sigma(1)$. Since the image of 1 can be any member of $\{1, 2, \ldots, n\}$, we see that there are n choices. Without loss of generality, let us say that $\sigma(1) = k$. Now how many choices are there for the image of 2? Since k has already been used, there are only $n - 1$ possibilities. In a similar manner, we may reason that there are $n - 2$ choices for the image of 3, $n - 3$ for the image of 4, and so on until there is only one choice left for the image of n. Thus we have derived the following theorem.

Theorem 5.5

The symmetric group S_n has $n!$ elements.

Our first goal is to develop a 'slick' notation to describe these $n!$ elements of S_n. Let $\sigma \in S_n$. If σ takes 1 to $\sigma(1)$, 2 to $\sigma(2)$, etc., we will write σ as

$$\begin{pmatrix} 1 & 2 & 3 & \cdots & n \\ \sigma(1) & \sigma(2) & \sigma(3) & \cdots & \sigma(n) \end{pmatrix}$$

For example, the elements of S_3 are

$$\begin{pmatrix} 1 & 2 & 3 \\ 1 & 2 & 3 \end{pmatrix}, \quad \begin{pmatrix} 1 & 2 & 3 \\ 2 & 3 & 1 \end{pmatrix}, \quad \begin{pmatrix} 1 & 2 & 3 \\ 3 & 1 & 2 \end{pmatrix}, \quad \begin{pmatrix} 1 & 2 & 3 \\ 1 & 3 & 2 \end{pmatrix}, \quad \begin{pmatrix} 1 & 2 & 3 \\ 3 & 2 & 1 \end{pmatrix},$$

and $\begin{pmatrix} 1 & 2 & 3 \\ 2 & 1 & 3 \end{pmatrix}$

This notation may seem somewhat cumbersome, for we have listed the complete domain and image of σ, but we will soon see that this form of σ lends itself to rapid calculation of products of permutations (permutation composition). Since permutations are bijections, a product of two permutations refers to the composite of the two bijections. Let

$$\tau = \begin{pmatrix} 1 & 2 & 3 & \cdots & n \\ \tau(1) & \tau(2) & \tau(3) & \cdots & \tau(n) \end{pmatrix} \in S_n$$

Then

$$\tau \circ \sigma = \begin{pmatrix} 1 & 2 & \cdots & n \\ \tau(1) & \tau(2) & \cdots & \tau(n) \end{pmatrix} \begin{pmatrix} 1 & 2 & \cdots & n \\ \sigma(1) & \sigma(2) & \cdots & \sigma(n) \end{pmatrix}$$

$$= \begin{pmatrix} 1 & 2 & \cdots & n \\ \tau(\sigma(1)) & \tau(\sigma(2)) & \cdots & \tau(\sigma(n)) \end{pmatrix}$$

since $(\tau \circ \sigma)(1) = \tau(\sigma(1))$, etc.

Consider the example with

$$\sigma = \begin{pmatrix} 1 & 2 & 3 & 4 \\ 2 & 3 & 4 & 1 \end{pmatrix}, \quad \tau = \begin{pmatrix} 1 & 2 & 3 & 4 \\ 3 & 4 & 2 & 1 \end{pmatrix} \in S_4$$

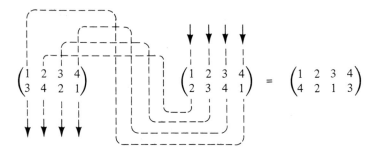

Figure 5.1

Then $(\tau \circ \sigma)(1) = \tau(\sigma(1)) = 4$, $(\tau \circ \sigma)(2) = \tau(\sigma(2)) = 2$, $(\tau \circ \sigma)(3) = \tau(\sigma(3)) = 1$, and $(\tau \circ \sigma)(4) = \tau(\sigma(4)) = 3$, so that

$$\tau \circ \sigma = \begin{pmatrix} 1 & 2 & 3 & 4 \\ 4 & 2 & 1 & 3 \end{pmatrix}$$

We diagram this composition in Figure 5.1. The key to composition of permutations is "follow the arrows right to left." With just a little practice, you should be able to learn to calculate products of permutations quickly.[1]

Our second goal is to modify our "big" notation into something more compact. Toward that objective we make the following definition:

Definition 5.6

A permutation $\sigma \in S_n$ is said to be a **cycle of length** m, or an m-**cycle** if there exists a nonempty subset $\{i_1, i_2, \ldots, i_m\} \subseteq \{1, 2, \ldots, n\}$ such that

$$\sigma(i_1) = i_2, \quad \sigma(i_2) = i_3, \quad \ldots, \quad \sigma(i_{m-1}) = i_m, \quad \sigma(i_m) = i_1,$$

and $\sigma(j) = j$

for all $j \in \{1, 2, \ldots, n\} - \{i_1, i_2, \ldots, i_m\}$. A cycle of length 2 is also known as a **transposition**.

Example 5.5

The permutation

$$\sigma = \begin{pmatrix} 1 & 2 & 3 \\ 2 & 3 & 1 \end{pmatrix}$$

---

[1] Some authors prefer to define multiplication of permutations by the rule "follow the arrows left to right." (Many English mathematicians use this method not only for permutations but also for multiplication of matrices acting as mappings.) Their treatment produces a different result. Our method is motivated by our desire to be consistent with our treatment of composition of mappings and our view of permutations as bijective mappings.

is a 3-cycle, since $\sigma(1) = 2$, $\sigma(2) = 3$, and $\sigma(3) = 1$. The permutation

$$\tau = \begin{pmatrix} 1 & 2 & 3 & 4 & 5 & 6 & 7 & 8 \\ 1 & 3 & 2 & 4 & 5 & 6 & 7 & 8 \end{pmatrix}$$

is a 2-cycle, since $\tau(2) = 3$, $\tau(3) = 2$, and $\tau(k) = k$ for all other elements. The permutation

$$\gamma = \begin{pmatrix} 1 & 2 & 3 & 4 & 5 & 6 & 7 & 8 \\ 8 & 2 & 5 & 4 & 7 & 6 & 1 & 3 \end{pmatrix}$$

is a 5-cycle, since $\gamma(1) = 8$, $\gamma(8) = 3$, $\gamma(3) = 5$, $\gamma(5) = 7$, $\gamma(7) = 1$, and $\gamma(j) = j$ for $j = 2, 4, 6$.

A permutation $\sigma \in S_n$ is said to **fix** $i \in \{1, 2, \ldots, n\}$ if $\sigma(i) = i$. If $\sigma(i) \neq i$, σ is said to **move** i. Two permutations $\sigma, \tau \in S_n$ are said to be **disjoint** if every element moved by σ is fixed by τ and every element moved by τ is fixed by σ. It can happen that an element can be fixed by both σ and τ. The permutations

$$\alpha = \begin{pmatrix} 1 & 2 & 3 & 4 & 5 & 6 & 7 & 8 \\ 1 & 4 & 3 & 6 & 5 & 2 & 7 & 8 \end{pmatrix} \quad \text{and}$$

$$\beta = \begin{pmatrix} 1 & 2 & 3 & 4 & 5 & 6 & 7 & 8 \\ 3 & 2 & 5 & 4 & 1 & 6 & 7 & 8 \end{pmatrix}$$

are disjoint. The permutations τ and γ of Example 5.5 are **not** disjoint since $\tau(3) = 2$ and $\gamma(3) = 5$.

It is not easy to tell when two permutations are disjoint in the form we are presently using to represent permutations, but there is another way to write permutations that allows for immediate determination. If σ is a cycle, it is possible to describe σ quite simply. Again consider the cycle

$$\gamma = \begin{pmatrix} 1 & 2 & 3 & 4 & 5 & 6 & 7 & 8 \\ 8 & 2 & 5 & 4 & 7 & 6 & 1 & 3 \end{pmatrix}$$

and focus attention on the subset $\{1, 8, 3, 5, 7\}$, consisting of elements moved by γ. Write $\gamma = (1 \quad 8 \quad 3 \quad 5 \quad 7)$. Then γ acts in the following way:

$$1 \overset{\gamma}{\to} 8 \overset{\gamma}{\to} 3 \overset{\gamma}{\to} 5 \overset{\gamma}{\to} 7 \overset{\gamma}{\to} 1$$

Another way to depict the action of γ is given in Figure 5.2. If we simply remember that $\gamma = (1 \quad 8 \quad 3 \quad 5 \quad 7)$ takes any of the numbers 1, 8, 3, or 5 to the next and the number 7 back to the 1, then we have a compact way to represent the cycle γ. With this representation, it is quite simple to determine whether or not two cycles are disjoint. For example, the cycles

$$\tau = (1 \quad 3 \quad 5 \quad 4) \quad \text{and} \quad \sigma = (2 \quad 6 \quad 7)$$

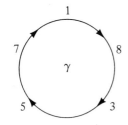

Figure 5.2

are disjoint while

$$\gamma = (2 \quad 4 \quad 6 \quad 5) \qquad \text{and} \qquad \sigma = (1 \quad 4 \quad 3 \quad 7)$$

are not disjoint.

The next problem is how to extend this compact notation to permutations that are not cycles. Consider the next example.

Let

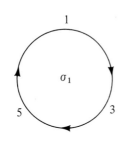

$$\sigma = \begin{pmatrix} 1 & 2 & 3 & 4 & 5 & 6 & 7 & 8 & 9 \\ 3 & 6 & 5 & 9 & 1 & 8 & 2 & 7 & 4 \end{pmatrix} \in S_9.$$

Then

$$\sigma: 1 \to 3 \to 5 \to 1$$

$$\sigma: 2 \to 6 \to 8 \to 7 \to 2$$

$$\sigma: 4 \to 9 \to 4$$

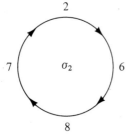

Consider the elements σ_1, σ_2, and σ_3 given by

$$\sigma_1 = \begin{pmatrix} 1 & 2 & 3 & 4 & 5 & 6 & 7 & 8 & 9 \\ 3 & 2 & 5 & 4 & 1 & 6 & 7 & 8 & 9 \end{pmatrix} = (1 \quad 3 \quad 5)$$

$$\sigma_2 = \begin{pmatrix} 1 & 2 & 3 & 4 & 5 & 6 & 7 & 8 & 9 \\ 1 & 6 & 3 & 4 & 5 & 8 & 2 & 7 & 9 \end{pmatrix} = (2 \quad 6 \quad 8 \quad 7)$$

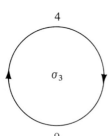

$$\sigma_3 = \begin{pmatrix} 1 & 2 & 3 & 4 & 5 & 6 & 7 & 8 & 9 \\ 1 & 2 & 3 & 9 & 5 & 6 & 7 & 8 & 4 \end{pmatrix} = (4 \quad 9)$$

We leave it to you to verify that $\sigma = \sigma_1 \sigma_2 \sigma_3$. We will write

$$\sigma = (1 \quad 3 \quad 5)(2 \quad 6 \quad 8 \quad 7)(4 \quad 9)$$

Figure 5.3

Note that the action of σ is represented in Figure 5.3.

The nice thing about cycles is that they can be used to rid ourselves of the clumsy two-row notation that we have used to describe permutations. To see this, recall our earlier example of permutation multiplication.

$$\begin{pmatrix} 1 & 2 & 3 & 4 \\ 3 & 4 & 2 & 1 \end{pmatrix} \begin{pmatrix} 1 & 2 & 3 & 4 \\ 2 & 3 & 4 & 1 \end{pmatrix} = \begin{pmatrix} 1 & 2 & 3 & 4 \\ 4 & 2 & 1 & 3 \end{pmatrix}$$

$$(1 \quad 3 \quad 2 \quad 4)(1 \quad 2 \quad 3 \quad 4) = (1 \quad 4 \quad 3)$$

Beneath each permutation, we have listed its representation as a product of cycles. We multiply the cycles in much the same way we composed the permutations, that is, follow the path of each number from right to left.

In $(1 \quad 3 \quad 2 \quad 4)(1 \quad 2 \quad 3 \quad 4)$, $1 \to 2 \to 4$, $2 \to 3 \to 2$, $3 \to 4 \to 1$, $4 \to 1 \to 3$. The product is thus

$$\begin{pmatrix} 1 & 2 & 3 & 4 \\ 4 & 2 & 1 & 3 \end{pmatrix} = (1 \quad 4 \quad 3)$$

Notice that the product on the left in our example has the same cycle representation as that on the right.

By the way, the cycle representation of a cycle is not necessarily unique. Clearly,

$$(1 \quad 2 \quad 3 \quad 4), \qquad (2 \quad 3 \quad 4 \quad 1), \qquad (3 \quad 4 \quad 1 \quad 2), \qquad (4 \quad 1 \quad 2 \quad 3)$$

are different representations of the same permutations. The set of numbers that appears in each cycle, however, is the same.

There are two other difficulties encountered in using the cyclical representation of a permutation. The first is that we must remember that the action of a cycle is read left to right inside the parentheses and that a cycle takes the last element in its cyclical representation back to the first. The second problem is that the cycle

$$(1 \quad 2 \quad 3) \in S_3 \qquad \text{and} \qquad (1 \quad 2 \quad 3) = \begin{pmatrix} 1 & 2 & 3 & 4 \\ 2 & 3 & 1 & 4 \end{pmatrix} \in S_4$$

are indistinguishable. Hence, we must keep in mind the symmetric group in which we are working. These difficulties are not overwhelming, however, and the notational ease that the use of cycles yields is one advantage that usually outweighs these problems.

Recall that two cycles are disjoint when all elements moved by either one are left fixed by the other. In the discussion involving Figure 5.3, we demonstrated how a particular permutation could be written as a product of other permutations each of which was a cycle. In fact, these cycles were disjoint. Interestingly, this property is true in general.

Theorem 5.6

Every permutation in S_n can be written as either a cycle or as a product of disjoint cycles.

Proof

Let σ be a nonidentity element of S_n. Our proof follows our example. Let i_1 be the first element of $\{1, 2, \ldots, n\}$ not fixed by σ. If $\sigma(i_1) = i_2$, $\sigma(i_2) = i_3$, $\ldots$, $\sigma(i_{m-1}) = i_m$, and $\sigma(i_m) = i_1$, set $\sigma_1 = (i_1 \quad i_2 \quad \cdots \quad i_m)$. If σ fixes every element of $\{1, 2, \ldots, n\} - \{i_1, i_1, \ldots, i_m\}$, then σ is a cycle and $\sigma = \sigma_1$. If not, then $\sigma(j_1) \neq j_1$ for some element in $\{1, 2, \ldots, n\} - \{i_1, i_2, \ldots, i_m\}$. Suppose

$$\sigma(j_1) = j_2 \qquad \sigma(j_2) = j_3 \qquad \cdots, \qquad \sigma(j_{k-1}) = j_k, \qquad \text{and} \qquad \sigma(j_k) = j_1$$

Set $\sigma_2 = (j_1 \quad j_2 \quad \cdots \quad j_k)$. Since σ is an injective mapping

$$\{i_1, i_2, \ldots, i_m\} \cap \{j_1, j_2, \ldots, j_k\} = \varnothing$$

Now $\sigma_1\sigma_2$ and σ act the same way on $\{i_1, i_2, \ldots, i_m, j_1, j_2, \ldots, j_k\}$. If the remaining elements of $\{1, 2, \ldots, n\}$ are fixed by σ, then $\sigma = \sigma_1\sigma_2$. Otherwise, we can proceed as above and obtain another cycle $\sigma_3 \in S_n$. Eventually this process must stop, and σ will be a product of disjoint cycles.

To complete the proof, we need only point out that the identity permutation can be written as the trivial cycle (1). □

If two permutations are each expressed as products of disjoint cycles, then the permutations are disjoint if no number appears in both. For example,

$$\sigma = (1 \quad 4 \quad 7 \quad 8)(3 \quad 5) \qquad \text{and} \qquad \tau = (2 \quad 6 \quad 9)$$

are disjoint, but σ and $\delta = (1 \quad 6 \quad 7)$ are not.

In writing a permutation σ as a product of disjoint cycles, you obtain a factorization of σ similar to the factorization of a natural number into a product of primes. While the disjoint cycles that appear are unique, the order in which they appear is not. So just as $6 = 2 \cdot 3 = 3 \cdot 2$, so too is $(1 \quad 2 \quad 3)(4 \quad 5) = (4 \quad 5)(1 \quad 2 \quad 3)$. This follows from the important observation that disjoint cycles commute. We leave the proof of this result as Exercise 7.

If $\sigma \in S_n$ is written as a product of disjoint cycles, say $\sigma = \sigma_1 \sigma_2 \cdots \sigma_t$, then σ induces a partition of $\{1, 2, \ldots, n\}$ in the following way. Each of the cycles σ_i contains elements in a subset of $\{1, 2, \ldots, n\}$. Since the cycles are disjoint, the subsets are disjoint. These subsets, together with singleton subsets of elements not moved by σ, are called the **orbits** of σ. These orbits form a partition of $\{1, 2, \ldots, n\}$.

For example, the permutation

$$n = \begin{pmatrix} 1 & 2 & 3 & 4 & 5 & 6 & 7 & 8 & 9 \\ 4 & 5 & 1 & 3 & 2 & 8 & 7 & 9 & 6 \end{pmatrix} = (1 \quad 4 \quad 3)(2 \quad 5)(6 \quad 8 \quad 9)$$

has orbits $\{1, 4, 3\}$, $\{2, 5\}$, $\{6, 8, 9\}$, and $\{7\}$.

In the next section, we will show how to factor a permutation into a product of transpositions or 2-cycles. But there the goal will be to discover information about the nature of permutations and to identify an important subgroup.

Exercises 5.2

1. List the elements of S_2 and S_4.

2. Write each element of S_3 in its cyclic representation, and construct the Cayley table for S_3.

3. Let

$$\sigma = \begin{pmatrix} 1 & 2 & 3 & 4 \\ 4 & 3 & 2 & 1 \end{pmatrix} \qquad \tau = \begin{pmatrix} 1 & 2 & 3 & 4 \\ 3 & 4 & 2 & 1 \end{pmatrix}$$

   a. Compute $\sigma\tau$ and $\tau\sigma$.
   b. Find σ^{-1} and τ^{-1}.
   c. Compute $\sigma^{-1}\tau^{-1}$ and $\tau^{-1}\sigma^{-1}$.

4. Let

$$\alpha = \begin{pmatrix} 1 & 2 & 3 & 4 & 5 & 6 & 7 & 8 \\ 2 & 1 & 5 & 6 & 4 & 3 & 8 & 7 \end{pmatrix} \quad \beta = \begin{pmatrix} 1 & 2 & 3 & 4 & 5 & 6 & 7 & 8 \\ 1 & 3 & 6 & 8 & 5 & 7 & 2 & 4 \end{pmatrix}$$

$$\gamma = \begin{pmatrix} 1 & 2 & 3 & 4 & 5 & 6 & 7 & 8 \\ 3 & 6 & 8 & 2 & 7 & 4 & 1 & 5 \end{pmatrix}$$

be elements of S_8.
   a. Write α, β, and γ as products of disjoint cycles.
   b. Compute $\alpha\beta$, $\beta\gamma$, and $\alpha\gamma$.
   c. Find α^{-1} and write it as a product of disjoint cycles. Do the same for β and γ.

5. Let

$$\alpha = \begin{pmatrix} 1 & 2 & 3 & 4 & 5 & 6 & 7 & 8 \\ 2 & 3 & 7 & 4 & 1 & 8 & 6 & 5 \end{pmatrix} \quad \beta = \begin{pmatrix} 1 & 2 & 3 & 4 & 5 & 6 & 7 & 8 \\ 2 & 1 & 3 & 5 & 7 & 8 & 4 & 6 \end{pmatrix}$$

$$\gamma = \begin{pmatrix} 1 & 2 & 3 & 4 & 5 & 6 & 7 & 8 \\ 6 & 8 & 2 & 5 & 1 & 4 & 7 & 3 \end{pmatrix}$$

Follow the instructions in Exercise 4.

6. Let

$$\alpha = \begin{pmatrix} 1 & 2 & 3 & 4 & 5 & 6 \\ 6 & 5 & 4 & 3 & 2 & 1 \end{pmatrix} \quad \beta = \begin{pmatrix} 1 & 2 & 3 & 4 & 5 & 6 \\ 2 & 1 & 4 & 5 & 6 & 3 \end{pmatrix}$$

$$\gamma = \begin{pmatrix} 1 & 2 & 3 & 4 & 5 & 6 \\ 1 & 3 & 2 & 4 & 6 & 5 \end{pmatrix}$$

Proceed as in Exercise 4.

7. Prove that if σ_1, $\sigma_2 \in S_n$ are disjoint, then $\sigma_1\sigma_2 = \sigma_2\sigma_1$.
8. Determine the orbits of α, β, and γ in Exercise 4.
9. Verify that $(1 \ \ 2 \ \ 3 \ \ 4 \ \ 5) = (1 \ \ 5)(1 \ \ 4)(1 \ \ 3)(1 \ \ 2)$.
10. Suppose $\sigma \in S_n$ is a k-cycle. Prove $|\langle \sigma \rangle| = k$.
11. Let $\sigma \in S_n$ be a cycle and $1 \leqslant i \leqslant n$. Suppose $\sigma(i) \neq i$. Show

$$\{j \mid \sigma(j) \neq j\} = \{j \mid j = \sigma^m(i), \ m \in \mathbb{N}\}$$

12. Let σ and τ be the disjoint cycles of length k and m, respectively. Use Exercises 7 and 10 to show that the order of $\sigma\tau$ is $km/(k, m)$, where (k, m) is the greatest common divisor of k and m.
13. Show that if σ, $\tau \in S_n$, then $o(\sigma\tau) = o(\tau\sigma)$.
14. If $\sigma \in S_n$ is written as a product of disjoint cycles, then how are the disjoint cycles in such representations of σ^2 and σ^{-1} related to those of σ?
15. Suppose $\sigma \in S_n$ is a p-cycle where p is prime. Prove that every power of σ is also a p-cycle or (1).
16. Prove that if $n > 2$, then S_n is a nonabelian group.
17. Show that $K = \{(1), (1 \ \ 2)(3 \ \ 4), (1 \ \ 3)(2 \ \ 4), (1 \ \ 4)(2 \ \ 3)\}$ is a group.
18. Use induction to prove $|S_n| = n!$.

*19. Write a computer program that takes a permutation of S_5 and returns its cycle decomposition. Use as input the second row of the standard two-row representation of the permutation.

*20. Write a computer program that takes two permutations in S_5 and computes their product.

5.3

Odd and Even Permutations

The major result of the last section stated that every permutation in S_n can be written as a product of disjoint cycles. The cycles in such a factorization have various lengths. If we do not require the cycles to be disjoint, then we can also write any permutation as a product of 2-cycles or transpositions.

Theorem 5.7

Every cycle can be written as a product of transpositions.

Proof

Let $\sigma = (k_1 \quad k_2 \quad \cdots \quad k_m)$ be a cycle in S_n. Consider $\omega \in S_n$ defined by $\omega = (k_1 \quad k_m)(k_1 \quad k_{m-1}) \cdots (k_1 \quad k_2)$. Then $\omega(k_1) = k_2$, $\omega(k_2) = k_3$, ..., $\omega(k_{m-1}) = k_m$, $\omega(k_m) = k_1$. Moreover $\omega(k') = k'$ for all $k' \in \{1, 2, \ldots, n\} - \{k_1, k_2, \ldots, k_m\}$. Thus $\sigma = \omega$. $\square$

By Theorem 5.6, every permutation can be written as a product of cycles. By Theorem 5.7, every cycle can be written as a product of transpositions. Thus we have the following result:

Corollary 5.3

Every permutation can be written as a product of transpositions.

Example 5.6

We saw in the last section that

$$\sigma = \begin{pmatrix} 1 & 2 & 3 & 4 & 5 & 6 & 7 & 8 & 9 \\ 3 & 6 & 5 & 9 & 1 & 8 & 2 & 7 & 4 \end{pmatrix} = (1 \quad 3 \quad 5)(2 \quad 6 \quad 8 \quad 7)(4 \quad 9)$$

To write a cycle as a product of transpositions, take the first (from the left) element of the cycle as the first element of each transposition, and then write transpositions using elements from right to left in the cycle. Therefore

$$(1 \quad 3 \quad 5) = (1 \quad 5)(1 \quad 3) \qquad \text{and}$$

$$(2 \quad 6 \quad 8 \quad 7) = (2 \quad 7)(2 \quad 8)(2 \quad 6)$$

This gives us

$$\sigma = (1 \quad 3 \quad 5)(2 \quad 6 \quad 8 \quad 7)(4 \quad 9)$$

$$= (1 \quad 5)(1 \quad 3)(2 \quad 7)(2 \quad 8)(2 \quad 6)(4 \quad 9)$$

Corollary 5.3 tells us that every permutation can be written as a product of transpositions. Is this decomposition unique? Is the number of these transpositions determined by the permutation? The answer to both these questions is no! The identity of S_n for $n > 3$ can be written as

$$(1) = (1 \quad 2)(1 \quad 2) \qquad \text{or} \qquad (1) = (3 \quad 4)(3 \quad 4)$$

while the element $(1 \quad 2)$ may also be written as $(1 \quad 2)(3 \quad 4)(3 \quad 4)$. In addition,

$$(1 \quad 2 \quad 3) = (2 \quad 1)(2 \quad 3) = (1 \quad 3)(1 \quad 2)$$

There is, however, a useful property of permutations that does not change regardless of the method of representation. This property is termed an **invariant** and is extremely important in the decomposition of a permutation into a product of transpositions.

Definition 5.7

Let $\sigma \in S_n$. We say that σ is an **even permutation** if it is possible to write σ as a product of an even number of transpositions. We say that σ is an **odd permutation** if it is possible to write σ as a product of an odd number of transpositions.

Such a definition is acceptable only when we show that a given permutation is either odd or even, that is, that no permutation can be both. For example, we have mentioned (and will investigate further in Section 5.4) the fact that the set of all permutations on a given set forms a group with the operation of composition of mappings. In this regard, we recall that the identity for the group is the permutation $e = (1)$ defined by $e(n) = n$ for all $n \in \{1, 2, \ldots, n\}$.

If $n > 1$, then we note that the identity of S_n seems to be an even permutation, for $(1) = (1 \quad 2)(1 \quad 2)$. Is the identity also an odd permutation? In fact, if σ is an arbitrary permutation, can σ be both an even and an odd permutation? The answer is no, and the next result will demonstrate this fact.

Theorem 5.8

Parity Theorem
No permutation is both even and odd.

Proof

Suppose $\sigma \in S_n$ is both even and odd. We seek a contradiction. Toward that end, suppose $\sigma = \tau_1 \tau_2 \cdots \tau_m$, where the τ_i's are transpositions and m is even, and $\sigma = \sigma_1 \sigma_2 \cdots \sigma_q$, where the σ_j's are transpositions and q is odd. Then

$$(1) = \sigma \sigma^{-1} = (\tau_1 \tau_2 \cdots \tau_m)(\sigma_1 \sigma_2 \cdots \sigma_q)^{-1}$$

$$= (\tau_1 \tau_2 \cdots \tau_m)(\sigma_q^{-1} \sigma_{q-1}^{-1} \cdots \sigma_1^{-1})$$

By Theorem 3.2 and the fact that any transposition is its own inverse, we have

$$(1) = (\tau_1 \tau_2 \cdots \tau_m)(\sigma_q \sigma_{q-1} \cdots \sigma_1)$$

Thus (1) can be written as a product of $m + q$ transpositions. Hence, if σ is both an even and odd permutation, then (1) can be written as an odd permutation. We use mathematical induction to show that this is not true.

Assume $(1) = \tau_1 \tau_2 \cdots \tau_p$, where the τ_i's are transpositions and p is the smallest such odd integer. Then $p \neq 1$, for in that case $(1) = (a \ b)$, which contradicts the fact that the identity permutation (1) fixes a. Consider $\tau_{p-1} \tau_p$, where $\tau_p = (a \ b)$. There are four possible cases to consider.

Case 1: $\tau_{p-1} \tau_p = (a \ b)(a \ b)$. Since any transposition is its own inverse, we have

$$(1) = \tau_1 \tau_2 \cdots \tau_{p-2}$$

contrary to our choice of p. Therefore, this case cannot occur.

Case 2: $\tau_{p-1} \tau_p = (c \ d)(a \ b)$. Then $\tau_{p-1} \tau_p = (a \ b)(c \ d)$ since disjoint transpositions commute.

Case 3: $\tau_{p-1} \tau_p = (a \ d)(a \ b)$. Then $\tau_{p-1} \tau_p = (a \ b)(b \ d)$. (Verify.)

Case 4: $\tau_{p-1} \tau_p = (b \ d)(a \ b)$. Then $\tau_{p-1} \tau_p = (a \ d)(b \ d)$. (You should check this, also.)

Cases 2–4 show that we can replace any factorization of (1) ending in $(a \ b)$ with one in which a appears in the second to last transposition without adding to the number of transpositions present.

Make the appropriate change in the representation and repeat the argument used with $\tau_{p-1} \tau_p$ to the second to last and third to last transpositions. We again encounter the four cases just considered, the first allowing us to terminate our argument and the other three permitting us to replace the factorization we have by one in which a appears in the third to last transposition. Continuing in this manner, if Case 1 never occurs we conclude that the only transposition that contains a is τ_1. This means that either Case 1 occurs or (1) moves the element a, both of which are contradictions. Thus the identity permutation cannot be written as a product of an odd number of transpositions and, hence, (1) must be an even permutation. As a consequence, any permutation is either even or odd but not both.   □

This is but one of many approaches to the proof of the Parity Theorem. See Spitznagel's article in the *American Mathematical Monthly* for another version.[2] We outline two traditional approaches to the proof of the Parity Theorem in the exercises.

---

[2]E. L. Spitznagel, Jr., "Note on the alternating group," *Amer. Math. Monthly*, Vol. 75, No. 1 (1968), pp. 68–69.

| Corollary 5.4 |

If $\sigma, \tau \in S_n$, then

    a. $\sigma\tau$ is even if σ and τ are both even or both odd, and

    b. $\sigma\tau$ is odd when one permutation is odd and the other is even.

Proof

This result follows immediately from the observation that if σ can be written as a product of m transpositions and τ can be written as a product of q transpositions, then $\sigma\tau$ can be represented as a product of $m+q$ transpositions. □

We next wish to investigate a mapping from S_n to the set $\{1, -1\}$. If $\sigma \in S_n$, define sign$(\sigma) = 1$ if σ is even, and sign$(\sigma) = -1$ if σ is odd. Since no permutation is both even and odd, sign is well-defined; that is, sign(σ) is 1 or -1 but not both. The mapping sign appears in linear algebra where it is utilized in the definition of *determinant*.

A number of statements can be made about the mapping sign:

| Corollary 5.5 |

If $\sigma, \tau \in S_n$, then

    a. sign$(\sigma\tau) = ($sign $\sigma)($sign $\tau)$;

    b. sign$(\sigma) = $sign$(\sigma^{-1})$; and

    c. if σ can be written as a product of k transpositions, then sign$(\sigma) = (-1)^k$.

Proof

Part a is a restatement of Corollary 5.4. Since $\sigma\sigma^{-1} = (1)$, we have $1 = $ sign$((1)) = ($sign $\sigma)($sign $\sigma^{-1})$ by part a. But then this equation shows that sign σ and sign σ^{-1} must have the same value for their product to equal 1. This proves part b. Finally, part a can be extended by induction to prove that the sign of the composite of any number of permutations is the product of the signs. Then part c follows immediately. □

In Section 5.4, we will apply our results to study the subgroup structure of S_n and also to see how groups of the form S_n relate to the study of all finite groups. However, before we end our discussion of even and odd permutations, we want to mention an interesting application to the world of puzzles.

Over one hundred years ago, Sam Loyd, a noted creator of puzzles, invented what is known as the "15"-puzzle. The game enjoyed instant success, similar to that of Rubik's cube. Its popularity is attested to by the fact that a game, or variations of it, is still found among Christmas "stocking stuffers", in toy stores, and every so often is included as a prize in boxes of breakfast cereal.

The puzzle, depicted in Figure 5.4, consists of fifteen numbered tiles contained in a frame with sixteen spaces. The empty space is used to slide the tiles and alter the arrangement of the numbers on the tiles. Note that the order of the tile numbered 15 is reversed with the tile numbered 14. The

1	2	3	4
5	6	7	8
9	10	11	12
13	15	14	

Figure 5.4

object of the game is to slide the tiles around so that numbers on the tiles appear in order from 1 to 15 with the empty space again in the lower right-hand corner. Of course, breaking the case that holds the tiles, or prying out the tiles and putting them back in the correct order, is not allowed.

Part of the instant success of this puzzle was, no doubt, due to the fact that Sam Loyd offered $1000 to the first person to solve the puzzle. A number of persons tried to claim the prize, but none of them was able to recreate the sequence of moves or slides he or she used. The reason for this is that no solution exists! Through the use of our knowledge of permutations, it is not difficult to figure out why this is so. First, note that whenever a succession of moves is made so that the empty space appears in the lower-right corner, then an even permutation has been performed. For example, Figure 5.5 is the puzzle after tile 12 is slid down, the third row slid to the right, tile 13 is slid up, and the fourth row is slid to the left. In essence we have the permutation

1	2	3	4
5	6	7	8
13	9	10	11
15	14	12	

Figure 5.5

$$\begin{pmatrix} 9 & 10 & 11 & 12 & 13 & 15 & 14 \\ 13 & 9 & 10 & 11 & 15 & 14 & 12 \end{pmatrix} = (9 \quad 13 \quad 15 \quad 14 \quad 12 \quad 11 \quad 10)$$

$$= (9 \quad 10)(9 \quad 11)(9 \quad 12)(9 \quad 14)(9 \quad 15)(9 \quad 13)$$

which is even. Secondly, observe that the original configuration is just the transposition (14   15). Therefore, part b of Corollary 5.4 guarantees that any sequence of slides will produce another odd permutation. Since the identity permutation or arrangement is even, it can never be obtained by a sequence of "legal" moves. This puzzle gives you an example to rebuff anyone who might scoff that "abstract algebra isn't good for anything."[3]

Exercises 5.3

1. Determine the parity of each of the following permutations by writing them as products of transpositions.
   a. (1   2   3   4)
   b. (1   3   2   6)(4   5)
   c. (1   2   3)(2   3   4)
   d. (1   3   4   6)(1   3   4   6)
   e. (1   2   9   8)(7   6   3   4)
   f. (2   3   6   5   7   9   1   4)(6   8   9   1   5   4   3   2)

2. Write each of the permutations in Exercise 4 of Section 5.2 as a product of transpositions.

3. Determine the parity of each of the permutations in Exercise 5 of Section 5.2 by writing them as products of transpositions.

[3] For excellent descriptions of other puzzles, we recommend Martin Gardner, *Mathematical Puzzles and Diversions*, New York: Simon and Schuster, 1959.

4. Write each of the permutations in Exercise 6 of Section 5.2 as a product of transpositions.

5. Prove that if k is even then a k-cycle is odd, and that if k is odd then a k-cycle is even.

6. Determine the sign or parity of each permutation in S_4.

7. List all even permutations of S_4.

8. Is it possible for a subgroup of S_n to contain only odd permutations? If so, give an example. If not, why?

9. Prove that if $(i \ \ j)$ and $(i \ \ k)$ are transpositions in S_n, then there exists $\sigma \in S_n$ such that $\sigma(i \ \ j)\sigma^{-1} = (i \ \ k)$.

10. If σ is an even permutation of S_n and $\tau \in S_n$, prove that $\tau \sigma \tau^{-1}$ is even.

11. Let $i \in \{1, 2, \ldots, n\}$ and set $H_i = \{\sigma \in S_n | \sigma(i) = i\}$. Prove $H_i \leqslant S_n$. (H_i is known as the **stabilizer of** i.)

12. Let $\sigma \in S_n$. Define $\mathrm{sgn}(\sigma) = \displaystyle\prod_{i<j} \frac{\sigma(j) - \sigma(i)}{j-i}$ for $1 \leqslant i < j \leqslant n$.

    a. Calculate $\mathrm{sgn}(\sigma)$ for $\sigma = (1 \ \ 2 \ \ 3) \in S_3$ and $\sigma = (1 \ \ 3 \ \ 2 \ \ 4) \in S_4$.
    b. Prove $\mathrm{sgn}(\tau) = -1$ if τ is a transposition.
    c. Verify $\mathrm{sgn}(\sigma\tau) = (\mathrm{sgn} \ \sigma)(\mathrm{sgn} \ \tau)$.
    d. Prove $\mathrm{sgn}(\sigma) = \mathrm{sign}(\sigma)$.

13. Let $n \in \mathbb{N}$ and $p = p(x) = \prod(x_j - x_i)$, where the product runs over all i, j such that $i \leqslant i < j \leqslant n$. For $\sigma \in S_n$, define

$$\sigma(p) = \prod_{1 \leqslant i < j \leqslant n} (x_{\sigma(j)} - x_{\sigma(i)})$$

    a. For $(1 \ \ 2 \ \ 3) \in S_3$ and $\sigma \in S_4$, calculate $\sigma(p)$.
    b. Verify that $\tau(p) = -p$ if τ is a transposition.
    c. Prove $\sigma(p) = p$ if and only if σ is even.

14. Prove that if $\sigma \in S_n$ can be written as $\tau_1 \tau_2 \cdots \tau_k$, where each τ_i is a transposition, then $\sigma^{-1} = \tau_k^{-1} \tau_{k-1}^{-1} \cdots \tau_1^{-1} = \tau_k \tau_{k-1} \cdots \tau_1$.

15. The following approach to the proof of the Parity Theorem was first given by Hans Liebeck in the June 1969 issue of the *American Mathematical Monthly*. Let $(1) = \tau_1 \tau_2 \cdots \tau_k$, where each τ_i is a transposition. An element a that is moved by one transposition must be moved by a second transposition. Thus the number of transpositions that move a must be even. Since this is true for each element that is moved once, the total number of transpositions must be even. Discuss the validity of this "simple" proof of the Parity Theorem.

*16. Use Exercise 5 to modify the program of Exercise 19 of Section 5.2 to determine if a permutation is odd or even.

5.4

Symmetric Groups

The subject of this chapter is permutation groups. We have not, as yet, formally defined what is meant by this term. A **permutation group** is simply a group of permutations. Since these permutations must be permutations of some set, a permutation group is a subgroup of the set of all permutations of that set. In other words, H is a permutation group if $H \leqslant S_X$ for some set X.

This brings up the question as to what subgroups does S_n possess. Are there any subgroups of S_n besides S_n and $\{(1)\}$? For $n > 2$, the answer is yes. In fact, in the next chapter, we will present a classic result of Cayley that states that, in one sense, *every* finite group is a permutation group on some set. This means that if you know the subgroup structure of S_n for all n, then you know the structure of *all* finite groups, because we will see that if $|X| = n$, then S_X and S_n have identical subgroup structure.

Now, obviously S_n contains cyclic subgroups. But there are many other noncyclic subgroups of S_n. For example, by Exercise 17 of Section 5.2, the set

$$K = \{(1), (1\quad 2)(3\quad 4), (1\quad 3)(2\quad 4), (1\quad 4)(2\quad 3)\}$$

is a subgroup of S_4. We now wish to investigate what is probably the most important subgroup of S_n, namely, the alternating group. We will show in a later chapter that the structure of this subgroup for $n = 5$ is intimately related to the question of whether there exists a result analogous to the quadratic formula for finding the roots of a polynomial of degree 5.

Let us, therefore, consider A_n, the set of all even permutations of S_n. Recall that $(1) = (1\quad 2)(1\quad 2)$ is an even permutation of S_n. We saw in Corollary 5.4 that the composite of two even permutations is another even permutation. This means that A_n is a nonempty subset of S_n closed under composition. By Theorem 4.5, A_n is a subgroup. As a consequence, we now have another proof that the inverse of an even permutation is even. This last observation also follows from part b of Corollary 5.5.

Our results about A_n are contained in Theorem 5.9.

Theorem 5.9

For each $n \in \mathbb{N}$, $n > 1$, the set A_n, consisting of all even permutations in S_n, is a subgroup of S_n called the **alternating group on n letters**.

The group A_3 consists of the permutations $\{(1), (1\quad 2\quad 3), (1\quad 3\quad 2)\}$. Note that this is exactly half of the number of permutation in S_3. The same is true for S_n, $n > 3$.

Theorem 5.10

In S_n, $n > 1$, the number of even permutations is equal to the number of odd permutations.

Proof

Let $\Lambda = \{$even permutations in $S_n\}$ and $\Gamma = \{$odd permutations in $S_n\}$. We wish to show that $|\Lambda| = |\Gamma|$. Let $\delta \in \Gamma$, and consider the mapping $\phi_\delta \colon \Lambda \to \Gamma$ defined by $\phi_\delta(\sigma) = \sigma\delta$ for $\sigma \in \Lambda$. Since δ is odd and σ is even, $\sigma\delta$ is odd by Corollary 5.4. If $\phi_\delta(\sigma_1) = \phi_\delta(\sigma_2)$, then $\sigma_1\delta = \sigma_2\delta$; thus $\sigma_1 = \sigma_2$ by right cancellation. We conclude that ϕ_δ is injective and $|\Lambda| \leqslant |\Gamma|$.

Now consider the mapping $\Psi_\delta: \Gamma \to \Lambda$ defined by $\Psi_\delta(\tau) = \tau\delta$ for $\tau \in \Gamma$. Since τ and δ are both odd, $\tau\delta$ is even by Corollary 5.4. Again this mapping is injective, so that $|\Gamma| \leqslant |\Lambda|$. Therefore, we conclude $|\Lambda| = |\Gamma|$. $\square$

Since $|S_n| = n!$ by Theorem 5.5 we have the following corollary.

Corollary 5.6

$$|A_n| = \frac{n!}{2}$$

The symmetric group S_4 is large enough to possess a number of interesting properties and contain many useful examples of permutation groups. For that reason, we have given the Cayley table for S_4 in Figure 5.6.

Let us remind you that in order to determine the computation $(1\ \ 2\ \ 4)(2\ \ 4\ \ 3)$, you should find the element in the $(1\ \ 2\ \ 4)$ row (horizontal line) and the $(2\ \ 4\ \ 3)$ column (vertical line). Thus

$$(1\ \ 2\ \ 4)(2\ \ 4\ \ 3) = (1\ \ 2)(3\ \ 4) \qquad \text{while}$$
$$(2\ \ 4\ \ 3)(1\ \ 2\ \ 4) = (1\ \ 4)(2\ \ 3)$$

There are times when permutations are quite useful in computer science. Sometimes there is a need to determine if a graph possesses some sort of property or configuration or if a certain path exists between numbered vertices. In other instances, data must be examined in order to rearrange input. Permutations can be used to enumerate all possibilities. In algorithm analysis, permutations can be utilized to check all possible branches in a program. In each case, time can be saved by examining the problem in terms of permutations.

Permutations used in computer science are seldom, if ever, stored in computer memory. Just to store S_{10} would require that $10! = 3,628,800$ permutations be assigned storage locations. Instead a subroutine is incorporated into a computer program that generates the permutations one at a time. As each permutation is generated, the graph or data is tested, a new permutation is generated, and the old one discarded. This procedure is repeated until all permutations have been used or until the answer to the original question is obtained.

When we say that a permutation is **generated**, we mean that a rearrangement of $\{1, 2, \ldots, n\}$ is produced. This rearrangement is the second row of the two-row description of permutations. There are quite a few intriguing schemes that have been devised for generating permutations. Many of them can be found in the *Communications of the A.C.M.* (See, for example, Vol. 5, pp. 434–435; Vol. 10, pp. 298–299, 452; and Vol. 11, p. 117.)

In the next chapter, we will discover two interesting facts about S_n. The first is that any symmetric group on a set X with $|X| = n$ can be identified with S_n. The second is that *any* finite group is, in one way, a subgroup of S_n for some n. These will result from our investigation into the question of when are two groups of the same order identical.

S_4	(1)	(1234)	(13)(24)	(1432)	(123)	(1342)	(243)	(14)	(132)	(34)	(124)	(1423)
(1)	(1)	(1234)	(13)(24)	(1432)	(123)	(1342)	(243)	(14)	(132)	(34)	(124)	(1423)
(1234)	(1234)	(13)(24)	(1432)	(1)	(1324)	(143)	(12)	(234)	(14)	(123)	(1342)	(243)
(13)(24)	(13)(24)	(1432)	(1)	(1234)	(142)	(23)	(134)	(1243)	(234)	(1324)	(143)	(12)
(1432)	(1432)	(1)	(1234)	(13)(24)	(34)	(124)	(1423)	(132)	(1243)	(142)	(23)	(134)
(123)	(123)	(1342)	(243)	(14)	(132)	(34)	(124)	(1432)	(1)	(1234)	(13)(24)	(1432)
(1342)	(1342)	(243)	(14)	(123)	(24)	(14)(23)	(13)	(12)(34)	(1423)	(132)	(34)	(124)
(243)	(243)	(14)	(123)	(1342)	(143)	(12)	(234)	(1324)	(12)(34)	(24)	(14)(23)	(13)
(14)	(14)	(123)	(1342)	(243)	(1234)	(13)(24)	(1432)	(1)	(1324)	(143)	(12)	(234)
(132)	(132)	(34)	(124)	(1423)	(1)	(1234)	(13)(24)	(1432)	(123)	(1342)	(243)	(14)
(34)	(34)	(124)	(1423)	(132)	(1243)	(142)	(23)	(134)	(1432)	(1)	(1234)	(13)(24)
(124)	(124)	(1423)	(132)	(34)	(14)(23)	(13)	(12)(34)	(24)	(134)	(1243)	(142)	(23)
(1423)	(1423)	(132)	(34)	(124)	(1342)	(243)	(14)	(123)	(24)	(14)(23)	(13)	(12)(34)
(12)	(12)	(234)	(1324)	(143)	(23)	(134)	(1243)	(142)	(13)	(12)(34)	(24)	(14)(23)
(234)	(234)	(1324)	(143)	(12)	(13)(24)	(1432)	(1)	(1234)	(142)	(23)	(134)	(1243)
(1324)	(1324)	(143)	(12)	(234)	(14)	(123)	(1342)	(243)	(1234)	(13)(24)	(1432)	(1)
(143)	(143)	(12)	(234)	(1324)	(12)(34)	(24)	(14)(23)	(13)	(243)	(14)	(123)	(1342)
(23)	(23)	(134)	(1243)	(142)	(13)	(12)(34)	(24)	(14)(23)	(12)	(234)	(1342)	(143)
(134)	(134)	(1243)	(142)	(23)	(124)	(1423)	(132)	(34)	(14)(23)	(13)	(12)(34)	(24)
(1243)	(1243)	(142)	(23)	(134)	(1432)	(1)	(1234)	(13)(24)	(34)	(124)	(1423)	(132)
(142)	(142)	(23)	(134)	(1243)	(234)	(1324)	(143)	(12)	(13)(24)	(1432)	(1)	(1234)
(13)	(13)	(12)(34)	(24)	(14)(23)	(12)	(243)	(14)	(123)	(1342)	(143)	(23)	(134)
(12)(34)	(12)(34)	(24)	(14)(23)	(13)	(243)	(14)	(123)	(1342)	(143)	(12)	(234)	(1324)
(24)	(24)	(14)(23)	(13)	(12)(34)	(1423)	(132)	(34)	(124)	(1342)	(243)	(14)	(123)
(14)(23)	(14)(23)	(13)	(12)(34)	(24)	(134)	(1243)	(142)	(23)	(124)	(1423)	(132)	(34)

S_4	(12)	(234)	(1324)	(143)	(23)	(134)	(1243)	(142)	(13)	(12)(34)	(24)	(14)(23)
(1)	(12)	(234)	(1324)	(143)	(23)	(134)	(1243)	(142)	(13)	(12)(34)	(24)	(14)(23)
(1234)	(134)	(1243)	(142)	(23)	(124)	(1423)	(132)	(34)	(14)(23)	(13)	(12)(34)	(24)
(13)(24)	(1423)	(132)	(34)	(124)	(1342)	(243)	(14)	(123)	(24)	(14)(23)	(13)	(12)(34)
(1432)	(243)	(14)	(123)	(1342)	(143)	(12)	(234)	(1324)	(12)(34)	(24)	(14)(23)	(13)
(123)	(13)	(12)(34)	(24)	(14)(23)	(12)	(234)	(1324)	(143)	(23)	(134)	(1243)	(142)
(1342)	(234)	(1324)	(143)	(12)	(13)(24)	(1432)	(1)	(1234)	(142)	(23)	(134)	(1243)
(243)	(1432)	(1)	(1234)	(13)(24)	(34)	(124)	(1423)	(132)	(1243)	(142)	(23)	(134)
(14)	(124)	(1423)	(132)	(34)	(14)(23)	(13)	(12)(34)	(24)	(134)	(1243)	(142)	(23)
(132)	(23)	(134)	(1243)	(142)	(13)	(12)(34)	(24)	(14)(23)	(12)	(234)	(1324)	(143)
(34)	(12)(34)	(24)	(14)(23)	(13)	(243)	(14)	(123)	(1342)	(143)	(12)	(234)	(1324)
(124)	(14)	(123)	(1342)	(243)	(1234)	(13)(24)	(1432)	(1)	(1324)	(143)	(12)	(234)
(1423)	(13)(24)	(1432)	(1)	(1234)	(142)	(23)	(134)	(1243)	(234)	(1324)	(143)	(12)
(12)	(1)	(1234)	(13)(24)	(1432)	(123)	(1342)	(243)	(14)	(132)	(34)	(124)	(1423)
(234)	(1342)	(243)	(14)	(123)	(24)	(14)(23)	(13)	(12)(34)	(1423)	(132)	(34)	(124)
(1324)	(14)(23)	(13)	(12)(34)	(24)	(134)	(1243)	(142)	(23)	(124)	(1423)	(132)	(34)
(143)	(1243)	(142)	(23)	(134)	(1432)	(1)	(1234)	(13)(24)	(34)	(124)	(1423)	(132)
(23)	(132)	(34)	(124)	(1423)	(1)	(1234)	(13)(24)	(1432)	(123)	(1342)	(243)	(14)
(134)	(1234)	(13)(24)	(1432)	(1)	(1324)	(143)	(12)	(234)	(14)	(123)	(1342)	(243)
(1243)	(143)	(12)	(234)	(1324)	(12)(34)	(24)	(14)(23)	(13)	(243)	(14)	(123)	(1342)
(142)	(24)	(14)(23)	(13)	(12)(34)	(1423)	(132)	(34)	(124)	(1342)	(243)	(14)	(123)
(13)	(123)	(1342)	(243)	(14)	(132)	(34)	(124)	(1423)	(1)	(1234)	(13)(24)	(1432)
(12)(34)	(34)	(124)	(1423)	(132)	(1243)	(142)	(23)	(134)	(1432)	(1)	(1234)	(13)(24)
(24)	(142)	(23)	(134)	(1243)	(234)	(1324)	(143)	(12)	(13)(24)	(1432)	(1)	(1234)
(14)(23)	(1324)	(143)	(12)	(234)	(14)	(123)	(1342)	(243)	(1234)	(13)(24)	(1432)	(1)

Figure 5.6

Exercises 5.4

1. Give the Cayley table for A_3 and A_4.

2. Label the vertices of an equilateral triangle and describe each symmetry in terms of permutations. For example, $\rho \sim$ 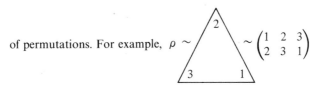 $\sim \begin{pmatrix} 1 & 2 & 3 \\ 2 & 3 & 1 \end{pmatrix}$

3. Follow the instructions in Exercise 2 for the symmetries of a rectangle.

4. Follow the instructions in Exercise 2 for the symmetries of a square.

5. Find the Cayley table for the group of permutations in Exercise 2.

6. Give the Cayley table for the permutation group in Exercise 3.

7. Construct the Cayley table for the group of permutations in Exercise 4.

8. Let $\sigma \in S_n$. Set $k(\sigma)$ equal to the number of interchanges of adjacent numbers in the bottom row of a two-row representation of σ, which are necessary to return the rearrangement to natural order. For example, for

$$\sigma = \begin{pmatrix} 1 & 2 & 3 & 4 \\ 2 & 4 & 1 & 3 \end{pmatrix}$$

we have $2413 \rightarrow 2143 \rightarrow 2134 \rightarrow 1234$, so that $k(\sigma) = 3$. Prove that $\text{sign}(\sigma) = (-1)^{k(\sigma)}$.

9. Examine the table for S_4 (Figure 5.6). Find:
   a. Cyclic subgroups of order 2, 3, and 4
   b. A noncyclic subgroup of order 4
   c. A subgroup of order 6
   d. A subgroup of order 8
   e. A subgroup of order 12

10. With the notation in Exercise 19 of Section 4.3, determine
    a. $Z(S_3)$, $Z(A_3)$, $Z(S_4)$, $Z(A_4)$
    b. $N_{S_4}(A_4)$, $N_{S_4}(S_3)$

11. Show that A_4 contains no subgroup of order 6.

12. Show that if a subgroup of S_n contains all 3-cycles, then it contains all even permutations. Use the fact that any even permutation is the product of an even number of transpositions.

13. Verify that
    a. $(2\ \ 3)(1\ \ 2)(2\ \ 3) = (1\ \ 3)$
    b. $(3\ \ 4)(1\ \ 3)(3\ \ 4) = (1\ \ 4)$
    c. $(k\ \ k+1)(1\ \ k)(k\ \ k+1) = (1\ \ k+1)$

14. Let $\sigma = (1\ \ 2\ \ 3\ \ \cdots\ \ n-1\ \ n)$ and $\tau = (1\ \ 2)$. Show:
    a. $\sigma\tau\sigma^{-1} = (2\ \ 3)$
    b. $\sigma^2\tau\sigma^{-2} = (3\ \ 4)$
    c. $\sigma^k\tau\sigma^{-k} = (k+1\ \ k+2)$ for $k < n-1$

15. Use Exercises 13 and 14 to show that if $H \leqslant S_n$ and $(1\ \ 2\ \ 3\ \ \cdots\ \ n), (1\ \ 2) \in H$, then $H = S_n$.

16. Verify that if σ is a k-cycle and τ is a transposition, then $\tau\sigma\tau^{-1}$ is a k-cycle. Use Theorem 5.7 to show that $\gamma \in S_n$ then $\gamma\sigma\gamma^{-1}$ is a k-cycle.

17. If H is a subgroup of S_n that contains all transpositions of the form $(1 \quad s)(1 \quad t)$, for $s, t \in \{1, 2, \ldots, n\}$, show that $H = S_n$ or A_n.

18. If K is a subgroup of S_n that contains all permutations for the form $(1 \quad t)$ for $t = 1, 2, \ldots, n$, show that $K = S_n$.

6

Group

Homomorphisms

Recall that when we discussed cyclic groups, we stated that cyclic groups could be divided into two classes—finite cyclic groups, which are essentially the groups $(\mathbb{Z}_n, \oplus)$, and infinite cyclic groups, which act like the integers. In Chapter 5, we made the statement that any finite group could be viewed as a permutation group. In particular, in Exercises 2, 3, and 4 of Section 5.4, we asked you to show how elements of certain groups of rigid motions of geometric shapes can be described by permutations. In this chapter we will explore these concepts in more detail and make explicit what we mean by *essentially*, *act like*, and *can be viewed as*. We will do this first by considering homomorphisms, that is, mappings between groups which *preserve* operations, and then by examining isomorphisms, that is, homomorphisms which are also bijections. In the last section, we will prove Cayley's Theorem, the result that states that every group is isomorphic to a permutation group. In Chapter 8, we will examine more closely the connection between the subgroup structure of a group G and the set of homomorphisms on G.

6.1

Group Homomorphisms

Let us start by considering two groups. Suppose G is the group of rotations of an equilateral triangle and G' is the group of integers modulo 3; that is, $G' = \mathbb{Z}_3$. (See Figures 6.1 and 6.2, respectively.)

G	e	ρ	ρ^2
e	e	ρ	ρ^2
ρ	ρ	ρ^2	e
ρ^2	ρ^2	e	ρ

Figure 6.1

Let us associate G with G' in the following way:

$$e \leftrightarrow 0 \qquad \rho \leftrightarrow 1 \qquad \rho^2 \leftrightarrow 2$$

Then $\rho^2 = \rho \cdot \rho \leftrightarrow 1 + 1 = 2$. You may wish to check that all the other products of corresponding elements are related in the same way. Thus, you might suspect that there is actually only one group involved here and that G and G' are but two different representations of that group.

G'	0	1	2
0	0	1	2
1	1	2	0
2	2	0	1

Figure 6.2

Consider another example. Let G be the group of integers under addition and G the group consisting of 1 and -1 under multiplication. Define a mapping $\Gamma\colon G \to G'$ by $\Gamma(n) = 1$ if n is even and $\Gamma(n) = -1$ if n is odd. Then note that

$$\text{odd} + \text{odd} = \text{even and } (-1)(-1) = 1$$

$$\text{odd} + \text{even} = \text{odd and } (-1)(1) = -1$$

$$\text{even} + \text{odd} = \text{odd and } (1)(-1) = -1$$

$$\text{even} + \text{even} = \text{even and } (1)(1) = 1$$

Thus, we see that to every integer there corresponds a number 1 or -1 such that sums of integers in G correspond to products of the corresponding numbers in G'. Note that this correspondence is no longer injective, but, nevertheless, there still seems to be some sort of relationship between these two groups.

The relationships involved in these examples are made explicit by the following definition:

Definition 6.1

Let $(G, \circ)$ and $(G', *)$ be groups. A mapping $\phi\colon G \to G'$ is a **group homomorphism** if

$$a, b \in (G, \circ) \qquad \text{implies that} \qquad \phi(a \circ b) = \phi(a) * \phi(b)$$

Both mappings in the two previous examples are group homomorphisms. In the first case, the mapping was described by a correspondence. To verify that this correspondence is a homomorphism involves checking each composition in the table. In the second case, the formula for Γ and the four lines illustrating the relationship between addition of integers and multiplication in G' provide the verification that $\Gamma(m + n) = \Gamma(m) \cdot \Gamma(n)$.

Homomorphisms are given special names due to the nature of the mapping. For example, suppose $\phi: G \to G'$ is a homomorphism. If ϕ is also a surjection, then ϕ is termed an **epimorphism**. If ϕ is an injection, then ϕ is called a **monomorphism**. Of special interest is the situation where a homomorphism $\phi: G \to G'$ is a bijection. In this case, ϕ is said to be an **isomorphism**, and $(G, \circ)$ and $(G', *)$ are said to be **isomorphic groups** (denoted $G \cong G'$). A group isomorphism from G to itself is termed an **automorphism** of G.

Example 6.1

Let $G = G'$ be a group, and $\phi = 1_G$, the identity mapping on G. Then ϕ is an isomorphism. (It is an automorphism, as well.)

Example 6.2

Let $z \in \mathbb{Z}$. Define $\theta(z)$ to be the remainder of z upon division by 5. Then $\theta: \mathbb{Z} \to \mathbb{Z}_5$, the group of integers modulo 5, is a mapping. We leave it to you to verify that θ is a homomorphism. (See Exercise 7.)

Example 6.3

Let G and G' be groups, and let $1'$ be the identity element of G'. Define $\tau(a) = 1'$ for all $a \in G$. For $a, b \in G$, we then have

$$\tau(ab) = 1' = 1'1' = \tau(a)\tau(b)$$

so that τ is a group homomorphism.

Example 6.4

Let $G = \mathbb{Z}$ and $G' = 2\mathbb{Z}$, the group of even integers. Define $\rho: G \to G'$ by $\rho(a) = 2a$ for $a \in G$. Since $\rho(a + b) = 2(a + b) = 2a + 2b = \rho(a) + \rho(b)$, ρ is a group homomorphism. If $\rho(a) = \rho(b)$, then $2a = 2b$ and, by multiplicative cancellation, $a = b$. Thus ρ is injective. If $x \in G'$, then $x = 2a$ for some $a \in G$. Thus $\rho(a) = 2a = x$ and ρ is surjective. Therefore, G is isomorphic to a proper subgroup of itself.

Example 6.5

Let $G = S_n$, $n > 1$, and G' be the group on $\{1, -1\}$ under the operation of multiplication. Our work in the last chapter, in particular Corollary 5.5, shows that the mapping sign: $S_n \to \{1, -1\}$ is a group homomorphism.

Example 6.6

Let G be the group of symmetries of an equilateral triangle and $G' = S_3$. Then the correspondence

$$(1) \leftrightarrow e \qquad (1\ \ 2) \leftrightarrow \tau_A \qquad (1\ \ 2\ \ 3) \leftrightarrow \rho$$

$$(2\ \ 3) \leftrightarrow \tau_C \qquad (1\ \ 3) \leftrightarrow \tau_B \qquad (1\ \ 3\ \ 2) \leftrightarrow \rho^2$$

is actually an isomorphism between these two groups.

Example 6.7

If $X = \{x_1, x_2, \ldots, x_n\}$, then S_X is isomorphic to S_n. We leave it to you to verify that if $\alpha \in S_X$ and $\alpha(x_i) = x_{\alpha(i)}$, then the correspondence

$$\begin{pmatrix} x_1 & x_2 & \cdots & x_n \\ x_{\alpha(1)} & x_{\alpha(2)} & \cdots & x_{\alpha(n)} \end{pmatrix} \rightarrow \begin{pmatrix} 1 & 2 & \cdots & n \\ \alpha(1) & \alpha(2) & \cdots & \alpha(n) \end{pmatrix}$$

is the required isomorphism. (See Exercise 12.)

	1	a
1	1	a
a	a	1

Figure 6.3

Many important questions in group theory involve the idea of group isomorphism. For example, given two groups of the same order, how can you determine whether or not the groups are isomorphic? How many nonisomorphic groups of the same order are there? If two groups have the property that each proper subgroup of one is isomorphic to a proper subgroup of the other, are the two isomorphic?

While we will not attempt to answer these questions now, we will use Cayley tables to determine all nonisomorphic groups G where $|G| < 5$. Clearly if $|G| = 1$, then G consists of just the identity 1. If $|G| = 2$, then there must be an element a in addition to 1. This element a must be its own inverse. We therefore have derived the Cayley table in Figure 6.3.

	1	a	b
1	1	a	b
a	a		
b	b		

(a)

Suppose $|G| = 3$ and $G = \{1, a, b\}$. We can start with the partial Cayley table in Figure 6.4a. Consider the element ab. Since neither a nor b is the identity, $ab = 1$ (Figure 6.4b). Because of the cancellation property, the Cayley table of a group contains each element exactly once in each row and each column. Thus $aa = b$, $bb = a$, and $ba = 1$. We have shown that any group of order 3 has a Cayley table isomorphic to that in Figure 6.4c. This means that we can speak now of "the" group of order 3.

	1	a	b
1	1	a	b
a	a		1
b	b		

(b)

	1	a	b
1	1	a	b
a	a	b	1
b	b	1	a

(c)

Figure 6.4

If $G = \{1, a, b, c\}$, then let us proceed as before. The element ab is neither a nor b. Suppose $ab = 1$. We then obtain the table fragment in Figure 6.5a. Then ac is not a, c, nor 1. Thus $ac = b$ and aa must then equal c (Figure 6.5b). Consider the column containing ba. Now $ba \neq b$, $ba \neq a$, and $ba \neq c$. Thus $ba = 1$ and $ca = b$ (Figure 6.5c). Filling in the table, we see that $bc = a$, $bb = c$, $cb = a$, and $cc = 1$. This gives us Figure 6.5d. You should be able to show that this group is isomorphic to $(\mathbb{Z}_4, \oplus)$. (See Exercise 13.)

There was another possibility in the preceding paragraph. If $ab = c$, then $ac = 1$ or $ac = b$. These cases lead to the tables in Figures 6.6a and b. In the first, $bc \neq b$, $bc \neq c$, and $bc \neq 1$ leads us to conclude that $bc = a$, and we

	1	a	b	c
1	1	a	b	c
a	a		1	
b	b			
c	c			

(a)

	1	a	b	c
1	1	a	b	c
a	a	c	1	b
b	b			
c	c			

(b)

	1	a	b	c
1	1	a	b	c
a	a	c	1	b
b	b	1		
c	c	b		

(c)

	1	a	b	c
1	1	a	b	c
a	a	c	1	b
b	b	1	c	a
c	c	b	a	1

(d)

Figure 6.5

	1	a	b	c
1	1	a	b	c
a	a	b	c	1
b	b	c		
c	c	1		

(a)

	1	a	b	c
1	1	a	b	c
a	a	1	c	b
b	b	c		
c	c	b		

(b)

	1	a	b	c
1	1	a	b	c
a	a	b	c	1
b	b	c	1	a
c	c	1	a	b

(c)

	1	a	b	c
1	1	a	b	c
a	a	1	c	b
b	b	c	1	a
c	c	b	a	1

(d)

	1	a	b	c
1	1	a	b	c
a	a	1	c	b
b	b	c	a	1
c	c	b	1	a

(e)

Figure 6.6

obtain the table in Figure 6.6c. In the second case, two possibilities occur. We leave it to you to show that these two instances are given in Figures 6.6d and e. The group defined by Figure 6.6d was encountered before. It is isomorphic to the group of motions of a rectangle and also to the group K in Exercise 17 of Section 5.2. (Also see Exercise 14 in this section.) This group is known as the **Klein Four-Group**.

Now how many groups of order 4 are there? We claim there are only two, and that the groups in Figures 6.5d, 6.6c, and 6.6e are isomorphic. The verification of this statement and the assertion that none of these is isomorphic to the group of Figure 6.6d is left to the exercises. (See Exercise 13.)

Once we have stated and proved a result in the next chapter that connects the order of an element in a group G to the order of G itself, we will be able to dispose easily of the case $|G| = 5$. However, this table approach to groups of order greater than 5 can get quite difficult, and we stop our analysis of groups of small order at this point. In Section 6.2, we turn instead to the question of what group properties are *preserved* by group homomorphisms.

Exercises 6.1

1. Let G consist of the complex numbers $\{1, i, -1, -i\}$. Show $(G, \cdot) \cong (\mathbb{Z}_4, \oplus)$.

2. Prove that the function $f: \mathbb{R} \to \mathbb{R}^+$ defined by $f(x) = 10^x$ is an isomorphism between the additive group $(\mathbb{R}, +)$ and the multiplicative group $(\mathbb{R}^+, \cdot)$.

3. Let $\phi: \mathbb{Z}_{12} \to \mathbb{Z}_{12}$ be defined by $\phi(a) = 3a$. Verify that ϕ is a homomorphism. Is ϕ a monomorphism or an epimorphism?

4. a. Prove that if $\phi_1: G_1 \to G_2$ and $\phi_2: G_2 \to G_3$ are group homomorphisms, then $\phi_2 \circ \phi_1: G_1 \to G_3$ is a group homomorphism.
   b. Prove that if $\phi_1: G_1 \to G_2$ and $\phi_2: G_2 \to G_3$ are group isomorphisms, then $\phi_2 \circ \phi_1: G_1 \to G_3$ is a group isomorphism.

5. Suppose $\phi: G \to G'$ is an isomorphism. Verify that the mapping $\phi^{-1}: G' \to G$ is also an isomorphism.

6. Let $(\mathbb{Q}^+, \cdot)$ be the group of positive rational numbers under the operation of multiplication. Does $\phi(a/b) = b/a$ define an isomorphism?

7. Verify that the mapping of Example 6.2 is a homomorphism.

8. Show that every proper subgroup of S_3 is isomorphic to a proper subgroup of $(\mathbb{Z}_6, \oplus)$. Is $S_3 \cong \mathbb{Z}_6$?

9. Let $\phi: G \rightarrow G$ be defined by $\phi(g) = g^{-1}$. Prove that ϕ is an isomorphism if and only if G is an abelian group.

10. Let G be an abelian group and $H = \{g^2 | g \in G\}$. Show $H \leq G$. If $\phi: G \rightarrow H$ is defined by $\phi(g) = g^2$, show that ϕ is an epimorphism.

11. a. Prove that S_2 is isomorphic to a subgroup of S_3.
    b. Prove that S_2 is isomorphic to a subgroup of S_4.
    c. Generalize the proofs of parts a and b to prove that if k and n are positive integers, $k < n$, then S_k is isomorphic to a subgroup of S_n.

12. Verify the statement in Example 6.7 that if $|X| = n$, then $S_X \cong S_n$.

13. Complete the verification that there are but two groups of order 4. (The noncyclic group is termed the *Klein Four-Group*.)

14. Suppose $\phi: G \rightarrow G'$ is a group epimorphism. If G is abelian, prove that G' is abelian. Is the converse of this result true? Is the result still true if *abelian* is replaced by *cyclic*?

15. Let $\mathbb{Q}' = \mathbb{Q} - \{1\}$. Define $\#$ on $\mathbb{Q}$ by $a \# b = a + b - ab$. Show that $\#$ restricts to an operation on $\mathbb{Q}'$ and that $(\mathbb{Q}', \#)$ is a group isomorphic to $(\mathbb{Q} - \{0\}, \cdot)$.

16. Prove that the set of all matrices of the form

$$\begin{bmatrix} a & 0 \\ 0 & a \end{bmatrix}$$

where a is a rational number not equal to zero, forms a group under matrix multiplication and that this group is isomorphic to the group $(\mathbb{Q} - \{0\}, \cdot)$.

6.2

Properties

One of the most useful tools in group theory is the group homomorphism. The fact that a mapping $\phi: G \rightarrow G'$ is a group homomorphism allows you to draw conclusions about the structure of G' from the structure of G and vice-versa. There are a number of results that describe the connection between the subgroup structures of G and G'. The first of these shows how a group homomorphism not only preserves products but also identities and inverses.

Theorem 6.1

If $\phi: G \rightarrow G'$ is a group homomorphism and $1, 1'$ are the identities of G, G', respectively, then

   i. $\phi(1) = 1'$ and
   ii. $\phi(a)^{-1} = \phi(a^{-1})$ for all $a \in G$.

Proof

By the definition of group homomorphism, we have

$$\phi(1)1' = \phi(1) = \phi(1 \cdot 1) = \phi(1)\phi(1)$$

By using cancellation in G', we conclude $\phi(1) = 1'$. This proves part i.

Let $a \in G$. Then $1' = \phi(1) = \phi(aa^{-1}) = \phi(a)\phi(a^{-1})$, so that $\phi(a^{-1})$ must be the inverse of $\phi(a)$. Thus, part ii holds. □

Probably the most important property of a group homomorphism between two groups G and G' is the fact that group homomorphisms are structure preserving; that is, images of subgroups of G are subgroups of G' and preimages of subgroups of G' are subgroups of G. Therefore, the existence of a group homomorphism and knowledge of the structure of either the domain or range often allows us to deduce much about the structure of the other.

We will summarize these results in Theorem 6.2, but to do so in a convenient manner, we want to use a notational convention first introduced in Section 2.1 and studied in Exercises 6 and 7 of that section.

Definition 6.2

If $f: X \to Y$, then $f^{\to}: \mathscr{P}(X) \to \mathscr{P}(Y)$ is defined by $f^{\to}(A) = \{f(x) | x \in A\}$, where $A \in \mathscr{P}(X)$. Similarly, $f^{\leftarrow}: \mathscr{P}(Y) \to \mathscr{P}(X)$ is defined by $f^{\leftarrow}(B) = \{x \in X | f(x) \in B\}$, where $B \in \mathscr{P}(Y)$.

We have to define the mappings $f^{\to}$ and $f^{\leftarrow}$, since f acts on elements of X, and we wish to discuss images or preimages of entire subsets (or in our case, subgroups). It might be argued that this is just a case of semantics but, in our opinion, it is a required one. Many authors write $f(A)$ for our $f^{\to}(A)$ and $f^{-1}(B)$ for our $f^{\leftarrow}(B)$. This latter notation is confusing and misleading, since f^{-1} denotes the inverse of the mapping f. Recall that the mapping f^{-1} exists only when f is both injective and surjective. Neither need be true to define $f^{\to}$ or $f^{\leftarrow}$.

Theorem 6.2

Suppose G and G' are groups, and $\phi: G \to G'$ is a group homomorphism. If $H \leqslant G$ and $H' \leqslant G'$, then

    i. $\phi^{\to}(H)$ is a subgroup of G' and
    ii. $\phi^{\leftarrow}(H')$ is a subgroup of G.

Proof

We will leave the proof of part i for Exercise 1. To prove part ii, recall that $\phi^{\leftarrow}(H') = \{g \in G | \phi(g) \in H'\}$. By Theorem 6.1, $\phi(1) = 1' \in H'$ so that $1 \in \phi^{\leftarrow}(H')$

and $\phi^{\leftarrow}(H') \neq \varnothing$. Let $g_1, g_2 \in \phi^{\leftarrow}(H')$. Then $\phi(g_1), \phi(g_2) \in H'$ and, since H' is a subgroup of G', we have $\phi(g_1 g_2) = \phi(g_1)\phi(g_2) \in H'$. Thus $g_1 g_2 \in \phi^{\leftarrow}(H')$. Since H' is a subgroup, $\phi(g_1)^{-1} \in H'$, and so $\phi(g_1^{-1}) = \phi(g_1)^{-1} \in H'$. Therefore $g_1^{-1} \in \phi^{\leftarrow}(H')$. By Theorem 4.4, $\phi^{\leftarrow}(H') \leqslant G'$. $\square$

In the preceding result, $\phi^{\rightarrow}(H)$ is termed the **homomorphic image** of H under ϕ, and $\phi^{\leftarrow}(H')$ is called the **homomorphic preimage** of H' under ϕ.

A natural question that arises in the study of groups and their homomorphic images is, "How many distinct (that is, nonisomorphic) homomorphic images can a given group have?" This question is difficult to answer. Although we must postpone a complete description of all homomorphic images to the next chapter, we can begin to address the question now.

By Theorem 6.2, we know that, since $\{1'\}$ is a subgroup of G', $\phi^{\leftarrow}(\{1'\})$ is a subgroup of G. This subgroup of G is of such importance in our study that we will give it a special name.

Definition 6.3

> If G and G' are groups, with $\phi: G \rightarrow G'$ being a group homomorphism and $1'$ the identity for G', then $\phi^{\leftarrow}(\{1'\})$ is called the **kernel of ϕ** (denoted ker ϕ).[1]

Example 6.8

If $\phi: G \rightarrow G'$ is an isomorphism, then ker $\phi = \{1\}$ since ϕ must be injective and 1 can have but one preimage.

Example 6.9

If $\phi: G \rightarrow G'$ is defined by $\phi(g) = 1'$ for all $g \in G$, then ker $\phi = G$.

Example 6.10

If $\phi: \mathbb{Z} \rightarrow \mathbb{Z}_5$ is the homomorphism of Example 6.2 such that $\phi(z)$ is the remainder upon division of z by 5, then ker $\phi = 5\mathbb{Z}$, the set of integral multiples of 5.

Example 6.11

If sign: $S_n \rightarrow \{1, -1\}$ is the homomorphism discussed in Chapter 5, then ker(sign) $= A_n$, the subgroup of all even permutations in S_n.

The following results show the usefulness of the kernel in actually determining the images of the mapping. They illustrate the reason why the kernel is such an important subgroup.

---

[1] The term "kernel" was coined by P. S. Alexandroff, a noted Soviet topologist, and first appeared in print in 1935 in the algebraic supplement to the text: P. S. Alexandroff and H. Hopf, *Topologie I*, Berlin: Springer-Verlag, 1935, p. 557.

Lemma 6.1

If G and G' are groups, $\phi: G \to G'$ is a group homomorphism, and $K = \ker \phi$, then $xK = Kx$ for all $x \in G$.

Proof

Let $x \in G$, and consider $xK = \{xk \mid k \in K = \ker \phi\}$. If $k \in K$, then

$$\phi(xkx^{-1}) = \phi(x)\phi(k)\phi(x^{-1}) = \phi(x)1'\phi(x^{-1}) = \phi(x)\phi(x^{-1})$$

$$= \phi(xx^{-1}) = \phi(1) = 1'$$

so that $xkx^{-1} \in \ker \phi = K$. Set $xkx^{-1} = k_0 \in K$. Then $xk = k_0x \in Kx$. Therefore $xK \subseteq Kx$.

A similar argument considering the element $x^{-1}kx$ shows that $Kx \subseteq xK$. The result follows. $\qquad\square$

Theorem 6.3

Under the hypothesis of Lemma 6.1, we have $\phi(x) = \phi(y)$ if and only if $xK = yK$.

Proof

Suppose $\phi(x) = \phi(y)$ for some $x, y \in G$. Then consider $x^{-1}y$.

$$\phi(x^{-1}y) = \phi(x^{-1})\phi(y) = \phi(x^{-1})\phi(x) = \phi(x^{-1}x) = \phi(1) = 1'$$

Thus $x^{-1}y = k \in K = \ker \phi$, and $y = xk \in xK$. Then $yK = (xk)K = x(kK) = xK$ since the subgroup K is closed with respect to composition. Conversely, suppose $xk_1 = yk_2$ for $k_1, k_2 \in K$. Then

$$\phi(x) = \phi(x)1' = \phi(x)\phi(k_1) = \phi(xk_1) = \phi(yk_2) = \phi(y)\phi(k_2)$$

$$= \phi(y)1' = \phi(y) \qquad\qquad\square$$

The preceding results actually characterize the subgroups that are potential kernels of homomorphisms. These subgroups, which are called *normal subgroups*, will be studied in more detail in Chapter 7. There we will show how a group has a homomorphic image for each subgroup of this type.

We will give one more result that describes how the type of homomorphism relates to the subgroup structure of the group, and then we will conclude this section with the basic classification of cyclic groups that we can now obtain through the use of group isomorphisms.

Corollary 6.1

A group homomorphism $\phi: G \to G'$ is a monomorphism if and only if $\ker \phi = \{1\}$.

Proof

If ϕ is a monomorphism, then ϕ is injective and $\phi(x) = \phi(y)$ if and only if $x = y$. If $x \in \ker \phi$, then $\phi(x) = 1' = \phi(1)$ so $x = 1$. Thus $\ker \phi = \{1\}$.

Conversely, if $\ker \phi = \{1\}$, then Theorem 6.3 implies that $\phi(x) = \phi(y)$ if and only if $x\{1\} = y\{1\}$ or $x = y$. $\qquad\square$

Theorem 6.4

Let $G = \langle a \rangle$ be a cyclic group. If $|G| = n$, then $G \cong (\mathbb{Z}_n, \oplus)$. If $|G|$ is infinite, then $G \cong (\mathbb{Z}, +)$.

Proof

Suppose G is finite and $G = \{a^0, a, a^2, \ldots, a^{n-1}\}$. Define $\phi: G \to \mathbb{Z}_n$ by $\phi(a^k) = k$. If $a^s, a^t \in G$ and $s + t < n$, then $s + t = s \oplus t$ and

$$\phi(a^s a^t) = \phi(a^{s+t}) = s + t = s \oplus t = \phi(a^s) \oplus \phi(a^t)$$

If $s + t = n + r$, where $0 \leqslant r < n$, then $s \oplus t = r$ and

$$\phi(a^s a^t) = \phi(a^{s+t}) = \phi(a^{n+r}) = \phi(a^n a^r) = \phi(1 a^r) = \phi(a^r) = r = s \oplus t$$

$$= \phi(a^s) \oplus \phi(a^t)$$

Therefore ϕ is a homomorphism. Clearly a^k is the preimage of $k \in \mathbb{Z}_n$ so ϕ is surjective. If $\phi(a^s) = \phi(a^t)$ for $s \leqslant t < n$, then $s = t$ so that $a^s = a^t$. This shows ϕ is injective and, hence, an isomorphism.

We leave it to you to show that if G is an infinite group, the $\phi(a^m) = m$ defines an isomorphism from G to $(\mathbb{Z}, +)$. (See Exercise 16.)   □

Exercises 6.2

1. Prove Theorem 6.2i: If $\phi: G \to G'$ is a group homomorphism and if $H \leqslant G$, then $\phi^{\to}(H) \leqslant G'$.

2. Determine the kernel of the homomorphism $\phi: \mathbb{Z}_{12} \to \mathbb{Z}_{12}$ defined by $\phi(a) = 3a$.

3. Let $\phi: G \to G'$ be a group homomorphism. If $n \in \mathbb{Z}$, prove that $\phi(a^n) = \phi(a)^n$ for all $a \in G$.

4. If $\phi: G \to G'$ is a group homomorphism, $a \in G$, and $|\langle a \rangle| = n$, prove that $|\langle \phi(a) \rangle| \leqslant n$.

5. If $\phi: G \to G'$ is a group homomorphism and $|G|$ is finite, prove that $|\phi^{\to}(G)|$ is finite.

6. If $\phi: G \to G'$ is a group homomorphism and H' is a finite subgroup of G', does it follow that $|\phi^{\leftarrow}(H')|$ is a finite group? Explain.

7. Determine all the subgroups of $\mathbb{Z}_4$. How many of these can be kernels? What is the maximum number of distinct (nonisomorphic) images of $\mathbb{Z}_4$?

8. If $(G, *)$ and $(H, \circ)$ are finite groups, prove that they are isomorphic only if $|G| = |H|$.

9. a. Prove that if $\phi: G \to G'$ is a homomorphism and G is abelian, then $\phi^{\to}(G)$ is abelian. Is G' abelian?
   b. If $\phi: G \to G'$ is a homomorphism and G is cyclic, prove that $\phi^{\to}(G)$ is cyclic. How are the generators of $\phi^{\to}(G)$ related to the generators of G?

10. If $\phi: G \to G'$ is an epimorphism, prove $G \cong G'$ if and only if $\ker \phi = \{1\}$.

11. a. Suppose $\phi: G \to G'$ is a monomorphism and $H \leqslant G$ with $|H| = k$. Prove that $|\phi^{\to}(H)| = k$.
    b. If $\phi: G \to G'$ is an epimorphism and $H' \leqslant G'$ with $|H'| = k$, does it follow that $|\phi^{\leftarrow}(H')| = k$?

12. If G is a group and $a \in G$, define $\phi_a: G \to G$ by $\phi_a(g) = aga^{-1}$ for $g \in G$. Prove that ϕ_a is an isomorphism.

13. Suppose $G \cong G'$ and $H \cong H'$. Prove $G \times H \cong G' \times H'$.

14. If G is a cyclic group of order n and $\phi: G \to G'$ is a homomorphism, prove that $o(b)$ divides n for each $b \in \phi^{\to}(G)$.

15. If $\phi: G \to G'$ is a homomorphism, verify that $\phi^{\to}(Z(G)) \subseteq Z(\phi^{\to}(G))$.

16. Complete the proof of Theorem 6.4 by showing that if $G = \langle a \rangle$ is an infinite cyclic group, then $\phi(a^n) = n$ defines an isomorphism between G and $(\mathbb{Z}, +)$.

6.3

The Theorem of Cayley

We treat the final result of this chapter in a section by itself because of its fundamental importance in the study of groups. In the introduction to Chapter 5, we mentioned that historically permutation groups were among the first to be studied when the field of mathematics turned from almost total immersion in applications to the investigation of mathematics for its own sake. This change occurred in the middle 1800's and brought about much of the mathematics we know today. Prior to that time, mathematics was generally created from a need and for a specific purpose. A good deal of the time, proofs were not of the rigorous nature that we know today.

Therefore, we conclude this chapter with an important result that links the concept of group isomorphism with our discussion of symmetric groups. The Theorem of Cayley shows that a knowledge of all symmetric groups and their subgroups, the permutation groups, would imply a knowledge of all groups. The beauty of this result is that it is true for both finite and infinite groups. While we have not studied infinite symmetric groups at any length, our proof of the main theorem is valid for them as well. You might find the theorem easier to understand, however, if you think of the group under consideration as being finite.

Theorem 6.5

Cayley's Representation Theorem

Every group is isomorphic to a group of permutations.

	e	ρ	α	β
e	e	ρ	α	β
ρ	ρ	e	β	α
α	α	β	e	ρ
β	β	α	ρ	e

Figure 6.7

Before we give a proof of this result, we feel it might be helpful to see the construction of an actual example. Recall the group of symmetries G of a rectangle given in Figure 6.7. Consider the following permutations on the set $\{e, \rho, \alpha, \beta\}$:

$$\phi_e = \begin{pmatrix} e & \rho & \alpha & \beta \\ e & \rho & \alpha & \beta \end{pmatrix} \qquad \phi_\rho = \begin{pmatrix} e & \rho & \alpha & \beta \\ \rho & e & \beta & \alpha \end{pmatrix}$$

$$\phi_\alpha = \begin{pmatrix} e & \rho & \alpha & \beta \\ \alpha & \beta & e & \rho \end{pmatrix} \qquad \phi_\beta = \begin{pmatrix} e & \rho & \alpha & \beta \\ \beta & \alpha & \rho & e \end{pmatrix}$$

These permutations are defined from the table by using the row to the

	ϕ_e	ϕ_ρ	ϕ_α	ϕ_β
ϕ_e	ϕ_e	ϕ_ρ	ϕ_α	ϕ_β
ϕ_ρ	ϕ_ρ	ϕ_e	ϕ_β	ϕ_α
ϕ_α	ϕ_α	ϕ_β	ϕ_e	ϕ_ρ
ϕ_β	ϕ_β	ϕ_α	ϕ_ρ	ϕ_e

Figure 6.8

right of the group element as the image of the permutation. Using permutation composition, we obtain the table of the group G' (Figure 6.8). Obviously, these permutations form a group G.

It should be clear to you that these Cayley tables are isomorphic by the mapping $\phi: G \to G'$ defined by $\phi(g) = \phi_g$. This mapping is the main tool in the proof of Cayley's Theorem.

Proof of Cayley's Representation Theorem

Let G be a group and $g \in G$. Define $\phi_g: G \to G$ by $\phi_g(x) = gx$, where x is an element of G. We wish to show first that ϕ_g is a bijection for each $g \in G$. Suppose $\phi_g(x) = \phi_g(y)$ for $x, y \in G$. Then $gx = gy$. By left cancellation, $x = y$. Therefore ϕ_g is injective. Let $y \in G$. Choose $z = g^{-1}y$ in G. Then $\phi_g(z) = gz = g(g^{-1}y) = (gg^{-1})y = y$. Thus ϕ_g is a surjection, so that ϕ_g is a bijection.

We have now shown that for each $g \in G$, ϕ_g is a permutation of G. (Be careful! We are not saying that ϕ_g is necessarily an isomorphism.) Now consider the mapping $\phi: G \to S_G$ defined by $\phi(g) = \phi_g$. If we can show that ϕ is a monomorphism between G and the set of all permutations of the form ϕ_g, then Theorem 6.2 guarantees that the set of such permutations forms a subgroup of S_G. Therefore, ϕ is in fact an isomorphism between G and that group of permutations.

Let $g, h \in G$. Then $\phi_{gh}(x) = (gh)x = g(hx) = \phi_g(hx) = \phi_g\phi_h(x)$ for any $x \in G$. Thus, $\phi_{gh} = \phi_g\phi_h$. If we now consider $\phi(gh)$, then we see that $\phi(gh) = \phi_{gh} = \phi_g\phi_h = \phi(g)\phi(h)$, so that ϕ is a homomorphism from G to S_G. What about the kernel of ϕ? The identity permutation ϕ_1 is the only permutation of the form ϕ_g such that $\phi_g(x) = x$ for all $x \in G$. Therefore, the kernel of ϕ consists of all $g \in G$ such that $\phi_g = \phi_1$. But $\phi_g = \phi_1$ if and only if $g = 1$. Thus, the kernel of ϕ is $\{1\}$, and ϕ is a monomorphism by Corollary 6.1. The result follows. $\qquad\square$

Another way of stating Cayley's Theorem is to say that any group can be **embedded** in S_G. From the proof, we see that if $|G| = n$, then G is isomorphic to a subgroup of S_n. This says that any group of order 4 is isomorphic to a subgroup of S_4. Thus we could determine all groups of order 4 by studying S_4. Similarly, any group of order 8 is isomorphic to a subgroup of S_8. However, since $|S_8| = 8! = 40,320$, it is probably easier to study the groups of order 8 directly!

Suppose G is a group of order n. How many such groups are there? Could there be an infinite number of nonisomorphic groups G for some value of n? Cayley's Theorem provides an immediate answer to this question. Since G can be viewed as a subgroup of S_G and any finite group can have but a finite number of subgroups, the number of groups of order n is finite.

In the Theorem of Cayley, the map ϕ is termed the **regular left representation** of G. Representation theory, the idea of studying some algebraic object by finding a structure-preserving homomorphism from it to a group of matrices or operators, is one of the more applicable topics in algebra. Its results are utilized in such areas as X-ray spectroscopy and quantum mechanics.

Exercises 6.3

1. Using the proof of the Theorem of Cayley, define the mappings ϕ_g for each $g \in \mathbb{Z}_4$.

2. Find a subgroup of S_4 that is isomorphic to $\mathbb{Z}_4$. Is there more than one?

3. Determine all subgroups of order 4 in S_4. How many different (nonisomorphic) subgroups are there?

4. Let G be a group and $g \in G$. Define $\alpha_g: G \to G$ by $\alpha_g(x) = xg$ for $x \in G$. Show that α_g is a permutation of the elements of G. Is the map $\alpha: G \to \{\alpha_g | g \in G\}$ a homomorphism?

5. We saw in Section 6.1 that there is but one group of order 3. Use Cayley's Theorem and the structure of S_3 to give an alternate proof of this fact.

6. Recall that an isomorphism from a group G to itself is termed an **automorphism**. Let Aut(G) denote the set of all automorphisms of G.
   a. Prove that Aut(G) is a group under the operation of composition of mappings.
   b. Show that $|\text{Aut}(\mathbb{Z}_3)| = 2$.

7. This exercise uses Exercise 6. An automorphism $\phi: G \to G$ is called an **inner automorphism** if there exists $g \in G$ such that $\phi(x) = gxg^{-1}$ for all $x \in G$. Show that Inn(G), the set of all inner automorphisms of G, is a subgroup of Aut(G) and that Inn(G) = $\{1_G\}$ if and only if $Z(G) = G$. (See Exercise 15 of Section 4.3.)

8. By considering the rows of a Cayley table as permutations of the ground set of a group G, give an alternate proof of the fact that there exists but a finite number of nonisomorphic groups of order $|G|$.

9. Let G and G' be permutation groups on the sets X and X', respectively. G and G' are said to be **permutation group isomorphic** if there exist an isomorphism $\phi: G \to G'$ and a bijection $f: X \to X'$ such that $f(gx) = \phi(g)(f(x))$ for all $g \in G$ and $x \in X$. Show that S_4 has two subgroups of order 3 that are group isomorphic but not permutation group isomorphic.

10. Prove that the group of all nonzero complex numbers under multiplication is isomorphic to the group under the operation of matrix multiplication of all 2×2 matrices with real number entries of the form

$$\begin{bmatrix} a & b \\ -b & a \end{bmatrix}$$

where a and b are not both equal to 0.

11. Show that a cyclic group of order p, where p is prime, has exactly $p - 1$ distinct automorphisms.

12. Find Aut($(\mathbb{Z}_4, \oplus)$) and Aut(K), where K is the Klein-Four Group. (See Exercise 6.)

Normal Subgroups

Twice in the preceding chapters, we have made use of a special subset notation in our investigation of subgroup properties. In Chapter 4 we proved that a nonempty subset H of a finite group G is a subgroup if the subset is closed under the group operation. In the proof we introduced the notation $Ha = \{ha \mid h \in H\}$. Then in Chapter 6, we saw that if K is the kernel of a homomorphism on G, then K has the property that $Kx = xK$ for all $x \in G$. These two ideas relate to the study of a particular type of subgroup, the normal subgroup. The study of these subgroups will provide us with important information about the homomorphic images of a group. In order to develop the theory of normal subgroups, we need to deal with a basic idea in algebra, the equivalence relation. We have delayed the introduction of equivalence relations to this point so that we could apply them immediately.

The term *equivalence* means equality, and it is a concept that you have used before. In the study of rational fractions, you conveniently determined when two fractions were equal by the following rule:

$$\frac{a}{b} = \frac{c}{d} \qquad \text{if and only if} \quad ad = bc$$

This idea of equality is really that of an equivalence relation. There is an entire class of numbers that *represents* the same rational number, and two of these numbers are equivalent if they satisfy the preceding condition. The idea of equivalence relation will be very important in our study.

7.1

Equivalence Relations

In order to put our definition and discussion of this concept on as elementary a level as possible, we will begin in the following manner.

Definition 7.1

A **relation** R on a set X is a nonempty subset of $X \times X$. If $(a, b) \in R$, then we will write $a \, R \, b$.

We have already studied many examples of relations, although we did not call them that at the time. For example, "is a subset of" is a relation on the collection of subsets of a set X. "Is equal to" is a relation on the set of all mappings from a set X into a set Y.

We are interested in three special properties of relations:

Definition 7.2

A relation R on a set X is said to be

a. **reflexive** if $a \, R \, a$ for all $a \in X$;
b. **symmetric** if $b \, R \, a$ whenever $a \, R \, b$;
c. **transitive** if $a \, R \, c$ whenever $a \, R \, b$ and $b \, R \, c$.

Example 7.1

"Is a subset of" on a nonempty set is a relation that is reflexive and transitive but not symmetric. It is reflexive since every set is a subset of itself. It is transitive since $A \subseteq B$ and $B \subseteq C$ imply $A \subseteq C$. However, it is not symmetric since the empty set is a subset of every set, but no nonempty set is a subset of the empty set.

Example 7.2

The relation $<$ on the set of real numbers is transitive, but neither reflexive nor symmetric. The last two statements are obvious since no number is less than itself, and a number cannot be both less than and greater than any other number. However, the relation "less than or equal to" is both reflexive and transitive.

The preceding two examples indicate to us that there is a distinction between transitive and the other two properties. We will leave it as an exercise for you to show that these ideas are completely different. See Exercise 3.

There are relations that satisfy all three of these conditions. As an example, consider the relation we defined on the rational numbers: $a/b = c/d$ if and only if $ad = bc$.

1. First, the relation is reflexive; that is, $a/b = a/b$ since $ab = ba$ for all real numbers.
2. Secondly, the relation is symmetric. If $a/b = c/d$, then $ad = bc$. This means that $da = cb$ and $cb = da$, which, in turn, is equivalent to the statement $c/d = a/b$.
3. Finally, the relation is transitive. If $a/b = c/d$ and $c/d = e/f$, then $ad = bc$ and $cf = de$, so

$$af = \frac{adf}{d} = \frac{bcf}{d} = \frac{bde}{d} = be$$

It therefore follows that $a/b = e/f$, which is what we intended to prove.

With a bit of thought (or maybe it seems quite obvious), we see that equality of rational numbers is a relation that satisfies these three properties. From our discussion, we can see that by saying two rational numbers are equal, we really mean they are related by a special type of relation. Thus we are led to the following definition.

Definition 7.3

A relation R on X is an **equivalence relation** if it is reflexive, symmetric, and transitive.

Many equivalence relations are found throughout mathematics. In geometry, both congruence and similarity are equivalence relations on the set of all triangles. In calculus, if we define two functions to be related when they have the same derivative, then we obtain an equivalence relation. One of the most important examples of an equivalence relation is connected to $\mathbb{Z}_n$, the set of integers modulo n, which we first discussed in Chapter 2.

Example 7.3

Let n be a positive integer and $a, b \in \mathbb{Z}$. Define

$$a \equiv b \ (\text{modulo } n) \qquad \text{or} \qquad a \equiv b \ (\text{mod } n)$$

for short, if $a - b$ is a multiple of n. This relation is termed **congruence modulo** n. For example, $6 \equiv 1 \ (\text{mod } 5)$, $147 \equiv 15 \ (\text{mod } 3)$, and $15 \equiv -7 \ (\text{mod } 11)$. For each positive integer n, congruence modulo n is an equivalence relation. To verify this, note that 0 is a multiple of n, so that $a - a = 0 = 0 \cdot n$ and $a \equiv a \ (\text{mod } n)$ for any $a \in \mathbb{Z}$. If $a \equiv b \ (\text{mod } n)$, then $a - b = nq$ for some $q \in \mathbb{Z}$. Then $b - a = n(-q)$ and $b \equiv a \ (\text{mod } n)$. Thus, congruence modulo n is a symmetric relation. Finally, if $a \equiv b \ (\text{mod } n)$ and $b \equiv c \ (\text{mod } n)$, then $a - b = n(q_1)$ and $b - c = n(q_2)$. Therefore,

$$a - c = (a - b) + (b - c) = n(q_1 + q_2)$$

so that $a \equiv c \pmod{n}$ and the relation is transitive. This shows that congruence modulo n is an equivalence relation.

The establishment of an equivalence relation on a set X (by whatever means) immediately induces a partition of the set X. This is a consequence of the next result.

Theorem 7.1

Let R be an equivalence relation on a set X. For $x \in X$, set $A_x = \{y \in X \mid y \, R \, x\}$. Then the collection of subsets $\{A_x \mid x \in X\}$ forms a partition of X.

Proof

We must show that we have a collection of nonempty subsets that are mutually disjoint and whose union is X. Since R is reflexive, $x \in A_x$ and, hence, $A_x \neq \emptyset$. Moreover

$$X = \bigcup_{x \in X} \{x\} \subseteq \bigcup_{x \in X} A_x \subseteq X \qquad \text{implies} \qquad X = \bigcup_{x \in X} A_x$$

It remains to be shown that if $A_x \neq A_y$, then $A_x \cap A_y = \emptyset$. So suppose $z \in A_x \cap A_y$. Then $z \, R \, x$ and $z \, R \, y$. By symmetry we have $x \, R \, z$. Since $z \, R \, y$, the transitive property implies $x \, R \, y$.

Now let $w \in A_x$. Then $w \, R \, x$ and, since $x \, R \, y$, we have $w \, R \, y$ and $w \in A_y$. This says that $A_x \subseteq A_y$. Now $y \, R \, x$, by symmetry and a similar argument, shows that $A_y \subseteq A_x$. But then $A_x = A_y$, contradicting our assumption that $A_x \neq A_y$. We must conclude that $A_x \cap A_y = \emptyset$. The result follows. $\square$

These sets A_x are called the **equivalence classes** of R. An equivalence class for the relation "is similar to" on the set of triangles consists of all triangles similar to a given triangle. For equivalence of rational numbers, the class of the number $1/2$ consists of all rational numbers of the form $n/2n$ for $n \in \mathbb{Z}$, $n \neq 0$.

Example 7.4

To find the partition of the integers determined by the relation congruence modulo 5, let us calculate the equivalence classes. For notational sake, let us use $\bar{x}$ to indicate A_x. For $x = 0$, we have

$$\bar{0} = \{y \in \mathbb{Z} \mid y \equiv 0 \pmod{5}\} = \{\ldots, -5, 0, 5, 10, \ldots\}$$

Similarly,

$$\bar{1} = \{\ldots, -4, 1, 6, 11, \ldots\}$$
$$\bar{2} = \{\ldots, -3, 2, 7, 12, \ldots\}$$
$$\bar{3} = \{\ldots, -2, 3, 8, 13, \ldots\}$$
$$\bar{4} = \{\ldots, -1, 4, 9, 14, \ldots\}$$

Note that if $x = 5q + r$ for $0 \leqslant r < 5$, we have that $\bar{x} = \bar{r}$. See Exercise 5. The partition of $\mathbb{Z}$, therefore, consists of the classes $\{\bar{0}, \bar{1}, \bar{2}, \bar{3}, \bar{4}\}$. But these correspond to the elements of $\mathbb{Z}_5$, the set of integers modulo 5.

Analogous statements can be made for congruence modulo n and the set $\mathbb{Z}_n$. In the future, we will use $\mathbb{Z}_n$ to denote the set of equivalence classes represented by the elements $0, 1, 2, \ldots, n-1$.

An important result concerning equivalence relations and partitions is that not only does an equivalence relation give rise to a partition but also that any partition generates an equivalence relation. This is the content of the following theorem.

Theorem 7.2

If $\{A_\alpha | \alpha \in \Gamma\}$ is a partition of X, then there is an equivalence relation R on X for which the resulting equivalence classes are the subsets of the partition.

Proof

Let us define R on X by $a \, R \, b$ if and only if there is an A_α such that both $a \in A_\alpha$ and $b \in A_\alpha$. We first need to show that R is an equivalence relation.

Since, for any a, there exists an A_α such that $a \in A_\alpha$, we have $a \, R \, a$. Therefore R is reflexive.

If $a \, R \, b$ for two elements a and b, then there is a subset A_α such that $a \in A_\alpha$ and $b \in A_\alpha$. Consequently, it follows that $b \, R \, a$ and R is symmetric.

If $a \, R \, b$ and $b \, R \, c$, then there are A_α, A_β such that $a, b \in A_\alpha$, and $b, c \in A_\beta$. Since $b \in A_\alpha$ and $b \in A_\beta$, $A_\alpha \cap A_\beta \neq \varnothing$. Therefore, since the sets of the form A_α make up a partition, it follows that $A_\alpha = A_\beta$. Thus, by the definition of R, $a \, R \, c$ and R is transitive.

Therefore R is an equivalence relation. Because of the definition of R, it follows immediately that the members of the partition are the equivalence classes for the equivalence relation R. $\square$

The duality between equivalence relations and partitions expressed in Theorems 7.1 and 7.2 is sometimes called the **Fundamental Theorem of Equivalence Relations**. It is a very useful mathematical tool, one that we will apply later in this chapter.

Exercises 7.1

1. a. Show that "has the same parents as" is an equivalence relation on the set of people in the world.
   b. Do the same as in part a for "has the same mother as."
   c. Do the equivalence relations of parts a and b determine the same equivalence classes?

2. Determine whether "is the brother of" on the set of people of the world is reflexive or symmetric.

3. Describe a relation on a set that is:
   a. Reflexive, but neither symmetric nor transitive

    b. Symmetric, but neither reflexive nor transitive

    c. Transitive, but neither reflexive nor symmetric

4. Construct a relation on a set that is:

    a. Reflexive and symmetric, but not transitive

    b. Transitive and reflexive, but not symmetric

5. Suppose $x \in \mathbb{Z}$ and $x = 5q + r$ for q, $r \in \mathbb{Z}$ and $0 \leqslant r < 5$. Verify that the equivalence class of x mod 5 is the same as the equivalence class of r mod 5.

6. Let $X = \{1, 2, 3, 4, 5, 6\}$ and

$$R = \{(1, 3), (1, 1), (5, 4), (4, 4), (4, 5), (3, 1), (5, 5), (3, 3), (2, 2), (6, 6)\}$$

    Show that R is an equivalence relation on X, and list the corresponding partition.

7. Let $X = \{1, 2, 3, 4, 5, 6\}$. Then $\{\{1, 3, 5\}, \{2, 6\}, \{4\}\}$ is a partition of X. List the elements of the corresponding equivalence relation.

8. For each of the following relations on $\mathbb{Z}$, determine whether they are reflexive, symmetric, or transitive.

    a. $a R_1 b$ if $a + b > 0$ b. $a R_2 b$ if a divides b

    c. $a R_3 b$ if $|a| = |b|$ d. $a R_4 b$ if $|a| \leqslant |b|$

9. For each of the following relations on $\mathscr{P}(X)$, where X is a finite set, determine whether they are reflexive, symmetric, or transitive.

    a. $A \cong B$ if $|A| = |B|$ b. $A @ B$ if $|A| \neq |B|$

    c. $A < \sim B$ if $|A| < |B|$ d. $A \leqslant B$ if $|A| \leqslant |B|$

    e. $A \sim B$ if $A \subseteq B$ f. $A \# B$ if $A - B = \varnothing$

    g. $A \& B$ if $A \cup B = X$ h. $A * B$ if $A - B = A$

10. A relation R on a set X is termed **circular** if $a R b$ and $b R c$ imply $c R a$. Prove that R is an equivalence relation on X if and only if R is reflexive and circular.

11. Let R be defined on $\mathbb{N} \times \mathbb{N}$ by $(a, b) R (c, d)$ if $a + d = b + c$. Show that R is an equivalence relation.

12. Let G be a group, and define a relation R on G by $a R b$ if there exists $g \in G$ such that $b = gag^{-1}$. Prove that R is an equivalence relation on G.

13. Let $f: X \to Y$ be a mapping, and define R on X by $x_1 R x_2$ if $f(x_1) = f(x_2)$.

    a. Prove that R is an equivalence relation on X.

    b. Does this mean that $\{f^{\leftarrow}(y) | y \in Y\}$ is a partition of X?

14. If R is a symmetric and transitive relation on a set X and $a R b$, then $b R a$ by symmetry. Then transitivity forces $a R a$. Does this argument show that every symmetric and transitive relation is automatically reflexive?

Cosets and the Theorem of Lagrange

A question that arises naturally when considering groups and their homomorphic images is just how many distinct (that is, nonisomorphic) homomorphic images can a given group have? The remainder of this chapter and the next is devoted to answering this question. If $\phi: G \to G'$ is a group homomorphism, recall that we showed in Theorem 6.2 that $\phi^{\leftarrow}\{1'\}$ is a subgroup of G called the kernel of ϕ. Moreover if $K = \ker \phi$, we also saw a number of results concerning the subsets Kx for $x \in G$. Let us investigate these subsets in more detail.

Definition 7.4

> If G is a group and $H \leqslant G$, then for $a \in G$ the set $Ha = \{ha \mid h \in H\}$ is termed the **right coset of H by a**. Similarly $aH = \{ah \mid h \in H\}$ is called the **left coset of H by a**.

Example 7.5

If $G = \mathbb{Z}$ and $H = 2\mathbb{Z}$, the even integers, then the coset $1 + H = \{1 + 2m \mid m \in \mathbb{Z}\}$ is the set of all odd integers. Note that this set is **not** a subgroup; moreover, the coset $3 + H = \{3 + 2m \mid m \in \mathbb{Z}\}$ also consists of all odd integers.

In the preceding example, $1 + H = n + H$ for any odd integer n. At times we will select a particular member of a coset to represent the coset. This element will be termed the **coset leader** or **coset representative**.

Example 7.6

Let $G = S_3$ and $H = \langle (1 \quad 2) \rangle$. Then

$$H(1 \quad 3) = \{(1)(1 \quad 3), (1 \quad 2)(1 \quad 3)\} = \{(1 \quad 3), (1 \quad 3 \quad 2)\} \qquad \text{while}$$

$$(1 \quad 3)H = \{(1 \quad 3)(1), (1 \quad 3)(1 \quad 2)\} = \{(1 \quad 3), (1 \quad 2 \quad 3)\}$$

This last example is interesting in that it shows that not every right coset is necessarily equal to the corresponding left coset. This concept is connected to the question of how many distinct homomorphic images a group can possess.

Recall Lemma 6.1, which said that if $\phi: G \rightarrow G'$ is a group homomorphism and $K = \ker \phi$, then $xK = Kx$ for all $x \in G$. In the last example, we see that H could never be the kernel of a group homomorphism. Moreover, by Theorem 6.3, $\phi(x) = \phi(y)$ if and only if $xK = yK$. This says that the number of distinct elements in the homomorphic image of a group depends upon the number of distinct left cosets. Also the number of distinct homomorphic images of a group can never be more than the number of those subgroups N of G that possess the additional property that $xN = Nx$ for all $x \in G$.

We now have the two main problems before us. The first is to investigate the type of subgroup that could be the kernel of a homomorphism. The second is to determine when such a subgroup is indeed a kernel of a homomorphism. We will deal with the former in the next section and the latter in the next chapter. In order to do so, however, we need to study the relationships that exist between a group and the set of all possible cosets of one of its subgroups.

Suppose $H \leqslant G$ and $a, b \in G$. Let us define $a \, R \, b$ if $a \in bH$. Since $a = a1 \in aH$, we have $a \, R \, a$, and R is a reflexive relation on G. Suppose $a \, R \, b$, so that $a \in bH$. Then $a = bh$ for some $h \in H$. Since H is a subgroup, $h^{-1} \in H$ and $b = ah^{-1} \in aH$; so $b \in aH$ and $b \, R \, a$. This says that R is symmetric. Finally suppose $a \, R \, b$ and $b \, R \, c$. Then $a = bh_1$ and $b = ch_2$ for $h_1, h_2 \in H$, so that $a = (ch_2)h_1 = c(h_2 h_1) \in cH$ and $a \, R \, c$. This says that R is transitive. Summing up the discussion, we obtain Theorem 7.3.

Theorem 7.3

Let $H \leqslant G$, and define relation R on G by $a \, R \, b$ if $a \in bH$. Then R is an equivalence relation on G whose equivalence classes are the left cosets of H in G.

The last statement follows from the fact that a is contained in the equivalence class of b if and only if $a \in bH$. While the theorem deals with left cosets, there is an analogous development using right cosets. We refer you to the exercises for these details.

Using the basic duality between equivalence relations and partitions, we obtain an important consequence.

Corollary 7.1

If $H \leqslant G$, the left coset of H in G form a partition of G.

Returning to Example 7.6, we list all the left cosets of $H = \langle (1 \quad 2) \rangle$ in S_3.

$$(1)H = \{(1)(1), (1)(1 \quad 2)\} = \{(1), (1 \quad 2)\}$$

$$(1 \quad 2)H = \{(1 \quad 2)(1), (1 \quad 2)(1 \quad 2)\} = \{(1 \quad 2), (1)\}$$

$$(1 \quad 3)H = \{(1 \quad 3)(1), (1 \quad 3)(1 \quad 2)\} = \{(1 \quad 3), (1 \quad 2 \quad 3)\}$$

$$(2 \quad 3)H = \{(2 \quad 3)(1), (2 \quad 3)(1 \quad 2)\} = \{(2 \quad 3), (1 \quad 3 \quad 2)\}$$

$$(1 \quad 2 \quad 3)H = \{(1 \quad 2 \quad 3)(1), (1 \quad 2 \quad 3)(1 \quad 2)\} = \{(1 \quad 2 \quad 3), (1 \quad 3)\}$$

$$(1 \quad 3 \quad 2)H = \{(1 \quad 3 \quad 2)(1), (1 \quad 3 \quad 2)(1 \quad 2)\} = \{(1 \quad 3 \quad 2), (2 \quad 3)\}$$

Note that there are but three **distinct** left cosets of H in G.

The number of distinct left cosets of a subgroup H in a group G is termed the **index of H in G** and is denoted $[G:H]$. In the example above $[G:H] = 3$. There are only two cosets of the even integers $2\mathbb{Z}$ in $\mathbb{Z}$; namely, $2\mathbb{Z}$ itself and the coset $1 + 2\mathbb{Z}$ consisting of all odd integers. Thus $[\mathbb{Z}:2\mathbb{Z}] = 2$.

If G is a finite group, then there is an important result that relates the index of a subgroup and the group order. To present it, we require a lemma.

Lemma 7.1

If H is a finite subgroup of a group G, then $|H| = |aH|$ for all $a \in \mathbf{G}$.

Proof

Suppose $|H| = m$ and $H = \{h_1, h_2, \ldots, h_m\}$. Then $aH = \{ah_1, ah_2, \ldots, ah_m\}$. The cancellation law implies that these elements are distinct, so that $|aH| = |H|$. $\qquad \square$

Now suppose G is finite and $H \leqslant G$. If $[G:H] = m$, then the m left cosets form a partition of G. Therefore

$$|G| = |a_1 H| + |a_2 H| + \cdots + |a_m H|$$

By the preceding lemma, $|a_i H| = |H|$, so that

$$|G| = |H| + |H| + \cdots + |H| = m \cdot |H|$$

This proves the following theorem.

| Theorem 7.4 |

Lagrange's Theorem[1]
Let H be a subgroup of the finite group G. Then $|G| = [G:H] \cdot |H|$.

As an example recall that

$$[S_3 : \langle (1 \quad 2) \rangle] = 3 = \frac{|S_3|}{|\langle (1 \quad 2) \rangle|}$$

While Lagrange's Theorem appears quite simple, it has a host of applications. For one thing, it can be used to determine the possible orders of subgroups of a finite group G. A group of order 36 could never contain a subgroup of order 8. If $a \in G$ has order k, then $|\langle a \rangle| = k$. Since $\langle a \rangle$ is a subgroup of G, we obtain the following corollary.

| Corollary 7.2 |

If G is a finite group, then the order of each element of G divides $|G|$.

If G is a group of prime order, then the structure of G is determined:

| Corollary 7.3 |

If G is a group of prime order, then G is cyclic.

Proof

Let $a \in G$, $a \neq 1$, and $H = \langle a \rangle$. Then $|H|$ divides $|G|$. Since $|G|$ is a prime and $|H| \neq 1$, we must have $|H| = |G|$. But then $G = H = \langle a \rangle$. □

Lagrange's Theorem can also be used to obtain a classical result from number theory:

---

[1] In "Reflexions sur la resolution algebrique des equations," published in 1771, Lagrange established that certain groups of permutations of n objects divide $n!$. Note that Galois first defined *group* over sixty years later.

| Corollary 7.4 |

Fermat's "Little" Theorem[2]

If $a \in \mathbb{Z}$ and p is a prime integer, then $a^p \equiv a$ (modulo p).

Proof

If $a = 0$, the result is clear. If $a \neq 0$, then $a \in \mathbb{Z}_p^*$, which is a group of order $p - 1$. If $o(a) = k$, then k divides $p - 1$ by Corollary 7.2 and $p - 1 = kt$. But then

$$a^{p-1} = a^{kt}$$
$$= (a^k)^t$$
$$\equiv 1^t \text{ (modulo } p)$$
$$\equiv 1 \text{ (modulo } p)$$

and $a^p \equiv a$ (modulo p). $\square$

Therefore, in $\mathbb{Z}_{17}$, the element $2^{17} = 2$. Another way to state this fact is to note that 2^{17} has remainder 2 upon division by 17.

We will give a few more applications of Lagrange's Theorem in the exercises. In the chapters ahead, we will make use of this result again and again. It is amazing that such an important theorem can be put in such a "small package."

We have now developed some information about orders of possible subgroups. For further information about subgroups that are kernels of group homomorphism, we must look at a special type of subgroup. We do this in the next section.

| Exercises 7.2 |

1. If $H \leqslant G$ and $a \in G$, prove that $aH = H$ and $Ha = H$ if and only if $a \in H$.

2. Determine all left cosets of H in G if
   a. $H = \langle 4 \rangle$, $G = \mathbb{Z}_{24}$,
   b. $H = A_3$, $G = S_3$
   c. $H = \langle (2 \quad 3) \rangle$, $G = S_3$
   d. $H = S_3$, $G = S_4$
   e. $H = \langle (1 \quad 2 \quad 3) \rangle$, $G = A_4$
   f. $H = 5\mathbb{Z}$, $G = \mathbb{Z}$
   g. $H = \{(1), (1 \quad 2)(3 \quad 4), (1 \quad 3)(2 \quad 4), (1 \quad 4)(2 \quad 3)\}$, $G = A_4$

3. Determine all right cosets of H in G for the groups in Exercise 2.

4. Suppose $H \leqslant G$. Define R_1 on G by $a R_1 b$ if $a \in Hb$. Prove that R_1 is an equivalence relation on G. What are the equivalence classes?

5. Use Exercise 4 to prove that the right cosets of a subgroup form a partition of the group.

---

[2]Pierre-Simon de Fermat (1601–1665) was the French mathematician who was the founder of analytic geometry and modern number theory. Together with Blaise Pascal (1623–1662), he developed the theory of probability.

6. Suppose $H \leqslant G$. Show that $\theta(aH) = Ha^{-1}$ defines a bijection from the set of left cosets of H in G to the set of right cosets of H in G. (This shows that the number of left cosets equals the number of right cosets.)

7. Find all possible orders of subgroups of a group G if
    a. $|G| = 17$ b. $|G| = 12$ c. $|G| = 24$

8. Suppose G is a finite group and $K \leqslant H$, $H \leqslant G$. Prove that

    $$[G:K] = [G:H] \cdot [H:K]$$

9. Verify directly that the orders of the elements of $\mathbb{Z}_{24}$ divide 24.

10. Let G be a group, $H \leqslant G$, and $K \leqslant G$.
    a. Verify that if $x \in G$, then $x(H \cap K) = xH \cap xK$.
    b. Prove that if $[G:H]$ and $[G:K]$ are finite, then $[G:H \cap K]$ is finite.
    c. Prove Poincare's[3] Theorem: The intersection of a finite number of subgroups of finite index is a subgroup of finite index.

11. Prove that a group of order 4 must either be cyclic or have the property that every element is its own inverse. Conclude that any group of order 4 is abelian.

12. Suppose a group G has an element of order p, where p is a prime. Prove that G has at least $p - 1$ distinct elements of order p.

13. If H and K are subgroups of G of orders m and n, respectively, where $(m, n) = 1$, show that $H \cap K = \{1\}$.

14. Suppose H and K are distinct subgroups of G of order p, where p is a prime. Verify that $H \cap K = \{1\}$. Show that if p is not prime, then the result is no longer true.

15. Suppose $\phi: G \to G'$ is a group isomorphism, and H is a subgroup of finite index in G. Prove that $[G:H] = [G':\phi^{\to}(H)]$.

16. If G is a noncyclic group of order 27, why must $x^9 = 1$ for all $x \in G$.

17. Let G be a group of order p^2, where p is a prime. Prove that every proper subgroup of G must be cyclic. Is the result still true if p^2 is replaced by p^3?

18. If G is a group of order pq, where p and q are distinct primes, verify that every proper subgroup of G is cyclic.

19. Determine all abelian groups of order 8.

20. Let G be the set of all 2×2 matrices of the form

    $$\begin{bmatrix} a & b \\ c & d \end{bmatrix}$$

    with entries in $(\mathbb{Z}_2, \oplus)$ and having the additional property that $ad - bc = 1$. Prove that G is a group if the operation is matrix multiplication using addition and multiplication in $\mathbb{Z}_2$. What is the order of G?

21. Let H and K be subgroups of G. Define a relation $\approx$ on G by writing $a \approx b$ if there exist $h \in H$, $k \in K$ such that $a = hbk$. Prove that $\approx$ is an equivalence relation

---

[3] Henri Poincare (1854–1912) was a contemporary and chief rival of David Hilbert. Poincare published extensively on numerous topics that included differential equations, non-euclidean geometry, topology, and probability.

on G. (The equivalence classes are known as **double cosets** and have the form HbK.)

7.3

Normal Subgroups

We saw in Lemma 6.1 that if K is a kernel of a homomorphism on a group G, then $xK = Kx$ for all $x \in G$. If we hope to determine which subgroups of G can be kernels of homomorphisms, we should investigate subgroups with exactly this property.

Definition 7.5

A subgroup N of a group G is **normal** in G if $xN = Nx$ for all $x \in G$, that is, every left coset is a right coset. We will denote the fact that N is a normal subgroup of G by writing $N \trianglelefteq G$.

Example 7.7

Let G be the set of motions of an equilateral triangle and $H = \{e, \rho, \rho^2\}$, the subgroup of G consisting of the rotations. (See Example 4.7). H has two left cosets in G: one is

$$H = eH = \rho H = \rho^2 H = \{e, \rho, \rho^2\}$$

and the other left coset is

$$\tau_A H = \tau_B H = \tau_C H = \{\tau_A, \tau_B, \tau_C\}$$

The right cosets of H in G are

$$H = He = H\rho = H\rho^2 = \{e, \rho, p^2\}$$

and

$$H\tau_A = H\tau_B = H\tau_C = \{\tau_A, \tau_B, \tau_C\}$$

Therefore, each left coset equals the corresponding right coset, and H is a normal subgroup of G.

Example 7.8

Let $(G, +)$ be an abelian group, and let H be any subgroup of G. Then H is normal in G since for any $x \in H$,

$$x + H = \{x + h \mid h \in H\}$$
$$= \{h + x \mid h \in H\} \qquad \text{since } G \text{ is abelian}$$
$$= H + x$$

Example 7.9

For any group G, G, and $\{1\}$ are normal. This follows since $xG = G = Gx$, for all $x \in G$, and $x1 = x = 1x$, for all $x \in G$.

Example 7.10

Consider S_3. An investigation of the Cayley table for S_3 shows that its subgroups are

$$E = \langle (1) \rangle, \qquad A = \langle (1 \quad 2) \rangle, \qquad B = \langle (1 \quad 3) \rangle,$$
$$C = \langle (2 \quad 3) \rangle, \qquad D = \langle (1 \quad 2 \quad 3) \rangle, \qquad \text{and} \qquad S_3$$

By the preceding example, $E \trianglelefteq S_3$ and $S_3 \trianglelefteq S_3$. The left coset of $A = \{(1), (1 \quad 2)\}$ in S_3 are

$$(1)A = (1 \quad 2)A = \{(1), (1 \quad 2)\}$$
$$(1 \quad 3)A = (1 \quad 2 \quad 3)A = \{(1 \quad 3), (1 \quad 2 \quad 3)\}$$
$$(2 \quad 3)A = (1 \quad 3 \quad 2)A = \{(2 \quad 3), (1 \quad 3 \quad 2)\}$$

The right cosets of A in S_3 are

$$A(1) = A(1 \quad 2) = \{(1), (1 \quad 2)\}$$
$$A(1 \quad 3) = A(1 \quad 3 \quad 2) = \{(1 \quad 3), (1 \quad 3 \quad 2)\}$$
$$A(2 \quad 3) = A(1 \quad 2 \quad 3) = \{(2 \quad 3), (1 \quad 2 \quad 3)\}$$

Since not every left cosets equals the corresponding right coset, for example, $(1 \quad 3)A \neq A(1 \quad 3)$, A is not normal in S_3. Similarly B and C are not normal.

By Lagrange's Theorem, D has two left cosets and two right cosets. One is

$$(1)D = (1 \quad 2 \quad 3)D = (1 \quad 3 \quad 2)D = D = D(1 \quad 3 \quad 2) = D(1 \quad 2 \quad 3)$$
$$= D(1)$$

The other is

$$(1 \quad 2)D = (1 \quad 3)D = (2 \quad 3)D = \{(1 \quad 2), (1 \quad 3), (2 \quad 3)\}$$
$$= D(2 \quad 3) = D(1 \quad 3) = D(1 \quad 2)$$

Therefore $D \trianglelefteq S_3$.

Recall that we showed in Section 6.3 that S_3 and the group of motions of an equilateral triangle are isomorphic. Under this isomorphism, the subgroup H is isomorphic to the subgroup D. Note that both are normal. This is not a coincidence; in fact, homomorphic images of normal subgroups are normal in the homomorphic image. We leave the proof of this fact to you. (See Exercise 4.)

If $H \trianglelefteq G$ and $h \in H$, $x \in G$, then it is not necessarily true that $xh = hx$, but rather that $xh = h'x$, where $h' \in H$. The element h' may or may not equal h. In Example 7.7, $\tau_A H = H\tau_A$, although $\tau_A \rho = \tau_C$ and $\rho\tau_A = \tau_B$.

In Example 7.10, we had $D \trianglelefteq S_3$ and $[S_3:D] = 2$. The following shows that such a result is to be expected.

Theorem 7.5

If $H \leqslant G$ and $[G:H] = 2$, then $H \trianglelefteq G$.

Proof

If $x \in H$, then $xH = H = Hx$ by Exercise 1 of Section 7.2. Since there are but two left cosets and one is H, the other is $G - H$; that is, $xH = G - H$ for $x \notin H$. The same is true for the right cosets, so $Hx = G - H$ for all $x \notin H$. Therefore $xH = Hx$ for all $x \in G$ and $H \trianglelefteq G$. $\quad\square$

Since the alternating group A_n is a subgroup of index 2 in S_n, we have the following corollary.

Corollary 7.5

For $n > 1$, A_n is a normal subgroup of S_n.

Example 7.11

Let us work through the details in showing that the subgroup

$$K = \{(1), (1\ \ 2)(3\ \ 4), (1\ \ 3)(2\ \ 4), (1\ \ 4)(2\ \ 3)\}$$

is normal in S_4. We know that $(1)K = K(1) = K$, since $(1) \in K$. If we hope to find other left or right cosets of K in S_4, we must take as coset leaders elements not in K. Try the coset $(1\ \ 2)K$. We obtain the left coset

$$
\begin{aligned}
(1\ \ 2)K = \{&(1\ \ 2)(1), (1\ \ 2)(1\ \ 2)(3\ \ 4), \\
&(1\ \ 2)(1\ \ 3)(2\ \ 4), (1\ \ 2)(1\ \ 4)(2\ \ 3)\} \\
= \{&(1\ \ 2), (3\ \ 4), (1\ \ 3\ \ 2\ \ 4), (1\ \ 4\ \ 2\ \ 3)\}
\end{aligned}
$$

This left coset is certainly different from the first one we found. Since cosets as equivalence classes are disjoint and any group element lies in the equivalence class for which it is a coset leader, we must continue our search for additional cosets by looking for coset leaders among those group elements not already listed. Consider $(1\ \ 3)K$. We have

$$
\begin{aligned}
(1\ \ 3)K = \{&(1\ \ 3)(1), (1\ \ 3)(1\ \ 2)(3\ \ 4), \\
&(1\ \ 3)(1\ \ 3)(2\ \ 4), (1\ \ 3)(1\ \ 4)(2\ \ 3)\} \\
= \{&(1\ \ 3), (1\ \ 2\ \ 3\ \ 4), (2\ \ 4), (1\ \ 4\ \ 3\ \ 2)\}
\end{aligned}
$$

Since $(2\ \ 3)$ is not one of the elements in the cosets already calculated, we can take $(2\ \ 3)K$ as our next coset:

$$
\begin{aligned}
(2\ \ 3)K = \{&(2\ \ 3)(1), (2\ \ 3)(1\ \ 2)(3\ \ 4), \\
&(2\ \ 3)(1\ \ 3)(2\ \ 4), (2\ \ 3)(1\ \ 4)(2\ \ 3)\} \\
= \{&(2\ \ 3), (1\ \ 3\ \ 4\ \ 2), (1\ \ 2\ \ 4\ \ 3), (1\ \ 4)\}
\end{aligned}
$$

The element $(1 \quad 2 \quad 3)$ is one possibility for our next coset leader.

$$(1 \quad 2 \quad 3)K = \{(1 \quad 2 \quad 3)(1), (1 \quad 2 \quad 3)(1 \quad 2)(3 \quad 4),$$
$$(1 \quad 2 \quad 3)(1 \quad 3)(2 \quad 4), \ (1 \quad 2 \quad 3)(1 \quad 4)(2 \quad 3)\}$$
$$= \{(1 \quad 2 \quad 3), (1 \quad 3 \quad 4), (2 \quad 4 \quad 3), (1 \quad 4 \quad 2)\}$$

By Lagrange's Theorem, the four elements that remain must constitute the last coset. Since $(1 \quad 3 \quad 2)$ is one of these elements, we can take $(1 \quad 3 \quad 2)K$ as our last coset:

$$(1 \quad 3 \quad 2)K = \{(1 \quad 3 \quad 2)(1), (1 \quad 3 \quad 2)(1 \quad 2)(3 \quad 4),$$
$$(1 \quad 3 \quad 2)(1 \quad 3)(2 \quad 4), \ (1 \quad 3 \quad 2)(1 \quad 4)(2 \quad 3)\}$$
$$= \{(1 \quad 3 \quad 2), (2 \quad 3 \quad 4), (1 \quad 2 \quad 4), (1 \quad 4 \quad 3)\}$$

To verify that K is normal in S_4, we need only calculate the corresponding right cosets:

$$K(1 \quad 2) = \{(1)(1 \quad 2), (1 \quad 2)(3 \quad 4)(1 \quad 2),$$
$$(1 \quad 3)(2 \quad 4)(1 \quad 2), (1 \quad 4)(2 \quad 3)(1 \quad 2)\}$$
$$= \{(1 \quad 2), (3 \quad 4), (1 \quad 4 \quad 2 \quad 3), (1 \quad 3 \quad 2 \quad 4)\}$$
$$K(1 \quad 3) = \{(1)(1 \quad 3), (1 \quad 2)(3 \quad 4)(1 \quad 3),$$
$$(1 \quad 3)(2 \quad 4)(1 \quad 3), (1 \quad 4)(2 \quad 3)(1 \quad 3)\}$$
$$= \{(1 \quad 3), (1 \quad 4 \quad 3 \quad 2), (2 \quad 4), (1 \quad 2 \quad 3 \quad 4)\}$$
$$K(2 \quad 3) = \{(1)(2 \quad 3), (1 \quad 2)(3 \quad 4)(2 \quad 3),$$
$$(1 \quad 3)(2 \quad 4)(2 \quad 3), (1 \quad 4)(2 \quad 3)(2 \quad 3)\}$$
$$= \{(2 \quad 3), (1 \quad 2 \quad 4 \quad 3), (1 \quad 3 \quad 4 \quad 2), (1 \quad 4)\}$$
$$K(1 \quad 2 \quad 3) = \{(1)(1 \quad 2 \quad 3), (1 \quad 2)(3 \quad 4)(1 \quad 2 \quad 3),$$
$$(1 \quad 3)(2 \quad 4)(1 \quad 2 \quad 3), (1 \quad 4)(2 \quad 3)(1 \quad 2 \quad 3)\}$$
$$= \{(1 \quad 2 \quad 3), (2 \quad 4 \quad 3), (1 \quad 4 \quad 2), (1 \quad 3 \quad 4)\}$$
$$K(1 \quad 3 \quad 2) = \{(1)(1 \quad 3 \quad 2), (1 \quad 2)(3 \quad 4)(1 \quad 3 \quad 2),$$
$$(1 \quad 3)(2 \quad 4)(1 \quad 3 \quad 2), (1 \quad 4)(2 \quad 3)(1 \quad 3 \quad 2)\}$$
$$= \{(1 \quad 3 \quad 2), (1 \quad 4 \quad 3), (2 \quad 3 \quad 4), (1 \quad 2 \quad 4)\}$$

Since each left coset equals the corresponding right coset, we conclude that $K \trianglelefteq S_4$.

There are other ways of defining the concept of normal subgroup. Let G

be a group and $H \leqslant G$. For $x \in G$, define $x^{-1}Hx = \{x^{-1}hx \mid h \in H\}$ to be the **conjugate of H in G by** x.

Theorem 7.6

Let $H \leqslant G$. Then the following are equivalent:
  i. $H \trianglelefteq G$,
 ii. $x^{-1}Hx \subseteq H$ for all $x \in G$,
iii. $x^{-1}Hx = H$ for all $x \in G$.

Proof

We first show that part i implies part ii. If $H \trianglelefteq G$, then $xH = Hx$ for all $x \in G$, so that $hx = xh'$ for some $h' \in H$. But then $x^{-1}hx = h'$ is an element of H for all $h \in H$, $x \in G$. Thus, $x^{-1}Hx \subseteq H$ for all $x \in G$.

We next show part ii implies part iii. If $x^{-1}Hx \subseteq H$ for all $x \in G$, then $x(x^{-1}Hx)x^{-1} \subseteq xHx^{-1}$ for all $x \in G$. (Verify.) But

$$x(x^{-1}Hx)x^{-1} = (xx^{-1})H(xx^{-1}) = H$$

We have $H \subseteq x^{-1}Hx = x^{-1}H(x^{-1})^{-1}$, and so $H = x^{-1}Hx$ for all $x \in G$.

Finally, we verify that part iii implies part i. If $x^{-1}Hx = H$ for all $x \in G$, then for any $h \in H$ and $x \in G$, we have $x^{-1}hx = h' \in H$; that is, $hx = xh'$ so $Hx \subseteq xH$. Similarly, $xH \subseteq Hx$. Therefore $Hx = xH$ for all $x \in G$. □

Normal subgroups are the building blocks of group theory. They allow examination of a group by considering "layers" of subgroups. We will present this idea to you in Chapter 11. Some groups, however, do not possess any normal subgroups other than the group itself and the identity subgroup. These groups are termed **simple**. For example, a group of prime order was shown in Section 7.2 to possess only the identity subgroup as a proper subgroup. Therefore there can be no proper nontrivial normal subgroups. The problem of determining all simple groups has been the major question in group theory and has recently been solved. Chapter 11 contains some further information on this problem.

Let us return now to our original question, namely, of determining how many homomorphic images a group possesses by finding out which subgroups can be kernels of homomorphisms. We have shown that any kernel of a homomorphism of a group G must be a normal subgroup of G. But is the converse true? Is every normal subgroup the kernel of a group homomorphism? Let us look at some special cases.

Theorem 7.7

If G is a group, then G and $\{1\}$ are kernels of homomorphisms.

Proof

Consider the identity mapping $\phi: G \to G$ given by $\phi(g) = g$ for all $g \in G$. This is an isomorphism whose kernel is $\{1\}$. If H is any group with identity 1_H, then

G is the kernel of the trivial homomorphism $\theta\colon G \to H$ defined by $\theta(x) = 1_H$ for all $x \in G$. $\square$

Note that $A_n \trianglelefteq S_n$ by Corollary 7.5 and A_n is the kernel of the homomorphism sign: $S_n \to \{1, -1\}$ defined in Chapter 5. In the next chapter we will show that *every* normal subgroup is the kernel of a group homomorphism and discover some additional connections between normal subgroups and group homomorphisms.

Exercises 7.3

1. Determine all normal subgroups of the following groups.
   a. $\mathbb{Z}_{10}$ b. $\mathbb{Z}_{19}$
   c. Motions of rectangle   d. Motions of square

2. Why is no subgroup of order 2 in S_3 the kernel of a group homomorphism? Is the same true for subgroups of S_n for $n > 3$?

3. Suppose G is a group, $N_1 \trianglelefteq G$ and $N_2 \trianglelefteq G$. Prove that $N_1 \cap N_2 \trianglelefteq G$.

4. Suppose $\phi\colon G \to G'$ is a group homomorphism.
   a. If $N \trianglelefteq G$, show that $\phi^{\to}(N) \trianglelefteq \phi^{\to}(G)$.
   b. If $N' \trianglelefteq G'$, verify that $\phi^{\leftarrow}(N') \trianglelefteq G$.

5. Let G be a group, $H \leqslant G$, and $K \trianglelefteq G$. Prove that $H \cap K \trianglelefteq H$.

6. Suppose that G is a group, $N \trianglelefteq G$ and $n \in N$. Show that $gng^{-1} \in N$ for all $g \in G$.

7. Let G be a group and $K \leqslant G$. From Exercise 19 of Section 4.3 recall $N_G(K)$ is a subgroup of G. Prove that $N_G(K)$ is the largest subgroup of G in which K is normal.

8. Let G be a group and $K \leqslant G$. K is said to be **characteristic** in G if $\phi^{\to}(K) \leqslant K$ for all homomorphisms on G. Prove that every characteristic subgroup is normal..

9. If A_4 contained a subgroup H of order 6, then it would be normal in A_4 by Theorem 7.5. Therefore if $x \in H$, then $axa^{-1} \in H$ for all $a \in A_4$. Use this idea to show A_4 contains no subgroup of order 6. (This example shows that the converse to Lagrange's Theorem is false!)

10. Examine the subgroups of the group of rigid motions of a square. Find a normal subgroup of index 2 that itself contains a normal subgroup of index 2. Use this example to show that "is a normal subgroup of" is **not** a transitive relation on the set of groups.

11. Let $D_n = \langle a, b \,|\, a^n = b^2 = 1, bab = a^{n-1} \rangle$ be the dihedral group of order $2n$. Prove that every subgroup of $\langle a \rangle$ is normal in D_n.

12. Suppose $H \trianglelefteq G$ and $H' \trianglelefteq G'$. Show that $H \times H' \trianglelefteq G \times G'$.

13. If $H \trianglelefteq K$ and $K \leqslant G$, show that K is a subgroup of $N_G(H)$, the normalizer of H in G.

14. Verify that $Z(G) \trianglelefteq G$ for any group G.

15. Suppose $K \leqslant G$ and $\Omega = \{a_1K, a_2K, \ldots, a_nK\}$ is the set of distinct left cosets of K in G. For each $x \in G$, define $\phi_x\colon \Omega \to S_n$, the symmetric group on n letters, by $\phi_x(a_iK) = xa_iK$. Let $\phi\colon G \to S_\Omega \cong S_n$ be given by $\phi(x) = \phi_x$. Prove that ϕ is a homomorphism with

$$\ker \phi = \bigcap_{b \in G} bKb^{-1}$$

a. Conclude that any group containing a proper subgroup of finite index contains a proper normal subgroup of finite index.

b. Suppose the finite group G contains a subgroup K with $[G:K] = p$, the smallest prime divisor of $|G|$. Prove $K \trianglelefteq G$.

16. Let G be a group and $a \in G$ have order 2. Set $N = \langle a \rangle$. If $N \trianglelefteq G$, prove $a \in Z(G)$.

17. Set G equal to the set of all 2×2 matrices of the form

$$\begin{bmatrix} a & b \\ c & d \end{bmatrix}$$

whose entries are rational numbers and which have the property that $ad - bc \neq 0$. Prove that the set of all such matrices with the property $ad - bc = 1$ forms a normal subgroup of G.

18. Let G be a group, and suppose $|G| = 97$.
a. Describe the subgroups of G.
b. Describe the normal subgroups of G.
c. If $\phi \colon G \to G'$ is a homomorphism from G to a group G', then describe ϕ.

19. Show that if G is a finite group, then G cannot be isomorphic to a proper subgroup of itself.

20. Give an example to show that an infinite group can be isomorphic to a proper subgroup of itself.

21. Let $K = \{(1), (1 \ \ 2)(3 \ \ 4), (1 \ \ 3)(2 \ \ 4), (1 \ \ 4)(2 \ \ 3)\}$. We demonstrated in this section that $K \trianglelefteq S_4$.
a. How many subgroups of S_4 are isomorphic to K?
b. Does there exist an epimorphism from S_4 with kernel K?

8

Factor Groups and

Homomorphisms

We have learned from our study of group homomorphisms and their kernels that if G is a group and K is the kernel of a group homomorphism of group G, then K is a normal subgroup of G. Is the converse true? Are there as many group homomorphisms as normal subgroups? Does each normal subgroup give rise to a group homomorphism and, if so, how is such a homomorphism defined?

Let us consider a specific example, say S_3. From Example 7.10, we know that the subgroups of S_3 have orders 1, 2, 3, and 6. We also have seen that $\{(1)\}$ and S_3 are kernels of homomorphisms whose homomorphic images are isomorphic to S_3 and $\{(1)\}$, respectively. An analysis of the subgroups of S_3 showed that A_3 is the only other normal subgroup of S_3. Thus, no subgroup of order 2 can be the kernel of a homomorphism of S_3; it may be possible, however, that A_3 can be the kernel of a homomorphism. Can we find such a mapping? The answer is yes; we have already seen in Section 6.1 that the mapping $\Gamma: S_3 \to \{1, -1\}$ is a homomorphism and its kernel is A_3, the set of even permutations.

In this example, we have verified that each normal subgroup does indeed give rise to a homomorphism. Note, too, that the order of each homomorphic image is the index of the normal subgroup. These observations hold true in general for all groups. The objective of this chapter is to verify these assertions and establish additional relationships between normal subgroups and group homomorphisms.

8.1

Factor Groups

Our investigation of the kernel of a homomorphism and of those subgroups that might be kernels led us to study normal subgroups. Combining this inquiry with Theorem 7.3, we are led to the following development.

Let G be a group with the operation denoted by juxtaposition and N a normal subgroup of G (where N is a candidate for the kernel of a homomorphism). By Theorem 6.3, we know that the number of distinct left cosets of the kernel of a group homomorphism equals the number of distinct elements in the homomorphic image of the group. Thus let us denote by G/N the set of all left cosets of N in G; that is, $G/N = \{xN \mid x \in G\}$. In the case that G is a finite group, Lagrange's Theorem tells us that $|G/N| = |G|/|N|$. G/N then denotes the **set** (not necessarily the **group**) that we will eventually show is isomorphic to the homomorphic image of G.

Since homomorphisms preserve the operation in a group, let us attempt to define a binary operation $*$ on G/N by

$$xN * yN = xyN$$

Does this make sense? If $x_1 \in xN$, then $x_1 N = xN$. If our definition is to be meaningful, it should be the case that the composite of two cosets is the same no matter which element x or x_1 (called **coset leaders**) we use to name the coset. Therefore we must show that if $xN = x_1 N$ and $yN = y_1 N$, then $xN * yN$ is the same as $x_1 N * y_1 N$. In mathematical terminology, we need to check that our operation $*$ is **well defined**.

To do this, note that $x = x1 \in xN = x_1 N$ implies that $x = x_1 n_1$ for some $n_1 \in N$. Similarly if $yN = y_1 N$, then $y = y_1 n_2$ for some $n_2 \in N$. Since $N \trianglelefteq G$, we know that $y_1^{-1} n_1 y_1 = n_3 \in N$, so that $n_1 y_1 = y_1 n_3$. Therefore

$$xN * yN = xyN$$
$$= (x_1 n_1)(y_1 n_2)N$$
$$= x_1 (n_1 y_1)n_2 N$$
$$= x_1 (y_1 n_3)n_2 N$$
$$= (x_1 y_1)(n_3 n_2)N$$
$$= x_1 y_1 N = x_1 N * y_1 N$$

and we do have a well-defined operation.

Since N is a normal subgroup of G, we have $xN = Nx$ for all $x \in G$. Therefore we could have defined $*$ by $Nx * Ny = Nxy$ and obtained the same operation. If N is **not** a normal subgroup, however, then our definition $xN * yN = xyN$ does not produce, in general, an operation on the cosets N in G. We leave examples of this to Exercises 1 and 2.

Note that the coset $1N$ is an identity for the operation $*$ on G/N since

$1N * yN = 1yN = yN$. Then the coset $x^{-1}N$ is the inverse of $xN \in G/N$ because

$$x^{-1}N * xN = (x^{-1}x)N = 1N$$

Finally the equalities

$$(xN * yN) * zN = xyN * zN$$
$$= (xy)zN$$
$$= x(yz)N$$
$$= xN * (yz)N$$
$$= xN * (yN * zN)$$

show that $*$ is an associative operation on G/N. We have proved an important result.

Theorem 8.1

If $N \trianglelefteq G$, then $(G/N, *)$ is a group.

G/N is termed the **factor group** or **quotient group of G by N**. G/N is read "G mod N."

Recall that we want to show that a normal subgroup N of a group G gives rise to a group homomorphism. The existence of the group G/N is just the key we need.

Theorem 8.2

Let $N \trianglelefteq G$. Then there exists a group homomorphism $\tau : G \to G/N$ with $\ker \tau = N$.

Proof

Define $\tau : G \to G/N$ by $\tau(a) = aN$ where $a \in G$. Then, if $a, b \in G$,

$$\tau(ab) = (ab)N = aN * bN = \tau(a) * \tau(b)$$

We still must show that $\ker \tau = N$. If $x \in \ker \tau$, then $\tau(x) = 1N = N$. But $\tau(x) = xN$. Thus $xN = N$, so that $x = x1 \in N$. Conversely, if $y \in N$, then $\tau(y) = yN$; but $y \in N$ implies $yN = N$. Thus $\tau(y) = N$ and $y \in \ker \tau$. We conclude $\ker \tau = N$.   □

Example 8.1

In Example 7.11, we showed that the subgroup

$$K = \{(1), (1 \quad 2)(3 \quad 4), (1 \quad 3)(2 \quad 4), (1 \quad 4)(2 \quad 3)\}$$

is normal in S_4. In that example we denoted the cosets

$$(1)K = K$$

$$(1 \quad 2)K = \{(1 \quad 2), (3 \quad 4), (1 \quad 3 \quad 4 \quad 2), (1 \quad 4 \quad 2 \quad 3)\}$$

$$(1 \quad 3)K = \{(1 \quad 3), (1 \quad 2 \quad 3 \quad 4), (2 \quad 4), (1 \quad 4 \quad 3 \quad 2)\}$$
$$(2 \quad 3)K = \{(2 \quad 3), (1 \quad 3 \quad 4 \quad 2), (1 \quad 2 \quad 4 \quad 3), (1 \quad 4)\}$$
$$(1 \quad 2 \quad 3)K = \{(1 \quad 2 \quad 3), (2 \quad 4 \quad 3), (1 \quad 3 \quad 4), (1 \quad 4 \quad 2)\}$$
$$(1 \quad 3 \quad 2)K = \{(1 \quad 3 \quad 2), (1 \quad 2 \quad 4), (2 \quad 3 \quad 4), (1 \quad 4 \quad 3)\}$$

With this representation, Theorems 8.1 and 8.2 tell us that there is a homomorphism with kernel K. The definition of the factor group operation $(xK * yK = xyK)$ and the fact that the elements we have chosen to represent the equivalence classes are the elements of S_3 leads us to conjecture that the factor group S_4/K is isomorphic to S_3. This will be substantiated by Theorem 8.3 (The First Isomorphism Theorem) in the next section.

Example 8.2

	1	a	b	ab
1	1	a	b	ab
a	a	1	ab	b
b	b	ab	1	a
ab	ab	b	a	1

Figure 8.1

Let $G = \{1, a, b, ab\}$ with the operation of juxtaposition given by the Cayley table in Figure 8.1. This is the table of motions of a rectangle. Let $H_1 = \{1, a\}$, and $H_2 = \{1, b\}$. Then H_1 and H_2 are distinct (though isomorphic) normal subgroups of G (check) and $G/H_1 \cong G/H_2$.

Example 8.3

	H	$1 + H$
H	H	$1 + H$
$1 + H$	$1 + H$	H

Figure 8.2

Let $G = \mathbb{Z}_{12}$, the group of integers modulo 12 under modular addition. Since $\mathbb{Z}_{12}$ is abelian, every subgroup is normal. What do these factor groups look like? $H = \langle 2 \rangle = \{0, 2, 4, 6, 8, 10\}$ has two cosets: H and $1 + H = \{1, 3, 5, 7, 9, 11\}$. Thus $|G/H| = 2$. The mapping $\tau(a) = a + H$ can be expressed as

$$\tau(a) = H \quad \text{if} \quad a \in H \qquad \text{or} \qquad \tau(a) = 1 + H \quad \text{if} \quad a \notin H$$

The Cayley table of $\mathbb{Z}_{12}/H$ is in Figure 8.2. Clearly this group is isomorphic to $\mathbb{Z}_2$. (Why?)

The subgroup $K = \langle 4 \rangle = \{0, 4, 8\}$ has index 4, and $\sigma(a) = a + K$ is the desired homomorphism. The Cayley table is in Figure 8.3. Note that $(3 + K) + (3 + K) = 6 + K = 2 + K$. This group is cyclic and, hence, isomorphic to $\mathbb{Z}_4$.

Similar results can be obtained for $N = \langle 3 \rangle$ and $M = \langle 6 \rangle$.

Figure 8.3

	K	$1 + K$	$2 + K$	$3 + K$
K	K	$1 + K$	$2 + K$	$3 + K$
$1 + K$	$1 + K$	$2 + K$	$3 + K$	K
$2 + K$	$2 + K$	$3 + K$	K	$1 + K$
$3 + K$	$3 + K$	K	$1 + K$	$2 + K$

Example 8.4

	$2\mathbb{Z}$	$1+2\mathbb{Z}$
$2\mathbb{Z}$	$2\mathbb{Z}$	$1+2\mathbb{Z}$
$1+2\mathbb{Z}$	$1+2\mathbb{Z}$	$2\mathbb{Z}$

Figure 8.4

Let $G = \mathbb{Z}$ and $N = 2\mathbb{Z}$, the even integers. Then $\mathbb{Z}/2\mathbb{Z}$ consists of just two elements: one coset is $2\mathbb{Z}$; the other is $1+2\mathbb{Z}$, the odd integers. The Cayley table for $\mathbb{Z}/2\mathbb{Z}$ is in Figure 8.4. Note that this factor group is isomorphic to $\mathbb{Z}_2$.

We said that our goal in this chapter was to determine the connections between normal subgroups and group homomorphisms. At last, you might say, we have reached our goal. The kernel of a homomorphism is a normal subgroup, and every normal subgroup is the kernel of a homomorphism. Not only that, but we showed by our construction of factor groups exactly how these homomorphisms can be defined.

But wait! There is more. We will show in Section 8.2 how an arbitrary homomorphism from one group G to another can be interpreted as a homomorphism to a factor group of G. We also will investigate factor groups in more detail and see what form subgroups of factor groups have.

Exercises 8.1

1. Let $N = \langle (1 \quad 2) \rangle$. Define $*$ on the cosets of N in S_3 by $xN * yN = xyN$. Construct a Cayley table for $*$. Do you obtain a group? Why? Is the operation $*$ well defined?

2. Follow the instructions in Exercise 1 for the subgroup $N = \langle (1 \quad 2 \quad 3) \rangle$ in A_4.

3. Suppose N is a subgroup of G and $(xN)(yN) = (xy)N$ for all $x, y \in G$. Prove that $N \trianglelefteq G$.

4. Let $K = \{(1), (1 \quad 2)(3 \quad 4), (1 \quad 3)(2 \quad 4), (1 \quad 4)(2 \quad 3)\}$. Show directly that $K \trianglelefteq A_4$, and determine the Cayley table for A_4/K. What is the order of this group? (We showed in Chapter 7 that $K \trianglelefteq S_4$.)

5. Determine the Cayley table for G/K if $G = (\mathbb{Z}_{24}, \oplus)$ and $K = \langle 6 \rangle$. Is this group abelian? This group is isomorphic to either $\mathbb{Z}_6$ or S_3? Which group is it and what is the isomorphism?

6. Determine the Cayley table for G/K if $G = (\mathbb{Z}_{24}, \oplus)$ and $K = \langle 4 \rangle$. Is this group abelian? Is this group isomorphic to $\mathbb{Z}_4$ or the Klein Four-Group? What is the isomorphism?

7. Suppose $a \in G$, $N \trianglelefteq G$, and $a^m = 1$. Prove that $(aN)^m = N$.

8. Prove that if $N \trianglelefteq G$, then
    a. G abelian implies G/N is abelian, and
    b. G cyclic implies G/N is cyclic.

9. Suppose $N \trianglelefteq G$. If G is finite, prove that G/N is finite. If G/N is finite, does G have to be finite? Explain.

10. If $N \trianglelefteq G$, $K \trianglelefteq G$, then G/N and G/K are both groups. Verify that $N \cap K \trianglelefteq G$. Is $G/(N \cap K)$ isomorphic to the group $(G/N) \times (G/K)$ defined in Exercise 15 of Section 3.2?

11. Let G be an infinite abelian group and $N = \{g \in N \,|\, o(g) \text{ is finite}\}$. Prove
    a. $N \trianglelefteq G$, and
    b. if $aN \in G/N$, then aN has infinite order.

12. If $|G| = 97$, determine all factor groups of G.

13. Suppose $K \trianglelefteq G$ and $K \leqslant N \leqslant G$. Prove that N/K is a subgroup of G/K.

14. Let $G = (\mathbb{Z}_{13}^*, \odot)$, $K = \langle 2 \rangle$, and $M = \langle 3 \rangle$. Calculate the Cayley tables of the factor groups G/K and G/M.

15. Suppose $N \trianglelefteq G$ and $ghg^{-1}h^{-1} \in N$ for all g, $h \in G$. Prove that G/N is abelian.

16. Let G be a group and G' be the subgroup of G generated by all elements of the form $aba^{-1}b^{-1}$, where a, $b \in G$.
    a. Prove that $G' \trianglelefteq G$. (This subgroup is known as the **commutator subgroup of** G.)
    b. Find G' for $G =$
        i. $(\mathbb{Z}_8, \oplus)$ ii. S_3 iii. A_4 iv. S_4
    c. Verify that if $K \trianglelefteq G$ and G/K is abelian, then $G' \trianglelefteq K$.

The Isomorphic Connection

In the preceding section, we saw how we could define a homomorphism of a group G for each normal subgroup of G. The image of this homomorphism is a factor group. In fact, *every* homomorphic image can be interpreted as arising in this way. This idea is made explicit by the following result.

Theorem 8.3

First Isomorphism Theorem

Let $\phi: G \to G'$ be a group epimorphism with kernel K. Then $G/K \cong G'$.

Proof

By Theorem 8.2, the mapping $\tau: G \to G/K$ defined by

$$\tau(a) = aK$$

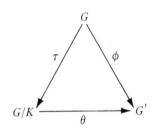

is a group homomorphism with kernel K. Clearly, τ is an epimorphism. We will define a map $\theta: G/K \to G'$ and show that θ is an isomorphism. (See Figure 8.5.)

Let $aK \in G/K$. Define $\theta(aK) = \phi(a)$. Does this definition make sense? In other words, is $\theta: G/K \to G'$ a well-defined mapping? Recall that this means: if $aK = bK$, is it true that $\theta(aK) = \theta(bK)$? Well, if $aK = bK$, then $a = bk$ for some $k \in K$ and

$$\theta(aK) = \phi(a) = \phi(bk) = \phi(b)\phi(k) = \phi(b)1' = \phi(b) = \theta(bK)$$

since $k \in K = \ker \phi$. Thus, θ is indeed a mapping.

If aK, $bK \in G/K$, then

$$\theta(aK * bK) = \theta(abK) = \phi(ab) = \phi(a)\phi(b) = \theta(aK)\theta(aK)$$

Hence, θ is a group homomorphism.

Let $aK \in \ker \theta$. Then $\phi(a) = \theta(aK) = 1'$. Therefore, a is an element of K, the kernel of ϕ; but then $aK = K$. By Corollary 6.1, θ is a monomorphism since the kernel of θ is the identity coset of G/K.

Figure 8.5

It remains to be shown that θ is an epimorphism. To do this, let $a' \in G'$. Since ϕ is an epimorphism, there exists an $a \in G$ such that $\phi(a) = a'$. Consider aK. $\theta(aK) = \phi(a) = a'$. Therefore, for all $a' \in G'$, there exists an $aK \in G/K$ such that $\theta(aK) = a'$. Thus, θ is an epimorphism and, hence, the desired isomorphism. □

Since any group homomorphism $\phi: G \to G'$ can be considered as a group epimorphism from G to $\phi^{\to}(G)$, the image of G (denoted $\operatorname{Im} \phi$), we obtain the following result.

Corollary 8.1

Let $\phi: G \to G'$ be a group homomorphism. Then $G/\ker \phi \cong \operatorname{Im} \phi$.

Example 8.5

Let $\phi: \mathbb{Z} \to \mathbb{Z}_n$ be defined by setting $\phi(a)$ equal to the remainder obtained by dividing a by n. Then ϕ is a homomorphism. (Check.) This image of ϕ is $\mathbb{Z}_n$. Theorem 8.3 asserts, therefore, that $\mathbb{Z}_n$ is isomorphic to a factor group of $\mathbb{Z}$. Which factor group? To answer this question, we need only compute the kernel of ϕ. Since 0 is the identity of $\mathbb{Z}_n$ and since the integers with remainders (upon division by n) equal to 0 are precisely $n\mathbb{Z}$, the multiples of n, we have $\mathbb{Z}/n\mathbb{Z} \cong \mathbb{Z}_n$. This isomorphism is described as follows:

$$0 + n\mathbb{Z} \leftrightarrow 0$$

$$1 + n\mathbb{Z} \leftrightarrow 1$$

$$\vdots$$

$$(n-1) + n\mathbb{Z} \leftrightarrow n-1$$

Example 8.6

Let

$$D = \{(1), (1 \ 2 \ 3 \ 4), (1 \ 3)(2 \ 4), (1 \ 4 \ 3 \ 2),$$
$$(1 \ 2)(3 \ 4), (2 \ 4), (1 \ 4)(2 \ 3), (1 \ 3)\}$$

Then D is a subgroup of S_4. Its Cayley table is given in Figure 8.6. Note that

	(1)	(1 2 3 4)	(1 3)(2 4)	(1 4 3 2)	(1 2)(3 4)	(2 4)	(1 4)(2 3)	(1 3)
(1)	(1)	(1 2 3 4)	(1 3)(2 4)	(1 4 3 2)	(1 2)(3 4)	(2 4)	(1 4)(2 3)	(1 3)
(1 2 3 4)	(1 2 3 4)	(1 3)(2 4)	(1 4 3 2)	(1)	(1 3)	(1 2)(3 4)	(2 4)	(1 4)(2 3)
(1 3)(2 4)	(1 3)(2 4)	(1 4 3 2)	(1)	(1 2 3 4)	(1 4)(2 3)	(1 3)	(1 2)(3 4)	(2 4)
(1 4 3 2)	(1 4 3 2)	(1)	(1 2 3 4)	(1 4)(2 3)	(1 3)(2 4)	(2 4)	(1 3)	(1 2)(3 4)
(1 2)(3 4)	(1 2)(3 4)	(2 4)	(1 4)(2 3)	(1 3)	(1)	(1 2 3 4)	(1 3)(2 4)	(1 4 3 2)
(2 4)	(2 4)	(1 4)(2 3)	(1 3)	(1 2)(3 4)	(1 4 3 2)	(1)	(1 2 3 4)	(1 3)(2 4)
(1 4)(2 3)	(1 4)(2 3)	(1 3)	(1 2)(3 4)	(2 4)	(1 3)(2 4)	(1 4 3 2)	(1)	(1 2 3 4)
(1 3)	(1 3)	(1 2)(3 4)	(2 4)	(1 4)(2 3)	(1 2 3 4)	(1 3)(2 4)	(1 4 3 2)	(1)

Figure 8.6

this group is isomorphic to the group of symmetries of a square. (See Exercise 18.)

The normal subgroups of D have possible orders 1, 2, 4, and 8. We know that $\{(1)\}$ and D are normal subgroups, and so $D \cong D/\{(1)\}$ and $\{(1)\} \cong D/D$ are isomorphic to homomorphic images of D. Any other homomorphic images have orders 4 (kernel of order 2) or 2 (kernel of order 4). The subgroups of order 2 are

$$\{(1), (1 \quad 3)(2 \quad 4)\}, \qquad \{(1), (1 \quad 2)(3 \quad 4)\}, \qquad \{(1), (2 \quad 4)\},$$

$$\{(1), (1 \quad 4)(2 \quad 3)\}, \qquad \text{and} \qquad \{(1), (1 \quad 3)\}$$

$K_1 = \{(1), (1 \quad 3)(2 \quad 4)\}$ is the only normal subgroup of order 2 and

$$D/K_1 = \{K_1, (1 \quad 2 \quad 3 \quad 4)K_1, (1 \quad 4)(2 \quad 3)K, (1 \quad 3)K_1\}$$

Since there are only two groups (up to isomorphism) of order 4, either $D/K_1 \cong K_4$ (the Klein Four-Group) or $D/K_1 \cong \mathbb{Z}_4$ (the cyclic group of order 4). Since no element of D/K_1 has order 4, we know that $D/K_1 \cong K_4$. Thus for D there is only one homomorphism with a homomorphic image of order 4.

The subgroups of D of order 4 are

$$\{(1), (1 \quad 3), (2 \quad 4), (1 \quad 3)(2 \quad 4)\}$$

$$\{(1), (1 \quad 2 \quad 3 \quad 4), (1 \quad 3)(2 \quad 4), (1 \quad 4 \quad 3 \quad 2)\}$$

$$\{(1), (1 \quad 3)(2 \quad 4), (1 \quad 2)(3 \quad 4), (1 \quad 4)(2 \quad 3)\}$$

Since each subgroup has index 2 in D, each is normal. Thus there are three homomorphic images of order 2, and these images are all isomorphic to $\mathbb{Z}_2$.

There are some other isomorphism theorems that may prove useful, but first we need a definition.

Definition 8.1

If $A \subseteq G$ and $B \subseteq G$, then $AB = \{ab \mid a \in A, b \in B\}$.

Even if A and B are subgroups, the set AB might not be a subgroup. For example, in S_3, if $A = \{(1), (1 \quad 2)\}$ and $B = \{(1), (1 \quad 3)\}$, then

$$AB = \{(1), (1 \quad 3), (1 \quad 2), (1 \quad 3 \quad 2)\}$$

Since $|S_3| = 6$ and $|AB| = 4$, then Lagrange's Theorem guarantees that AB is not a subgroup of G.

Theorem 8.4

If $A \leqslant G$, $B \leqslant G$, and either $A \trianglelefteq G$ or $B \trianglelefteq G$, then $AB = BA$ and $AB \leqslant G$.

Proof

Let us assume that $B \trianglelefteq G$. Then

$$AB = \{ab \mid a \in A, b \in B\}$$

$$= \bigcup_{a \in A} aB$$

$$= \bigcup_{a \in A} Ba$$

$$= \{ba \mid b \in B, a \in A\}$$

$$= BA$$

To complete the proof, we must show that AB is a subgroup of G. Since $AB \neq \varnothing$ (why?), we need only verify that, for $a_1 b_1,\ a_2 b_2 \in AB$, we have $(a_1 b_1)(a_2 b_2)^{-1} \in AB$ (Theorem 4.6). Now $(a_2 b_2)^{-1} = b_2^{-1} a_2^{-1} \in BA = AB$, so $(a_2 b_2)^{-1} = a_3 b_3$ for some $a_3 \in A$, $b_3 \in B$. Thus

$$(a_1 b_1)(a_2 b_2)^{-1} = (a_1 b_1)(a_3 b_3)$$

$$= a_1 (b_1 a_3) b_3$$

$$= a_1 (a_4 b_4) b_3 \qquad \text{since } b_1 a_3 \in BA = AB$$

$$= (a_1 a_4)(b_4 b_3) \in AB \qquad \qquad \square$$

If $A \leqslant G$, $B \trianglelefteq G$, then AB is a subgroup of G, but not necessarily normal in G. For example, if A is not normal and $B = \{1\}$, then $AB = A$. It is true, however, that $B \trianglelefteq AB$, since $B \trianglelefteq G$ and $B \leqslant AB \leqslant G$, and also that $A \cap B \trianglelefteq A$. The connection between these groups is given in the next result.

Theorem 8.5

Second Isomorphism Theorem

If $A \leqslant G$ and $B \trianglelefteq G$, then $A \cap B \trianglelefteq A$ and $A/(A \cap B) \cong AB/B$.

Proof

Recall from Exercise 10 of Section 7.2 that if $a \in A$, then $a(A \cap B) = aA \cap aB$. But $aA = A = Aa$ since $a \in A$, and $aB = Ba$ since $B \trianglelefteq G$. Thus

$$a(A \cap B) = aA \cap aB = Aa \cap Ba = (A \cap B)a \qquad \text{and} \qquad A \cap B \trianglelefteq A$$

The elements of AB/B are the cosets aB where $a \in A$. (See Exercise 19.) Define $\phi: A \to AB/B$ by $\phi(a) = aB$. This is clearly a surjection. If $a_1,\ a_2 \in A$, then

$$\phi(a_1 a_2) = (a_1 a_2)B = (a_1 B) * (a_2 B) = \phi(a_1) * \phi(a_2)$$

shows that ϕ is an epimorphism. What is $\ker \phi$? Since $B = 1B$ is the identity coset of AB/B,

$$\ker \phi = \{a \in A \,|\, \phi(a) = B\}$$
$$= \{a \in A \,|\, aB = B\}$$
$$= \{a \in A \,|\, a \in B\}$$
$$= A \cap B$$

The result now follows immediately from Theorem 8.3. $\square$

Corollary 8.2

If A and B are finite subgroups of G and $B \trianglelefteq G$, then

$$|AB| = \frac{|A| \cdot |B|}{|A \cap B|}$$

Proof

By the Second Isomorphism Theorem, we have $|AB/B| = |A/A \cap B|$. We obtain our conclusion by applying Lagrange's Theorem to note that

$$\left| \frac{AB}{B} \right| = \frac{|AB|}{|B|} \quad \text{and} \quad \left| \frac{A}{A \cap B} \right| = \frac{|A|}{|A \cap B|}$$ $\square$

Example 8.7

Let $G = \mathbb{Z}_{12}$, $A = \langle 4 \rangle$, and $B = \langle 6 \rangle$. Since $\mathbb{Z}_{12}$ is abelian, both A and B are normal in $\mathbb{Z}_{12}$. Then $AB = A + B$ is a subgroup of order

$$\frac{|A| \cdot |B|}{|A \cap B|}$$

The order of $A \cap B$ must divide both $|A| = 3$ and $|B| = 2$. Thus $|A \cap B| = 1$ and $|AB| = 6$. We leave the determination of these 6 elements as Exercise 17.

Example 8.8

Let $G = S_4$, $C = \langle (1 \quad 2 \quad 3) \rangle$ and

$$K = \{(1), (1 \quad 2)(3 \quad 4), (1 \quad 3)(2 \quad 4), (1 \quad 4)(2 \quad 3)\}$$

We showed $K \trianglelefteq S_4$ in Chapter 7. Now $|C \cap K| = 1$ since the order of $C \cap K$ must divide both $|C| = 3$ and $|K| = 4$. Therefore, $|CK| = 12$. If you note that every element of CK is a product of two even permutations, you should now be able to conclude that $CK = A_4$.

Recall that in Theorem 6.2, we showed how a group homomorphism $\phi : G \to G'$ takes subgroups of G to subgroups of G'. Moreover, the inverse images of subgroups of G' are subgroups of G. That result can be extended to a bijection between certain subgroups of G and subgroups of Im ϕ, the image of ϕ. We leave the proof of the result expressed in Theorem 8.6 for Exercise 12.

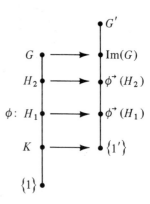

Figure 8.7

| Theorem 8.6 |

Correspondence Theorem

If $\phi\colon G \to G'$ is a group homomorphism with $K = \ker \phi$, then there is a bijection between the set of subgroups of G containing K and subgroups of $\operatorname{Im} \phi$.

As in the case of Theorem 6.2, the desired mapping assigns to each subgroup $H \leqslant G$ the image $\phi^{\to}(H)$. $\phi^{\leftarrow}\colon G' \to G$ assigns to each $H' \leqslant \operatorname{Im} \phi$ the subgroup $\phi^{\leftarrow}(H')$. This relationship is pictured in Figure 8.7.

If $K \trianglelefteq G$, then we can apply the Correspondence Theorem to the natural homomorphism $\tau\colon G \to G/K$ to obtain:

| Corollary 8.3 |

Suppose $K \trianglelefteq G$. If $K \leqslant H \leqslant G$, then $H/K \leqslant G/K$; moreover, if $N \leqslant G/K$, then there exists a subgroup H of G containing K such that $H/K \cong N$.

In the result of Corollary 8.3, we can say more if $H \trianglelefteq G$. In that case $H/K \trianglelefteq G/K$. To verify this, we must show that $(gK)(hK)(gK)^{-1} \in H/K$ for any $g \in G$, $h \in H$. But

$$(gK)(hK)(gK)^{-1} = (ghg^{-1})K = h_1 K \in H/K$$

since $H \trianglelefteq G$. Therefore $H/K \trianglelefteq G/K$.

Let us consider a concrete example. If

$$K = \{(1), (1\ \ 2)(3\ \ 4), (1\ \ 3)(2\ \ 4), (1\ \ 4)(2\ \ 3)\}$$

then

$$\frac{A_4}{K} = \{K, (1\ \ 2\ \ 3)K, (1\ \ 3\ \ 2)K\} \qquad \text{and}$$

$$\frac{S_4}{K} = \{K, (1\ \ 2\ \ 3)K, (1\ \ 3\ \ 2)K, (1\ \ 2)K, (1\ \ 3)K, (2\ \ 3)K\}$$

Figure 8.8

	A_4/K	$(1\ \ 2)(A_4/K)$
A_4/K	A_4/K	$(1\ \ 2)(A_4/K)$
$(1\ \ 2)(A_4/K)$	$(1\ \ 2)(A_4/K)$	A_4/K

Figure 8.9

	A_4	$(1\ \ 2)A_4$
A_4	A_4	$(1\ \ 2)A_4$
$(1\ \ 2)A_4$	$(1\ \ 2)A_4$	A_4

Now A_4/K is a normal subgroup of S_4/K. (Check.) Therefore we can consider the factor group of S_4/K by A_4/K. What does this group look like? Well, the left cosets of A_4/K in S_4/K are

$$A_4/K = \{K, (1\ \ 2\ \ 3)K, (1\ \ 3\ \ 2)K\} \quad \text{and}$$

$$(1\ \ 2)(A_4/K) = \{(1\ \ 2)K, (2\ \ 3)K, (1\ \ 3)K\}$$

The Cayley table of this group is given in Figure 8.8. Compare this table with that of S_4/A_4 given in Figure 8.9. It is clear that these two groups are isomorphic. (Why?) That this situation occurs in general is the content of the last result of this chapter.

Theorem 8.7

Third Isomorphism Theorem

If $K \trianglelefteq G$ and $K \leqslant H \trianglelefteq G$, then $H/K \trianglelefteq G/K$. Moreover,

$$\frac{G/K}{H/K} \cong G/H$$

Proof

We have already proved the first part of this result. To complete the proof, define $\phi: G/K \to G/H$ by $\phi(gK) = gH$. This is clearly an epimorphism. What is the kernel of this mapping? $gK \in \ker \phi$ if and only if $\phi(gK) = H$, the identity coset of G/H. But $\phi(gK) = gH$ and $gH = H$ if and only if $g \in H$. Therefore $\ker \phi$ consists of all cosets gK such that $g \in H$. But this is H/K. The result now follows immediately from the First Isomorphism Theorem. $\square$

Exercises 8.2

1. Consider the epimorphism $\phi: \mathbb{Z} \to 2\mathbb{Z}$ given by $\phi(n) = 2n$. What is $\ker \phi$? Is $\mathbb{Z}/\ker \phi \cong \text{Im } \phi$?

2. Find a homomorphism ϕ from $\mathbb{Z}$ onto $\mathbb{Z}_{12}$ such that $\mathbb{Z}/\ker \phi \cong \mathbb{Z}_{12}$.

3. If $N \trianglelefteq G$ and $K \trianglelefteq G$ and $|N \cap K| = 1$, prove that $nk = kn$ for all $n \in N$, $k \in K$.

4. Prove for any integer $n > 1$ that S_2 is the homomorphic image of S_n.

5. For $A = \langle (1 \quad 2) \rangle$, $B = \langle (1 \quad 2 \quad 3) \rangle$, construct AB.

6. For $A = \langle (1 \quad 2 \quad 3) \rangle$, $B = \langle (1 \quad 2 \quad 3 \quad 4) \rangle$, show that neither A nor B is normal in S_4. Construct AB. Is AB a group?

7. Let $A = S_3$, $B = A_4$. Verify that $A \cap B \trianglelefteq A$ and that $A/A \cap B \cong AB/B$.

8. Consider the group D of Example 8.6. If $K_1 = \{(1), (1 \quad 3)(2 \quad 4)\}$, construct D/K_1 and N/K_1 where $N = \{(1), (1 \quad 2)(3 \quad 4), (1 \quad 3)(2 \quad 4), (1 \quad 4)(2 \quad 3)\}$. Verify that $N/K_1 \trianglelefteq D/K_1$ and $(D/K_1)/(N/K_1) \cong D/N$.

9. Prove that if AB is a group and $|A| = r$ and $|B| = s$ where $(r, s) = 1$, then $|AB| = |A| \cdot |B|$.

10. Let $G = \mathbb{Z}_{24}$, $A = \langle 3 \rangle$, $B = \langle 4 \rangle$. Compute AB. What is $|A \cap B|$?

11. Give an example of a group G containing a normal subgroup A and a nonnormal subgroup B such that $AB \trianglelefteq G$.

12. Complete the proof of Theorem 8.6, the Correspondence Theorem.

13. The idea of **inner automorphism** was developed in Exercise 7 of Section 6.3. Use the First Isomorphism Theorem to prove that $\text{Inn}(G) \cong G/Z(G)$.

14. Suppose that $A \trianglelefteq G$ and $B \leqslant C \leqslant G$. Prove that $AB \leqslant AC$.

15. Define $\phi : \mathbb{Z}_{24} \to \mathbb{Z}_6$ by setting $\phi(a)$ equal to the equivalence class of a in $\mathbb{Z}_6$. Determine $\ker \phi$ and show $\mathbb{Z}_{24}/\ker \phi \cong \mathbb{Z}_6$.

16. Does there exist a group epimorphism from $\mathbb{Z}_{24}$ to $\mathbb{Z}_5$? Why?

17. In Example 8.7 determine the six elements of $AB = A + B$.

18. Show that the group of Example 8.6 is isomorphic to the group of symmetries of a square.

19. Complete the proof of the Second Isomorphism Theorem (Theorem 8.5) by showing that if $A \leqslant G$, $B \trianglelefteq G$, then the elements of AB/B are the cosets aB where $a \in A$.

20. Extend Theorem 8.6 to show that if $\phi : G \to G'$ is a homomorphism and $K' \trianglelefteq G'$, then $\phi^{\leftarrow}(K') \trianglelefteq G$.

21. Let K be the set of all complex numbers of the form $a + bi$ where $a^2 + b^2 = 1$.
    a. Prove that K is a normal subgroup of $(\mathbb{C} - \{0\}, \cdot)$.
    b. Verify that the mapping $\phi : \mathbb{R} \to \mathbb{C} - \{0\}$ given by $\phi(x) = \cos(2\pi x) + i \sin(2\pi x)$ is a homomorphism.
    c. Prove that the factor group of $\mathbb{R}$ by $\mathbb{Z}$ is isomorphic to K.

9

The Sylow Theorems

Group theory has long been important in the study of mathematics; nevertheless, in the last twenty years we have seen a great resurgence in activity, especially in the field of finite group theory. It is still one of the more significant and beautiful subject areas of algebra, wherein number theory, combinatorics, and, more recently, modern high-speed computers have joined to uncover whole families of previously unknown groups and to provide answers to questions first posed almost a century ago.

Unlike some algebraic systems involving a number of operations, finite groups have an advantage in that deep results on their internal structure can often be established with a minimum of preliminaries. We will concentrate on these finite groups and provide some answers to questions about the existence of certain subgroups. Along the way, we will develop the classical theorems of Sylow,[1] study a special classification of groups, and prove that the alternating groups of the form A_n are simple groups.

Throughout this chapter, G signifies a finite group whose operation is denoted by juxtaposition.

---

[1]Ludwig Sylow (1832–1918) was a Norwegian mathematician who worked primarily with substitution groups.

Conjugacy Classes

In studying a given finite group G, we often want to know the number of subgroups of G and the order of each of these subgroups. The Theorem of Lagrange provides us with a set of possible orders of subgroups, but it does not guarantee that a subgroup of one of these orders actually exists. The Theorem of Lagrange, therefore, is a starting point. While it is sometimes impossible to determine all subgroups of a group, often one can find enough subgroups to deduce significant information on the group itself. We will provide you with examples of this process.

First we develop a counting procedure that enables us to work with a group's elements and subgroups. As is so often the case, this investigation utilizes an equivalence relation. The primary definition of this section is based upon a concept we first encountered in studying normal subgroups.

Definition 9.1

If a, $b \in G$, then a is said to be **conjugate to b in G**, denoted $a \sim b$, if there exists some $x \in G$ such that $b = xax^{-1}$. In this case, a and b are said to be **conjugates**.

In an abelian group, each element is conjugate only to itself, since $xax^{-1} = xx^{-1}a = 1a = a$. However, in a nonabelian group, distinct elements can be conjugate. For example, in S_3, the element (1 2 3) is conjugate to (1 3 2) because

$$(1 \quad 2)(1 \quad 3 \quad 2)(1 \quad 2)^{-1} = (1 \quad 2)(1 \quad 3 \quad 2)(1 \quad 2) = (1 \quad 2 \quad 3)$$

Theorem 9.1

Conjugation is an equivalence relation on G.

Proof

If $a \in G$, then $a = 1a1 = 1a1^{-1}$, so that $a \sim a$ and conjugation is reflexive.

If $a \sim b$, then there exists $x \in G$ such that $b = xax^{-1}$. But then

$$a = x^{-1}bx = (x^{-1})b(x^{-1})^{-1}$$

so that $b \sim a$ and conjugation is symmetric.

Finally, if $a \sim b$ and $b \sim c$, then $b = xax^{-1}$ and $c = yby^{-1}$ for some x, $y \in G$. But then

$$c = yby^{-1}$$
$$= y(xax^{-1})y^{-1}$$
$$= (yx)a(x^{-1}y^{-1})$$
$$= (yx)a(yx)^{-1}$$

Thus $a \sim c$ and conjugation is transitive.   □

The elements of G are partitioned into equivalence classes by the conjugation relation. These classes are termed **the conjugate classes of** G. The conjugate class of a particular element $a \in G$ is denoted by cl(a).

Example 9.1

The conjugate classes of S_3 are

$$\text{cl}((1)) = \{(1)\}$$

$$\text{cl}((1 \quad 2)) = \{(1 \quad 2), (1 \quad 3), (2 \quad 3)\}$$

$$\text{cl}((1 \quad 2 \quad 3)) = \{(1 \quad 2 \quad 3), (1 \quad 3 \quad 2)\}$$

In this example, observe that the number of elements in each equivalence class of S_3 divides $|S_3|$. To prove that this property holds in general, we need the following definition.

Definition 9.2

If $a \in G$, then the **centralizer of** a **in** G, denoted $C_G(a)$, is defined by

$$C_G(a) = \{g \in G \mid ag = ga\}$$

This definition states that the centralizer of a in G is the set of all elements of G that commute with a. It is a straightforward exercise to prove that $C_G(a)$ is a subgroup of G. (See Exercise 7.) In fact, if G is an abelian group, then all elements commute with one another; that is, $C_G(a) = G$ for all $a \in G$. If G is a nonabelian group, then G must contain centralizers that are proper subgroups. In S_3, we calculate that

$$C_{S_3}((1)) = S_3 \qquad\qquad C_{S_3}((1 \quad 2)) = \{(1), (1 \quad 2)\}$$

$$C_{S_3}((1 \quad 3)) = \{(1), (1 \quad 3)\} \qquad C_{S_3}((2 \quad 3)) = \{(1), (2 \quad 3)\}$$

$$C_{S_3}((1 \quad 2 \quad 3)) = C_{S_3}((1 \quad 3 \quad 2)) = \{(1), (1 \quad 2 \quad 3), (1 \quad 3 \quad 2)\}$$

Recall that the Theorem of Lagrange asserts that if H is a subgroup of the finite group G, then $|G| = |H| \cdot [G:H]$ where $[G:H]$, the index of H in G, is the number of left cosets of H in G. We can use this result to tie together the definitions of centralizers and conjugacy classes.

Theorem 9.2

Let G be a group, $a \in G$. The number of conjugates of a in G is equal to the index in G of the centralizer of a; that is,

$$|\text{cl}(a)| = [G : C_G(a)]$$

Hence the cardinality of a conjugacy class must divide the group order.

Proof

Suppose $x, y \in G$ and $xC_G(a) = yC_G(a)$. Then $x^{-1}yC_G(a) = C_G(a)$, so that $x^{-1}y \in C_G(a)$. Therefore $x^{-1}ya = ax^{-1}y$, and hence, $yay^{-1} = xax^{-1}$. We conclude that identical left cosets yield identical conjugates. The steps in this argument are reversible so identical conjugates lead to the same left cosets. The result follows.   □

Corollary 9.1

If J is a subset of G consisting of one element from each conjugate class of G, then

$$|G| = \sum_{a \in J} [G : C_G(a)]$$

Proof

Since conjugation is an equivalence relation, we have

$$G = \bigcup_{a \in J} \mathrm{cl}(a)$$

But distinct equivalence classes are disjoint so

$$|G| = \sum_{a \in J} |\mathrm{cl}(a)|$$

The conclusion now follows from Theorem 9.2.   □

Corollary 9.1 is termed the **class equation of** G and is one of the more useful tools in the study of finite groups. However, there is another form of the class equation that is easier to apply. This other form uses a subgroup first encountered in Example 4.10. For the sake of completeness, we repeat the definition here.

Definition 9.3

If G is a group, then the **center of** G, denoted $Z(G)$, is defined by

$$Z(G) = \{z \in G \mid zg = gz \text{ for all } g \in G\}$$

In other words, the center of a group is the set of elements that lie in the centralizer of all elements in G. We showed in Example 4.10 that $Z(G)$ is a subgroup of G. It is not hard to verify that the condition $G = Z(G)$ is equivalent to G being abelian. (See Exercise 10.)

Theorem 9.3

Let $z \in G$. Then $z \in Z(G)$ if and only if $\mathrm{cl}(z) = \{z\}$.

Proof

If $z \in Z(G)$ and $y \in \mathrm{cl}(z)$, then there exists some $x \in G$ such that $y = xzx^{-1}$. Since $z \in Z(G)$, we have $y = xzx^{-1} = xx^{-1}z = z$. Thus $\mathrm{cl}(z) = \{z\}$. Conversely, not that $xzx^{-1} \in \mathrm{cl}(z)$ for all $x \in G$. If $\mathrm{cl}(z) = \{z\}$, then $xzx^{-1} = z$ for all $x \in G$. But then $xz = zx$ for all $x \in G$ and $z \in Z(G)$.   □

Since there are precisely $|Z(G)|$ conjugacy classes consisting of a single element, we can restate the class equation as follows.

Corollary 9.2

If J' is a subset of G consisting of one representative from each conjugate class containing more than one element of G, then

$$|G| = |Z(G)| + \sum_{a \in J'} [G : C_G(a)]$$

The knowledge of the conjugate classes of a group provides a great deal of information about the group itself. In representation theory, the number of conjugacy classes of a group G is related to the number of distinct representations of G. (A **representation** of a group G can be thought of as a homomorphism from G onto a group of matrices.) A more accessible result utilizing conjugate classes relates to the existence of normal subgroups in a group.

Suppose N is a normal subgroup of a group G and $a \in N$. By the definition of normal subgroup, we know that $gag^{-1} \in N$ for all $g \in G$. But this says that

$$N = \bigcup_{a \in N} \{a\} = \bigcup_{a \in N} cl(a) \subseteq N$$

We have proved the following theorem.

Theorem 9.4

Let $N \trianglelefteq G$. Then N is a union of conjugate classes of elements of G.

Recall that when we first encountered Lagrange's Theorem we said that, while the order of a subgroup of a finite group G divides $|G|$, the converse does not always hold. To verify this, let $G = A_4$, the alternating group on four letters. You are asked in Exercise 2 to verify that the conjugate classes of A_4 are

$$\{(1)\}$$

$$\{(1 \ 2)(3 \ 4), (1 \ 3)(2 \ 4), (1 \ 4)(2 \ 3)\}$$

$$\{(1 \ 2 \ 3), (1 \ 4 \ 2), (1 \ 3 \ 4), (2 \ 4 \ 3)\}$$

$$\{(1 \ 3 \ 2), (1 \ 2 \ 4), (1 \ 4 \ 3), (2 \ 3 \ 4)\}$$

If $N \trianglelefteq A_4$ and $|N| = 6$, then N would be a union of some of these disjoint classes, and 6 could be written as a sum of their cardinalities. But 6 cannot be written as the sum of the numbers 1, 3, 4, and 4. Therefore, although 6 divides $12 = |A_4|$, A_4 contains no *normal* subgroup of order 6. But *any* subgroup of order 6 would have index 2 in A_4 and, hence, be normal in A_4 by Theorem 7.5. Thus, A_4 contains no subgroup of order 6.

So far we have concentrated on conjugate classes of **elements** of G. It is not too difficult to extend our methods to investigate conjugate classes of **subgroups** of G.

Definition 9.4

Let H_1 and H_2 be subgroups of G. H_1 is said to be **conjugate to H_2 in G**, denoted $H_1 \approx H_2$, if there exists some $x \in G$ such that

$$H_2 = xH_1x^{-1} = \{xhx^{-1} | h \in H_1\}$$

In this definition, H_1 and H_2 are termed **conjugate subgroups**.

Example 9.2

Let $H_1 = \{(1), (1 \quad 2 \quad 3), (1 \quad 3 \quad 2)\}$ and $H_2 = \{(1), (1 \quad 2 \quad 4), (1 \quad 4 \quad 2)\}$ be subgroups of S_4. Then $(3 \quad 4)^{-1} = (3 \quad 4)$ and

$(3 \quad 4)(1)(3 \quad 4) = (1), \qquad (3 \quad 4)(1 \quad 2 \quad 3)(3 \quad 4) = (1 \quad 2 \quad 4), \qquad$ and

$(3 \quad 4)(1 \quad 3 \quad 2)(3 \quad 4) = (1 \quad 4 \quad 2)$

Thus H_1 and H_2 are conjugate in S_4.

The proof of the following result is similar to that of Theorem 9.1 and is left to you as Exercise 13.

Theorem 9.5

Subgroup conjugation $\approx$ is an equivalence relation on the set of subgroups of G.

The equivalence classes of subgroups of G under $\approx$ are also known as **conjugate classes**. It is a straightforward exercise to show that conjugate classes of subgroups of A_4 are

$\{(1)\}, \qquad \{\langle(1 \quad 2 \quad 3)\rangle, \langle(1 \quad 2 \quad 4)\rangle, \langle(1 \quad 3 \quad 4)\rangle, \langle(2 \quad 3 \quad 4)\rangle\},$

$\{\{(1), (1 \quad 2)(3 \quad 4), (1 \quad 3)(2 \quad 4), (1 \quad 4)(2 \quad 3)\}\}, \qquad$ and $\qquad A_4$, itself

As in the case of element conjugation, the number of subgroups in a class is a divisor of the group order. The proof of this result involves a certain type subset of G first encountered in Exercise 19 of Section 4.3.

Definition 9.5

If $H \leqslant G$, then the **normalizer of H in G**, denoted $N_G(H)$, is defined by

$$N_G(H) = \{x \in G | xH = Hx\}$$

Example 9.3

In Exercise 15 you are asked to verify that

$$N_{S_4}(\langle (1\ \ 2\ \ 3) \rangle) = \{(1), (1\ \ 2), (1\ \ 3), (2\ \ 3), (1\ \ 2\ \ 3), (1\ \ 3\ \ 2)\}$$

This subgroup is S_3.

From the definition of normalizer, one can easily prove that $N_G(H)$ is a subgroup of G (see Exercise 19 of Section 4.3) and that every subgroup of G is contained in its normalizer. Furthermore, H is normal in G if and only if $N_G(H) = G$ since a subgroup of G is normal in G if $xH = Hx$ for all $x \in G$. It is possible to extend this remark to show that $N_G(H)$ is the largest subgroup of G in which H is normal. The proofs of the preceding statements are left to you as Exercise 14.

With the definition of normalizer, we can now prove that for a group G the number of subgroups conjugate to a given subgroup of G divides $|G|$. This result is similar in statement and proof to Theorem 9.2. (See Exercise 16.)

Theorem 9.6

Let H be a subgroup of G. The number of conjugates of H in G is equal to the index in G of the normalizer of H; that is, the number of conjugates of H in G equals $[G:N_G(H)]$. Hence, the number of conjugates of H in G is a divisor of $|G|$.

The preceding result will be very useful in the next two sections, wherein we analyze a special type of group called a "p-group" and determine information about its existence as a subgroup.

Exercises 9.1

1. Determine the conjugate classes of D, the group in Example 8.6.
2. Verify that the conjugate classes of A_4 are:

    $\{(1)\}$, $\{(1\ \ 2\ \ 3), (1\ \ 4\ \ 2), (1\ \ 3\ \ 4), (2\ \ 4\ \ 3)\}$,

    $\{(1\ \ 3\ \ 2), (1\ \ 2\ \ 4), (2\ \ 3\ \ 4)\}$,   and

    $\{(1\ \ 2)(3\ \ 4), (1\ \ 3)(2\ \ 4), (1\ \ 4)(2\ \ 3)\}$

3. Determine the five conjugate classes of S_4.
4. Are H_1 and H_2 of Example 9.2 conjugate in A_4?
5. List the twenty-four 5-cycles of A_5. Show that they fall into two distinct conjugate classes of twelve elements each. Conclude that it is not necessarily true that group elements of the same order are conjugate.
6. Is $(1\ \ 2\ \ 3)$ conjugate to $(1\ \ 3\ \ 2)$ in A_4? Are these elements conjugate in S_4?
7. If $a \in G$, prove that $C_G(a)$ is a subgroup of G.
8. Suppose $a \in H \leqslant G$ and $C_H(a) = \{h \in H \mid ah = ha\}$.
    a. Prove $C_H(a) \leqslant C_G(a)$.
    b. If $C_H(a) = C_G(a)$, prove that the number of conjugates of a in H is a divisor of the number of conjugates of a in G.

9. Compute the following:

a. $C_{A_4}((1 \quad 2)(3 \quad 4))$
b. $C_{S_4}((1 \quad 2)(3 \quad 4))$
c. $C_{A_5}((1 \quad 2 \quad 3))$
d. $C_{A_4}((1 \quad 2 \quad 3))$
e. $C_{A_4}((1 \quad 3 \quad 2))$
f. $C_{A_5}((1 \quad 2)(3 \quad 4))$
g. $C_{A_5}((1 \quad 2 \quad 3 \quad 4 \quad 5))$

10. In Exercise 14 of Section 7.3, you were asked to prove that $Z(G) \trianglelefteq G$.
    a. Verify that $G = Z(G)$ if and only if G is abelian.
    b. Show that $Z(G) = \bigcap_{a \in G} C_G(a)$ where $C_G(a)$ is defined as in Exercise 8.

11. Verify Corollary 9.2 for the group D of Exercise 1.

12. Determine the conjugacy classes of:
    a. S_3 b. A_4 c. S_4

13. Prove that subgroup conjugation is an equivalence relation on the set of subgroups of a group G.

14. Let $H \leqslant G$. Prove that $N_G(H)$ is a subgroup of G containing H, and that it is the largest subgroup of G in which H is normal.

15. Verify that $N_{S_4}(\langle (1 \quad 2 \quad 3) \rangle) = \{(1), (1 \quad 2), (1 \quad 3), (2 \quad 3), (1 \quad 2 \quad 3), (1 \quad 3 \quad 2)\}$.
    Compute the following:

a. $N_{S_3}(\langle (1 \quad 2 \quad 3) \rangle)$
b. $N_{A_4}(\langle (1 \quad 2 \quad 3) \rangle)$
c. $N_{A_4}(\langle (1 \quad 2)(3 \quad 4) \rangle)$
d. $N_{S_4}(\langle (1 \quad 2)(3 \quad 4) \rangle)$
e. $N_{S_4}((1))$
f. $N_{S_4}(A_4)$

16. Prove that if $H \leqslant G$, then the number of subgroups of G conjugate to H in G is equal to $[G : N_G(H)]$.

17. Suppose $H \leqslant N \trianglelefteq G$. Show that $\bigcup_{x \in G} xHx^{-1} \subseteq N$.

18. If $H \leqslant G$, then the **centralizer of H in G**, denoted $C_G(H)$, is defined by

$$C_G(H) = \{x \in G \mid xh = hx \text{ for all } h \in H\}$$

a. Prove that $C_G(H)$ is a subgroup of G.
b. Show $C_G(H) \leqslant N_G(H)$.
c. Prove that $C_G(H)$ is a normal subgroup of $N_G(H)$.
d. Compute $C_{S_3}(A_3)$, $C_{S_3}(\langle (1 \quad 2) \rangle)$, and $C_{S_4}(S_3)$.
e. Is it true that $H \leqslant G$ implies $H \leqslant C_G(H)$?

19. This is a continuation of Exercise 18. Let $H \leqslant G$ and $n \in N_G(H)$. Define a mapping

$$\Psi_n: H \to H \qquad \text{by} \quad \Psi_n(h) = nhn^{-1} \quad \text{for} \quad h \in H$$

a. Prove that Ψ_n is an automorphism of H.
b. Show that $\Psi_n = 1_H$, the identity mapping, if and only if $n \in C_G(H)$.

20. This is a continuation of Exercise 19. Define $\phi: N_G(H) \to \text{Aut}(H)$, the group of all automorphisms of H, by $\phi(n) = \Psi_n$. Show that ϕ is a group homomorphism with kernel $C_G(H)$. Conclude that $N_G(H)/C_G(H)$ is isomorphic to a subgroup of $\text{Aut}(H)$.

21. Verify that the converse to Lagrange's Theorem holds for S_4.

*22. Write a computer program to input an element of S_n for $n \leqslant 9$, and output all conjugates of that element.

*23. Extend the program in Exercise 22 to determine all conjugate classes in S_n for $n \leqslant 9$.

9.2

p-Groups

Certain types of groups have exceedingly interesting properties. After the abelian groups, the p-groups are probably the most accessible classes of groups. Not only are their properties easily investigated by the methods we have developed, but they also provide a wealth of examples. In Section 9.3 we will consider p-groups as they relate to the problem of determining the subgroups of an arbitrary group. We will devote this section to investigating p-groups themselves.

First we state the basic definition of this section:

Definition 9.6

Let p be a prime integer. A p-**group** P is a group with the property that each $x \in P$ has order equal to some integral power of p. (The identity element 1 has order $p^0 = 1$.)

Example 9.4

A_3, the alternating group on three letters, consists of the elements (1), (1   2   3), and (1   3   2). Each of these elements has order 1 or 3. Thus A_3 is a 3-group.

Example 9.5

The group $K = \{(1), (1 \quad 2 \quad 3 \quad 4), (1 \quad 3)(2 \quad 4), (1 \quad 4 \quad 3 \quad 2)\}$ is a 2-group, since every element has order 1, 2, or 4.

Definition 9.6 for p-groups does not lend itself to easy application. It does not seem reasonable to require one to check the order of each and every element of a given group G to determine if the group is a p-group. In order to give an alternate definition, however, we need a classical result of Cauchy. We first prove the result in the case that G is an abelian group.

Lemma 9.1

Let G be an abelian group and p a prime divisor of $|G|$. Then G contains an element of order p.

Proof

We must find $a \in G$, $a \neq 1$, such that $a^p = 1$. If $G = \{1\}$, then the result is true trivially. Let us proceed by induction, assuming that the result is true for all abelian groups of order less than $|G|$.

Let $x \in G - \{1\}$ and m be the order of x. If $m = pk$, then x^k is an element of order p. Without loss of generality, we can assume p does not divide m. Let $N = \langle x \rangle$. Since G is abelian, $N \trianglelefteq G$ and G/N is an abelian group of order n/m.

Inasmuch as p divides (n/m) and $(n/m) < n$, we can apply our induction hypothesis to G/N and conclude that there exists some $yN \in G/N$ such that $(yN)^p = N$ and $yN \neq N$. But $(yN)^p = y^p N$. Thus $y^p \in N$ so $(y^m)^p = (y^p)^m = 1$. If $y^m = 1$, then $(yN)^m = y^m N = N$ and p divides m by Exercise 13 of Section 4.3, contradicting our assumption that p does not divide m. Therefore $y^m \neq 1$ and $a = y^m$ is an element of order p.   □

Now that we have proved this result in the case where G is abelian, we are in a better position to prove the same result for an arbitrary group G. This "divide and conquer" technique, coupled with mathematical induction, is at the heart of a great many proofs in finite group theory.

| Theorem 9.7 |

Cauchy's Theorem

If p is a prime divisor of $|G|$, then G contains an element of order p.

Proof

We again use induction on $|G|$. Since the theorem is true if $|G| = 1$, we can proceed and assume the result true for all groups of order less than $|G|$.

If some proper subgroup of $|G|$ has order divisible by p, then that subgroup, and hence G, contains an element of order p. We thus assume that *no proper* subgroup of G has order divisible by p.

By the class equation (Corollary 9.2), we know that

$$|G| = |Z(G)| + \sum_{a \in J'} [G : C_G(a)]$$

If $a \in G - Z(G)$, then Theorem 9.3 assures us that $C_G(a) < G$. Thus the condition that p does not divide $|C_G(a)|$ and Lagrange's Theorem imply that p divides $[G : C_G(a)]$ for all $a \in J'$.

Since p divides $|G|$, the class equation forces $|Z(G)|$ to be divisible by p, also. But then $Z(G)$ is an abelian group whose order is a multiple of p. By Lemma 9.1, $Z(G)$, and hence G, contains an element of order p.   □

Often in mathematics some proofs are more aesthetically satisfying than others. The proof of Cauchy's Theorem is one of these. Its beauty lies in the subtlety in reducing the problem to one solved by an earlier result. Contrast this procedure with some of our previous endeavours utilizing more brute force than style. By reviewing proofs of this type, you can gain some appreciation for the beauty of algebra and possibly experience the same satisfaction as one gains from assembling a complicated device or restoring Rubik's cube for the first time.

One of the first consequences of Cauchy's Theorem is a characterization of all simple abelian groups. Recall that a group is simple if it contains no proper nontrivial normal subgroups.

Corollary 9.3

Let G be a nontrivial abelian simple group. Then G is cyclic of prime order.

Proof

Let p be a prime divisor of $|G|$. By Cauchy's Theorem there exists some $x \in G$ of order p. If $T = \langle x \rangle$, then $T \trianglelefteq G$ since G is abelian. But the fact that G is simple then forces $G = T$ and therefore $|G| = |T| = p$. $\qquad\square$

Cauchy's Theorem allows us to give an alternate characterization of *p*-groups that concerns itself with the number of elements in the group rather than with the order of its elements.

Corollary 9.4

P is a *p*-group if and only if $|P|$ power of p.

Proof

If $|P|$ is a power of p, then Corollary 7.2 assures us that every element of P has order that is a power of p.

Conversely, suppose P is a *p*-group with order divisible by a prime q where $q \neq p$. Then Theorem 9.7 asserts that P contains an element whose order is q, not a power of p. This contradicts the assumption that P is a *p*-group. $\qquad\square$

Recall Examples 9.4 and 9.5 wherein the order of the 3-group A_3 was 3 and the order of the 2-group K was $4 = 2^2$. On the other hand, since $|S_3| = 6$, S_3 is not a *p*-group for any prime p.

The proof of Cauchy's Theorem depended heavily on the use of the class equation. The next result does also.

Theorem 9.8

If P is a nontrivial *p*-group, then P contains a nontrivial center; that is, $|Z(P)| > 1$.

Proof

Consider the class equation:

$$|P| = |Z(P)| + \sum_{a \in J'} [P : C_P(a)]$$

If $a \in P$ and $[P : C_P(a)] = 1$, then $C_P(a) = P$ and $a \in Z(P)$. Thus, for $a \in J'$, $[P : C_P(a)] > 1$. Since $|P|$ is a power of p, we have p divides $[P : C_P(a)]$ for all $a \in J'$. Therefore, in the class equation, p divides each index in the summation on the right. This, together with the fact that p divides $|P|$, implies that p must also divide $|Z(P)|$. Thus $|Z(P)| > 1$. $\qquad\square$

In Chapter 6, we discovered through the use of Cayley tables that any group of order 4 is abelian. Theorem 9.8 can be used to extend this result to groups of order p^2 for any prime p.

Corollary 9.5

Let p be a prime and P a group of order p^2. Then P is abelian.

Proof

By Theorem 9.8, $|Z(P)| > 1$. Therefore $|Z(P)| = p$ or p^2. If $|Z(P)| = p^2$, then $Z(P) = P$ and P is abelian. We can therefore assume that $|Z(P)| = p$. If $x \in P - Z(P)$, then $C_P(x)$ is a subgroup of P that properly contains $Z(P)$. Thus $|C_P(x)|$ must equal p^2 and $P = C_P(x)$. But then $x \in Z(P)$, contrary to the choice of x. We conclude that the case $|Z(P)| = p$ cannot occur, and P must be abelian.   □

If P is a group of order p^n, where p is a prime, we know by Cauchy's Theorem that P has a subgroup of order p. The structure of a p-group is such, however, that it is possible to show that P has a subgroup of order p^k for **every** k such that $0 \leqslant k < n$.

Corollary 9.6

Let P be a p-group of order p^n with $n \geqslant 1$. If $0 \leqslant k < n$, then P contains a subgroup of order p^k.

Proof

If $k = 0$ or $k = 1$, or if $|P| = p$, there is little to prove. Suppose that $k > 1$, and let us proceed by induction on $|P|$, assuming the result true for all p-groups of order less than $|P|$. Since $|Z(P)| > 1$ by Theorem 9.8, these exists an element $x \in Z(P)$ of order p. Let $N = \langle x \rangle$; then $|N| = p$ and $x \in Z(P)$ implies that $N \trianglelefteq P$. Since $|P/N| = p^{n-1}$, our induction hypothesis and Theorem 8.6 imply that P/N contains a subgroup H/N of order p^{k-1}. Then H is a subgroup of P of order

$$|H| = |H/N| \cdot |N| = p^{k-1}p = p^k$$

The result now follows immediately.   □

In this section, we have but sampled the many interesting results that have been obtained on the structure of finite p-groups. Notice how many of the proofs utilized mathematical induction and Lagrange's Theorem. In addition to the abelian groups and the p-groups, there are numerous classes of groups in group theory. We will encounter some of them in Chapter 11. The p-groups are contained in almost all of the important classes of groups, and the proofs that demonstrate this fact usually involve the results of this section.

In the next section, these same propositions and techniques will be used to determine important information about the subgroup structure of an arbitrary finite group.

1. List all elements of order 2 in A_4. Can you find an element of order 6 in A_4, in S_4, or in S_5?

2. Let D be the group of Example 8.6. Find $Z(D)$.

3. Let p be a prime. Show that Corollary 9.4 is not true if we replace p^2 by p^3.

4. Compute $Z(S_3)$. Compare your answer with Theorem 9.8.

5. Let P be a p-group and $H \trianglelefteq P$. Prove that H and P/H are p-groups.

6. Let p be a prime and $H \trianglelefteq G$. If both H and G/H are p-groups, then prove that G is a p-group.

7. Determine $C_D(a)$ for all $a \in D$, the group of Example 8.6.

8. Is it true that if p is a prime dividing the order of a group G and p^2 divides $|G|$, then G contains an element of order p^2?

9. Let p and q be distinct primes dividing the order of the abelian group G.
   a. Prove that G contains an element of order pq.
   b. If $|G| = pq$, then show that G is cyclic.

10. Let G be a group of order p^2 where p is a prime. Show that any subgroup of G is normal in G.

11. If P is a p-group of order p^n, prove that any subgroup of P of order p^{n-1} is normal in P.

12. Let G be a group and $x \in Z(G)$. Prove that if $N = \langle x \rangle$, then $N \trianglelefteq G$.

13. Suppose P is a nonabelian group of order p^3 where p is a prime. Prove that $|Z(P)| = p$.

14. Let G be a group and $N \trianglelefteq G$.
    a. Prove $Z(G)N \trianglelefteq G$ and $Z(G)N/N \trianglelefteq G/N$.
    b. Show that $Z(G)N/N \leqslant Z(G/N)$.

15. Let G be a group, not necessarily of finite order, and H be a subgroup of G of finite index.
    a. Prove that $\text{Core}(H) = \bigcap_{a \in G} aHa^{-1}$ is a subgroup of G.
    b. Verify that G contains a normal subgroup of finite index.
    c. If G is also a p-group, prove that $[G:H] = p^n$ for some n.

16. Let G be a group and p any prime. Prove that G contains a unique largest normal p-subgroup and a unique smallest normal subgroup K such that G/K is a p-group.

17. The **class number** of a finite group G, denoted $k(G)$, is defined to be the number of conjugacy classes of G.
    a. If G is abelian, verify that $k(G) = |G|$.
    b. Prove that $k(G) \geqslant 1 + |Z(G)|$ if and only if G is nonabelian.
    c. Suppose G is a nontrivial group and p is the smallest prime divisor of $|G|$. If $k(G) > |G|/p$, then prove $|Z(G)| > 1$.
    d. Suppose G is a nonabelian group of order p^3, where p is a prime integer. Prove that $|Z(G)| = p$ and $k(G) = p^2 + p - 1$.

*18. If p is a prime, then we know that any group of order p or p^2 is abelian. Write a program to list the orders of all such groups G of order n for $n < 10{,}000$ when n is a prime or the square of a prime.

9.3

The Theorems of Sylow

Few results have had such a great impact on a particular mathematical discipline as have the Sylow Theorems on finite group theory. Since the appearance of the first theorem of Sylow in 1872, group theorists have concerned themselves with applying and generalizing consequences of these classical results. In this section, we will state and prove these theorems and try to indicate some of the many ways these results can be applied. Throughout this section, p will denote a prime integer.

Lagrange's Theorem asserts that the order of any subgroup of a finite group G divides the group order. If G is a p-group, then Corollary 9.6 states that the converse of Lagrange's Theorem holds for G; that is, given any integral divisor of $|G|$, there exists a subgroup of G of that order. However, if G is not a p-group, the converse of Lagrange's Theorem does not necessarily hold. For example, in Section 9.1 we showed that A_4, the alternating group on four letters, possesses no subgroup of order 6. Therefore, the question arises as to just exactly what divisors of the group order correspond to orders of subgroups of that group. It is to this question that the First Theorem of Sylow responds.

Recall that Cauchy's Theorem asserts that if the order of a group G is divisible by a prime p, then G possesses an element of order p. The cyclic group generated by that element is then a subgroup of G of order p. The First Theorem of Sylow extends this result.

Theorem 9.9

First Theorem of Sylow

Let G be a group of order $p^n m$ where $(p, m) = 1$. Then G possesses a subgroup of order p^n.

Proof

Let us induct on the order of G, the theorem obviously true in the case that $|G| = 1$. Thus, we can assume that the result holds in all groups of order less than $|G|$.

First suppose that p divides $|Z(G)|$. Then by Cauchy's Theorem, there exists an $x \in Z(G)$ such that $x^p = 1$. Let $N = \langle x \rangle$. Since $x \in Z(G)$, we have $N \trianglelefteq G$. (See Exercise 12 of Section 9.2.) Consider $G' = G/N$. G' is a group of order $p^{n-1}m$. By our induction hypothesis, G' possesses a subgroup P' of order p^{n-1}. By Corollary 8.3, $P' = P/N$ where P is a subgroup of G. But then

$$|P| = |P/N| \cdot |N| = |P'| \cdot |N| = p^{n-1} \cdot p = p^n$$

Thus the results holds if p divides $|Z(G)|$. Now suppose that p does not divide $|Z(G)|$. Consider the class equation for G.

$$|G| = |Z(G)| + \sum_{a \in J'} [G : C_G(a)]$$

where J' is a subset of G consisting of one element from each conjugate class of G that contains more than one element. If every term in the summand on the right side of the equation is divisible by p, then, since p divides $|G|$, we could conclude that p divides $|Z(G)|$, contrary to our assumption. Thus there must exist some $a \in G - Z(G)$ such that p does not divide $[G : C_G(a)]$. By Lagrange's Theorem, this forces $|C_G(a)|$ to be divisible by p^n. Now $a \notin Z(G)$ implies that $C_G(a)$ is a proper subgroup of G. We can therefore apply our induction hypothesis to $C_G(a)$ and conclude that $C_G(a)$, and hence G, contains a subgroup of order p^n.   □

Let us apply this result to S_4. The First Theorem of Sylow asserts that S_4 has subgroups of order 3 and $8 = 2^3$. Clearly $H = \langle (1 \quad 2 \quad 3) \rangle$ is a subgroup of S_4 of order 3, while the group D of Example 8.6 is a subgroup of S_4 of order 8.

In the last section, we studied the structure of p-groups and showed in Corollary 9.6 that if P is a p-group of order p^n, then P has a subgroup of order p^k, where $0 \leqslant k \leqslant n$. If G is a group of order $p^n m$ where $(p, m) = 1$, and P is a subgroup of G of order p^n, then every subgroup of P is also a subgroup of G. We can therefore combine the preceding result with Corollary 9.6 to obtain the following corollary.

Corollary 9.7

Let G be a group of order $p^n m$, where $(p, m) = 1$. Then G possesses a subgroup of order p^k where $0 \leqslant k \leqslant n$.

The subgroups of order p^n in Theorem 9.10 are given a special name.

Definition 9.7

Let G be a group of order $p^n m$ where $(p, m) = 1$. If P is a subgroup of G of order p^n, then P is termed a p-**Sylow subgroup of** G. The set of p-Sylow subgroups of G is denoted by $\text{Syl}_p(G)$.

Therefore, the First Theorem of Sylow can be restated to assert that every finite group G possesses a p-Sylow subgroup for each prime p dividing $|G|$.

The next corollary concerns the nature of the normalizer of a p-Sylow subgroup. Recall that if $H \leqslant G$, then the very definition of normalizer implies $H \trianglelefteq N_G(H)$.

Corollary 9.8

Let P be a p-Sylow subgroup of G. Then the identity is the only element in $N_G(P)/P$ whose order is a power of p.

Proof

Let $|G| = p^n m$, where $(p, m) = 1$. Then $|P| = p^n$ and Lagrange's Theorem implies that

$$m = [G:P] = [G:N_G(P)][N_G(P):P]$$

Therefore $(p, m) = 1$ implies that

$$p \text{ does not divide } [N_G(P):P] = |N_G(P)/P|$$

Since the order of any element in $N_G(P)/P$ divides $|N_G(P)/P|$, this shows that no nonidentity element of $N_G(P)/P$ has order divisible by p.   □

The following is a consequence of Corollary 9.7, and its proof is left as Exercise 6.

Corollary 9.9

If P is a p-Sylow subgroup of G, and P' is a p-group contained in $N_G(P)$, then $P' \leqslant P$.

Let P' be a p-**subgroup** of G; that is, a subgroup of G that is itself a p-group. The Second Theorem of Sylow states that if P' is not a p-Sylow subgroup of G, then there is at least one p-Sylow subgroup of G containing P'. In other words, the p-Sylow subgroups are not only the largest p-subgroups of G in terms of order, but also in terms of containment. In order to prove this result, it is necessary that we investigate further the conjugates of a p-Sylow subgroup.

Let P be a p-Sylow subgroup of G. For $g \in G$, define P^g by

$$P^g = gPg^{-1} = \{gxg^{-1} \mid x \in P\}$$

We leave it to you to show that P^g is another subgroup of G and that the mapping $\phi \colon P \to P^g$ defined by $\phi(x) = gxg^{-1}$ is a group isomorphism. (See Exercise 3.) Therefore $|P^g| = |P|$, and P^g is another p-Sylow subgroup of G. In the language of Definition 9.3, any conjugate of a p-Sylow subgroup of G is another p-Sylow subgroup of G. The conjugates of a p-Sylow subgroup of G play an important part in the proofs of the remaining theorems of Sylow. To utilize them, we need to develop some machinery.

Let G be a group of permutations on a set Ω. Define a relation on Ω in the following way: if $\alpha, \beta \in \Omega$, we write $\alpha \approx \beta$ if there exists some $g \in G$ such that $g(\alpha) = \beta$. We leave it to you to show that $\approx$ is an equivalence relation on Ω. (See Exercise 4.) An equivalence class of this relation is termed an **orbit**. The orbit of $\alpha \in \Omega$ consists of the elements $g(\alpha)$ for all $g \in G$. Let G_α denote the set of all $g \in G$ such that $g(\alpha) = \alpha$. Then G_α is a subgroup of G called **the stabilizer of** α.

Example 9.6

Let $G = \langle(1 \quad 2 \quad 3)(5 \quad 6)\rangle$. Then G is a group of permutations of

$$\Omega = \{1, 2, 3, 4, 5, 6\}$$

The orbits of Ω under G are $\{1, 2, 3\}$, $\{4\}$, and $\{5, 6\}$.

From this example, we see that our new definition of orbit is consistent with the one given in our discussion of permutation groups. If G is not cyclic, our new definition is an extension of our previous concept.

The stabilizer of $\alpha \in \Omega$ and the orbit of α are related in the following way.

Lemma 9.2

Let G be a finite group of permutations on a set Ω. Then the number of elements in an orbit of $\alpha \in \Omega$ is $[G : G_\alpha]$.

Proof

The orbit of α is the set $\{g(\alpha) | g \in G\}$. If $g_1(\alpha) = g_2(\alpha)$ for $g_1, g_2 \in G$, then $g_1^{-1}g_2 \in G_\alpha$ and $g_2 \in g_1 G_\alpha$. Thus $g_2 G_\alpha = g_1 G_\alpha$. If $g_1(\alpha) \neq g_2(\alpha)$, then $g_1^{-1}g_2 \notin G_\alpha$. Therefore $g_2 \notin g_1 G_\alpha$ and $g_2 G_\alpha \neq g_1 G_\alpha$. Thus the elements in the orbit of α and the cosets of G_α are in one-to-one correspondence.   □

Lemma 9.3

Let P be a p-group and $\Psi: P \to S_n$ be a homomorphism. Then $\Psi^{\rightarrow}(P)$ is a group of permutations of $\{1, 2, \ldots, n\}$ and the size of each orbit of $\Psi^{\rightarrow}(P)$ is a power of p (possibly $p^0 = 1$).

Proof

The result is an immediate consequence of the fact that $\Psi^{\rightarrow}(P)$ is a subgroup of S_n and, if $\alpha \in \{1, 2, \ldots, n\}$, then $[P : P_\alpha]$ is a divisor of $|P|$.   □

Example 9.7

$P = \langle(1 \quad 2 \quad 3)\rangle$ is a 3-Sylow subgroup of S_4. Its conjugates are

$$P_1 = P$$

$$P_2 = \langle(1 \quad 2 \quad 4)\rangle = (3 \quad 4)P_1(3 \quad 4)$$

$$P_3 = \langle(1 \quad 4 \quad 3)\rangle = (2 \quad 4)P_1(2 \quad 4)$$

$$P_4 = \langle(2 \quad 3 \quad 4)\rangle = (1 \quad 3)P_1(1 \quad 3)$$

We now proceed to develop a homomorphism Ψ from P to the symmetric group on $\Omega = \{P_1, P_2, P_3, P_4\}$. For $x \in P$, define $\phi_x: \Omega \to \Omega$ by $\phi_x(P_i) = P_i^x = xP_ix^{-1}$. If $\phi_x(P_i) = \phi_x(P_j)$, then $xP_ix^{-1} = xP_jx^{-1}$ and $P_i = P_j$. (See Exercise 7.) Thus ϕ_x is an injection on a finite set and, hence, a permutation of Ω. In permutation notation,

$$\phi_x = \begin{pmatrix} P_1 & P_2 & P_3 & P_4 \\ P_1^x & P_2^x & P_3^x & P_4^x \end{pmatrix}$$

If $\Psi\colon P \to S_\Omega$ is defined by $\Psi(x) = \phi_x$, then

$$\phi_{xy}(P_i) = (xy)P_i(xy)^{-1} = (xy)P_i(y^{-1}x^{-1}) = x(yP_iy^{-1})x^{-1}$$

$$= \phi_x(yP_iy^{-1}) = \phi_x(\phi_y(P_i)) = (\phi_x \circ \phi_y)(P_i)$$

implies that $\Psi(xy) = \Psi(x)\Psi(y)$, so that Ψ is a homomorphism from P to S_Ω, which is isomorphic to S_4.

Now two subgroups P_i, P_j of Ω are in the same orbit if there exists some $x \in P$ such that $P_j = xP_ix^{-1}$. It is a straightforward matter to show that the orbit of P_1 is $\{P_1\}$ and the orbit of P_2 is $\{P_2, P_3, P_4\}$. (See Exercise 8.) Thus the length of each orbit is a power of 3.

We wish to repeat the ideas of this example to analyze the p-Sylow subgroups of an arbitrary group.

Theorem 9.10

Second Theorem of Sylow
Every p-subgroup of G is contained in a p-Sylow subgroup of G.

Proof

Let P' be a p-group of G and P be a p-Sylow subgroup of G. Set Ω equal to the set of all conjugates of P in G, $\Omega = \{P = P_1, P_2, \ldots, P_n\}$. For $x \in G$ define $\phi_x\colon \Omega \to \Omega$ by $\phi_x(P_i) = xP_ix^{-1}$. If $\phi_x(P_i) = \phi_x(P_j)$, then $P_i = P_j$ (see Exercise 7), so that ϕ_x is an injection. Since Ω is a finite set, ϕ_x must also be surjective. Thus ϕ_x is a permutation on Ω.

We next show that the mapping $\Psi\colon G \to S_\Omega \cong S_n$ defined by $\Psi(x) = \phi_x$ is a homomorphism. For $P_i \in \Omega$ and $x, y \in G$, we have $\phi_{xy}(P_i) = (\phi_x \circ \phi_y)(P_i)$ by an argument identical to that in the preceding example. Thus $\phi_{xy} = \phi_x\phi_y$ and $\Psi(xy) = \Psi(x)\Psi(y)$.

Now let us restrict Ψ to P'. Then $\Psi_{P'}$ is still a homomorphism from P' to S_Ω. By Lemma 9.3, the size of each orbit of $\Psi^\rightarrow(P')$ is a power of p. By Lemma 9.2, the cardinality of the orbit that fixes an element of Ω, say P, is $[P' : P'_P]$ where P'_P is the set of all elements $x \in P'$ such that $\phi_x(P) = xPx^{-1} = P$.

Now suppose the cardinality of some orbit of $\Psi^\rightarrow(P')$ is 1; that is, suppose, for some $P_i \in \Omega$, we have $xP_ix^{-1} = P_i$ for all $x \in P'$. Then $P' \leqslant N_G(P_i)$. By Corollary 9.9 this can only occur if $P' \leqslant P_i$ and P' is contained in a p-Sylow subgroup of G.

To complete the proof, we need only show that some orbit of Ω under $\Psi^\rightarrow(P')$ has length 1. Let $k_1, k_2, \ldots, k_r$ denote the lengths of the distinct orbits of Ω under the action of $\Psi^\rightarrow(P')$. Then

$$|\Omega| = k_1 + k_2 + \cdots + k_r$$

If no orbit has length 1, then each k_i is a nontrivial power of p. Thus p divides $k_1 + k_2 + \cdots + k_r = |\Omega|$. But Theorem 9.6 implies that Ω, the number of conjugates of P, is equal to $[G:N_G(P)]$, which is relatively prime to p by combining Lagrange's Theorem with the fact that P is a p-Sylow subgroup of G. This contradiction forces the length of at least one orbit to be equal to 1.

$\square$

The preceding theorem is of great help in determining the subgroups of a group. For example, the theorem guarantees that every subgroup of order 2 or 4 in S_4 is contained in some subgroup of S_4 of order 8. Therefore to find a subgroup of order 8, we could find a subgroup H of order 4 and look for subgroups that contain H. Can we go the other way? Does a subgroup of S_4 of order 8 contain **every** type of subgroup of S_4 of order 2 or 4? The answer is yes, but we need to know that all p-Sylow subgroups are conjugate to one another to prove it.

If P is a p-Sylow subgroup of G, then we know that any conjugate of P in G is another p-Sylow subgroup of G. But do there exist other p-Sylow subgroups of G that are **not** conjugates of P in G? The Third Theorem of Sylow answers this question. In addition, it gives a strong clue to determining just how many p-Sylow subgroups G contains.

Theorem 9.11

Third Theorem of Sylow

If p is a prime divisor of the order of G, then all p-Sylow subgroups of G are conjugate. The number of p-Sylow subgroups of G divides $|G|$ and is congruent to 1 modulo p.

Proof

Let P be a p-Sylow subgroup of G, and again let $\Omega = \{P_1 = P, P_2, \ldots, P_n\}$ be the set of all conjugates of P in G. By Theorem 9.6, we know that $|\Omega| = [G:N_G(P)]$ is a divisor of $|G|$, so that it remains to be shown that every p-Sylow subgroup of G is in Ω and that $|\Omega| \equiv 1 \pmod{p}$.

Let $\Psi: G \to S_\Omega$ be the homomorphism developed in the proof of Theorem 9.10 and restrict Ψ to P. By Lemma 9.3, the length of each orbit under the action of $\Psi^{\to}(P)$ is a power of p. Since $xPx^{-1} = P$ for all $x \in P$, the length of the orbit containing P is 1. Now can the length of any other orbit be 1? If $xP_ix^{-1} = P_i$ for all $x \in P$, $i \neq 1$, then $P \leqslant N_G(P_i)$. By Corollary 9.9, $P \leqslant P_i$, contradicting the fact that $P \neq P_i$. Therefore, the answer is no; and Ω has exactly one orbit of length 1, and all orbits under the action of $\Psi^{\to}(P)$ have length, which is a nontrivial power of p. Thus

$$|\Omega| = 1 + \sum_{i=1}^{s} p^{a_i} \equiv 1 \text{ (modulo } p)$$

Now suppose that Q is a p-Sylow subgroup of G not in Ω. Restrict $\Psi: G \to S_\Omega$ to Q, and write

$$|\Omega| = t_1 + t_2 + \cdots + t_v$$

where the t_i's are the lengths of the distinct orbits of Ω under the action of $\Psi^{\rightarrow}(Q)$. By an argument similar to that used in the preceding paragraph, no t_i can equal 1, for then Q would be equal to $P_i \in \Omega$.

Therefore, Lemma 9.3 implies that each t_i, $i = 1, 2, \ldots, v$, is a nontrivial power of p, and $|\Omega|$ is congruent to 0 modulo p. But we have just shown that $|\Omega|$ is congruent to 1 modulo p. This contradiction implies that every p-Sylow subgroup of G must be in Ω. $\square$

We now know by this theorem that the p-Sylow subgroups of a group G form a single conjugacy class, which we denoted $\text{Syl}_p(G)$. The numerical implications of this result are often of more immediate application. Consider the next two examples.

Example 9.8

If $P = \langle (1 \quad 2 \quad 3) \rangle$, then P is a 3-Sylow subgroup of S_4. By Theorem 9.11, the number k of 3-Sylow subgroups of S_4 must divide $24 = |S_4|$ and be congruent to 1 modulo 3. The only possibilities are $k = 1$ or $k = 4$. Since $P_2 = \langle (1 \quad 2 \quad 4) \rangle$ is another 3-Sylow subgroup of S_4, $k \neq 1$. Therefore, $k = 4$. These 3-Sylow subgroups appear in Example 9.7. Moreover, since $4 = [S_4 : N_{S_4}(P)]$, we can conclude that $|N_{S_4}(P)| = 6$.

How many 2-Sylow subgroups does S_4 possess? By Theorem 9.11, we know that S_4 has one or three 2-Sylow subgroups. If the number were 1, then the 2-Sylow subgroup of S_4 would be normal in S_4. The subgroup D of S_4 in Example 8.6 has order 8, however, and is not normal in S_4. (See Exercise 2.) Thus S_4 has three 2-Sylow subgroups.

Example 9.9

Let G be a group of order 15. The number of 3-Sylow subgroups of G must divide 15 and be congruent to 1 modulo 3. The only possibility is 1, so that G has a normal 3-Sylow subgroup C_3. Similarly, G must possess exactly one 5-Sylow subgroup C_5, which must also be normal. But then $C_3 C_5$ is a subgroup of G of order 15, forcing $G = C_3 C_5$. By Exercise 9 of Section 9.2, G is a cyclic group of order 15.

Example 9.10

Suppose G is a group of order 28. The number of 7-Sylow subgroups of G must divide 28 and be congruent to 1 modulo 7. But 1 is the only possibility since 8 does not divide 28 and 15 does not divide 28. Therefore G must have a normal 7-Sylow subgroup.

Even when the Sylow Theorems do not allow us to determine the complete structure of a group, they often provide us with enough information to make important conclusions about the nature of the group. For example, in the case $|G| = 28$ in Example 9.10, we were able to conclude that G had a normal 7-Sylow subgroup. This is especially important if we are trying to find groups containing no normal subgroups other than the identity and the group itself. We will have more to say about this in Chapter 11. In the next chapter, however, we utilize the Sylow Theorems to aid our analysis of finite abelian groups.

Exercises 9.3

1. Prove that G has a normal p-Sylow subgroup P if and only if P is the only p-Sylow subgroup of G.

2. List the three 2-Sylow subgroups of S_4.

3. If P is a p-Sylow subgroup of G and $g \in G$, let $P^g = gPg^{-1} = \{gxg^{-1} | x \in P\}$. Prove that:
   a. P^g is a subgroup of G.
   b. If $\phi: P \to P^g$ is defined by $\phi(x) = gxg^{-1}$, then ϕ is an isomorphism.
   c. P^g is p-Sylow subgroup of G.

4. Let G be a group of permutations on a set Ω. If α, $\beta \in \Omega$, define $\alpha \approx \beta$ if there exists some $g \in G$ such that $g(\alpha) = \beta$. Prove that $\approx$ is an equivalence relation on Ω. If $G_\alpha = \{g \in G | g(\alpha) \approx \alpha\}$, verify $G_\alpha \leqslant G$.

5. Determine the orbits of $\Omega = \{1, 2, 3, 4\}$ under the action of the group

$$G = \{(1), (1 \quad 2), (3 \quad 4), (1 \quad 2)(3 \quad 4)\}$$

6. Let P be a p-Sylow subgroup of G and P' be a subgroup of G contained in $N_G(P)$. Prove that P' is a subgroup of P.

7. Let P_1 and P_2 be subgroups of G and $x \in G$. If $xP_1x^{-1} = xP_2x^{-1}$, then show that $P_1 = P_2$.

8. In Example 9.7, prove that the orbit of P is $\{P\}$ and the orbit of P_2 is $\{P_2, P_3, P_4\}$.

9. In Theorem 9.10, what is the stabilizer of P?

10. Suppose H is a subgroup of A_4 of order 6.
    a. Prove that $H \trianglelefteq A_4$.
    b. Verify that the number of 3-Sylow subgroups of A_4 is 4.
    c. Let P be a 3-Sylow subgroup of H. Then prove that P is also a 3-Sylow subgroup of A_4.
    d. Prove that A_4 has a normal 2-Sylow subgroup.
    e. Show that $N_{A_4}(P) = P$.
    f. Show that $P \trianglelefteq H$, contrary to the fact that $N_{A_4}(P) = P$ implies that no element of $H - P$ normalizes P.
    g. Conclude that A_4 contains no subgroup of order 6.

11. If $N \trianglelefteq G$ and P is a p-Sylow subgroup of G, then prove that $N \cap P$ is a p-Sylow subgroup of N.

12. If $N \trianglelefteq G$ and P is a p-Sylow subgroup of G, then verify that PN/N is a p-Sylow subgroup of G/N.

13. Show that any group of order 35 must be cyclic.

14. Determine all possible subgroup structures of a group of order 21.

15. If P is a p-Sylow subgroup of G and $N_G(P) \leqslant H \leqslant G$, verify that

$$[H : H_G(P)] \equiv 1 \text{ (modulo } p)$$

16. Prove that $N_G(P^g) = (N_G(P))^g$.

17. Let P be a p-Sylow subgroup of G. If $N_G(P) \leqslant H \leqslant G$, then show that $N_G(H) = H$. Conclude that $N_G(N_G(P)) = N_G(P)$.

18. Let p be a prime dividing the order of a group G and $N \trianglelefteq G$. Prove that the number of p-Sylow subgroups of N and G/N are less than or equal to the number of p-Sylow subgroups of G. Find examples of groups in which one or the other equality holds.

19. Suppose P is a p-Sylow subgroup of a group G and H is a subgroup of G such that $N_G(P) \leqslant H$. Verify that $[G : H] \equiv 1$ (modulo p).

20. If G is a p-group and N is a nontrivial normal subgroup of G, prove that $N \cap Z(G) \neq \{1\}$.

21. Suppose G is a group of order pq, where p and q are distinct primes such that q is not congruent to 1 modulo p. Prove that G contains a normal p-Sylow subgroup.

22. Suppose G is a finite p-group.
    a. If G is cyclic, prove that G contains exactly one subgroup of order p^k for all p^k that divide $|G|$.
    b. If G contains exactly one subgroup for all p^k dividing $|G|$, must G be cyclic?

23. Let G be a group of permutations of a finite set Ω. The element $a \in \Omega$ is **invariant** under the permutation $\sigma \in G$ if $\sigma(a) = a$. Set

$$\text{Inv}(\sigma) = |\{x \in \Omega | \sigma(a) = a\}|$$

    a. Define an operation $\approx$ on Ω by $a \approx b$ if there exists $\sigma \in G$ such that $\sigma(a) = b$. Prove that $\approx$ is an equivalence relation on Ω, called the **equivalence relation on Ω induced by** G.
    b. Prove Burnside's Theorem:[2] The number of equivalence classes in the equivalence σ is given by

$$\frac{1}{|G|} \sum_{\sigma \in G} \text{Inv}(\sigma)$$

24. This is a continuation of Exercise 23. Let Ω be a set. **A coloring of** Ω is a function $f : \Omega \to R$ where R is a set of colors. Let $C(\Omega, R)$ be the set of all colorings of Ω using the set of colors R. If G is a group of permutations of Ω and $\sigma \in G$, define a mapping σ^* on $C(\Omega, R)$ by $[\sigma^*(f)](a) = f(\sigma(a))$ for $a \in \Omega$.
    a. If $G^* = \{\sigma^* | \sigma \in G\}$, verify that G^* is a group of permutations of $C(\Omega, R)$ and that $G^* \cong G$.
    b. Prove that if $G = \{\sigma_1, \sigma_2, \ldots, \sigma_n\}$ is a group of permutations on Ω, then the number of distinct colorings in $C(\Omega, R)$ where $|R| = m$ is given by

$$\frac{1}{|G|} [m^{\text{cyc}(\sigma_1)} + m^{\text{cyc}(\sigma_2)} + \cdots + m^{\text{cyc}(\sigma_n)}]$$

    where $\text{cyc}(\sigma_i)$ is the number of cycles in the cyclic decomposition of σ_i.

---

[2] William Burnside (1852–1927) made numerous contributions to group theory. One of his major results states that any group of order $p^\alpha q^\beta$, where p and q are primes, is a solvable group. His 1902 question as to whether a finitely-generated group all of whose elements have finite order must necessarily be a finite group has still not been answered completely.

25. This is a continuation of Exercise 24. Let

$$\Omega = \{a_1, a_2, \ldots, a_n\} \quad \text{and} \quad R = \{r_1, r_2, \ldots, r_m\}$$

A function $w: R \to \mathbb{Z}^+$ assigns a **weight** $w(r_i)$ to each $r_i \in R$. The **weight of a coloring** is the product of the weights assigned to the elements of Ω; that is, if $f \in C(\Omega, R)$, then

$$W(f) = \prod_{i=1}^{n} w(f(a_i))$$

If $K \subseteq C(\Omega, R)$, the **inventory** of K is given by

$$\text{inv } K = \sum_{f \in K} W(f)$$

a. Two colorings $f, g \in C(\Omega, R)$ are **equivalent** if there exists $\sigma \in G$ such that $g = f \circ \sigma$. Verify that equivalent colorings have the same weight.
b. Suppose $\sigma \in G$ has $\ell_k(\sigma)$ cycles of length k in the representation of σ as a product of disjoint cycles. Prove $n = 1 \cdot \ell_1(\sigma) + 2 \cdot \ell_2(\sigma) + \cdots + n \cdot \ell_n(\sigma)$.
c. If $x_1, x_2, \ldots, x_n$ denote variables and $\sigma \in G$, define the **cycle index** of σ to be the expression

$$x^{\ell_1(\sigma)} x^{\ell_2(\sigma)} \cdots x^{\ell_n(\sigma)}$$

The **cycle index** of G is the polynomial

$$P_G(x_1, x_2, \ldots, x_n) = \frac{1}{|G|} \sum_{\sigma \in G} x_1^{\ell_1(\sigma)} x_2^{\ell_2(\sigma)} \cdots x_n^{\ell_n(\sigma)}$$

Prove Polya's[3] Theorem:

$$\text{inv}(G) = P_G \left(\sum_{i=1}^{m} r_i, \sum_{i=1}^{m} r_i^2, \ldots, \sum_{i=1}^{m} r_i^n \right)$$

*26. Let G be a group. Write a computer program that takes a value of $n = |G|$ for $n \leqslant 1000$ and:
a. Prints out the orders of the various Sylow subgroups of G.
b. Checks the possible values of the number of Sylow subgroups of G to determine if G has a normal Sylow subgroup.

---

[3] George Polya (1887–1985) was a Hungarian-born mathematician whose mathematical activity included work in complex variables, number theory, probability, applied mathematics, and combinatorics. His *Collected Papers* (edited by R. Boas, MIT Press, Cambridge, Mass., 1972, 1975) fills two volumes, and he authored a number of popular books on mathematics and problem solving.

10

Group Products and

Abelian Groups

Up to this point, we have concerned ourselves with investigating certain group classifications and analyzing the internal subgroup structure of arbitrary groups. We have considered cyclic groups, abelian groups, groups of symmetries, permutation groups, and others. The theorems of Lagrange and Sylow have allowed us to deduce information about the subgroups of groups of certain orders. In this chapter we first want to construct new groups from those we already know. We then wish to turn around and show how a similar analysis can be used to decompose a given group into its parts. By this we mean to show how a group can be viewed as a product of certain of its subgroups. In the next chapter, we will see that this decomposition is vital to an understanding of important classes of groups whose internal structure is related to the problem of solving polynomial equations.

Also in this chapter our work in decomposing a group into its parts will be used to determine the nature or structure of all finite abelian groups. It will be shown that a finite abelian group has unique subgroups of prime power order that fit together in a fairly straightforward manner to make up the group. This will allow us to state with assurance whether or not two finite abelian groups are isomorphic. Later in this text, the results of this chapter concerning finite cyclic groups will be used to investigate how the speed of computer addition can be increased.

10.1

External and Internal Products

If $(H, \cdot)$ and $(K, \circ)$ are groups, then the Cartesian product $H \times K$ can be made into a group if we define $*$ on $H \times K$ by

$$(h_1, k_1) * (h_2, k_2) = (h_1 \cdot h_2, k_1 \circ k_2)$$

The identity of $G = H \times K$ is $(1_H, 1_K)$, while the inverse of an element (h, k) is given by $(h, k)^{-1} = (h^{-1}, k^{-1})$. The group $G = H \times K$ is called the **external direct product of H and K**. The elements of the group $G = H \times K$ are ordered pairs of the form (h, k). For this reason, neither H nor K is actually a subgroup of G. Instead $H \times \{1_K\}$ and $\{1_H\} \times K$ are subgroups of G isomorphic to H and K, respectively. In fact, these two subgroups are normal in G, and they have only the identity $(1_H, 1_K)$ in common. (We leave it to you to check the details of these assertions.) Nevertheless, it is common practice to identify H and K with their isomorphic counterparts and refer to them as normal subgroups of $G = H \times K$. Henceforth, we will do the same.

| Example 10.1 | Let $H = K = (\mathbb{R}, +)$. Then $\mathbb{R} \times \mathbb{R}$ is an abelian group of ordered pairs of real numbers. Its elements constitute the points in the Cartesian plane. |

| Example 10.2 | Let $H = K = (\mathbb{Z}, +)$. $\mathbb{Z} \times \mathbb{Z}$ is another abelian group; in fact, it is a subgroup of $\mathbb{R} \times \mathbb{R}$. The elements of $\mathbb{Z} \times \mathbb{Z}$ are known as **lattice points in the plane**. |

| Example 10.3 | Let $H = \mathbb{Z}_2$, $K = \mathbb{Z}_3$. Then $\mathbb{Z}_2 \times \mathbb{Z}_3$ is a group of order 6. It is easy to show that $(1, 1)$ is a generator of the group. Hence $\mathbb{Z}_2 \times \mathbb{Z}_3$ is a cyclic group of order 6. But we have seen in Theorem 6.4 that any cyclic group of order 6 is isomorphic to $\mathbb{Z}_6$. Thus $\mathbb{Z}_6 \cong \mathbb{Z}_2 \times \mathbb{Z}_3$. |

| Example 10.4 | Let $H = K = \mathbb{Z}_2$. Then $\mathbb{Z}_2 \times \mathbb{Z}_2$ is a noncyclic group of order 4. This group can be shown to be isomorphic to the group of symmetries of a rectangle. |

If H and K are abelian groups, then so is the direct product $H \times K$. In additive notation, the direct product becomes a **direct sum**, and the group $H \times K$ is often written $H \oplus K$.

If H and K are finite groups, then $H \times K$ has $|H| \cdot |K|$ elements. If H_1, $H_2, \ldots, H_m$ are all groups, then a similar construction can be made to show that $H_1 \times H_2 \times \cdots \times H_m$ is a group. In this case we note that since

$$H_i \times H_j \cong H_j \times H_i \qquad \text{and} \qquad H_i \times (H_j \times H_k) \cong (H_i \times H_j) \times H_k$$

(see Exercise 3), we have the fact that

$$H_1 \times H_2 \times \cdots \times H_m = (H_1 \times H_2 \times \cdots \times H_{m-1}) \times H_m$$

is a group. For example, the set of binary numbers from 0 to 15 can be identified with the elements of the group $(\mathbb{Z}_2)^4 = \mathbb{Z}_2 \times \mathbb{Z}_2 \times \mathbb{Z}_2 \times \mathbb{Z}_2$. The addition of binary numbers, however, is not the same as the addition in this group.

The foregoing discussion shows how you can construct new groups from groups you already know. But now we want to go the other way. Suppose G is a group containing subgroups H and K such that

1. $G = HK$; that is, every element $g \in G$ can be written in the form hk, for some $h \in H$, $k \in K$;
2. the elements of H commute with the elements of K; and
3. $H \cap K = \{1\}$.

In this case we claim that G is actually isomorphic to the group $H \times K$. To see this, let us define a mapping $\phi \colon G \to H \times K$ in the following way. For $g \in G$, since $g = hk$ for some $h \in H$, $k \in K$, define $\phi(g) = (h, k)$. Is this mapping well-defined? That is, if g can also be written as $g = h'k'$ for $h' \in H$ and $k' \in K$ so that $\phi(g) = (h', k')$, is $(h, k) = (h', k')$? The answer is yes, for if $g = hk = h'k'$, then $h^{-1}h' = k(k')^{-1} \in H \cap K = \{1\}$. Thus $h^{-1}h' = 1 = k(k')^{-1}$ and $h = h'$, $k = k'$. We have shown that conditions 1 and 3 imply that each element $g \in G$ is **uniquely** expressible in the form hk. Thus the mapping ϕ makes sense.

We still need to show that ϕ is an isomorphism. Clearly ϕ is surjective, for $g = hk$ maps onto (h, k). Also ϕ is injective since $\phi(g_1) = (h_1, k_1) = \phi(g_2) = (h_2, k_2)$ implies that $h_1 = h_2$, $k_1 = k_2$, and $g_1 = h_1 k_1 = h_2 k_2 = g_2$. Finally if $g_1 = h_1 k_1$ and $g_2 = h_2 k_2$, then condition 2 says

$$g_1 g_2 = (h_1 k_1)(h_2 k_2) = h_1(k_1 h_2)k_2 = h_1(h_2 k_1)k_2 = (h_1 h_2)(k_1 k_2)$$

Therefore

$$\phi(g_1 g_2) = (h_1 h_2, k_1 k_2) = (h_1, k_1)(h_2, k_2) = \phi(g_1)\phi(g_2)$$

and ϕ is a homomorphism. We have proved the following result:

Theorem 10.1   Suppose G is a group, $H \leqslant G$, $K \leqslant G$, $H \cap K = \{1\}$, $G = HK$, and the elements of H commute with the elements of K. Then $G \cong H \times K$.

A group G, satisfying the conditions of the hypothesis of this theorem, is said to be an **internal direct product of H and K**. This theorem really says that $G = HK$ is an internal direct product, $H \times K$ is an external direct product, and the two are isomorphic. We will follow the common practice of identifying the two and simply say that G is the **direct product of H and K**.

If G is an abelian group, then there is no need to check if the elements of two subgroups commute with one another. If G is not abelian, however, then the next corollary is often easier to apply.

Corollary 10.1

If G is a group, $H \trianglelefteq G$, $K \trianglelefteq G$, $G = HK$, and $H \cap K = \{1\}$, then G is the direct product of H and K.

Proof

We need only show that if H and K are both normal, then the elements of H commute with the elements of K. If $h \in H$, $k \in K$, let $g = hkh^{-1}k^{-1}$. Since $K \trianglelefteq G$, we have $g = (hkh^{-1})k^{-1} = k_2 k^{-1} \in K$. But $H \trianglelefteq G$ implies $g = h(kh^{-1}k^{-1}) = hh_2 \in H$. Thus $g \in H \cap K = \{1\}$, and $hkh^{-1}k^{-1} = 1$ means $hk = kh$. It follows that the elements of H commute with elements of K. $\square$

Example 10.5

Let $G = \mathbb{Z}_{10}$. Set $H = \langle 5 \rangle$ and $K = \langle 2 \rangle$. Then H and K are both normal subgroups of $\mathbb{Z}_{10}$ since $\mathbb{Z}_{10}$ is abelian. $H \cap K$ is a subgroup of both H and K. Thus $|H \cap K|$ divides both $|H| = 2$ and $|K| = 5$. This forces $|H \cap K| = 1$. Finally

$$|HK| = \frac{|H| \cdot |K|}{|H \cap K|} = \frac{2 \cdot 5}{1} = 10$$

by Corollary 8.2; hence $\mathbb{Z}_{10} = HK$. By the corollary, $\mathbb{Z}_{10}$ is the direct product of H and K.

Example 10.6

Let $G = \langle a, b \,|\, a^4 = b^2 = 1, ab = ba \rangle$. Set $H = \langle a \rangle$ and $K = \langle b \rangle$. We leave it to you to show that G is a group of order 8 and that H and K are both normal in G. (See Exercise 8.) Since $H \cap K = \{1\}$, $|HK| = |H| \cdot |K| = 4 \cdot 2 = 8$, and $G = HK$. Therefore $G \cong H \times K$.

We noted earlier that if G is an abelian group, then the direct product of H and K is often written as $H \oplus K$. In Example 10.5, $\mathbb{Z}_{10} = \langle 5 \rangle \oplus \langle 2 \rangle$. Similarly, an abelian group G that is a direct product of the groups $H_1, H_2, \ldots, H_m$ can be expressed as

$$G = \bigoplus_{i=1}^{m} H_i$$

Suppose $G = H \times K \times L$. We can consider $G = H \times K_1$, where $K_1 = K \times L = KL$. Then $H \cap KL = \{1\}$. Similarly $K \cap HL = \{1\}$. We can extend this result to a direct product of more than three groups in the following way:

Theorem 10.2

Suppose $G = H_1 \times H_2 \times \cdots \times H_m$ is the direct product of groups $H_1, H_2, \ldots, H_m$. Then each $g \in G$ can be uniquely expressed in the form $g = h_1 h_2 \cdots h_m$ for $h_i \in H_i$. Moreover, for each $1 \leqslant j \leqslant m$, we have

$$H_j \cap (H_1 H_2 \cdots H_{j-1} H_{j+1} \cdots H_m) = \{1\}$$

Proof

Suppose first that $g = h_1 h_2 \cdots h_m = h'_1 h'_2 \cdots h'_m$. Then

$$h'_m h_m^{-1} = (h'_1 h'_2 \cdots h'_{m-1})^{-1}(h_1 h_2 \cdots h_{m-1})$$

$$\in (H_1 \times H_2 \times \cdots \times H_{m-1}) \cap H_m = \{1\}$$

implying that $h'_m = h_m$ and $h_1 h_2 \cdots h_{m-1} = h'_1 h'_2 \cdots h'_{m-1}$. The result follows by induction. The second assertion follows from the definition of direct product and the observation that

$$G \cong (H_1 \times H_2 \times \cdots \times H_{j-1} \times H_{j+1} \times \cdots \times H_m) \times H_j \qquad \Box$$

While we have worked so far with direct products in this section, there are other types of products. Recall from Chapter 8 that when $N \trianglelefteq G$ and $K \leqslant G$, then NK is a subgroup of G. If $G = NK$ and $N \cap K = \{1\}$, then G is termed the **semi-direct product of N by K**.

Example 10.7

Let $G = S_n$, $N = A_n$, and K be the cyclic group generated by any 2-cycle, say, $K = \langle (1 \quad 2) \rangle$. Then $A_n \cap K = \{1\}$ since A_n contains no odd permutations and

$$|A_n K| = \frac{|A_n| \cdot |K|}{|A_n \cap K|} = \frac{(n!/2)(2)}{1} = n! = |S_n|$$

This implies that $S_n = A_n K$. In particular, we have $S_4 = A_4 K$.

Example 10.8

Suppose $G = S_4$ and

$$N = \langle (1), (1 \quad 2)(3 \quad 4), (1 \quad 3)(2 \quad 4), (1 \quad 4)(2 \quad 3) \rangle$$

Recall that we showed in Example 7.11 that $N \trianglelefteq S_4$. If $K = S_3$, then $N \cap K = N \cap S_3 = \{(1)\}$ and

$$|N S_3| = \frac{|N| \cdot |S_3|}{|N \cap S_3|} = \frac{(4)(6)}{1} = |S_4|$$

Thus $S_4 = N S_3$.

The examples show that a group can be realized in more than one way as a semi-direct product. Any direct product, of course, is also a semi-direct product, but these examples show that the converse of this statement is false. Therefore the concept of semi-direct product is a generalization of direct product. We can even extend this idea.

In Examples 10.7 and 10.8 we used Corollary 8.2 to conclude that if $N \trianglelefteq G$, $K \leqslant G$ and $|G|$ is finite, then

$$|NK| = \frac{|N| \cdot |K|}{|N \cap K|}$$

Now suppose that $A \leqslant G$, $B \leqslant G$, and consider $AB = \{ab \,|\, a \in A, b \in B\}$. Recall that this set may or may not be a subgroup of G. Nevertheless, we claim that if G is finite, then it is still true that $|AB:A| = |B:A \cap B|$, so that

$$|AB| = \frac{|A| \cdot |B|}{|A \cap B|}$$

To see this, suppose

$$B = (A \cap B)b_1 \cup \cdots \cup (A \cap B)b_m$$

is a decomposition of B into a union of disjoint cosets of the subgroup $A \cap B$. We want to show that the cosets $\{Ab_i\}_{i=1}^m$ form a partition of AB. So suppose $x \in Ab_i \cap Ab_j$. Then there exist a_1, $a_2 \in A$ such that $a_1 b_i = a_2 b_j$. This implies that $a_1^{-1} a_2 = b_i b_j^{-1} \in A \cap B$ and $(A \cap B)b_j = [(A \cap B)(b_i b_j^{-1})]b_j = (A \cap B)b_i$, contrary to the choice of b_i and b_j. Therefore the sets of the form Ab_i are disjoint. Since each $b \in B$ is in $(A \cap B)b_i$ for exactly one i, we see that the set AB can be written as a union of sets of the form Ab for $b \in B$. By Lagrange's Theorem, we obtain the following theorem.

Theorem 10.3

If $A \leqslant G$ and $B \leqslant G$ where G is finite, then

$$|AB| = \frac{|A| \cdot |B|}{|A \cap B|}$$

This result will prove quite useful in analyzing sets of group elements of the form AB. But let us proceed first with an example. Let $G = S_4$ and D be a Sylow 2-subgroup of S_4 of order 8. Set $C = \langle (1 \quad 2 \quad 3) \rangle$. We leave it to you to verify that neither D nor C is normal in S_4, but

$$|DC| = \frac{|D| \cdot |C|}{|D \cap C|} = \frac{8 \cdot 3}{1} = |S_4|$$

Therefore the set of elements $\{dc \,|\, d \in D, c \in C\}$ contains all the elements of S_4, so $S_4 = DC$. In this case we say that S_4 is **factorized** by D and C.

Theorem 10.4

If $A \leqslant G$, $B \leqslant G$, then $AB \leqslant G$ if and only if $AB = BA$.

Proof

Suppose $AB \leqslant G$, and let $ab \in AB$. Then $(ab)^{-1} = b^{-1}a^{-1} \in BA$. However, since $AB \leqslant G$ and every element of AB is the inverse of an element of AB, this implies that $AB \leqslant BA$. A similar argument can be used to show that $BA \leqslant AB$. Thus $AB = BA$.

Conversely, suppose $AB = BA$ and $a_1 b_1$, $a_2 b_2 \in AB$. If $b_1 a_2 = a_2' b_1'$, we have that

$$(a_1 b_1)(a_2 b_2) = a_1(b_1 a_2)b_2 = a_1(a_2' b_1')b_2 = (a_1 a_2')(b_1' b_2) \in AB$$

so AB is closed under composition. Moreover $(a_1 b_1)^{-1} = b_1^{-1} a_1^{-1} \in BA = AB$, so AB contains the inverse of each element. Since $1_G = 1_G 1_G \in AB$, we have that $AB \leqslant G$.   □

The main differences between semi-direct products and these products are that a semi-direct product requires that the factors have intersection $\{1\}$ and at least one of the factors be normal. If D is a 2-Sylow subgroup of S_4, you should be able to verify that $S_4 = S_3 D$, although neither subgroup is normal and $|S_3 \cap D| = 2$. Factorization of a group into more elementary subgroups often aids in understanding the nature of the group. This technique is especially useful in analyzing groups that contain no proper normal subgroups other than the identity subgroup. We will have more to say on this in Chapter 11, but we turn in the next section to the use of direct products in classifying finite abelian groups.

Exercises 10.1

1. If $(H, \cdot)$ and $(K, \circ)$ are groups, verify that $H \times \{1_K\}$ and $\{1_H\} \times K$ are normal subgroups of $H \times K$.

2. Prove that if H and K are abelian groups, then $H \times K$ is abelian.

3. Verify that if H, K, and L are abelian groups, then

   $$H \times K \cong K \times H \quad \text{and} \quad (H \times K) \times L \cong H \times (K \times L)$$

4. Show directly that $\mathbb{Z}_2 \times \mathbb{Z}_3 \cong \mathbb{Z}_6$.

5. Verify that $\mathbb{Z}_2 \times \mathbb{Z}_2$ is isomorphic to the group of symmetries of a rectangle. Is $\mathbb{Z}_2 \times \mathbb{Z}_2 \times \mathbb{Z}_2$ isomorphic to the group of symmetries of the square?

6. Give an example to show that addition of binary numbers is not the same as group addition in $(\mathbb{Z}_2)^4$.

7. Prove that if $G \cong \mathbb{Z}_{15}$, then $G \cong \mathbb{Z}_3 \oplus \mathbb{Z}_5$.

8. a. Let $G = \langle a, b | a^4 = b^2 = 1, ab = ba \rangle$, the group of Example 10.6 generated by a and b.
      i. Show $G = \{a^m b^n | a^4 = b^2 = 1, ab = ba, m, n \in \mathbb{Z}\}$.
      ii. Prove $|G| = 8$.
      iii. Verify that $H = \langle a \rangle$ and $K = \langle b \rangle$ are both normal subgroups of G.
   b. Let $G = \langle a, b | a^4 = b^2 = 1, ab = ba^3 \rangle$, and set $H = \langle a \rangle$ and $K = \langle b \rangle$. Is $H \trianglelefteq G$? Is $K \trianglelefteq G$?

9. Let G be the group of rigid motions of a rectangle as given in Section 2.3, $H_1 = \langle \rho \rangle$, $H_2 = \langle \alpha \rangle$, and $H_3 = \langle \beta \rangle$. Show that

   $$H_1 \cap H_2 = H_2 \cap H_3 = H_1 \cap H_3 = \{e\}$$

   does not guarantee that $G \cong H_1 \times H_2 \times H_3$.

10. Verify that if G is the semi-direct product of N by K and $K \trianglelefteq G$, then $G \cong N \times K$.

11. Suppose G is a group order of 21, $P_3 \in \text{Syl}_3(G)$, and $P_7 \in \text{Syl}_7(G)$. Show that $G \cong P_3 \times P_7$ or $G \cong P_7 P_3$, a semi-direct product of P_7 by P_3.

12. Find an example of a group G with $N \trianglelefteq G$, $K \leqslant G$ and both N and K are abelian groups, but $G = NK$ is not abelian.

13. Let $G = S_4$ and $P_2 \in \mathrm{Syl}_2(G)$. Show $G = A_4 P_2$.

14. Suppose $A = \langle (1 \ \ 2 \ \ 3 \ \ 4) \rangle$ and $B = A_4$. Verify that $|A \cap B| = 2$ and that S_4 is factorized by A and B.

15. Let $A = S_3$ and $B = \{(1), (1 \ \ 2), (3 \ \ 4), (1 \ \ 2)(3 \ \ 4)\}$. Prove that AB is not a subgroup of S_4.

16. If $C = \langle (1 \ \ 2 \ \ 3) \rangle$ and $K = \{(1), (1 \ \ 2)(3 \ \ 4), (1 \ \ 3)(2 \ \ 4), (1 \ \ 4)(2 \ \ 3)\}$, verify that $A_4 = KC$.

17. Let G be a finite abelian group and $\pi(G) = \{p_1, p_2, \dots, p_k\}$ be the set of distinct prime divisors of $|G|$. If $P_i \in \mathrm{Syl}_{p_i}(G)$, prove $G = P_1 P_2 \cdots P_k$.

18. Show that if A and B are subgroups of G and AB is finite, then $|AB| = |BA|$.

19. Suppose $N \trianglelefteq G$, $K \trianglelefteq G$. Prove that $G/(N \cap K)$ is isomorphic to a subgroup of $G/N \times G/K$.

20. Let G be a finite group with $N \trianglelefteq G$ and $K \trianglelefteq G$ and P a p-Sylow subgroup of G.
    a. Prove that $P(N \cap K) = (PN) \cap (PK)$.
    b. If $G = NK$, verify that $P = (P \cap N)(P \cap K)$.

21. Suppose $A \trianglelefteq G$ and $H \trianglelefteq N \leqslant G$. Prove that $AH \trianglelefteq AN$.

22. Let G be a finite group with the property that G has a normal p-Sylow subgroup for each prime p dividing $|G|$. (Such a group is termed **nilpotent**.) Prove that G is the direct product of its Sylow subgroups.

10.2

Finite Abelian Groups

In this section we will present a number of results that determine the structure of finite abelian groups. These results establish isomorphism classes for finite abelian groups of a given order, so that if we are given an abelian group A of order n, then we know that A must be isomorphic to one of the groups in the class corresponding to the number n. By using some additional information about the group A, we can determine which of the groups in the class is actually isomorphic to A.

We begin with a preliminary result.

Lemma 10.1

Let A be an abelian group. Suppose $a \in A$ has order p and $b \in A$ has order q where $(p, q) = 1$. Then $a + b$ has order pq.

Proof

By Exercise 13 of Section 4.3, we know that the order m of $a + b$ divides any integer k such that $k(a + b) = 0$. Since

$$(pq)(a + b) = (pq)a + (pq)b = q(pa) + p(qb) = q \cdot 0 + p \cdot 0 = 0$$

m divides pq. On the other hand, the statement $(p, q) = 1$ implies there exist

$s, t \in \mathbb{Z}$ such that $1 = sp + tq$. But then

$$sp(a + b) = spa + spb = 0 + spb = spb = (1 - tq)b = b - tqb = b$$

Similarly, $tq(a + b) = a$. Since $m(a + b) = 0$, this says

$$mb = m(sp)(a + b) = (sp)(m(a + b)) = sp \cdot 0 = 0$$

so q divides m. In a like manner, $ma = 0$ and p divides m. Since $(p, q) = 1$, pq divides m. (See Exercise 6.) We conclude $m = pq$. $\qquad\square$

In order to give an indication of the type of classification theorem we will encounter in this section, let us treat the problem of classifying finite cyclic (abelian) groups. From Chapter 6, we know that any finite cyclic group A of order n is isomorphic to $\mathbb{Z}_n$. Suppose $n = p_1^{e_1} p_2^{e_2} \cdots p_k^{e_k}$, where the p_i's are distinct primes and $e_i \geq 1$. If $a_i = n/p_i^{e_i}$, then it can be shown that the order of a_i in $\mathbb{Z}_n$ is $p_i^{e_i}$. (See Exercise 4.) Set $A_i = \langle a_i \rangle$. Then $A_i \cong \mathbb{Z}_{p_i^{e_i}}$. Since A is abelian, $A_i \trianglelefteq A$ for each i. Then

$$A_1 \times A_2 \times \cdots \times A_k = A_1 A_2 \cdots A_k \leq A$$

and

$$|A_1 A_2 \cdots A_n| = p_1^{e_1} p_2^{e_2} \cdots p_k^{e_k} = |A|$$

We conclude that $A = A_1 A_2 \cdots A_n$. This proves the following result that we state in additive form:

Theorem 10.5

Let A be a cyclic group of order n. If $n = p_1^{e_1} p_2^{e_2} \cdots p_k^{e_k}$, where the p_i's are distinct primes and $e_i \geq 1$, then

$$A \cong \mathbb{Z}_{p_1^{e_1}} \oplus \mathbb{Z}_{p_2^{e_2}} \oplus \cdots \oplus \mathbb{Z}_{p_k^{e_k}}$$

We wish to develop a similar result for noncyclic abelian groups; namely, that any finite abelian group is isomorphic to a direct sum of cyclic groups of prime power order. In the noncyclic case, the orders of the cyclic summands are not relatively prime. To give an illustration, let A be an abelian group of order 8. What can we say about A? If A is cyclic, then $A \cong \mathbb{Z}_8$. If A is not cyclic, then no element of A can have order 8. Since the order of any element divides $|A| = 8$ by Corollary 7.2, the nonidentity elements of A have order 2 or 4. Now $\mathbb{Z}_2 \oplus \mathbb{Z}_2 \oplus \mathbb{Z}_2$ and $\mathbb{Z}_4 \oplus \mathbb{Z}_2$ are both groups of order 8 and neither is isomorphic to $\mathbb{Z}_8$. (Why?) Thus A could be isomorphic to either of these groups. Are there any other possibilities? As we will see, the answer is no.

This example involves analyzing the structures of an abelian group of prime power order. The next result shows that a knowledge of the structure of such groups is sufficient to solve our classification problem. The theorem is a consequence of the Sylow Theorems and our work in the last section.

Theorem 10.6

Any finite abelian group is a direct sum of its Sylow subgroups.

Proof

Let A be an abelian group of order n and $\pi(A) = \{p_1, p_2, \ldots, p_k\}$ be the set of distinct prime divisors of n. If $P_i \in \mathrm{Syl}_{p_i}(A)$, then $P_i \trianglelefteq A$, since A is abelian. Consider the subgroup $P_1 P_2 \cdots P_k = P_1 \oplus P_2 \oplus \cdots \oplus P_k$. The orders of the subgroups P_i are relatively prime. Since these subgroups are Sylow subgroups,

$$|P_1 \oplus P_2 \oplus \cdots \oplus P_k| = |P_1| \cdot |P_2| \cdots |P_k| = |A|$$

Thus $A = P_1 \oplus P_2 \oplus \cdots \oplus P_k$, and the result follows. $\square$

This result allows us to focus our attention on the Sylow subgroups of A, themselves. Our goal, then, is to determine the nature of finite abelian p-groups. We will do this through the use of some preliminary results that are of interest in their own right.

Lemma 10.2

Suppose A is a finite abelian p-group, and p^k is the smallest positive integer such that, for all $a \in A$, $p^k a = 0$, the identity of A. Then there exists $b \in A$ such that $o(b) = p^k$.

Proof

This follows immediately from the fact that each element of a p-group has order some power of p and the observation that if no element had order p^k, then $p^{k-1} a = 0$ for all $a \in A$, contrary to the choice of p^k. $\square$

The **exponent** of an abelian group A is the smallest positive integer n such that $na = 0$ for all $a \in A$. We have just shown that the exponent of an abelian p-group A is also equal to the order of the largest cyclic subgroup of A. We will exploit this fact later. We next determine how this largest cyclic subgroup of an abelian p-group A is related to the structure of A itself. The intricately detailed argument is due to Fuchs.[1]

Lemma 10.3

If A is a finite abelian p-group and $B = \langle b \rangle$ is the subgroup corresponding to the exponent p^k of A, then B is a direct summand of A; that is, there exists $C \trianglelefteq A$ such that $A = B \oplus C$.

Proof

Suppose C is the subgroup of A of largest order such that $B \cap C = \{0\}$. Since A is abelian, both B and C are normal in A and $A_1 = B \oplus C$ is a subgroup of A. We claim that $A = A_1$. To show this, suppose not and choose $x \in A - A_1$.

---

[1] L. Fuchs, *Abelian Groups*, Oxford: Pergamon Press, 1968, p. 39.

Since $p^k x = 0$, there exists an integer j such that $p^j x \in A$, but $p^{j-1} x \notin A$. Set $y = p^{j-1} x$. Then $y \notin A_1$ by the choice of j, but $py \in A_1 = \langle b \rangle \oplus C$.

This means we can write $py = mb + c$, for $m \in \mathbb{N}$, $c \in C$, and

$$p^{k-1}(mb) + p^{k-1}c = p^{k-1}(py) = p^k y = 0$$

Since $p^{k-1}mb = -p^{k-1}c \in B \cap C = \{0\}$, the order of b, namely, p^k, must divide $p^{k-1}m$. Therefore we can conclude that $m = pm_1$ and

$$p(y - m_1 b) = py - pm_1 b = py - mb = c \in C$$

although $y - m_1 b \notin C$, since $y \notin A_1 = C \oplus \langle b \rangle$.

Consider $C_1 = C \oplus \langle y - m_1 b \rangle$. By the choice of C, $\langle b \rangle \cap C_1 \neq \{0\}$. This means that a multiple of b, say ub, is in C_1 and $ub = c' + v(y - m_1 b)$ for some $v \in \mathbb{N}$, $c' \in C$. Now $vy = (ub + vm_1 b) - c' \in \langle b \rangle + C = A_1$ and if $v = wp$, we would have

$$ub = c' + wp(y - m_1 b) = c' + wc \in C$$

contrary to the condition that $\langle b \rangle \cap C = \{0\}$. Therefore $(v, p) = 1$ and there exist $s, t \in \mathbb{Z}$ such that $1 = sv + tp$. But then $y = (sv + tp)y = s(vy) + t(py) \in A_1$. This contradicts the choice of y, and we must conclude that $A - A_1 = \varnothing$. Therefore, $A = A_1 = B \oplus C$. $\qquad\square$

This last result now allows us to determine the structure of finite abelian p-groups. We have shown that if a finite abelian p-group has exponent p^k, then $A = \langle b \rangle \oplus C \cong \mathbb{Z}_{p^k} \oplus C$. Moreover, since $p^k a = 0$ for all $a \in A$, we have $p^k c = 0$ for all $c \in C$. Therefore the exponent of C, say p^{k_2}, must divide p^k, and we can reapply Lemma 10.3 to the group C to show that $C \cong \mathbb{Z}_{p^{k_2}} \oplus D$, where D is another abelian p-group. The fact that A is a finite group says that, in continuing this procedure, we eventually reach a point where the direct summand is itself cyclic of prime power order.

We can summarize this discussion in the following way:

Theorem 10.7

If A is a finite abelian p-group, then

$$A \cong \mathbb{Z}_{p^{k_1}} \oplus \mathbb{Z}_{p^{k_2}} \oplus \cdots \oplus \mathbb{Z}_{p^{k_m}}$$

where $k_1 \geqslant k_2 \geqslant \cdots \geqslant k_m$ and

$$|A| = p^{k_1} p^{k_2} \cdots p^{k_m}$$

Combining our last two theorems, we have obtained the classical result on finite abelian groups that was first given in 1878 by G. Frobenius and L. Stickelberger.[2]

---

[2]G. Frobenius and L. Stickelberger, "Über Gruppen von vertauschbaren Elementen," *Journal reine u. angew. Math.*, Vol. 86 (1878), pp. 217–262.

Theorem 10.8

Basis Theorem

A finite abelian group is the direct sum of a finite number of cyclic groups of prime power order.

The prime powers that are the orders of the cyclic summands are called the **elementary divisors** of the group. The Basis Theorem is the first part of the Fundamental Theorem of Finite Abelian Groups. The Basis Theorem asserts there exists a decomposition of a finite abelian group into a direct sum of cyclic groups of prime power order. The second half of the Fundamental Theorem states that such a decomposition is unique in terms of the number and order of the summands. For the sake of completeness, we state the theorem but refer you to the texts of Fuchs or Rotman in the references for the proof of uniqueness.

Fundamental Theorem of Finite Abelian Groups

Let A be a finite abelian group. Then A is a direct sum of cyclic groups of prime power order. Moreover, any two decompositions of A into direct sums of cyclic groups of prime power order have the same number of summands of each order.

Let us illustrate this result by analyzing the possible representations of an abelian group A of order $72 = 2^3 3^2$. We list the possible sets of elementary divisors along with the corresponding direct sum decompositions.

Elementary Divisors	Direct Sum Representation of A
2^3, 3^2	$\mathbb{Z}_8 \oplus \mathbb{Z}_9$
2^3, 3, 3	$\mathbb{Z}_8 \oplus \mathbb{Z}_3 \oplus \mathbb{Z}_3$
2^2, 2, 3^2	$\mathbb{Z}_4 \oplus \mathbb{Z}_2 \oplus \mathbb{Z}_9$
2^2, 2, 3, 3	$\mathbb{Z}_4 \oplus \mathbb{Z}_2 \oplus \mathbb{Z}_3 \oplus \mathbb{Z}_3$
2, 2, 2, 3^2	$\mathbb{Z}_2 \oplus \mathbb{Z}_2 \oplus \mathbb{Z}_2 \oplus \mathbb{Z}_9$
2, 2, 2, 3, 3	$\mathbb{Z}_2 \oplus \mathbb{Z}_2 \oplus \mathbb{Z}_2 \oplus \mathbb{Z}_3 \oplus \mathbb{Z}_3$

By Theorems 10.5 and 10.6,

$$\mathbb{Z}_{72} \cong \mathbb{Z}_8 \oplus \mathbb{Z}_9 \qquad \mathbb{Z}_{36} \cong \mathbb{Z}_4 \oplus \mathbb{Z}_9$$
$$\mathbb{Z}_{24} \cong \mathbb{Z}_8 \oplus \mathbb{Z}_3 \qquad \mathbb{Z}_{18} \cong \mathbb{Z}_2 \oplus \mathbb{Z}_9$$
$$\mathbb{Z}_{12} \cong \mathbb{Z}_4 \oplus \mathbb{Z}_3 \qquad \mathbb{Z}_6 \cong \mathbb{Z}_2 \oplus \mathbb{Z}_3$$

In general, in a direct sum decomposition of an abelian group into a sum of cyclic summands of prime power order, we obtain another cyclic group if we collect together the cyclic summands of highest order, one for each prime. The order of this group is called the **first invariant factor** of the group. This order is equal to the exponent of the group. (See Exercise 17.) If we repeat this procedure with the remaining summands, we obtain another cyclic group whose order is called the **second invariant factor**. The continuation of this process produces the complete set of invariant factors of the group. We again analyze an abelian group A of order 72, this time listing the possible invariant factors and corresponding direct sum decompositions.

Elementary Divisors	Invariant Factors	Decomposition
2^3, 3^2	72	$\mathbb{Z}_{72}$
2^3, 3, 3,	24, 3	$\mathbb{Z}_{24} \oplus \mathbb{Z}_3$
2^2, 2, 3^2	36, 2	$\mathbb{Z}_{36} \oplus \mathbb{Z}_2$
2^2, 2, 3, 3	12, 6	$\mathbb{Z}_{12} \oplus \mathbb{Z}_6$
2, 2, 2, 3^2	18, 2, 2	$\mathbb{Z}_{18} \oplus \mathbb{Z}_2 \oplus \mathbb{Z}_2$
2, 2, 2, 3, 3	6, 6, 2	$\mathbb{Z}_6 \oplus \mathbb{Z}_6 \oplus \mathbb{Z}_2$

The Fundamental Theorem allows us to be certain that any abelian group of order 72 is isomorphic to one *and only one* group in each of the lists. Usually, additional information about the group is needed to determine the exact isomorphism class.

Our observations concerning the invariant factors are summarized in the following corollary.

Corollary 10.2

If A is a finite abelian group and $|A| > 1$, then there is a unique list $m_1, m_2, \ldots,$ m_t of positive integers, each a multiple of the next, such that

$$A \cong \mathbb{Z}_{m_1} \oplus \mathbb{Z}_{m_2} \oplus \cdots \oplus \mathbb{Z}_{m_t}$$

Moreover $|A| = m_1 m_2 \cdots m_t$ and m_1 is the exponent of A.

In this chapter, we have investigated products of groups and applied them to the problem of classifying finite abelian groups. Infinite abelian groups can be shown to be a direct sum of copies of $\mathbb{Z}$ and possibly a summand of finite order, called the **torsion subgroup**. Our work can then be applied to determine this finite abelian subgroup. We will let you consult the references for the details of this analysis. In the next chapter, we will comment on the classification problem as it relates to arbitrary finite groups.

Exercises 10.2

1. Verify that if $G = (\mathbb{Z}_{15},\ \oplus)$, then $o(5) = 3$, $o(3) = 5$, and $o(8) = o(3 + 5) = 5 \cdot 3 = 15$.

2. In $(\mathbb{Z}_{27,000},\ \oplus)$, what is $o(1027)$?

3. Prove directly that if G is an abelian group of order $3 \cdot 5 \cdot 7$, then G is cyclic.

4. Suppose $n = p_1^{e_1} p_2^{e_2} \cdots p_k^{e_k}$, where the p_i's are distinct primes and $e_i \geq 1$. If $a_i = n/p_i^{e_i}$, then prove in $\mathbb{Z}_n$ that $o(a_i) = p_i^{e_i}$.

5. Suppose that G is an abelian group of exponent p^n where p is a prime. Show that G is a p-group.

6. Let $p, q \in \mathbb{N}$ and $(p, q) = 1$.
   a. Suppose $m = pm_1$ for $m_1 \in \mathbb{Z}$ and $q|m$. Prove $q|m_1$.
   b. Suppose $p|m$ and $q|m$. Show that $pq|m$.

7. Suppose A is an abelian group with exponent n. Show that A has an element of order n.

8. If A is a finite abelian group and b is an element of maximal order n, then prove $na = 0$ for all $a \in A$.

9. Show that the elements of finite order in an abelian group A form a normal subgroup K and that A/K has no elements of finite order other than the identity coset.

10. Prove that if A is a finite group, p a prime, and $pa = 0$ for all $a \in A$, then

$$A \cong \mathbb{Z}_p \oplus \mathbb{Z}_p \oplus \cdots \oplus \mathbb{Z}_p$$

(Such a group is called **elementary abelian**.)

11. Determine the possible elementary divisors and cyclic decompositions for an abelian group A of order:
    a. 25   b. 225   c. 144   d. 720   e. 12,500   f. 16,875

12. Determine the invariant factors and corresponding cyclic decompositions for the groups in Exercise 11.

13. If A is an abelian group of order p^m with p a prime, how many possible sets of elementary divisors of A are there for $m = 3, 4, 5,$ and 6.

14. If A is an abelian group of order $p^m q^n$ where p and q are distinct primes, how is the set of elementary divisors of A related to the sets of elementary divisors of subgroups of A of order p^m and q^n?

15. Prove that if A is a finite abelian group, there are but a finite number of possible sets of elementary divisors of A.

16. Prove directly that $\mathbb{Z}_4 \oplus \mathbb{Z}_2$ is not isomorphic to $\mathbb{Z}_2 \oplus \mathbb{Z}_2 \oplus \mathbb{Z}_2$.

17. If A is an abelian group and m is equal to the first invariant factor of A, prove that $ma = 0$ for all $a \in A$.

18. Let A be a finite abelian group of order n, and suppose p is a prime divisor of n. Set $pA = \{pa | a \in A\}$. Prove $pA \leqslant A$.

19. Show that a finite abelian group A is cyclic if and only if every Sylow subgroup of A is cyclic.

20. A nontrivial subgroup K of a group G is termed **minimal normal** if $K \trianglelefteq G$ and if K contains no nontrivial normal subgroup of G. If A is a finite abelian group, prove that any minimal normal subgroup of A is cyclic of prime order.

*21. Write a computer program that takes an integer n and lists the possible elementary divisors and invariant factors for an abelian group of order n.

11 | Solvable Groups

During the past twenty-five years, group theory has experienced an enormous amount of activity. Surely, it has been the predominant discipline in traditional algebraic research, both in number of mathematicians involved and in quantity of results obtained. Classical conjectures of fundamental importance have been considered and solved. Many of these investigations concerned the ways that groups fit together, i.e., how new groups can be constructed from known groups. Other questions involve the search for nonabelian simple groups, groups containing no nontrivial proper normal subgroups. In this latter effort, major results have been obtained by amassing a considerable amount of work utilizing ingenious mathematical reasoning coupled with a great deal of computer calculation.

In this chapter, we will see how groups are analyzed through their normal subgroup structure. Through the concept of normal series of subgroups, we will investigate the idea of composition series and discuss the classical result of Jordan and Hölder. This material will then be applied to examine a special class of groups whose structure is related to the solution of polynomial equations. Finally, we will examine simple groups and give a historical perspective on the efforts to determine all such groups.

11.1

Series of Subgroups

Knowledge of the proper subgroups of a group G is usually not enough to determine the group itself. For example, each proper subgroup of $\mathbb{Z}_6$ is isomorphic to a proper subgroup of S_3, yet $\mathbb{Z}_6$ is not isomorphic to S_3. The way in which the subgroups of a group relate to one another characterizes the group itself. In other words, a group G is determined by the way its subgroups fit together. The following definition gives us one way to analyze the subgroup structure of a group. Throughout this chapter, G represents a finite group.

Definition 11.1

> Let H be a subgroup of G. Suppose there exists a set of subgroups $\{H_0, H_1, \ldots, H_n\}$ of G such that
>
> $$H = H_0 \trianglelefteq H_1 \trianglelefteq H_2 \trianglelefteq \cdots \trianglelefteq H_{n-1} \trianglelefteq H_n = G$$
>
> then this chain of subgroups is called a **normal series between H and G.**

Certain normal series have particularly descriptive names. A normal series between $\{1\}$ and G is termed a **normal series for G.** The group H_{i+1} is termed the **successor of H_i.** If $H_i < H_{i+1}$ for $i = 0, 1, \ldots, n-1$, then the series is called a **proper normal series.** The factor groups H_{i+1}/H_i, for $i = 0, 1, \ldots, n-1$, are the **factors** of the series. The definition for normal series does not require that $H_i \trianglelefteq G$ for each i, but just that each subgroup is normal in its successor. Normal series wherein each subgroup is also normal in G are sometimes called **invariant series.**

Example 11.1

Let $G = S_3$. Then $\{(1)\} \trianglelefteq S_3$ and $\{(1)\} \trianglelefteq A_3 \trianglelefteq S_3$ are the only two normal series for S_3.

Example 11.2

If $G = A_4$, set $K = \{(1), (1\ \ 2)(3\ \ 4), (1\ \ 3)(2\ \ 4), (1\ \ 4)(2\ \ 3)\}$. Then

$$\{(1)\} \trianglelefteq K \trianglelefteq A_4$$

is an invariant series for A_4, while

$$\{(1)\} \trianglelefteq \langle(1\ \ 2)(3\ \ 4)\rangle \trianglelefteq K \trianglelefteq A_4$$

is a normal, but not invariant, series for A_4.

Example 11.3

If $G = \mathbb{Z}_{12}$, then any normal series for G is an invariant series for G. (Why?) The following are two normal series for $\mathbb{Z}_{12}$:

$$\{0\} \trianglelefteq \langle 6 \rangle \trianglelefteq \langle 3 \rangle \trianglelefteq \mathbb{Z}_{12}$$

$$\{0\} \trianglelefteq \langle 4 \rangle \trianglelefteq \langle 2 \rangle \trianglelefteq \mathbb{Z}_{12}$$

Note that the factors in the first series are isomorphic to $\mathbb{Z}_2$, $\mathbb{Z}_2$, and $\mathbb{Z}_3$, while those in the second series are isomorphic to $\mathbb{Z}_3$, $\mathbb{Z}_2$, and $\mathbb{Z}_2$.

Groups are often analyzed in terms of their normal series. For example, a **solvable group** is defined to be one possessing a normal series whose factors are abelian groups. This type of series is called a **solvable series**. We will investigate these groups in the next section. For now, we wish to consider series themselves in more detail.

Suppose G is a simple group; that is, a group with no nontrivial normal subgroups. Then $\{1\} \trianglelefteq G$ is the only normal series for G. To see this, note that while each subgroup in a normal series is required to be normal in its successor, the subgroup having G as its successor must be normal in G. In this case, the only factor of this series is G, which is simple. We are interested in series wherein each and every factor is simple.

Definition 11.2

A normal series

$$\{1\} = H_0 \trianglelefteq H_1 \trianglelefteq \cdots \trianglelefteq H_n = G$$

is said to be a **composition series for** G if each H_i is a maximal proper normal subgroup of H_{i+1}; that is,

i. $H_i \trianglelefteq H_{i+1}$
ii. $H_i \neq H_{i+1}$,
iii. $H_i \trianglelefteq L \trianglelefteq H_{i+1}$ for $L \leqslant G$ implies $H_i = L$ or $L = H_{i+1}$.

Suppose H_{i+1}/H_i is a nontrivial factor of a composition series of a group G. If H_{i+1}/H_i is not simple, then there is a normal subgroup $N/H_i \trianglelefteq H_{i+1}/H_i$. The correspondence between a group and its factor groups allows us to conclude that N is a proper normal subgroup of H_{i+1}, contradicting the fact that H_i is a "maximal" proper subgroup of H_{i+1}. We have thus proved the following result.

Theorem 11.1

The factors of a composition series of a group G are simple groups.

Both series in Example 11.3 are composition series. Note that these two series have the same number of factors, which can be paired isomorphically.

Definition 11.3

Two normal series

$$\{1\} = H_0 \unlhd H_1 \unlhd \cdots \unlhd H_n = G \qquad \text{and}$$

$$\{1\} = G_0 \unlhd G_1 \unlhd \cdots \unlhd G_m = G$$

are **equivalent** if there is a one-to-one correspondence between the factors H_{i+1}/H_i and G_{j+1}/G_j such that corresponding factors are isomorphic.

In other words, the two series are equivalent if $n = m$ and if there exists a bijection between the two sets of factor groups. We leave it to you to show that this definition gives a rise to an equivalence relation on the set of normal series of a group. (See Exercise 12.)

Suppose $\{1\} = H_0 \unlhd H_1 \unlhd \cdots \unlhd H_n = G$ is a normal series of G. Another normal series

$$\{1\} = G_0 \unlhd G_1 \unlhd \cdots \unlhd G_m = G$$

is a **refinement** of the first series if it is the same series or one obtained from the first by the insertion of subgroups. In Example 11.2,

$$\{(1)\} \unlhd \langle (1\ \ 2)(3\ \ 4) \rangle \unlhd K \unlhd A_4$$

is a refinement of $\{(1)\} \unlhd K \unlhd A_4$. A refinement is **proper** if it involves more subgroups than the original.

Theorem 11.2

Every finite group has a composition series. A normal series is a composition series if and only if it has no proper refinements.

Proof

The first assertion is proved by induction on $|G|$. The case $|G| = 1$ is trivial. Assume that the result is true for groups of order less than $|G|$. If G is simple, then $\{1\} \unlhd G$ is a composition series for G. If G is not simple, then there exists a maximal proper normal subgroup $N \unlhd G$. By induction, N has a composition series

$$\{1\} = H_0 \unlhd H_1 \unlhd \cdots \unlhd H_n = N$$

But then

$$\{1\} = H_0 \unlhd H_1 \unlhd \cdots \unlhd H_n = N \unlhd G$$

is a composition series for G.

To prove the second assertion, note that if $H_i \unlhd H_{i+1}$ can be properly refined to $H_i \unlhd N \unlhd H_{i+1}$, then H_i cannot be a maximal normal subgroup of H_{i+1}. Conversely, if a series cannot be refined, then each subgroup must be a maximal normal subgroup of its successor.   □

The amazing thing about composition series is not that every group has a composition series but that the composition series is essentially unique. The key to this result is the following theorem published in 1928 by O. Schreier. The proof is left to more advanced texts.[1]

Theorem 11.3

Any two series of G have equivalent refinements.

Now suppose $\{1\} = H_0 \trianglelefteq H_1 \trianglelefteq \cdots \trianglelefteq H_n = G$ is a normal series for a group G and $\{1\} = G_0 \trianglelefteq G_1 \trianglelefteq \cdots \trianglelefteq G_m = G$ is a composition series for G. Schreier's Theorem asserts that both series have equivalent refinements. But a composition series has no proper refinements. Therefore the first normal series can be refined to obtain a series equivalent to a composition series of G. Since a series equivalent to a composition series is itself a composition series (see Exercise 11), we have proved the following:

Corollary 11.1

Any normal series can be refined to a composition series.

If you have two composition series for a group G, then Schreier's Theorem says that these series have equivalent refinements. Since a composition series has no proper refinement, these series must already be equivalent. This is the content of the classical Jordan-Hölder Theorem.

Theorem 11.4

Jordan-Hölder Theorem

Any two composition series for a group G are equivalent.

In 1869 C. Jordan showed that the orders of the factors of a composition series of a group G are unique in that they depend only on G. Twenty years later, O. Hölder proved that the composition factors themselves are unique. In the next section, we will focus on a particular type of group in terms of normal series.

Exercises 11.1

1. Prove that if G is an abelian group, then every normal series for G is an invariant series for G.

2. Find all normal series for the group D, which was described in Example 8.6. Which are invariant series? Which are composition series?

3. Determine all composition series for each of the following groups. Verify that all composition series for a particular group are equivalent.

---

[1] For this proof, see J. J. Rotman's *The Theory of Groups*, 2nd ed., Boston: Allyn and Bacon, 1973, p. 106.

        a. $\mathbb{Z}_{24}$ b. $\mathbb{Z}_8$ c. $\mathbb{Z}_{30}$

        d. $\mathbb{Z}_{32}$ e. $S_3 \times \mathbb{Z}_6$ f. $A_4 \times A_3$

4. What is the composition series for $G = \{1\}$?

5. Suppose $\{1\} = G_0 \trianglelefteq G_1 \trianglelefteq \cdots \trianglelefteq G_n = G$ is a composition series for G. Then n is called the **composition length** of G. Verify that G has composition length 1 if and only if G is simple. Prove that two finite isomorphic groups have the same composition length.

6. In Example 11.3, find the permutation that takes the composition factors of the first series to the corresponding factors in the second.

7. Prove that if N is a normal subgroup of G, then there is a composition series for G containing N as one of the subgroups. Is the statement true if the word *normal* is removed?

8. Suppose $\{H_0 = \{1\}, H_1, H_2, \ldots, H_k = N, H_{k+1}, \ldots, H_n = G\}$ forms a composition series for G and $N \trianglelefteq G$. Prove that $\{H_k/N, H_{k+1}/N, \ldots, H_n/N\}$ forms a composition series for G/N. Is the result still true if *composition* is replaced by *normal*?

9. Verify that if all the factors of a normal series for a group are nontrivial simple groups, then the normal series is actually a composition series.

10. Suppose that $\{G_0, G_1, \ldots, G_n\}$ forms a normal series for a group G and $H \leqslant G$. Set $H_i = G_i \cap H$. Prove that $\{H_0, H_1, \ldots, H_n\}$ forms a normal series for H.

11. Prove that a series for a group G equivalent to a composition series is a composition series itself.

12. Verify that **equivalence** is an equivalence relation on the normal series of a group G.

13. Verify that the series

$$\{0\} \trianglelefteq 24\mathbb{Z} \trianglelefteq 12\mathbb{Z} \trianglelefteq \mathbb{Z} \qquad \text{and} \qquad \{0\} \trianglelefteq 36\mathbb{Z} \trianglelefteq 6\mathbb{Z} \trianglelefteq \mathbb{Z}$$

have equivalent refinements.

14. Show that $\{0\} \trianglelefteq 16\mathbb{Z} \trianglelefteq 8\mathbb{Z} \trianglelefteq 4\mathbb{Z} \trianglelefteq 2\mathbb{Z} \trianglelefteq \mathbb{Z}$ cannot be refined to a composition series for $\mathbb{Z}$. (Not every infinite group has a composition series.)

15. Show that the composition factors of an abelian group must be cyclic of prime order.

16. Suppose that $\{H_0 = \{1\}, H_1, \ldots, H_n = G\}$ forms a normal series for the finite group G and set $h_i = |H_i/H_{i-1}|$. Prove that $|G| = h_1 h_2 \cdots h_n$.

17. Prove that the nontrivial composition factors of a group P of order p^n, p a prime, are isomorphic to $\mathbb{Z}_p$.

18. Verify that if $\{H_0 = \{1\}, H_1, H_2, \ldots, H_n\}$ and $\{G_0 = \{1\}, G_1, G_2, \ldots, G_m = G\}$ form composition series for groups H and G, respectively, then

$$\{H_0 \times G_0, H_1 \times G_0, \ldots, H_n \times G_0, H_n \times G_1, H_n \times G_2, \ldots, H_n \times G_m\}$$

forms a composition series for $H \times G$.

19. In Exercise 16 of Section 8.1 we defined the commutator subgroup G' of a group G by

$$G' = \langle aba^{-1}b^{-1} \,|\, a, b \in G \rangle$$

and showed that $G' \trianglelefteq G$. Set $G^{(1)} = G'$ and, for $i \geqslant 1$, $G^{(i)} = (G^{(i-1)})'$. $G^{(i)}$ is known as the *i*th **derived subgroup** of G. Prove that $G^{(i)} \trianglelefteq G$ for each i.

20. A subgroup H of a group G is called **subnormal in** G (denoted $H \trianglelefteq \trianglelefteq G$) if there exists a normal series between H and G.
    a. Give an example to show that a subnormal subgroup need not be normal.
    b. If $H \trianglelefteq \trianglelefteq G$ and $H \leqslant K \leqslant G$, then show $H \trianglelefteq \trianglelefteq K$.
    c. If $A \trianglelefteq \trianglelefteq G$ and $B \trianglelefteq \trianglelefteq G$, prove $A \cap B \trianglelefteq \trianglelefteq G$.

21. Let $G = \langle a, b \mid ba = ab^2 \rangle = \{a^m b^n \mid m, n \in \mathbb{Z}\}$. A **2-adic rational number** is one of the form $\pm 2^k$ where $k \in \mathbb{Z}$.
    a. Show that the set H_2 of all finite sums (including 0) of 2-adic rational numbers is a subgroup of $(\mathbb{Q}, +)$. (H_2 is called the **2-adic rational group**.)
    b. Verify that for any integer m, $b^{2^m} = a^{-m}ba^m$.
    c. Let $N = \langle b^{2^m} \mid m \in \mathbb{Z} \rangle$. Prove that N is isomorphic to a 2-adic rational group and $N \trianglelefteq G$.
    d. Prove that G/N is an infinite group.

11.2

Definition of Solvable Groups

A finite group possesses many properties that serve to aid in its characterization. We have already classified groups known as *abelian* and *cyclic*. There are other classifications that rely on a group's normal subgroup structure as it relates to normal series of the group. We will investigate one of the most general of such characterizations, *solvability*. Later in this text, we will utilize results on groups of this type to prove that there exists no general algebraic formula for determining the roots of an arbitrary polynomial of degree 5 or greater.

Definition 11.4

A finite group G is said to be **solvable** if there exists subgroups $G_i \leqslant G$ for $i = 0, 1, 2, \ldots, n$ such that

$$\{1\} = G_0 \trianglelefteq G_1 \trianglelefteq G_2 \trianglelefteq \cdots \trianglelefteq G_{n-1} \trianglelefteq G_n = G$$

and G_i/G_{i-1} is an abelian group for $i = 1, 2, \ldots, n$. The chain of subgroups

$$\{1\} = G_0 \trianglelefteq G_i \trianglelefteq G_2 \trianglelefteq \cdots \trianglelefteq G_{n-1} \trianglelefteq G_n = G$$

is termed a **solvable series of** G.

The definition above implies that G_1 is an abelian subgroup of G and $G_{n-1} \trianglelefteq G$. Now any normal series can be refined to a composition series. What does a composition series of a solvable group look like? It is sufficient to analyze the composition series between G_i and G_{i+1}. Since G_{i+1}/G_i is abelian in a solvable series. what form does a composition series for an abelian group take? By Theorem 11.1, the factors of a composition series are simple groups. According to Exercise 15 of Section 11.1, however, each factor must then be cyclic of prime order. Therefore, solvable groups can be alternately characterized in the following way:

Theorem 11.5

A group G is solvable if and only if its composition factors are cyclic of prime order.

The class of solvable groups is very large. Clearly any abelian group, and thus any cyclic group, is solvable. But there are many other examples.

Example 11.4

Consider $G = S_3$. If we set $G_1 = A_3$ and $G_2 = G = S_3$, then $\{(1)\} \trianglelefteq A_3 \trianglelefteq S_3$ and both $A_3 \cong A_3/\{(1)\}$ and $S_3/A_3 \cong \mathbb{Z}_2$ are abelian. Therefore S_3 is solvable.

Example 11.5

Set $G = S_4$, $G_1 = K = \{(1), (1\ \ 2)(3\ \ 4), (1\ \ 3)(2\ \ 4), (1\ \ 4)(2\ \ 3)\}$, $G_2 = A_4$, and $G_3 = G = S_4$. Then $\{(1)\} \trianglelefteq K \trianglelefteq A_4 \trianglelefteq S_4$ is an invariant series. Moreover, $K \cong K/\{(1)\}$, $A_4/K \cong \mathbb{Z}_3$, and $S_4/A_4 \cong \mathbb{Z}_2$ are abelian. Therefore S_4 is solvable.

Example 11.6

Suppose $G = S_3 \times A_4$. For K as in Example 11.5,

$$\{\{(1)\}, A_3, S_3\} \qquad \text{and} \qquad \{\{(1)\}, \langle (1\ \ 2)(3\ \ 4) \rangle, K, A_4\}$$

form composition series for S_3 and A_4, respectively. We leave it to you to verify that the subgroups in the set

$$\{\{(1)\} \times \{(1)\}, \{(1)\} \times \langle (1\ \ 2)(3\ \ 4) \rangle, A_3 \times \langle (1\ \ 2)(3\ \ 4) \rangle,$$
$$A_3 \times K, S_3 \times K, S_3 \times A_4\}$$

form a composition series for G whose factors are cyclic of prime order. Therefore G is solvable.

While we have shown that S_3 and S_4 are solvable, it happens that S_n is **not** solvable for $n \geq 5$. Before we prove this result, however, we wish to present some additional properties of solvable groups. The first shows that any factor group of a solvable group is also solvable.

Theorem 11.6

Let $N \trianglelefteq G$ where G is a solvable group. Then G/N is solvable.

Proof

Let $\{1\} = G_0 \trianglelefteq G_1 \trianglelefteq \cdots \trianglelefteq G_n = G$ be a solvable series for G. If $G_i \trianglelefteq G_{i+1}$, then $NG_i \trianglelefteq NG_{i+1}$ (see Exercise 8) and

$$N/N = NG_0/N \trianglelefteq NG_1/N \trianglelefteq \cdots \trianglelefteq NG/N = G/N$$

is a normal series for G/N. To show that this series is a solvable series for G/N, we must verify that each factor group is abelian. By the Third Isomorphism Theorem (Theorem 8.7), we have

$$(NG_{i+1}/N)/(NG_i/N) \cong NG_{i+1}/NG_i$$

But $G_i \trianglelefteq G_{i+1}$ implies $NG_{i+1} = (NG_i)(G_{i+1})$, so that

$$NG_{i+1}/NG_i \cong (NG_i)G_{i+1}/NG_i \cong G_{i+1}/(NG_i \cap G_{i+1})$$

by the Second Isomorphism Theorem (Theorem 8.5). Now G_{i+1}/G_i is abelian and $G_{i+1}/(NG_i \cap G_{i+1})$ is isomorphic to a factor group of G_{i+1}/G_i. (See Exercise 9.) Therefore $G_{i+1}/(NG_i \cap G_{i+1})$ is abelian for $i = 0, 1, \ldots, n-1$, and the result follows.   □

    If $N \trianglelefteq G$ and G is solvable, then we know that the series $\{1\} \trianglelefteq N \trianglelefteq G$ can be refined to a composition series of G. Since G is solvable, the composition series will have cyclic factors; but then N, as one of the terms of this composition series, is itself solvable. However, the condition that N is normal in G is not necessary, as the next result demonstrates.

Theorem 11.7

Let $H \leqslant G$ where G is a solvable group. Then H is solvable.

Proof

Let $\{1\} = G_0 \trianglelefteq G_1 \trianglelefteq \cdots \trianglelefteq G_n = G$ be a solvable series for G. Set $H_i = G_i \cap H$ for $i = 0, 1, 2, \ldots, n$. We claim that

$$\{1\} = H_0 \trianglelefteq H_1 \trianglelefteq H_2 \trianglelefteq \cdots \trianglelefteq H_n = H$$

is a solvable series for H.

    First, if $G_i \leqslant G$, then $H_i = G_1 \cap H \leqslant G \cap H = H$. Furthermore, if $h \in H_i = G_i \cap H$ and $x \in H_{i+1} = G_{i+1} \cap H$, then $G_i \trianglelefteq G_{i+1}$ implies that $xhx^{-1} \in G_i \cap H = H_i$. Thus $H_i \trianglelefteq H_{i+1}$ for $i = 0, 1, 2, \ldots, n-1$.

    We therefore only need show that H_{i+1}/H_i is abelian for $i = 0, 1, 2, \ldots, n-1$. Now

$$H_{i+1}/H_i = (G_{i+1} \cap H)/(G_i \cap H)$$

Since $H_i = G_i \cap H = (G_i \cap H) \cap G_{i+1}$, we can use the Second Isomorphism Theorem (Theorem 8.5) to obtain

$$H_{i+1}/H_i \cong (G_{i+1} \cap H)/((G_{i+1} \cap H) \cap G_i) \cong (G_{i+1} \cap H)G_i/G_i$$

But

$$(G_{i+1} \cap H)G_i/G_i \leqslant G_{i+1}/G_i$$

is an abelian group. Therefore H_{i+1}/H_i is abelian for $i = 0, 1, 2, \ldots, n-1$, and H is solvable.   □

We have now shown that if G is solvable, then so is every subgroup and factor group of G. In this context, solvable groups behave in the same manner as abelian and cyclic groups. The next result, however, does not hold for abelian or cyclic groups.

Theorem 11.8

If $N \trianglelefteq G$ and both N and G/N are solvable groups, then G is a solvable group.

Proof

Let $\phi: G \to G/N$ be the homomorphism defined by $\phi(g) = gN$. If

$$N/N = B_0 \trianglelefteq B_1 \trianglelefteq \cdots \trianglelefteq B_m = G/N \qquad \text{and}$$

$$\{1\} = A_0 \trianglelefteq A_1 \trianglelefteq \cdots \trianglelefteq A_n = N$$

are solvable series for G/N and N, respectively, then let $C_i = \phi^{\leftarrow}(B_i)$. By the Correspondence Theorem (Theorem 8.6), we have $C_i \leqslant G$, $B_i = C_i/N$, and $C_i \trianglelefteq C_{i+1}$, with $C_0 = N$ and $C_m = G$. But then we can easily see that

$$\{1\} = A_0 \trianglelefteq A_1 \trianglelefteq \cdots \trianglelefteq A_n = N = C_0 \trianglelefteq C_1 \trianglelefteq \cdots \trianglelefteq C_m = G$$

is a solvable series for G since $C_{i+1}/C_i \cong (C_{i+1}/N)/(C_i/N) = B_{i+1}/B_i$ is abelian for $i = 0, 1, 2, \ldots, n-1$. $\qquad\square$

Solvable groups have been investigated for over seventy years. However in 1962, two renowned mathematicians, Walter Feit and John G. Thompson, proved a long-standing conjecture of W. Burnside that every group of odd order is solvable. (Burnside had shown this result to be true for groups of order less than 40,000.) This result, stated so simply, is one of the outstanding accomplishments of modern mathematics. The proof of the Feit-Thompson Theorem extends to over 450 pages of deep mathematics.[2]

For a group to be nonsolvable, it is necessary that its order be divisible by 2. Just because a group has order divisible by 2, however, does not mean that the group is nonsolvable. The examples of solvable groups we gave in this section all had even order; indeed, if 2 is the only prime dividing the order of G, then G is solvable.

Theorem 11.9

Let p be a prime and P a p-group. Then P is solvable.

Proof

We will induct on $|P|$, with the case $|P| = 1$ being trivial. Assume that the result is true for all p-groups of order less than $|P|$. By Theorem 9.8, P

---

[2]Walter Feit and John G. Thompson, "Solvability of groups of odd order," *Pacific J. Math.*, Vol. 13, No. 3 (1963), pp. 775–1029.

contains a nontrivial center $Z(P)$. If $Z(P) = P$, then P is abelian and therefore solvable. If $Z(P) \neq P$, then both $Z(P)$ and $P/Z(P)$ are p-groups of order less than $|P|$. By our induction hypothesis, both $Z(P)$ and $P/Z(P)$ are solvable. The result follows immediately from Theorem 11.8. $\square$

After all these results, you might be tempted to think that every group is solvable, inasmuch as we have yet to give an example of a nonsolvable group.

We know that a nonabelian simple group could never be solvable, but again we need an example of such a group. It turns out that the set of alternating groups A_n, for $n > 4$, provides an infinite number of nonabelian simple groups.

To prove this result, let us first consider A_5, the smallest of these groups. A_5 has order $5!/2 = 60$. We want to show that A_5 has no normal subgroups other than $\{(1)\}$ and A_5. The major tool we utilize is Theorem 9.4, which states that a normal subgroup must be a union of conjugate classes.

A_5 consists of the following 60 elements:

   i. the identity (1);

   ii. $20 = \dfrac{5 \cdot 4 \cdot 3}{3}$ 3-cycles, such as $(1 \quad 2 \quad 3)$;

   iii. $15 = \dfrac{1}{2} \dfrac{(5 \cdot 4)}{2} \cdot \dfrac{(3 \cdot 2)}{2}$ products of disjoint 2-cycles, such as $(1 \quad 2)(3 \quad 4)$; and

   iv. $24 = \dfrac{5!}{5}$ 5-cycles, such as $(1 \quad 2 \quad 3 \quad 4 \quad 5)$.

We first claim that all twenty 3-cycles lie in the same conjugate class. This follows from Theorem 9.2 and the fact that

$$C_{A_5}((1 \quad 2 \quad 3)) = \langle (1 \quad 2 \quad 3) \rangle$$

(See Exercise 9c of Section 9.1.) One can also show that

$$C_{A_5}((1 \quad 2)(3 \quad 4)) = \{(1), (1 \quad 2)(3 \quad 4), (1 \quad 3)(2 \quad 4), (1 \quad 4)(2 \quad 3)\}$$

Therefore all products of disjoint 2-cycles are conjugate. Finally, since

$$C_{A_5}((1 \quad 2 \quad 3 \quad 4 \quad 5)) = \langle (1 \quad 2 \quad 3 \quad 4 \quad 5) \rangle$$

(See Exercise 9g of Section 9.1), we have

$$[A_5 : C_{A_5}((1 \quad 2 \quad 3 \quad 4 \quad 5))] = \frac{60}{5} = 12$$

Hence, A_5 contains two distinct conjugate classes of 5-cycles, each containing 12 elements.

Theorem 9.4 implies that every normal subgroup of A_5 is a union of conjugate classes. Thus the order of any normal subgroup of A_5 is a sum of numbers from $\{1, 12, 12, 15, 20\}$. Moreover, 1 must always appear in the sum. It is an easy computation to show that no such sum is a proper divisor of $|A_5| = 60$. We have therefore proved the following theorem.

Theorem 11.10

A_5 is a simple group.

Our next goal is to extend this theorem to show that A_n is simple for all $n \neq 4$.

Lemma 11.1

Let $N \trianglelefteq A_n$, $n \geqslant 5$. If N contains a 3-cycle, then $N = A_n$.

Proof

By suitable relabeling, we can assume that $(1 \quad 2 \quad 3) \in N$. Then

$$(1 \quad 3 \quad 2) = (1 \quad 2 \quad 3)(1 \quad 2 \quad 3) \in N$$

and so $N \trianglelefteq G$ implies that $g(1 \quad 3 \quad 2)g^{-1} \in N$ for all $g \in A_n$. Consider $g = (1 \quad 2)(3 \quad k)$ for $k > 3$. Since $g(1 \quad 3 \quad 2)g^{-1} = (1 \quad 2 \quad k)$, N contains $(1 \quad 2 \quad k)$ for all $k = 3, 4, \ldots, n$.

We claim that N contains **all** 3-cycles of A_n. First,

$$(1 \quad 2 \quad i)(1 \quad 2 \quad j)(1 \quad 2 \quad j) = (1 \quad j \quad i)$$

implies that N contains all 3-cycles of the form $(1 \quad j \quad i)$. Secondly,

$$(1 \quad j \quad k)(1 \quad j \quad i)(1 \quad k \quad j) = (i \quad j \quad k)$$

shows that N contains all 3-cycles of the form $(i \quad j \quad k)$, thus proving our claim.

Now let $(i \quad j)(k \quad \ell)$ be an element of A_n. Since

$$(i \quad j)(k \quad \ell) = (i \quad k \quad j)(i \quad k \quad \ell)$$

N contains all elements of the form $(i \quad j)(k \quad \ell)$. But **every** element of A_n is a product of elements of the form $(i \quad j)(k \quad \ell)$. Thus $N = A_n$.   □

The last two results have been proved by construction. A "brute force" argument of this type, while usually not elegant in style, is adequate and often the easiest available.

What follows is another proof by construction:

Theorem 11.11

Abel's Theorem

The alternating group A_n for $n \neq 4$ is a simple group.

Proof

We have already seen in Exercise 4 of Section 8.1 that A_4 has a normal subgroup of order 4. Moreover, A_1, A_2, and A_3 have orders 1, 1, and 3, respectively, and are therefore simple. We proceed by induction on n. By Theorem 11.10, we can assume $n > 5$ and that A_m is simple for all $m < n$, $m \neq 4$.

Suppose A_n is not simple and there exists $N \trianglelefteq A_n$ with $|N| \neq 1$. We first show that N contains a nontrivial element fixing some letter in $\{1, 2, \ldots, n\}$. Let $g \in N$, $g \neq (1)$. Now $g = c_1 c_2 \cdots c_t$, a product of disjoint cycles.

> **Case 1:** Suppose each c_i is a 2-cycle and $t = 2$. Then $n > 5$ implies that g moves only 4 letters and, thus, fixes at least 2 letters.
>
> **Case 2:** Suppose each c_i is a 2-cycle and $t > 2$, say $c_1 = (1 \quad 2)$, $c_2 = (3 \quad 4)$, $c_3 = (5 \quad 6)$, $\ldots$. Set $h = (1 \quad 2)(3 \quad 5)$. Since $N \trianglelefteq A_n$, $hgh^{-1} \in N$. But then $hgh^{-1}g^{-1}$ is an element of N fixing 1 and moving 3. Therefore $hgh^{-1}g^{-1}$ is the desired element.
>
> **Case 3:** Suppose c_1 is a 3-cycle, say $c_1 = (1 \quad 2 \quad 3)$, and c_i is a 2-cycle or a 3-cycle, say $c_2 = (4 \quad 5)$ or $c_2 = (4 \quad 5 \quad 6)$. Let $h = (1 \quad 3 \quad 4)$. Then $hgh^{-1}g^{-1}$ is an element of N fixing 6 but not 2.
>
> **Case 4.** If c_1 is a cycle of length 4 or more, say $c_1 = (1 \quad 2 \quad 3 \quad 4 \quad \cdots)$, set $h = (3 \quad 2 \quad 1)$. Then $hgh^{-1}g^{-1}$ is an element of N fixing 2 but not 1.

In any case, N contains a nontrivial element x fixing some letter, say $x(k) = k$.

Let $F(k) = \{g \in A_n \mid g(k) = k\}$. Then $F(k)$ is the set of all even permutations on $\{1, 2, \ldots, k-1, k+1, \ldots, n\}$ and is therefore isomorphic to A_{n-1}. Since $N \trianglelefteq A_n$ but $N \neq A_n$, we have $F(k) \cap N \trianglelefteq F(k) \cong A_{n-1}$.

By induction hypothesis, A_{n-1} is simple and the existence of $x \in F(k) \cap N$, $x \neq (1)$, forces $F(k) \cap N = F(k)$, so that $F(k) \leqslant N$. Since $n > 5$ and $F(k) \cong A_{n-1}$, we conclude that $F(k)$, and hence H, contains a 3-cycle. By Lemma 11.1, $N = A_n$, contrary to our choice of N. Thus A_n is simple, and the result follows. $\qquad \square$

Corollary 11.2

The alternating group A_n, $n \geqslant 5$, is not solvable.

We have now produced an infinite family of nonsolvable groups. We hasten to add, however, that it is not necessarily true that every nonsolvable group is simple.

Corollary 11.3

The symmetric group S_n for $n \geqslant 5$ is nonsolvable.

Proof

If S_n were solvable, then A_n would also be solvable by Theorem 11.7. Since A_n is not solvable, neither is S_n. $\qquad \square$

In Chapter 22, Corollary 11.3 will be used to prove a classical result of Abel on solutions to polynomial equations.

1. Verify the assertions in Example 11.5.

2. Is $\{1\} \leqslant A_3 \leqslant A_4 \leqslant S_4$ a solvable series for S_4? Why?

3. Prove directly that $S_3 \times S_3$ is a solvable group.

4. Verify the statement in Example 11.6 that the subgroups in the set

$$\{\{(1)\} \times \{(1)\}, \ \{(1)\} \times \langle (1\ 2)(3\ 4)\rangle, \ A_3 \times \langle (1\ 2)(3\ 4)\rangle, \ A_3 \times K, \ S_3 \times K, \ S_3 \times A_4\}$$

   form a composition series for $S_3 \times A_4$ with abelian factors. Determine all other composition series for this group.

5. Prove that if G_1 and G_2 are solvable, then $G_1 \times G_2$ is solvable.

6. Let D be the group described in Example 8.6. Prove directly that D is solvable.

7. A solvable group is sometimes defined as a group G possessing a normal series whose factors are cyclic of prime order. Show that this definition is equivalent to our Definition 11.4.

8. Suppose $N \trianglelefteq G$ and $G_1 \trianglelefteq G_2 \leqslant G$. Prove that $NG_1 \trianglelefteq NG_2$.

9. If $N \trianglelefteq G$, $G_1 \trianglelefteq G$, and $G_1 \leqslant G_2 \trianglelefteq G$, then prove that

$$NG_1 \cap G_2 \trianglelefteq G_2 \quad \text{and} \quad G_2/(NG_1 \cap G_2)$$

   is isomorphic to a factor group of G_2/G_1.

10. A classical theorem of W. Burnside (1904) says that if G is a group of order $p^\alpha q^\beta$, where p and q are primes, then G is solvable. Prove this theorem in the special case that G has a normal subgroup of the order p^α.

11. A class F of finite groups is called a **formation** if:
    a. $G \in F$ implies $G/N \in F$ for all $N \trianglelefteq G$, and
    b. $G/N_1 \in F$ and $G/N_2 \in F$ for $N_1 \trianglelefteq G$, $N_2 \trianglelefteq G$ imply $G/(N_1 \cap N_2) \in F$.
    Use Exercise 19 of Section 10.1 to show that the class of solvable groups is a formation. What about cyclic groups, abelian groups, and the p-groups?

12. Let G be a group, and define $G' = \langle ghg^{-1}h^{-1} | g, h \in G \rangle$; that is, G' is the subgroup of G generated by all elements of the form $ghg^{-1}h^{-1}$ for $g, h \in G$. In Exercise 16 of Section 8.1, you were asked to verify that $G' \trianglelefteq G$.
    a. If $N \trianglelefteq G$ and G/N is abelian, prove $G' \leqslant N$.
    b. If G is a finite solvable group, prove $G' \neq G$.
    c. Define $G^{(1)} = G$, $G^{(k+1)} = (G^{(k)})'$ for $k \geqslant 1$. If G is a finite solvable group, verify that there exists $n \in \mathbb{N}$ such that $G^{(n)} = \{1\}$.

13. Let N be a union of conjugate classes of G. Is it true that $N \trianglelefteq G$?

14. Verify the assertions contained in the various cases of Theorem 11.11.

15. If $H \trianglelefteq A_n$, $n \geqslant 5$, and $\alpha = (1\ 2)(3\ 4) \in H$, prove $H = A_n$. Is the statement true when we remove $n \geqslant 5$?

16. A finite group G is termed **nilpotent** if each Sylow subgroup of G is normal in G. Prove that any nilpotent group is solvable.

17. A group G is said to be **supersolvable** if it has an invariant series all of whose factors are cyclic. Prove that every supersolvable group is solvable. Verify that A_4 is an example of a solvable group that is not supersolvable.

18. Let G be a finite group, and let A and B be normal solvable subgroups of G.
    a. Verify that AB is another normal solvable subgroup of G.
    b. Prove that G contains a unique maximal normal solvable subgroup.

*19. If G is a group of order n and n is odd or $n = p^{\alpha}$ or $n = p^{\alpha}q^{\beta}$, where p and q are primes, then the Feit-Thompson Theorem, Theorem 11.9, and Burnside's Theorem (see Exercise 10) yield that G is solvable. Write a computer program to enumerate all values of such n for $n \leqslant 1000$.

11.3

The Search for Simple Groups

The Jordan-Hölder Theorem asserts that every finite group G has a unique set of composition factors associated with G. These factors must be simple groups. Although a group cannot always be constructed immediately from its composition factors (for example, $\mathbb{Z}_6$ and S_3 have the same composition factors), the composition factors can provide information on many properties of the group. In addition, there are methods in existence whereby, given groups K and H, all possible Cayley tables of a group G such that $G/K \cong H$ can be determined. In this way, the composition factors form the basic components or building blocks of group theory, and knowledge of all possible composition factors enables us to "determine" all possible finite groups. Since any simple abelian group must be cyclic of prime order (Corollary 9.3), the problem before us is to list all nonabelian simple groups.

This is an enormous problem and only recently has this classification been completed. The classification of finite simple groups has been a 30-year effort involving several hundred research papers and well over 5000 pages of long technical results. Although we have little hope of describing the details of this project in the next few pages, we do propose to discuss the history of the classification problem, consider some important milestones, and investigate some techniques and results that are interesting in their own right.

Simple groups first appeared on the mathematical scene when Galois gave the basic definition of simple group and proved that A_5 is simple. In 1870, C. Jordan considered five infinite families of simple groups. One of these families consisted of the alternating groups A_n, which we have shown to be simple nonabelian groups. The other four families involved groups of matrices whose entries were contained in an algebraic structure called a **finite field**. We will investigate finite fields in Chapter 21.

We will try to give an idea of what form a group in one of these families takes. We will see later that $\mathbb{Z}_p$, the integers modulo p for a prime p, is a finite field. Set

$$SL(2, 7) = \left\{ \begin{bmatrix} a & b \\ c & d \end{bmatrix} \middle| a, b, c, d \in \mathbb{Z}_7, \; ad - bc \equiv 1 \;(\text{mod } 7) \right\}$$

This set forms a group under matrix multiplication (see Exercise 2) and is called a **special linear group**. The center of this group, $Z(SL(2, 7))$, consists of all matrices of the form

$$\begin{bmatrix} a & 0 \\ 0 & a \end{bmatrix} \quad \text{where} \quad a = 1 \quad \text{or} \quad 6$$

Then the factor group

$$PSL(2, 7) = SL(2, 7)/Z(SL(2, 7))$$

is a group called a **projective special linear group**. This group is simple and nonabelian, and its order is 168. Additional simple groups can be obtained for other choices of the prime p.

This family and groups in Jordan's other three classes occur in transformational geometry and have been given names like **orthogonal**, **unitary**, and **symplectic groups**.

In 1892, O. Hölder determined all simple groups of order less than 200 by considering each order and utilizing numerical results including the Sylow Theorems. One of the earliest American group theorists, F. N. Cole, extended Hölder's work by determining all simple groups of order at most 600.[3]

Let us consider some of the methods used by Hölder and Cole. Suppose G is a group of order 56. The number of 7-Sylow subgroups of G must be 1 or 8 by Sylow's Third Theorem. If G has but one 7-Sylow subgroup, then that subgroup must be normal, and G is not simple. If G has eight 7-Sylow subgroups, let P_1 and P_2 be two distinct 7-Sylow subgroups. Then $P_1 \cap P_2 = \{1\}$ since $|P_1 \cap P_2|$ divides $|P| = 7$ by Lagrange's Theorem. Each 7-Sylow subgroup contains 6 nonidentity elements, each of order 7, that are not contained in any other 7-Sylow subgroup. This means that G contains $8 \cdot 6 = 48$ elements of order 7. Then there can exist only $56 - 48 = 8$ elements to form the 2-Sylow subgroup(s) of G. Because each 2-Sylow subgroup of G has order 8, we conclude that G has but one 2-Sylow subgroup R and $R \trianglelefteq G$. Therefore, G is not simple in any case.

Another technique for showing that a group of order n is not simple involves Cayley's Representation Theorem. It relies on the following result.

Theorem 11.12

Suppose H is a subgroup of G such that $|G|$ does not divide $[G:H]!$. Then G is not simple.

Proof

Suppose $[G:H] = n$ and $\Omega = \{g_1H, g_2H, \ldots, g_nH\}$ are the distinct left cosets of H in G. For $x \in G$, define $\phi_x \in S_\Omega$ by

---

[3] F. Cole "Simple groups from order 201 to order 500," *Amer. J. Math.*, Vol. 14 (1892), pp. 378–388 and "Simple groups as far as order 600," *Amer. J. Math.*, Vol. 15 (1893), pp. 303–315.

$$\phi_x = \begin{pmatrix} g_1 H & g_2 H & \cdots & g_n H \\ x g_1 H & x g_2 H & \cdots & x g_n H \end{pmatrix}$$

We leave it to you to verify that ϕ_x is indeed a permutation and the mapping $\phi: G \to S_\Omega \cong S_n$ defined by $\phi(x) = \phi_x$ is a homomorphism. (See Exercise 6.) Now $G/\ker \phi$ is isomorphic to $\phi^\to(G)$ by the First Isomorphism Theorem (Theorem 8.3). On the other hand, $\phi^\to(G)$ is isomorphic to a subgroup of S_n. Therefore

$$\frac{|G|}{|\ker \phi|} = |G/\ker \phi|$$

must divide $|S_n| = n!$. If $|G|$ does not divide $n!$, then $|\ker \phi| > 1$, and $\ker \phi$ is a nontrivial normal subgroup of G. Thus G is not simple. □

To see how to apply this result, consider a group G of order $600 = 2^3 \cdot 5^2$. The Sylow Theorems imply that G has either one or six 5-Sylow subgroups. (See Exercise 7.) If G is not simple and P is a 5-Sylow subgroup of G, then Sylow's Theorem 9.11 says that $[G : N_G(P)] = 6$. But $|G| = 600$ does not divide $6! = 720$. This contradiction forces $P \trianglelefteq G$, and G is not simple.

In his analysis, Cole discovered that there was a simple group of order 504. This group is $PSL(2, 8)$, a group of 2×2 matrices with entries in a finite field with 8 elements. In 1897, L. E. Dickson proved that all groups of the type $PSL(m, p^n)$, $m > 1$, $n \geqslant 1$, are simple groups except when $p^n = 2$ or 3.

In 1861, E. Mathieu discovered a family of transitive permutation groups. A permutation group G on a set Ω is termed **transitive** if, given any $a, b \in \Omega$, there exists $\sigma \in G$ such that $\sigma(a) = b$. G is **2-ply** or **doubly transitive on Ω** if, given distinct $a_1, a_2 \in \Omega$ and distinct $b_1, b_2 \in \Omega$, there exists at least one $\sigma \in G$ such that $\sigma(a_1) = b_1$ and $\sigma(a_2) = b_2$. Similar definitions can be made for 3-ply transitive, 4-ply transitive, and so on. Mathieu found five multiply-transitive permutation groups. The smallest of these **Mathieu groups** is M_{11} of order 7920, while the largest M_{24} has order 244,823,040. In his classification work, F. Cole showed that M_{11} is simple.[4] In 1900, G. A. Miller proved that the other four Mathieu groups are also simple. The five groups had an additional property: they did not appear to be contained in some infinite family of simple groups. Simple groups not contained in some infinite family of simple groups came to be known as **sporadic groups**, and the Mathieu groups were the first examples.

W. Burnside worked to extend the range of numbers used by Hölder and Cole and developed a good deal of theory relating to permutation groups. In 1901 he showed that no nonabelian group of odd order less than

---

[4] F. Cole, "List of the transitive substitution groups of ten and of eleven letters," *Quart. J. Pure Appl. Math.*, Vol. 27 (1895), pp. 39–50.

40,000 is simple. He conjectured that a nonabelian simple group must have even order. As we have already seen, this result was finally verified in 1963.

L. E. Dickson extended Jordan's other families of simple groups and investigated possible isomorphisms between groups of the same order lying in different families of simple groups. In 1905, he discovered a new infinite family of finite simple groups. However, his work, marked the start of a long period of relative inactivity in the search for simple groups. A number of infinite families of finite simple groups had been determined in addition to the five sporadic Mathieu groups. While work was done in extending the range of orders of simple groups, no new simple groups were found during the next fifty years.

However, in 1955, Claude Chevalley completed an investigation begun by Dickson and found several more families of simple groups termed **groups of Lie type**. Within the next five years, Chevalley's methods were adapted and more infinite families of simple groups were discovered. In 1960, M. Suzuki discovered a new infinite family while working with doubly-transitive groups. These groups could be looked at as groups of 4×4 matrices. A year later, R. Ree extended Suzuki's methods to find two more families. No additional infinite families of finite simple groups have been found since.

Despite this activity, the classification problem in the early 1960's appeared relatively settled. Then the dam burst! Beginning in 1966, new sporadic simple groups were discovered, one right after the other. A total of twenty-one new sporadic simple groups were found in the next ten years, ranging in size from Z. Janko's group of order

$$175,560 = 2^3 \cdot 3 \cdot 5 \cdot 7 \cdot 11 \cdot 19$$

to B. Fischer's "Monster" of order.[5]

$$808,017,424,794,512,875,886,459,904,961,710,757,005,754,368$$

What spurred this enormous amount of mathematical activity? Probably the Feit-Thompson Theorem provided the greatest impetus. It stated that every group of odd order is solvable. If G is a finite nonabelian simple group, then G must have even order. This allowed group theorists to focus attention on the 2-Sylow subgroups of G. For example, J. Walter in 1969 determined all simple groups with abelian 2-Sylow subgroups. Other examples can be given where the following approach was taken. Since the order of a simple group G is even, there exists $x \in G - \{1\}$ such that $x^2 = 1$.

---

[5] For more information on these sporadic groups, as well as the families of simple groups, we refer the reader to the excellent survey articles that have appeared in the September, 1976, issue of *Mathematics Magazine* and the December, 1985, issue of *Scientific American*. See Joseph A. Gallian, "The search for finite simple groups," *Math. Mag.*, Vol. 49, No. 4 (Sept. 1976), pp. 163–179, and Daniel Gorenstein, "The enormous theorem," *Sci. Amer.*, Vol. 253, No. 6 (Dec., 1985), pp. 104–115.

This element is termed an **involution**. Consider $C = C_G(\langle x \rangle)$, the centralizer of x in G. Suppose C has a certain form; say that C is isomorphic to the centralizer of an involution of a known simple group. Then G is shown to be one of a finite number of possibilities. Most results of this type utilized a great deal of computation and high-powered representation theory. When such an analysis led to an incomplete description of G, then researchers obtained candidates for new simple groups.

Other factors also contributed to the classification problem. John Thompson determined all simple groups with the property that all proper subgroups are solvable. These groups are termed **minimal simple groups**, and any simple group has at least one of these minimal simple groups related to it. The knowledge of these minimal simple groups allowed group theorists to eliminate many possible orders of simple groups. Connections were also found between group theory and graph and lattice theory. Major advances were made in representation and character theory, often aided by the use of large high-speed computers. All these factors contributed to the avid interest and activity in the field.

During the midst of the decade of sporadic group discovery, many group theorists surmised that the classification problem would never be solved. Are there an infinite number of sporadic groups? Do these sporadic groups themselves form some sort of infinite family? Some guessed that sporadic groups were similar to "fast guns" in the Old West—there would always be another to confront. This was not the case, however. By the late 1970's, some group theorists began to see the "light at the end of the tunnel."

In 1982, Daniel Gorenstein published the first of a projected three-volume series on the classification problem.[6] Not only does this volume give a description of all the infinite families and the twenty-six sporadic groups, but also contains an enormous amount of other related material. For example, it is now known that there exist at most two nonisomorphic simple groups of a given order n and all of these pairs of simple groups have been determined.

The solution to the classification problem does not bring an end to investigations in finite theory. Indeed there are still a vast number of questions to be resolved, not only concerning nonsimple groups but simple groups as well. The work involved in the classification problem has opened new areas of interest and research. Hopefully, activity in group theory will be as productive in the next hundred years as it has been in the last.

Exercises 11.3

1. Verify that S_3 and $\mathbb{Z}_6$ have the same composition factors.
2. Define matrix multiplication in $SL(2, 7)$ by using modular addition and multiplication with the corresponding formula for $M_2(\mathbb{Z})$ given in Exercise 15 of

---

[6]See Daniel Gorenstein, *Finite Simple Groups, An Introduction to Their Classification*, New York: Plenum Publishing, 1982.

Section 3.1. Prove that $SL(2, 7)$ is a group under multiplication. Verify that

$$Z(SL(2, 7)) = \left\{ \begin{bmatrix} 1 & 0 \\ 0 & 1 \end{bmatrix}, \begin{bmatrix} 6 & 0 \\ 0 & 6 \end{bmatrix} \right\}$$

3. If p is an odd prime, it can be proved that the group $PSL(2, p)$ has order

$$\frac{(p-1)p(p+1)}{2}$$

Show that $|PSL(2, 7)| = 168$ and $|PSL(2, 5)| = |A_5|$.

4. Use the techniques described in this section to show that G is a nonsimple group when $|G|$ has the value:
   a. 12   b. 48   c. 108
   d. 312   e. 2380   f. 392

5. Suppose H is a subgroup of G and $[G:H] \leqslant 4$. Prove that G is not simple.

6. Prove the assertions in Theorem 11.12 that ϕ_x is a permutation and that $\phi: G \to S_\Omega$ is a homomorphism.

7. Verify that if G is a group of order 600, then G has one or six 5-Sylow subgroups.

8. List all elements of $PSL(2, 3)$ and $PSL(2, 2)$. These groups are not simple. Can you identify these groups?

9. S_n is n-ply transitive on $\{1, 2, \ldots, n\}$. Find all permutations in S_4 that take 1 to 4 and 2 to 3.

10. Is A_4 doubly-transitive on $\{1, 2, 3, 4\}$? Is it 3-ply transitive? Is it 4-ply transitive?

11. Prove that

$$\begin{bmatrix} p-1 & 0 \\ 0 & p-1 \end{bmatrix}$$

is an involution in $SL(2, p)$ where p is an odd prime.

12. Suppose that G is simple and P is a 2-Sylow subgroup of G. Among all the involutions in G, prove that there exists an involution $x \in G$ such that $P \leqslant C_G(\langle x \rangle) < G$.

13. Let G be a group. The **upper central series of** G is defined by setting $Z_0(G) = \{1\}$, $Z_1(G) = Z(G)$, and for $n \geqslant 1$, $Z_{n+1}(G)$ equal to that subgroup of G such that

$$Z_{n+1}(G)/Z_n(G) = Z(G/Z_n(G))$$

   a. Prove that this series is well-defined in that $Z_n(G) \trianglelefteq G$ for all n.
   b. If P is a finite nontrivial p-group, then verify that there exists an $n \in \mathbb{N}$ such that $Z_n(P) = P$.
   c. If G is a finite group in which every Sylow subgroup is normal in G, then show that there exists an $n \in \mathbb{N}$ such that $Z_n(G) = G$.
   d. Find the upper central series for S_3 and the group D of Example 8.6.

14. Suppose G is a finite group and there exist $x, y \in G$ such that $yxy^{-1} = x^{-1}$. Prove that G has even order.

15. a. If G is a finite group containing a p-Sylow subgroup P such that $N_G(P) = C_G(P)$, prove that there exists $K \trianglelefteq G$ such that $G = KP$ and $K \cap P = \{1\}$.

    b. Use the result of part a to show that $\mathbb{Z}_2$ is the only finite simple group with a cyclic 2-Sylow subgroup.

16. Let

$$G = \left\{ \begin{bmatrix} a & b \\ c & d \end{bmatrix} \middle|\, a,\, b,\, c,\, d \in \mathbb{R},\; ad - bc = 1 \right\}$$

G is a group under the operation of matrix multiplication. Determine

$$H = \left\langle \begin{bmatrix} 0 & 1 \\ -1 & -1 \end{bmatrix}, \begin{bmatrix} 0 & 1 \\ 1 & 0 \end{bmatrix} \right\rangle.$$

*17. Extend Exercise 19 of Section 11.2 to include the Sylow Theorems and Theorem 11.9.

*18. Write a program to enumerate all the elements of $SL(2, 7)$.

12 Group Codes

One of the "boom" areas of modern technology is data communications, the transmission of information over communication channels, whether via telephone land line, microwave installations, ground-to-air satellites, or even connections in a computer from one component to another. Great importance is placed upon the accurate transmission and reception of large blocks of data. Natural and man-made phenomena, however, often cause "noise" along the communication channel resulting in incorrect reception of transmitted messages. Coding theory is an area of computer science whose goal is the detection and correction of such errors.

A basic model of a communication channel is given in Figure 12.1.

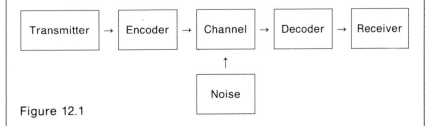

Figure 12.1

Information is normally transmitted as binary n-tuples, which are sequences of 0's and 1's. In other words, messages are written as binary numbers and sent out in strings or blocks of length n. In the encoder, certain **check digits** are added at the end of each block. These extended blocks are then sent along the communication channel where "noise" can sometimes cause a 0 to be changed to a

1 or a 1 to be changed to a 0. The decoder reads the message and, hopefully, detects and corrects errors and sends the message to the receiver where the message can be rewritten in original form.

    In this chapter, we wish to focus on two problems relating to coding. The first concerns the detection of errors in transmission. The second involves methods to correct a detected error. In error detection we will first use some basic probability theory as a means of comparing different coding procedures. Then we will utilize group theory to investigate error correction.

12.1

Binary Symmetric Channel

Since almost all computers use binary numbers within the CPU (Central Processing Unit), it makes sense to send out signals between computers in that form; then there is no need for conversion at either the transmitting end or the receiving end. However, with data originating at other sources, there first arises the problem of determining the simplest form in which to transmit the data while insuring that the number of errors occurring during transmission is minimal. It is obvious that the transmission format must contain at least two symbols, since with only one symbol everything would appear the same. That brings us to the computer binary system and the ASCII or American Standard Code for Information Interchange. In most microcomputers, this is a system by which each letter and/or operation has a special binary representation as a number between 0 and 127. For example, the upper-case letters A through Z are numbered 65 through 90 with A being 65 and Z being 90. Thus F has the ASCII code number of 71. The digits 0 through 9 are 48 through 57. Lower numbers are reserved for other operations relevant to computer functioning, at least in the case of home computers. We are not saying that the ASCII is *the* code used for all transmissions, but we use it as an example since it is a code familiar to those who use microcomputers.

    In considering how to transmit our data (including words, numbers, figures, pictures, etc.) in a "safe" manner, remember that no method of transmission is really 100% safe. When a signal is transmitted, two things can happen to an individual number or bit: it arrives either as a 0 or a 1; that is, it arrives intact or changed to the other bit.

    What are the chances of either happening? Instead of *chance*, mathematics uses the term **probability**. What is the probability of a bit being transmitted and received correctly? First, let us consider what is meant by *probability*. We are concerned with **experiments** (the transmission of a bit, for example) in which the **sample space** or the set of all possible outcomes of the experiment is finite. An **event** is the set of all members of the sample space that are considered to be favorable or successful outcomes of the experiment. For example, in the case of throwing or tossing a die (one half of a pair of dice) and obtaining a face with an odd number of dots, the sample space is $\{1, 2, 3, 4, 5, 6\}$ while the event is $\{1, 3, 5\}$. In the case of drawing an ace from a standard deck of playing cards, the sample space is all fifty-two cards (which for the sake of space we do not list here), while the event is the set $\{$ace of clubs, ace of diamonds, ace of hearts, ace of spades$\}$. Flipping a coin has sample space $\{$head, tail$\}$; the event of getting a head is $\{$head$\}$.

   The probability of a particular event is the cardinality (or number of elements) of the event divided by the cardinality of the sample space. Thus the probability of tossing an odd number on a die is $3/6 = 1/2$, the probability of tossing a head (or a tail, for that matter) is $1/2$, and the probability of drawing an ace is $4/52 = 1/13$.

   With this concept of probability, we might suspect at first that the probability of the reception of a particular bit in the manner in which it was transmitted is $1/2$. This is not correct because the probability of such an occurrence depends upon the media. To simplify things, let us say that the probability of the correct reception of the transmission of a bit is p in the case of a "1" and q in the case of a "0"—the values of p and q need not be the same. But then what is the probability of the incorrect transmission of a bit? Since there are only two elements in each sample space, we see that, in the case of a "1," it would be $1 - p$ and, in the case of a "0," $1 - q$.

   We are now prepared to consider the particular type of transmission media that we are going to study.

A **binary symmetric channel** consists of a transmitter sending signals 0 and 1, a receiver, and a probability p, $0 < p < 1$, of the incorrect transmission of a single digit.

   In this definition, the term **binary** refers, of course, to the fact that only one of two characters is being transmitted. **Symmetric** means that the probability of a 0 being sent and received incorrectly and the probability of a 1 being sent and received incorrectly are the same. In general, this is not always the case, since a 0 and a 1 may be transmitted differently and the format of one might be easier to convert than the other. For example, a 1 might be a peak in a signal while a 0 might be a plateau, and it might be easier to flatten a peak than to make a peak from a plateau. **Channel** simply refers to the method of transmission: radio signal, microwave, etc.

   Thus in a binary symmetric channel the probability of either bit being transmitted and received correctly is $1 - p$. Figure 12.2 represents the transmission probabilities in a binary symmetric channel.

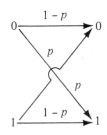

Figure 12.2

The probability p refers to a single digit or bit. If we assume that the probability of any one bit being transmitted incorrectly is independent of (does not depend or is not affected by) the incorrect or correct transmission of any other digit, and if we transmit a message of n bits, then the probability of the correct transmission of the message is $(1 - p)^n$. For example, suppose that the probability of an incorrect transmission is known to be .001. Then if we send an 8,000 bit message, the probability that it arrives correctly at the other end is $(1 - .001)^{8000} = .000334123$. This is not very good, you may say, but we are talking about an error occurring in any of the 8,000 bits of information transmitted. The probability of error .001 may be thought of as meaning that one out of each 1,000 bits transmitted is received incorrectly. That means that we can expect (but not necessarily get) eight errors in the signal transmission and any one of those can occur at any of the 8,000 positions in the message.

It is important to know how many different ways the errors can occur. For ease of discussion, let us consider a message that is transmitted in blocks, that is, n-digit binary words or n-place binary numbers consisting of 0's and 1's. If we consider the operation of place-addition modulo 2, then we have a group that we will denote by B_n. It is important to remember that by B_n we mean $(\mathbb{Z}_2)^n = \mathbb{Z}_2 \times \mathbb{Z}_2 \times \cdots \times \mathbb{Z}_2$, the direct product of n copies of the group $\mathbb{Z}_2$. For example, in the group B_5, $10010 + 11101 = 01111$. In Exercise 6, we ask you to prove directly that B_n is an abelian group under the operation of place-addition modulo 2.

In a similar manner, we consider the error pattern for the transmission of such a message. Let us denote by E the n-digit binary number having 1 in the ith place if an error occurred in the transmission of that bit and 0 in that position if there was no error in the transmission of that bit. How can E be determined? One method consists of finding where the errors occur and constructing the number E. Another method involves the comparison of the message transmitted and the message received. Let W be the transmitted message and R the received message; then the error pattern is $E = W + R$, where the addition is performed in B_n. To see this, note that, when the digits in the kth position in W and R agree, then the sum of those digits is 0 modulo 2, and there is no error in the transmission of the digit in that position. When the two digits disagree, indicating an error in transmission, then the sum of those two digits is 1.

Observing that each element in B_n is its own inverse, we see that the following equations hold in B_n:

    i. $E = W + R$
    ii. $R = W + E$
    iii. $W = R + E$

By utilizing the fact that each element in B_n is its own inverse, we can check the accuracy of a transmission. Each block of a message can be transmitted twice and the receiving station can compare the two inputs. If the sum is zero, then assume that the message is correct; otherwise, request a retransmission of that block.

This method, however, is not perfect. There is always the possibility that the same error is made twice. In order to analyze the probability of error in transmitting a message, we need to answer another question. Exactly how many different ways are there for k errors to occur in the transmission of a block of n characters?

What is needed here is a bit of combinatorics. The problem then is, "Given a set of n objects (the number of places or digits in the message), how many ways can you choose a subset of k elements (or how many different ways can k errors occur)?" It is a basic result in combinatorial set theory that the answer is the number

$$\binom{n}{k} = \frac{n!}{k!(n-k)!}$$

Recall that the number 0! is defined to be 1.

Theorem 12.1

A message block n units long is transmitted. The probability that there is an error in any bit is p, and the probability that there is no error is q. Then the probability that there are exactly k errors in the complete transmission is

$$\binom{n}{k} p^k q^{n-k}$$

Example 12.1

Suppose a message that consists of 200 digits is transmitted along a binary symmetric channel with probability of error $p = .01$ in the transmission of a single digit or bit. The probability of transmitting the message with no errors is

$$\binom{200}{0}(.01)^0(.99)^{200} = .13397967$$

while the probability of getting exactly one error is given by

$$\binom{200}{1}(.01)^1(.99)^{199} = .27066600$$

The probability of getting *at most* one error is the sum of these two exclusive or disjoint events; that is,

$$.13397967 + .27066600 = .40464567$$

In the next section, we will see how we use this method for detecting and/or correcting errors.

Exercises 12.1

1. Suppose that the probability that a digit 0 or 1 is transmitted correctly along a communication channel is .97. What is the probability that a message of 500 digits is transmitted:
   a. With no error?   b. With one error?   c. With at most one error?

2. If a communication channel has a probability $p = 1/10$ of error, then what is the probability that a message of 10 digits has:
   a. No errors?   b. Exactly two errors?   c. All errors?

3. If a binary symmetric channel has probability of error $p = .05$, then what is the probability that a message of 8 digits has:
   a. Exactly 3 errors?   b. At most 2 errors?

4. If a binary symmetric channel has probability $p = 1/8$ of error in the transmission of a single digit, what is the probability that a message of 6 digits has:
   a. Exactly 2 errors?   b. At most 2 errors?   c. At least 3 errors?

5. If a binary symmetric channel has probability of error $p = 1/10$, then what is the probability that a message of 10 digits has:
   a. Exactly 2 errors?   b. Exactly 3 errors?   c. At most 2 errors?
   d. At most 3 errors?   e. At least 4 errors?

6. Verify that for any natural number n, B_n is a commutative group.

7. Accepting the fact that the error message E for a transmitted message W and a received message R is determined by $E = R + W$, verify that:
   a. $R = W + E$ b. $W = R + E$

8. If the word 010111 is received as 010100, what is the error pattern E?

9. If a word W is received as $R = 100100$ with an error message or error pattern $E = 101011$, what is W?

10. If a transmitted word 101111 has an error message 010101, what word is received?

11. Suppose a message of 800 bits is transmitted along a communication channel where the probability of error in transmitting a 0 is 0.3 while the probability of correctly transmitting a 1 is 0.75. Find the probability that there are at most three errors in the complete transmission.

12. If a word of 5 bits is sent along a communication channel where the 0's are always transmitted correctly but the 1's are transmitted without error only 98% of the time, what is the probability that a message of 100 such words is transmitted correctly?

*13. Write a computer program that will determine the third of R, W, and E if the other two are known.

12.2

Elementary Coding Techniques

In this chapter we are considering what can happen to data sent over a communication channel. Before continuing this discussion, we pause briefly to discuss some different types of communication channels since we will see that some methods of data transmission and error detection are particularly suited for certain types of channels. A **simplex line** is a channel that permits the flow of information in only one way, comparable to a one-way street. A **half-duplex line** is a channel that permits communication both ways but only one way at a time, similar to the case of road work being done on a normally two-way street with traffic maintained but controlled. A **full-duplex line** is one that permits simultaneous communication, analogous to an actual two-way street.

In the actual transmission of information, there are basically two means of transmitting data serially. **Asynchronous**, or stop/start, transmission is a method by which an initial or "start" bit is transmitted and followed by message bits in an evenly-spaced pattern until a termination or "stop" bit is sent. The second method is **synchronous** transmission, which utilizes "hand-shaking." This means simply that the transmission rate follows a predeter-mined timing pattern, and thus has no need for start or stop bits.

With either of these basic means of transmission, we are concerned primarily with error detection and correction. Once an error is detected, there are several methods that may be used to correct the error. Of course, if a simplex line is being utilized, there is no way to notify the transmitting end of an encountered error. With a half-duplex line, a procedure called "stop and wait ARQ" can be used. In this method, a block of information is transmitted and then the sender pauses until receiving either an ACK (acknowledgement if no error is detected) or a NAK (negative acknowledgement in the case of error detection).

A more sophisticated procedure is employed if a full-duplex channel is being utilized; this one is called "go back N ARQ." The sender keeps transmitting blocks of data until receiving a NAK, at which point it backs up in its transmission N blocks and continues sending the message. With these aspects of transmission in mind, let us continue our examination of error detection.

| Example 12.2 |

Let us suppose that, in a specific communication channel, the probability that a digit is transmitted correctly is .998; so the probability of error in the transmission of a single digit is .002. Assume, furthermore, that a message of 1000 digits is transmitted in blocks of length 4. The probability of transmitting a block of 4 digits correctly is $(.998)^4 = .99202397$, so that the probability of sending the entire message of 250 blocks correctly is

$$[(.998)^4]^{250} = (.998)^{1000} = .13506453$$

Therefore the probability of receiving the message correctly is approximately 14%, meaning that there is about an 86% probability of getting a message containing one or more errors.

We would like to investigate a basic method for detecting errors, the **parity-check method**. To each block of four digits, let us add a fifth digit in the following way: if the sum of the first four digits is even, set the fifth digit equal to 0; if the sum of the first four digits is odd, then make the fifth digit a 1. For example, 0110 is transmitted as 01100, while 1011 is sent as 10111. Now every new block of five digits has even parity; that is, the sum of the digits in each block is even. If a single error occurs in the transmission of a block, the parity of the block will change from even to odd, so that the receiver will know an error has occurred.

How does this coding method affect the reliability of our transmission? Consider the following:

Example 12.3

In Example 12.2, if a block of 5 digits is sent, the probability of an error-free transmission is $(.998)^5 = .99003992$.

By Theorem 12.1 in Section 12.1, we know that the probability of getting a single error in the transmission of a 5-digit block is

$$\binom{5}{1}(.002)^1(.998)^4 = .00992024$$

Then the probability of transmitting a block with at most one error is

$$.99003992 + .00992024 = .99996016$$

If we repeat each block in which an error is detected until no error is detected, then the probability of getting a correct block transmitted can be shown to be .99995984 (see Exercise 2); so the probability of transmitting the original message of 1000 digits without error is now

$$(.99995984)^{250} = .99000018$$

which is a fantastic improvement over the previous .13506453.

There is a problem with the coding method just discussed. It requires two-way communication or a half-duplex channel. To correct an error, we had to have the block repeated. This may not always be possible or practical. Therefore we need to look for a real error-detecting and correcting coding scheme. One such method is called the **triple-repetition code**.

Suppose each digit x in our message is repeated three times as xxx. If the message xxx is received, then decode the message as x. If a single error occurs, the message received will be $xxy, xyx,$ or yxx. In this case we decode the message as the most frequently received digit, namely, x. Therefore, this method not only detects single and double errors, but corrects single errors immediately. One problem with this scheme is that a double error is decoded incorrectly. In other words, for a block xxx, a double error will be received as $xyy, yxy,$ or yyx and decoded as y. However, with this method, the probability of transmitting a digit without error along the channel of our last example is now $(.998)^3 = .99401199$. The probability of transmitting a digit with only one error is

$$\binom{3}{1}(.002)^1(.998)^2 = .00597602$$

so that the probability of transmitting a digit and decoding it correctly is

$$.99401199 + .00597602 = .99998801$$

Hence the probability that the original 1000 digit message is decoded correctly is

$$(.99998801)^{1000} = .98808152$$

This method is not quite as good as our parity-check model, but it does not require a two-way communication system.

There is, however, a drawback in the triple-repetition method. To use this procedure, we must transmit 3000 digits to receive a 1000 digit message, whereas we must transmit only 1250 digits plus repetitions of blocks with errors in the parity-check model. Thus the triple-repetition method is less efficient; nevertheless, the ability to correct and detect messages with only a one-way communication channel often makes up for lack of efficiency. In fact, if accuracy and not efficiency is desired, then one can extend the triple-repetition code to five-time and seven-time repetition codes to correct double and triple errors in transmission.

In order to see how group theory can be applied to coding theory, we need to analyze the errors that occur in the transmission of block or words. Let us assume that our words are blocks of five digits and consider them as elements of the group B_5, discussed in Section 12.1. Suppose the word $W = (1, 0, 1, 1, 0)$, or simply $W = 10110$, is received as $R = 10010$. Then a single error occurs in the third digit, and $E = 00100$ is termed the **error pattern** of the transmission.

The detection and correction of transmitted words is facilitated by the idea of *distance* as it relates to binary n-tuples.

Definition 12.2

> Suppose W_1 and W_2 are two code words. The **Hamming distance** $H(W_1, W_2)$ is the number of locations in which W_1 and W_2 differ.[1]

Thus $H(11001, 01101) = 2$, while $H(10001110, 10101011) = 3$. Note that if $E = W_1 + W_2$, where addition is performed in B_n, then $H(W_1, W_2)$ is equal to the number of 1's in E. The number of 1's in a word is its **weight**. For example, the weight of 11010 is 3. Therefore,

$$H(W_1, W_2) = \text{weight}(W_1 + W_2)$$

Suppose that a message in blocks of length n is encoded by the addition

---

[1] R.W. Hamming (1915–) pioneered much of the work in algebraic coding theory while working for Bell Labs.

of t digits into blocks of length $m = n + t$. Since $m > n$, not all possible blocks of length m appear in the dictionary of encoded words. If the distance between any two words in our encoded dictionary is at least $k + 1$, then no received word that differs from the transmitted word in at most k locations could ever be another word in our dictionary. This observation yields the following result:

Theorem 12.2

If the minimum distance between code words in an alphabet is $k + 1$, then the code can **detect** all errors of size k or less.

Example 12.4

Suppose the words

000	001	010	011	100	101	110	111

are encoded according to the following:

$$000 \rightarrow 0001001 \qquad 100 \rightarrow 1000111$$
$$001 \rightarrow 0011100 \qquad 101 \rightarrow 1010001$$
$$010 \rightarrow 0100011 \qquad 110 \rightarrow 1101101$$
$$011 \rightarrow 0111010 \qquad 111 \rightarrow 1110100$$

We leave it to you to verify that the minimum distance between the encoded words is three; so whenever a word is transmitted with one or two errors, it will be detected by the code.

A more important problem is how many errors can such a code **correct**. Suppose the word 0100001 is received. From the table in Example 12.4, we see that this word is only one digit away from 0100011 and more than one digit away from all the others. It seems logical to decode this word as 010 since the most probable error, a single error, would result in the received transmission. What about the received word 1100001? Note that both $H(1100001, 1101101) = 2$ and $H(1100001, 0100011) = 2$. The distance from 1100001 to the other code words is at least three. If our decoding rule is to choose the closest word in our alphabet to the word received, then we are faced with two choices and, therefore, cannot accurately decode the word.

This discussion leads to the following result whose proof is left to you in Exercises 9 and 10.

Theorem 12.3

If the minimum distance between code words in an alphabet is $2k + 1$, then the code can **correct** all words with k or fewer errors.

This result says that if $2k + 1 = 3$, then our code can correct all single errors; while if $2k + 1 = 5$, then the code can correct all double-error words.

Of course, if the minimum distance m between code words is even, then we must first set $2k + 1$ equal to $m - 1$ before solving for k.

On the basis of our two theorems, we want codes with a large enough minimum distance so as to accurately decode *almost all* of the errors. The number of errors in a transmitted word is related to the amount of *noise* on the channel. A noisy channel will require greater distance between code words to decode messages accurately than will a less noisy channel. On the other hand, if we add more check digits to a block, then the code is the less efficient from the viewpoint of the amount of digits that must be transmitted to accurately receive each word of the message.

We close this section with a historical note. The longest data transmission ever sent was on April 16–18, 1978, by the U.S. Post Office in a test of electronic mail. The total transmission was 56 hours, 21 minutes at 9600 characters per second. The total number of characters transmitted was in excess of 200,000,000.

Unfortunately, the Post Office was attempting to transmit 700 messages of 1000 characters each, which, because of internal problems, had to be retransmitted on an average of 286 times each!

Exercises 12.2

1. Does the parity-check code detect double errors? Triple errors? Quadruple errors? If the six digits are sent along a channel in which the probability of error is .05, what is the probability that there is:
   a. Exactly one error?   b. Exactly three errors?
   c. Exactly five errors?   d. An odd number of errors?
   e. An even number of errors?

2. Assume the probability of error in the transmission of a single digit along a communication channel is .002 and that messages are sent in blocks of five digits with the last digit being a parity-check digit. Suppose that whenever an error is detected in a block, that block is repeated until the parity-check code detects no error. Let x_i equal the number of errors in the ith transmission of a block and $\Pr(x_i = k)$ denote the probability that there are k errors in the ith transmission of a block.
   a. Verify the following:
       i. $\Pr(x_1 = 0) = .99003992$ ii. $\Pr(x_1 = 1) = .00992024$
       iii. $\Pr(x_1 = 2) = .00003976$ iv. $\Pr(x_1 = 3) = .00000008$
       v. $\Pr(x_1 = 4) = .00000000$ vi. $\Pr(x_1 = 5) = .00000000$
   b. Show that with the parity check code, the probability that an error is detected is .00992032 and the probability that an error occurs but is not detected is .00003976.
   c. From probability theory, the probability of finally receiving a correct block is

$$\Pr(x_1 = 0) + (\Pr(x_2 = 0 | x_1 = 1 \text{ or } 3) \cdot \Pr(x_1 = 1 \text{ or } 3))$$
$$+ (\Pr(x_3 = 0 | x_1 \text{ and } x_2 = 1 \text{ or } 3) \cdot \Pr(x_1 \text{ and } x_2 = 1 \text{ or } 3)) + \cdots$$
$$= .99003992 + (.99003992)(.00992024 + .00000008)$$
$$+ (.99003992)(.00992024 + .00000008)^2 + \cdots$$

Use the fact that

$$a(1 + p + p^2 + p^3 + \cdots) = \frac{a}{1 - p} \qquad \text{for} \quad 0 < p < 1$$

to show that the probability of sending a message of 250 blocks correctly is $(.99995984)^{250} = .99000018$.

3. Suppose that a triple repetition code is used with a channel whose probability of error is .01. Find the probability that a message of 500 digits is transmitted and decoded correctly.

4. Determine the detectability and correctability of the code $f: B_2 \to B_9$ defined by

     $00 \to 011001010$ $01 \to 100110011$

     $10 \to 010101100$ $11 \to 101110100$

5. Determine the detectability and correctability of the code $f: B_2 \to B_8$ given by

     $00 \to 00111011$ $01 \to 10010110$

     $10 \to 01001101$ $11 \to 11100000$

6. Determine the detectability and correctability of the following encodings:

     $000 \to 000000000$ $011 \to 011110010$

     $001 \to 001111101$ $101 \to 101100111$

     $010 \to 010101011$ $110 \to 110001101$

     $100 \to 100111110$ $111 \to 111011011$

7. Let $f: B_3 \to B_6$ be defined by

     $000 \to 000000$ $100 \to 100110$

     $001 \to 001011$ $101 \to 101101$

     $010 \to 010101$ $110 \to 110011$

     $011 \to 011110$ $111 \to 111000$

How many errors will the code (a) detect? (b) correct?

8. Show that the Hamming distance $H(W_1, W_2)$ satisfies the following properties for all W_1, W_2, and W_3 in $B_n = (\mathbb{Z}_2)^n$:
   a. $H(W_1, W_2) \geqslant 0$
   b. $H(W_1, W_2) = H(W_2, W_1)$
   c. $H(W_1, W_2) = 0$ if and only if $W_1 = W_2$
   d. $H(W_1, W_2) + H(W_2, W_3) \geqslant H(W_1, W_3)$

9. Prove Theorem 12.3: Suppose that the minimum distance between code words is $2k + 1$. Prove that the code can correct all words with k or fewer errors.

10. Prove the converse of Theorem 12.3; namely, if a code can correct all words with k or fewer errors, then the minimum distance between code words is at least $2k + 1$.

11. Find a code $f: B_3 \to B_7$ whose minimum distance between code words is four.

12.3

Group Codes

We wish to conclude our discussion by investigating one of the uses of group theory in coding. Recall our example in which blocks of length three were encoded as blocks of length 7 by adding certain check digits to obtain an alphabet of code words. In general, blocks of length n are augmented by check digits to obtain an alphabet C of code words of length m, a subset of the group B_m.

Definition 12.3

> A **group code** is a code in which the encoding function $f: B_n \to B_m$ is also a group homomorphism, so that the alphabet C of code words is a subgroup of B_m.

You should verify that code C given in Figure 12.3 is a group code. Since C is a subgroup of B_6, we know that the cosets of C form a partition of B_6. Any word received, therefore, occurs in one and only one coset of C. Furthermore, if the word W is sent along the channel and the word R is received, then we can exploit the fact that every element in the binary group B_6 is its own inverse to accurately decode R. If $C_1, C_2, \ldots, C_k$ are the cosets of C in B_6, let X_i be an element of the coset C_i with the smallest number of 1's. Then $C_i = X_i + C$, and every element of C_i can be written uniquely as $X_i + W'$ for some W' in C. If R is in the coset $C_i = X_i + C$, then we can write $R = X_i + W'$, and W' will be that code word in our alphabet whose distance from R is minimal.

Let us construct the cosets of C in B_6 by using as coset leaders those elements of B_6 with the fewest number of 1's, provided that these elements have not appeared already in a previously listed coset. (See Figure 12.4.)

Notice we cannot choose 110000 as our last coset leader since it already appears in the coset $000001 + C$. Among the possible coset leaders for the last coset, we choose 100100, although we could have selected 000011 or 011000.

The array given in Figure 12.4 can be used to decode messages. If the code word W is sent and the word R is received, simply locate the coset containing R and decode R as the word that appears at the top of R's column in Figure 12.4. Why does this work? The answer is clear if you remember that

$000 \to 000000$	$100 \to 100111$
$001 \to 001101$	$101 \to 101010$
$010 \to 010110$	$110 \to 110001$
$011 \to 011011$	$111 \to 111100$

Figure 12.3

Code Words

Cosets	000	001	010	011	100	101	110	111
$000000 + C$	000000	001101	010110	011011	100111	101010	110001	111100
$100000 + C$	100000	101101	110110	111011	000111	001010	010001	011100
$010000 + C$	010000	011101	000110	001011	110111	111010	100001	101100
$001000 + C$	001000	000101	011110	010011	101111	100010	111001	110100
$000100 + C$	000100	001001	010010	011111	100011	101110	110101	111000
$000010 + C$	000010	001111	010100	011001	100101	101000	110011	111110
$000001 + C$	000001	001100	010111	011010	100110	101011	110000	111101
$100100 + C$	100100	101001	110010	111111	000011	001110	010101	011000

Figure 12.4

we decode a word as the code word nearest to it in the sense of Hamming distance. If $R = E + W$ where E is the error pattern, then R lies in the coset $E + C$. Since we have used all the error patterns with the fewest number of 1's, W will be a code word whose distance from R is minimal.

Example 12.5

Suppose the message

110000   101010   100110   111111

is received. Our method says that we should conclude that

110001   101010   100111   011011

is the actual message sent and that we should decode the message as

110   101   100   011

Recall that since the minimum distance between our group code words is three, our code can correct all single errors. The last word differed from 011011 in two locations. Thus we cannot be sure that the actual word transmitted was 011011; in fact, if we had used 000011 as our coset leader, then we would have found that 111 appeared at the top of the column that contains the word 111111. Nevertheless, the decoded word 011011 was just as likely to be the actual word sent as was 111111, so that our choice is as good as any other. Our method does correct accurately all error patterns that appear as coset leaders, but not any others.

How can one implement this method in a computer? Suppose that the original message is given in blocks of length n and is encoded via a group code in blocks of length m. Then there exist 2^n code words and 2^m possible words that could be received. By Lagrange's Theorem, there exist $2^m/2^n = 2^{m-n}$ cosets of the code word subgroup C. If $n = 5$ and $m = 18$, this means that there are $2^{13} = 8192$ cosets. Certainly one would not wish to store all 8192 cosets, each containing $2^5 = 32$ words! Instead, simply store the code words and the coset leaders E, which represent error patterns with the least number of 1's as chosen in Example 12.5. Thus one needs to store only $2^n + 2^{m-n}$ elements. For $n = 5$ and $m = 18$, this means that one stores only $2^5 + 2^{13} = 32 + 3192 = 3224$, rather than all $2^{18} = 262,144$ words. If a word R is received, then $R = E + W$, where E is the error pattern for the actual word W transmitted. By exploiting the fact that every element of B_m is its own inverse, we can decode R by simply adding the coset leaders to R in turn until we obtain an element of W. This follows since $R = E + W$ implies

$$E + R = E + (E + W) = (E + E) + W = 0 + W = W$$

Here 0 represents the word that is the additive identity in B_m. Therefore, if $E + R$ is not a code word for any coset leader E, then we know that we cannot accurately decode R. If $E + R$ **is** a code word, then, to decode W, we only need to delete the check digits from $E + R = W$.

   We have seen how group theory, in particular, the structure of a binary group B_m and Lagrange's Theorem, can be used to implement an effective computerized method to encode and decode transmitted messages. This is only one example of how abstract algebra is used in applied mathematics and computer science. There are many other utilizations of algebra in computer science, and we recommend that you consult the references for additional material.

Exercises 12.3

1. Use the fragment of the group code table below to decode the words:

   011010   100001   010011   110001

   111001   000010   110111   001110

				Code Words			
Cosets	000	001	010	011	100	101	110	111
---	---	---	---	---	---	---	---	---
$000000 + C$	000000	001101	010010	011111	100110	101011	110100	111001
$100000 + C$	100000							
$010000 + C$	010000							
$001000 + C$	001000							
$000100 + C$	000100							
$000010 + C$	000010							
$000001 + C$	000001							
$100100 + C$	100100							

2. Is the code given in Example 12.4 a group code? Why or why not? Verify that the minimum distance between those code words is three.

3. Verify that the code given in Figure 12.3 is indeed a group code and that the minimum distance between code words is three.

4. Show that if B_n is encoded as a **group code** by the addition of check digits, then the digits added to $00000 \cdots 0$ must all be 0's.

5. Compare the efficiency between the triple repetition code and the group code given in the text in Figure 12.4.

6. a. Use the fact that a group homomorphism f satisfies $f(x + y) = f(x) + f(y)$ to complete the group code $f: B_3 \to B_6$ given by

   $000 \to 000000 \qquad 001 \to 001011 \qquad 010 \to 010101 \qquad 100 \to 100111$

   b. If the coset leaders for this code are

   $000000, \quad 100000, \quad 010000, \quad 001000,$

   $000100, \quad 000010, \quad 000001, \quad$ and $\quad 011000$

   then decode

   $011100 \qquad 111010 \qquad 100101 \qquad 110111$

7. Prove that if C is a group code in which the minimum distance between code words is k, then k is also the minimum number of 1's that appear in the nonzero code words of C.

8. How many words would one have to store in a computer to implement a group code $f: B_{21} \to B_{31}$?

9. a. As in Exercise 6, use the definition of group homomorphism to complete the group code $f: B_4 \to B_7$ given by

   $0001 \to 0001101 \qquad 0011 \to 0011010$

   $0111 \to 0111011 \qquad 1111 \to 1111101$

   b. How many errors can this code detect? correct?

10. Let a group code from B_2 to B_6 be given by

    $10 \to 101010 \qquad$ and $\qquad 01 \to 010101$

    a. How many errors can this code detect? correct?
    b. If the coset leaders include the words of weight one, find a complete set of coset leaders.
    c. Is this code equivalent to the triple-repetition code?

11. Does there exist a group code $f: B_2 \to B_6$ whose minimum distance between code words is four?

12. Suppose $f: B_m \to B_n$ is a group code.
    a. Verify that the code words of even weight form a subgroup of B_n.
    b. Give an example of a group code in which every code word has even weight.
    c. Prove that the subgroup of code words of even weight have index at most two in B_n.

13. If $f: B_m \to B_n$ is a group code, prove that the code cannot detect single errors unless f is a monomorphism.

Rings

So far in our study of mathematical structures, we have limited ourselves to considering those structures for which there was only one operation. However, our familiarity with the integers and practical everyday examples of many structures for which two or more operations are defined suggest that we should extend our horizons. The most general type of two-operation structure is called a *ring*. *Integral domains* and *fields* are two special types of rings that are quite important in abstract algebra. We will encounter all of these structures in this chapter.

The study of rings and fields constitutes an area that is in some ways older and in some ways younger than the study of groups. While the work of Abel and Galois utilized the concept of field, the precise definition was first given by Dedekind[1] in 1879. Early attempts to generalize the concept of integers led Gauss[2] to define the Gaussian integers, numbers of the form $a + bi$, where $a, b \in \mathbb{Z}$ and $i = \sqrt{-1}$. Another development was Dedekind's theory of *algebraic integers*, that is, numbers which satisfy a polynomial

---

[1]Julius Wilhelm Richard Dedekind (1831–1916) was the last pupil of Gauss. One of his noteworthy accomplishments was the extension of the rational numbers to the continuum of real numbers by a process now known as the method of Dedekind cuts.

[2]Johann Friedrich Gauss (1777–1855) is often called the "Prince of Mathematics" and is considered as the last genuine universal mathematician. One of the most prolific of all mathematicians, his list of achievements is too long to enumerate here since his research covered mathematics, physics, and astronomy.

equation with integral coefficients and leading coefficient 1. These generalizations created algebraic structures that did not require elements to have *unique factorization* in terms of *prime* elements. The desire to restore these familiar properties of $\mathbb{Z}$ motivated the definition of *ring* and *ideal* by Dedekind and Kummer.[3] These concepts, plus a number of other topics, await us in this and the following chapters.

---

[3]Ernst Eduard Kummer (1818–1893) introduced the concept of *ideal* into algebra. His great interest was in Fermat's Last Theorem and through his efforts to prove this result the concept of *ideal* arose. His attempts to prove Fermat's Last Theorem give another example of how the search for solutions to unsolved problems often leads to many other important discoveries along the way.

13.1

Definitions and Examples

When two operations are used concurrently on the same set, there is usually a desire to combine them in some fashion and have a means of interrelating these operations. This is accomplished by requiring one or more **distributive laws**. Therefore, a ring is defined by taking an abelian group, a second operation that is associative, and specifying how the second operation is related to the first. We make explicit the following definition.

Definition 13.1

A **ring** $(R, \oplus, \odot)$ is a nonempty set R, together with two binary operations $\oplus$ and $\odot$, such that

  i. $(R, \oplus)$ is an abelian group.
 ii. $\odot$ is associative on R:

$$a \odot (b \odot c) = (a \odot b) \odot c \qquad \text{for} \quad a, b, c \in R$$

iii. $\odot$ is both left and right distributive over $\oplus$:

$$a \odot (b \oplus c) = (a \odot b) \oplus (a \odot c)$$

and

$$(b \oplus c) \odot a = (b \odot a) \oplus (c \odot a) \qquad \text{for all} \quad a, b, c \in R$$

The rules contained in part iii are known as the **left and right distributive laws**. In a ring R, the first of the two binary operations on R is usually termed "addition" and the second "multiplication," although addition and multiplication do not have the same meaning as when applied to integers or real numbers.

A requirement that R be nonempty is unnecessary because any group must have an identity element. The second or "multiplicative" operation need not be commutative on R; in fact, we will see in some of the following examples that there exist many rings that do not have a commutative multiplication. We do require that the second operation be associative. Associativity is one of the most basic of properties; without it many simple calculations can become quite cumbersome.

Example 13.1

$(\mathbb{Z}, +, \cdot)$ is a ring. Here $+$ and $\cdot$ denote the familiar operations of addition and multiplication. This ring satisfies far more than just the basic requirements of the definition. In this ring, the operation $\cdot$ is commutative, and $\mathbb{Z}$ has an identity element for multiplication.

Example 13.2

Consider $(\mathbb{Z}_n, \oplus, \odot)$, the integers modulo n with modular addition and multiplication. $(\mathbb{Z}_n, \oplus, \odot)$ is also a ring in which the second operation is commutative and the ring has an identity for $\odot$. Furthermore, some of the elements in this ring have multiplicative inverses. (See Exercise 1.)

These two examples prompt us to introduce some definitions:

Definition 13.2

> A ring $(R, \oplus, \odot)$ is **commutative** if $\odot$ is a commutative operation on R.

Definition 13.3

> A ring $(R, \oplus, \odot)$ is a **ring with unity** if R contains an identity for the operation $\odot$.

The term **unity** is used to distinguish between the additive identity (denoted in our definition of **group** by e) and the multiplicative identity (sometimes denoted here by the symbol u). If $(R, +, \cdot)$ is a ring with unity u, then an element $a \in R$ is said to have a **multiplicative inverse** $b \in R$ if $ab = ba = u$.

From Examples 13.1 and 13.2, we observe that $(\mathbb{Z}, +, \cdot)$ and $(\mathbb{Z}_n, \oplus, \odot)$ are both commutative rings with unity.

Example 13.3

Suppose R represents the set of all real-valued functions on $\mathbb{R}$. If $f, g \in R$, set $(f + g)(x) = f(x) + g(x)$ and $(f \cdot g)(x) = f(x)g(x)$. Then $(R, +, \cdot)$ is a commutative ring with unity. (See Exercise 2.)

Example 13.4

The set $2\mathbb{Z}$ of even integers forms a commutative ring under the standard operations of integer addition and multiplication. This ring has no unity.

Example 13.5

Let $(G, *)$ be an abelian group with identity e. Define the operation $\#$ on G by $a \# b = e$ for all $a, b \in G$. Then $(G, *, \#)$ is a commutative ring.

Example 13.6

Let $M_2(\mathbb{Z})$ be the set of all matrices of the form

$$\begin{bmatrix} a & b \\ c & d \end{bmatrix}$$

where $a, b, c, d \in \mathbb{Z}$. If $+$ is defined on $M_2(\mathbb{Z})$ by

$$\begin{bmatrix} a & b \\ c & d \end{bmatrix} + \begin{bmatrix} e & f \\ g & h \end{bmatrix} = \begin{bmatrix} a+e & b+f \\ c+g & d+h \end{bmatrix}$$

we saw in Section 3.1 that $(M_2(\mathbb{Z}), +)$ is an abelian group. Recall the operation $\times$ on $M_2(\mathbb{Z})$ first defined in Section 2.2:

$$\begin{bmatrix} a & b \\ c & d \end{bmatrix} \times \begin{bmatrix} e & f \\ g & h \end{bmatrix} = \begin{bmatrix} ae+bg & af+bh \\ ce+dg & cf+dh \end{bmatrix}$$

We will leave to you the verification that $\times$ is associative and that $\times$ distributes over $+$. (See Exercise 4.) Therefore, $(M_2(\mathbb{Z}), +, \times)$ is a ring.

In this last example, note that we used a "$+$" sign to denote not only integer addition, but also the ring addition of matrices. This multiple utilization of symbols is quite common throughout algebra and other areas of mathematical investigation. When using symbols, be certain to be aware of the appropriate domain. In the following discussion in this chapter, we will try to alert you with special symbols for ring addition and multiplication whenever any confusion might occur.

We should warn you that the terminology in ring theory is not uniform. Some authors use the term *ring* for what we have termed a *ring with unity*; that is, they require all rings to have an identity for the second operation. Other authors use the expression *ring with identity* for what we call a *ring with unity*. Since all rings are rings with (additive) identity, we prefer our designation.

Example 13.7

Let $\mathbb{Z}[x]$ be the set of all polynomials of the form

$$\sum_{i=0}^{n} a_i x^i = a_0 + a_1 x^1 + \cdots + a_{n-1} x^{n-1} + a_n x^n$$

where $a_n, \ldots, a_1, a_0 \in \mathbb{Z}$ and $n \in \mathbb{Z}^+ \cup \{0\}$. Let

$$p(x) = \sum_{i=0}^{n} a_i x^i \qquad \text{and} \qquad q(x) = \sum_{i=0}^{m} b_i x^i$$

be elements of $\mathbb{Z}[x]$ with $m \leqslant n$. Define

$$p(x) + q(x) = \sum_{i=0}^{n} (a_i + b_i) x^i$$

$$= (a_0 + b_0) + (a_1 + b_1) x^1 + \cdots + (a_m + b_m) x^m + a_{m+1} x^{m+1}$$

$$+ \cdots + a_n x^n$$

and

$$p(x) \cdot q(x) = \sum_{k=0}^{m+n} c_k x^k$$

where

$$c_k = \sum_{j=0}^{k} a_j b_{k-j} = a_0 b_k + a_1 b_{k-1} + \cdots + a_k b_0$$

Then $(\mathbb{Z}[x], +, \cdot)$ is a ring, called the **ring of polynomials over** $\mathbb{Z}$. The polynomial $u(x) = 1$ is a unity for the second operation. Thus $\mathbb{Z}[x]$ is a commutative ring with unity.

We will investigate this last example more in Chapter 16.

Example 13.8

If, in the preceding example, we considered polynomials whose coefficients are equivalence classes in $\mathbb{Z}_n$, then we can define similar operations $+$ and $\cdot$ on this new set using modular addition and multiplication. For ease of notation, let us use the symbol $\bar{a}$ to denote the equivalence class of a in $\mathbb{Z}_n$. Set

$$\mathbb{Z}_n[x] = \{\bar{a}_0 + \bar{a}_1 x + \cdots + \bar{a}_n x^n \mid \bar{a}_i \in \mathbb{Z}_n\}$$

Then $\mathbb{Z}_n[x]$ becomes a commutative ring with unity $\bar{1}$.
    As an illustration, if $f(x) = \bar{2} + \bar{3}x + \bar{1}x^2$, $q(x) = \bar{4} + \bar{2}x^2 \in \mathbb{Z}_5[x]$, then

$$f(x) + g(x) = \bar{1} + \bar{3}x + \bar{3}x^2$$

and

$$f(x) \cdot g(x) = \bar{3} + \bar{2}x + \bar{3}x^2 + \bar{1}x^3 + \bar{2}x^4$$

Example 13.9

Let $G = \{a + bi \mid a, b \in \mathbb{Z}, i = \sqrt{-1}\}$. Define $+$ on G by

$$(a + bi) + (c + di) = (a + c) + (b + d)i$$

and define $\times$ on G by

$$(a + bi) \times (c + di) = (ac - bd) + (ad + bc)i$$

$(G, +, \times)$ is a commutative ring with unity. It is usually referred to as the **ring of Gaussian integers**.

We have seen various examples of rings. Some were commutative. Some had a unity. In the next section, we will see additional examples and continue to classify rings.

Exercises 13.1

1. Determine which elements of the following rings have multiplicative inverses.
   a. $(\mathbb{Z}, +, \cdot)$ b. $(\mathbb{Z}_6, \oplus, \odot)$
   c. $(\mathbb{Z}[x], +, \cdot)$ d. $(\mathbb{Z}_5, \oplus, \odot)$

2. Show that if R is the set of Example 13.3, then $(R, +, \cdot)$ is indeed a commutative ring with unity.

3. Prove the assertion in Example 13.5 that $(G, *, \#)$ is a commutative ring.

4. Verify that $M_2(\mathbb{Z})$ is a ring under the operations defined in Example 13.6. Prove that this ring is not commutative, but the matrix

$$\begin{bmatrix} 1 & 0 \\ 0 & 1 \end{bmatrix}$$

   is a unity. Which of the following elements have multiplicative inverses?

$$\begin{bmatrix} 1 & 2 \\ 2 & 4 \end{bmatrix} \quad \begin{bmatrix} 2 & 1 \\ 1 & 1 \end{bmatrix} \quad \begin{bmatrix} 1 & 0 \\ 0 & -1 \end{bmatrix} \quad \begin{bmatrix} 2 & 0 \\ 0 & 3 \end{bmatrix}$$

5. Let $M_2(\mathbb{Q})$ be the set of all 2×2 matrices with rational number entries. Define addition and multiplication as in Example 13.6. Show that this set is a noncommutative ring with unity. Prove that an element

$$\begin{bmatrix} a & b \\ c & d \end{bmatrix}$$

   has a multiplicative inverse if and only if $ad - bc \neq 0$.

6. In the ring $\mathbb{Z}_5[x]$ of Example 13.8, calculate the following:
   a. $(\bar{2} + \bar{3}x + \bar{4}x^2) + (\bar{1} + \bar{2}x + \bar{4}x^2)$
   b. $(\bar{2} + \bar{3}x + \bar{4}x^2) \cdot (\bar{1} + \bar{2}x + \bar{4}x^2)$
   c. $(x + x^3) \cdot (x + x^2 + \bar{2}x^3)$

7. Let $(R, \oplus, \odot)$ be a ring with unity 1. Show that the condition "$a \oplus b = b \oplus a$ for all $a, b \in R$" in the definition of a ring is redundant; that is, show that the other axioms for a ring with unity imply $a \oplus b = b \oplus a$.

8. Suppose R is a set and $*$ a binary operation on R. If $(R, *, *)$ is a ring, prove that $|R| = 1$.

9. Verify that if X is a set, then $(\mathscr{P}(X), \triangle, \cap)$ is a commutative ring with unity. Is $(\mathscr{P}(X), \triangle, \cup)$ a ring?

10. Let $R = \mathbb{Z} \times \mathbb{Z}$. This set is known as the set of all **lattice points in the plane**. Define

$$(a, b) \mathrel{\bar{+}} (c, d) = (a + c, b + d) \qquad \text{and} \qquad (a, b) \mathrel{\bar{\cdot}} (c, d) = (ac, bd)$$

   a. Show that $\bar{+}$ and $\bar{\cdot}$ are associative binary operations on R.
   b. Prove that R has identities for $\bar{+}$ and $\bar{\cdot}$.
   c. Show that R is a commutative ring.

11. Let R denote the set of all functions from $\mathbb{R}$ to $\mathbb{R}$. Define

$$(f + g)(x) = f(x) + g(x) \qquad \text{and} \qquad (f \circ g)(x) = f(g(x))$$

   Is $(R, +, \circ)$ a ring?

12. Suppose $(R, +, \cdot)$ and $(S, \bar{+}, \bar{\cdot})$ are rings. Let $T = \{(r, s) \mid r \in R, s \in S\}$. Define $\oplus$ and $\odot$ on T by

$$(r_1, s_1) \oplus (r_2, s_2) = (r_1 + r_2, s_1 \bar{+} s_2) \qquad \text{and}$$

$$(r_1, s_1) \odot (r_2, s_2) = (r_1 \cdot r_2, s_1 \bar{\cdot} s_2)$$

   Show that $(T, \oplus, \odot)$ is a ring. (Note that this exercise is an extension of Exercise 10. T is termed the **external direct product** of R and S.)

13. Determine the unity element of the ring of Gaussian integers. Show that if $a + bi$ has a multiplicative inverse, then $a^2 + b^2 = 1$. Determine all such elements and their inverses.

14. We know that the additive identity and additive inverses in a ring $(R, +, \cdot)$ must be unique since $(R, +)$ is an abelian group. If R has a unity element u, prove that u is unique.

15. Let $R = \mathbb{Z}$, the ring of integers, and define $\odot$ on R by $a \odot b = a + b - ab$. If $+$ refers to ordinary integer addition, is $(R, +, \odot)$ a commutative ring with unity?

16. Suppose $(R, +, \cdot)$ is a ring. Prove that the equality $(a + b)^2 = a^2 + 2(a \cdot b) + b^2$ holds for all $a, b \in R$ if and only if the operation $\cdot$ is commutative on R.

17. Let $(R, +, \cdot)$ be a commutative ring with unity u and $I(R)$ be the set of all invertible elements of R; that is,

$$I(R) = \{a \in R \mid a \cdot b = u \text{ for some } b \in R\}$$

   Prove that $(I(R), \cdot)$ is an abelian group.

18. Let $R = \mathbb{Z}$, and define $*$ on R by $a * b = b$. Is $(\mathbb{Z}, +, *)$ a ring?

19. If you have studied vectors in calculus, answer the following question. Does the set of all vectors in euclidean 3-space form a ring under the operations of vector addition and vector (cross) product?

20. A ring with unity $(R, +, \cdot)$ is a **Boolean ring** if every $a \in R$ satisfies the equation $a^2 = a \cdot a = a$. Show that
    a. $a = -a$ b. R is a commutative ring.

21. Determine the possible addition and multiplication tables for a ring with:
    a. Two elements   b. Three elements
    c. Four elements, if one is a unity

22. If $(R, +, \cdot)$ is a ring, prove that $(R, +, \otimes)$ is also a ring if $\otimes$ is defined by $a \otimes b = b \cdot a$.

23. Let R be the set of all 2×2 matrices of the form

$$\begin{bmatrix} a & 0 \\ 0 & b \end{bmatrix}$$

where $a, b \in \mathbb{Z}$. Prove that $(R, +, \times)$ is a commutative ring with unity.

13.2

Types of Rings and Their Properties

In Section 13.1, we presented the concepts of ring, commutative ring, and ring with unity. Here we will introduce other types of rings and investigate the relationships among them. Throughout, we will generally denote the additive identity of a ring by 0 and the unity element, if one exists, by 1.

It is often surprising to discover a ring that behaves contrary to the way one expects. For example, in the ring $(M_2(\mathbb{Z}), +, \times)$ of Example 13.6, we saw that

$$\begin{bmatrix} 1 & 0 \\ 1 & 0 \end{bmatrix} \times \begin{bmatrix} 0 & 0 \\ 1 & 1 \end{bmatrix} = \begin{bmatrix} 0 & 0 \\ 0 & 0 \end{bmatrix}$$

That is, the product of two nonzero (not equal to the additive identity) elements is

$$\begin{bmatrix} 0 & 0 \\ 0 & 0 \end{bmatrix}$$

which is the additive identity. In other words, zero can have nonzero factors in this ring. One immediate consequence of this is the loss of one of our familiar tools, the Cancellation Law. Note that

$$\begin{bmatrix} 1 & 0 \\ 1 & 0 \end{bmatrix} \times \begin{bmatrix} 0 & 0 \\ 1 & 1 \end{bmatrix} = \begin{bmatrix} 1 & 0 \\ 1 & 0 \end{bmatrix} \times \begin{bmatrix} 0 & 0 \\ 2 & 4 \end{bmatrix} = \begin{bmatrix} 0 & 0 \\ 0 & 0 \end{bmatrix}$$

and

$$\begin{bmatrix} 0 & 0 \\ 1 & 1 \end{bmatrix} \neq \begin{bmatrix} 0 & 0 \\ 2 & 4 \end{bmatrix}$$

Thus we are led to the following definitions:

Definition 13.4

Let $(R, +, \cdot)$ be a ring. An element $a \neq 0$ (where 0 is the identity for $+$) is a **left divisor of zero** or **left zero-divisor** if there is an element $b \in R - \{0\}$ such that $a \cdot b = 0$; and a is a **right divisor of zero** or **right zero-divisor** if there is an element $b \in R - \{0\}$ such that $b \cdot a = 0$.

Note that if a is a left divisor of zero and $a \cdot b = 0$ with $b \neq 0$, then b is a right divisor of zero.

Example 13.10

In the ring $(M_2, +, \times)$ of Example 13.5,

$$\begin{bmatrix} 0 & 1 \\ 0 & 2 \end{bmatrix}$$

is a left divisor of zero since

$$\begin{bmatrix} 0 & 1 \\ 0 & 2 \end{bmatrix} \times \begin{bmatrix} 2 & 1 \\ 0 & 0 \end{bmatrix} = \begin{bmatrix} 0 & 0 \\ 0 & 0 \end{bmatrix}$$

Note that

$$\begin{bmatrix} 2 & 1 \\ 0 & 0 \end{bmatrix} \times \begin{bmatrix} 0 & 1 \\ 0 & 2 \end{bmatrix} = \begin{bmatrix} 0 & 4 \\ 0 & 0 \end{bmatrix} \neq \begin{bmatrix} 0 & 0 \\ 0 & 0 \end{bmatrix}$$

although

$$\begin{bmatrix} 2 & 1 \\ 0 & 0 \end{bmatrix} \times \begin{bmatrix} -1 & 2 \\ 2 & -4 \end{bmatrix} = \begin{bmatrix} 0 & 0 \\ 0 & 0 \end{bmatrix}$$

In general, not every left zero-divisor is a right zero-divisor. In Exercise 8, we give an example of a ring in which a right zero-divisor is not a left zero-divisor.

We will complete our basic classification of rings with the following three definitions. These rings are the types most often encountered in an introductory study.

Definition 13.5

An **integral domain** is a commutative ring with unity that has no zero-divisors.

The standard example of an integral domain is the ring of integers since $ab = 0$ for $a, b \in \mathbb{Z}$ implies that either $a = 0$ or $b = 0$.

Definition 13.6

A ring $(R, +, \cdot)$ is a **division ring** (sometimes referred to as a **skew field**) if, in addition, $(R - \{0\}, \cdot)$ is a group.

Definition 13.7

A ring $(R, +, \cdot)$ is a **field** if $(R - \{0\}, \cdot)$ is also an abelian group.

These definitions require every division ring and every field to possess at least two elements since any group is nonempty. In talking about an abstract ring $(R, +, \cdot)$, we will follow our convention with groups and denote an additive inverse of $a \in R$ by $-a$. We will indicate the multiplicative inverse of an element $a \in R$, when such an element exists, by a^{-1}.

Example 13.11

The real number system with usual addition and multiplication forms a field as does the rationals with usual addition and multiplication.

Example 13.12

Consider the ring $(\mathbb{Z}_p, \oplus, \odot)$, where p is a prime. We saw in Example 3.5 that $(\mathbb{Z}_p^*, \odot)$ is an abelian group. Thus $(\mathbb{Z}_p, \oplus, \odot)$ is a field.

Example 13.13

The simplest example of a division ring which is not a field is the **quaternions** of Hamilton,[4] denoted $(Q, \oplus, \odot)$. Consider all constructs of the form $a \oplus b\mathbf{i} \oplus c\mathbf{j} \oplus d\mathbf{k}$, where a, b, c, and d are rational numbers and $\mathbf{i}$, $\mathbf{j}$, $\mathbf{k}$ are elements such that

$$\mathbf{ij} = \mathbf{k}, \quad \mathbf{jk} = \mathbf{i}, \quad \mathbf{ki} = \mathbf{j}, \quad \mathbf{ji} = -\mathbf{k}, \quad \mathbf{kj} = -\mathbf{i}, \quad \mathbf{ik} = -\mathbf{j},$$

and

$$\mathbf{i}^2 = \mathbf{j}^2 = \mathbf{k}^2 = -1$$

The diagram in Figure 13.1 can be used to recall these definitions. To find $\mathbf{jk} = \mathbf{j} \odot \mathbf{k}$, we move from $\mathbf{j}$ to $\mathbf{k}$ via a one-third turn in the **clockwise** direction. The answer is thus $+\mathbf{i}$. To find $\mathbf{kj}$, we move one-third turn in the **counterclockwise** direction. The answer then is $-\mathbf{i}$. The key to understanding multiplication from the diagram is to take the shortest path from the first of the two distinct factors to the second. Affix a "$+$" if you move in a clockwise direction; affix a "$-$" if you move in a counterclockwise direction. The diagram allows us to treat the elements of Q as polynomials. Thus, define

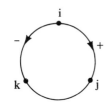

Figure 13.1

$$(a_1 \oplus a_2\mathbf{i} \oplus a_3\mathbf{j} \oplus a_4\mathbf{k}) \oplus (b_1 \oplus b_2\mathbf{i} \oplus b_3\mathbf{j} \oplus b_4\mathbf{k})$$

$$= (a_1 + b_1) \oplus (a_2 + b_2)\mathbf{i} \oplus (a_3 + b_3)\mathbf{j} \oplus (a_4 + b_4)\mathbf{k}$$

---

[4]Sir William Rowan Hamilton (1805–1865) was an Irishman whose contributions to the mathematical theory of crystals earned him a knighthood at age thirty. Seeking to expand the idea of complex numbers to ordered triples, he worked for ten years before determining that quadruples would suffice if he did not require that multiplication be commutative. This inspiration came to him during a Sunday afternoon walk with his wife. He was so happy at his realization that he carved the fundamental formula $\mathbf{i}^2 = \mathbf{j}^2 = \mathbf{k}^2 = \mathbf{ijk}$ on a stone of Brougham Bridge.

and define

$$(a_1 \oplus a_2\mathbf{i} \oplus a_3\mathbf{j} \oplus a_4\mathbf{k}) \odot (b_1 \oplus b_2\mathbf{i} \oplus b_3\mathbf{j} \oplus b_4\mathbf{k})$$
$$= (a_1b_1 - a_2b_2 - a_3b_3 - a_4b_4) \oplus (a_1b_2 + a_2b_1 + a_3b_4 - a_4b_3)\mathbf{i}$$
$$\oplus (a_1b_3 - a_2b_4 + a_3b_1 + a_4b_2)\mathbf{j} \oplus (a_1b_4 + a_2b_3 - a_3b_2 + a_4b_1)\mathbf{k}$$

Note that

$$r(a \oplus b\mathbf{i} \oplus c\mathbf{j} \oplus d\mathbf{k}) = (r \oplus 0\mathbf{i} \oplus 0\mathbf{j} \oplus 0\mathbf{k}) \odot (a \oplus b\mathbf{i} \oplus c\mathbf{j} \oplus d\mathbf{k})$$
$$= ra \oplus (rb)\mathbf{i} \oplus (rc)\mathbf{j} \oplus (rd)\mathbf{k}$$

We leave it as a simple exercise to show that $\odot$, so defined, distributes over $\oplus$ and that $(Q, \oplus, \odot)$ is a ring with unity $1 + 0\mathbf{i} + 0\mathbf{j} + 0\mathbf{k}$. Clearly $\mathbf{ij} \neq \mathbf{ji}$ so Q is not a field. However, to show that Q is a division ring, we must show that every element has an inverse. So what is $(a \oplus b\mathbf{i} \oplus c\mathbf{j} \oplus d\mathbf{k})^{-1}$? It happens that

$$(a \oplus b\mathbf{i} \oplus c\mathbf{j} \oplus d\mathbf{k})^{-1} = \frac{1}{a^2 + b^2 + c^2 + d^2}(a \oplus (-b)\mathbf{i} \oplus (-c)\mathbf{j} \oplus (-d)\mathbf{k})$$

provided that $a^2 + b^2 + c^2 + d^2 \neq 0$, or equivalently, that a, b, c, d are not all equal to 0. We leave the verification of this fact as Exercise 6. Thus we have shown that every nonzero element of Q has a multiplicative inverse. Therefore $(Q, \oplus, \odot)$ is a division ring.

If $(R, \oplus, \odot)$ is a ring, then $(R, \oplus)$ is a group, and this fact yields many consequences. Among these ring properties are the facts that the identity for $\oplus$ and inverses for $\oplus$ are unique and that R has a cancellation law for $\oplus$. Another result, which we will prove shortly, is that every field is an integral domain. (See Corollary 13.1.)

Figure 13.2 indicates the relationships between the various mathematical structures we have introduced thus far in this text with the most restricted (field) at the top and the most general (ring) at the bottom. Thus every integral domain is both a commutative ring and a ring with unity.

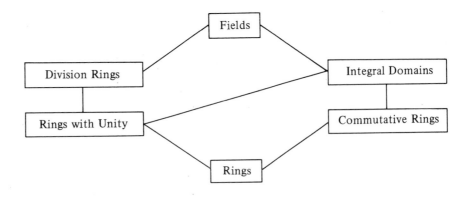

Figure 13.2

We will now try to develop some ways of determining whether a ring is an integral domain, a division ring, or a field.

Let $(R, +, \cdot)$ be a ring. Then:

    i. For all $a \in R$, $a \cdot 0 = 0 = 0 \cdot a$.

    ii. For all a, $b \in R$, $a \cdot (-b) = (-a) \cdot b = -(a \cdot b)$ and $(-a) \cdot (-b) = a \cdot b$.

    iii. If R is a ring with unity 1, then 1 is unique.

    iv. If R is a ring containing more than one element and R has a unity 1, then $1 \neq 0$.

    v. If R is a ring in which a left (right) cancellation law holds (for the second ring operation), then R has no left (right) zero-divisors.

Proof

To verify part i, note that

$$(a \cdot 0) + (a \cdot 0) = a \cdot (0 + 0) = a \cdot 0 = (a \cdot 0) + 0$$

Since there is a cancellation law in the group $(R, +)$, we have $a \cdot 0 = 0$. In a similar manner, we can show $0 \cdot a = 0$.

To prove part ii, we utilize the fact that inverses are unique in a group. We need only show that both $a \cdot (-b)$ and $(-a) \cdot b$ are both additive inverses of $a \cdot b$; then both must equal $-(a \cdot b)$. But

$$(a \cdot b) + (a \cdot (-b)) = a \cdot (b + (-b)) = a \cdot 0 = 0$$

Therefore $a \cdot (-b) = -(a \cdot b)$. A similar argument establishes $(-a) \cdot b = -(a \cdot b)$.

Using the preceding results,

$$(-a) \cdot (-b) = -(a \cdot (-b)) = a \cdot (-(-b)) = a \cdot b$$

Suppose 1 and v are both unity elements in R. Then $1 = 1 \cdot v$ since v is a unity element. On the other hand, $1 \cdot v = v$ since 1 is a unity. Hence $v = 1$, and part iii follows.

To prove part iv, suppose R has more than one element and 1 is the unity. If $r \in R$ and $r \neq 0$, then $r \cdot 1 = r$. If $1 = 0$, then part i implies $r = r \cdot 1 = r \cdot 0 = 0$. But $r \neq 0$ so we must conclude $1 \neq 0$.

Finally, to establish part v, let R be a ring in which there is a left cancellation law for $\cdot$. If $b \neq 0$ and $b \cdot a = 0$, then $b \cdot a = 0 = b \cdot 0$ and $a = 0$ by the left cancellation law. Thus there are no left divisors of zero. Similarly, if there is a right cancellation law, then there are no right divisors of zero.

                                                          $\square$

An immediate consequence of Theorem 13.1, Corollary 13.1, describes how field and integral domains are connected.

If $(F, +, \cdot)$ is a field, then $(F, +, \cdot)$ is an integral domain.

Proof

Let $(F, +, \cdot)$ be a field. We need only show that F contains no zero-divisors. To that end, suppose $a, b \in F$ and $a \cdot b = 0$. If $a \neq 0$, then there exists $a^{-1} \in F$ such that $a^{-1} \cdot a = 1$. Therefore,

$$b = 1 \cdot b = (a^{-1} \cdot a) \cdot b = a^{-1} \cdot (a \cdot b) = a^{-1} \cdot 0 = 0$$

by part i of Theorem 13.1. Thus F has no zero-divisors and $(F, +, \cdot)$ is an integral domain.   □

The integers $(\mathbb{Z}, +, \cdot)$ provide an example of an integral domain that is not a field. Part v of Theorem 13.1 can be used to obtain the following result:

Corollary 13.2

If $(R, +, \cdot)$ is a commutative ring with unity and cancellation laws for $\cdot$, then $(R, +, \cdot)$ is an integral domain.

A question to be asked at this point is, "Does an integral domain have cancellation laws?" In other words, does the converse of Corollary 13.2 hold?

If $(D, +, \cdot)$ is an integral domain and $a \cdot b = a \cdot c$, where $a \neq 0$, then $(a \cdot b) + (-(a \cdot c)) = 0$. Thus $0 = (a \cdot b) + (a \cdot (-c)) = a \cdot (b + (-c))$. Since D is an integral domain, there are no divisors of zero, and therefore either $a = 0$ or $b + (-c) = 0$. But $a \neq 0$ by hypothesis; so $b + (-c) = 0$ and, since additive inverses are unique, $b = c$. This discussion answers our question in the affirmative. This conclusion and Corollary 13.2 lead us to the following result:

Theorem 13.2

A commutative ring with unity is an integral domain if and only if there are (right and left) cancellation laws for the second operation.

Let us now turn our attention to questions concerning the relationship between a ring and its cardinality or size. Our simplest example of an integral domain that was neither a field nor a division ring was the integers, and our simplest example of a division ring that was not a field was the quaternions. Both of these rings are infinite. These examples lead us to the questions, "Are there finite integral domains that are not fields?" and "Are there finite division rings that are not fields?"

Rings of the form $(\mathbb{Z}_n, \oplus, \odot)$ are finite. We know that $(\mathbb{Z}_p, \oplus, \odot)$ is a field if p is a prime (Example 13.12). If n is not prime, then $(\mathbb{Z}_n, \oplus, \odot)$ is not an integral domain because $\odot$ is not an operation on $\mathbb{Z}_n - \{0\}$. What about other cases?

Theorem 13.3

Any finite integral domain with more than one element is a field.

Proof

To show that an integral domain with more than one element is a field, we only need to show the existence of a multiplicative inverse for each element other than 0. For $r \neq 0$, consider the set $\{r, r^2, \ldots, r^k, \ldots\}$. Since we have only a finite number of elements, there must be some repetition of elements. Let $W = \{n \in \mathbb{N} | r^n = r^k \text{ for some } k < n\}$. $W \neq \emptyset$, and by the Well-Ordering Property of the natural numbers, W has a least element, call it s. Then $r^k = r^s$, $k = s + t$, and therefore $r^k = r^{s+t} = r^s \cdot r^t$. But $r^k = r^s = r^s \cdot r^t$. By the cancellation law, $r^t = 1$. If $t = 1$, then $r = 1$, and r is its own inverse. If $t > 1$, then $r \cdot r^{t-1} = 1$, and therefore $r^{t-1} = r^{-1}$. In either case r has a multiplicative inverse. □

This result shows that the requirement that an integral domain be finite yields the additional property that every nonidentity element has a multiplicative inverse; so the integral domain becomes a field. It also provides an alternate proof of the result that $(\mathbb{Z}_n, \oplus, \odot)$ is a field when n is a prime. Does the similar restriction that a division ring be finite force the second operation to be commutative so that the division ring is a field? The answer, surprisingly, is yes. The fact that every finite division ring is a field is called Wedderburn's Theorem.[5] We refer you to the references for the proof of this result.

In the next chapter, we will explore structure-preserving mappings for rings and obtain results similar to those observed for groups. We will consider subsets of rings analogous to subgroups and normal subgroups and explore the connections between these mappings and these subsets.

Exercises 13.2

1. Verify that the subset of $(M_2(2\mathbb{Z}), +, \times)$, consisting of all matrices with only even integer entries, is a noncommutative ring without unity.

2. Verify that, in the ring $(\mathbb{Z}_n, \oplus, \odot)$, the second operation $\odot$ distributes over the first.

3. Is every division ring an integral domain? Why or why not?

4. Prove that if $(R, +, \cdot)$ is a ring with identity 0, then
   a. $0 \cdot a = 0$ for all $a \in R$,
   b. $(-a) \cdot b = -(a \cdot b)$ for all $a, b \in R$, and
   c. $a \cdot c = b \cdot c$ for $a, b, c \in R$, $c \neq 0$, implies $a = b$ when R has no right zero-divisors.

5. Does the ring of even integers under integer addition and multiplication have zero divisors? Is this ring an integral domain?

6. Verify the assertion of Example 13.13 that

$$(a \oplus b\mathbf{i} \oplus c\mathbf{j} \oplus d\mathbf{k})^{-1} = \frac{1}{a^2 + b^2 + c^2 + d^2}(a \oplus (-b)\mathbf{i} \oplus (-c)\mathbf{j} \oplus (-d)\mathbf{k})$$

Find the multiplicative inverses of the following elements:
   a. $2\mathbf{i}$ b. $\mathbf{i} \oplus \mathbf{j}$ c. $\mathbf{i} \oplus (-2)\mathbf{j} \oplus 3\mathbf{k}$

---

[5]This theorem was first proved by J. H. M. Wedderburn (1882–1948) in 1905 in an article: "A theorem on finite algebras," *Transactions of the Amer. Math. Soc.*, Vol. 6 (1905), pp. 349–52. A proof can be found in: I. N. Herstein, *Topics in Algebra*, 2nd ed., Lexington, Mass: Xerox College Publishing, 1975, p. 361.

7. Let $a + bi \in \mathbb{C}$, the set of complex numbers. If $a + bi \neq 0 + 0i$, prove that

$$\frac{a}{a^2 + b^2} - \frac{bi}{a^2 + b^2}$$

is the multiplicative inverse of $a + bi$; that is, show

$$(a + bi) \cdot \left(\frac{a}{a^2 + b^2} - \frac{bi}{a^2 + b^2} \right) = 1$$

Use this result to prove that $(\mathbb{C}, +, \cdot)$ is a field.

8. Let R be the set of all matrices in $M_2(\mathbb{Z})$ of the form

$$\begin{bmatrix} a & b \\ 0 & 0 \end{bmatrix}$$

a. Verify that R is a subring of $M_2(\mathbb{Z})$.
b. Show that

$$\begin{bmatrix} 1 & 0 \\ 0 & 0 \end{bmatrix}$$

is a right zero-divisor but never a left zero-divisor.

9. Find all zero-divisors in each of the following rings:
    a. $(\mathbb{Z}_4, \oplus, \odot)$ b. $(\mathbb{Z}_8, \oplus, \odot)$ c. $(\mathscr{P}(X), \triangle, \cap)$
    d. $\mathbb{Z}_4 \times \mathbb{Z}_4$, the external direct product of $\mathbb{Z}_4$ with itself
    e. $M_2(\mathbb{Z}_2)$, the ring of all 2×2 matrices with entries from $(\mathbb{Z}_2, \oplus, \odot)$

10. Prove that the ring of Gaussian integers has no zero divisors.

11. Is the ring consisting of the integer 0 with addition and multiplication an integral domain? Is it a field?

12. Prove that if $(\mathbb{Z}_n, \oplus, \odot)$ is an integral domain, then n is a prime integer.

13. a. For the ring of Example 13.9, compute $(3 + 2i) \times (3 - 2i)$.
    b. Show that $1 + i$ cannot be written as a product of two other Gaussian integers unless one is $1 - i$, $-1 - i$, or $-1 + i$.
    c. The Fundamental Theorem of Arithmetic asserts that every integer greater than 1 can be written as a product of prime integers. What can you conclude about similar factorization in the ring of Gaussian integers?

14. Is $(R, \mp, \dot{\cdot})$ in Exercise 10 of Section 13.1 an integral domain? Is it a field?

15. Show that the commutative ring in Exercise 12 of Section 13.1 is **not** an integral domain.

16. Give a proof to show that if $(D, +, \cdot)$ is an integral domain, then each element has at most one multiplicative inverse.

17. Show that the cancellation laws fail to hold in $\mathbb{Z}_6$.

18. Suppose $(D, +, \cdot)$ is an integral domain and $a \neq 0$ is an element of D. Define

$$a^1 = a, \ a^2 = a \cdot a, \ \dots, \ a^n = a^{n-1} \cdot a$$

Prove $a^n \neq 0$.

19. Prove that if $(F, +, \cdot)$ is a field, it is **not** generally true that $(a + b)^{-1} = a^{-1} + b^{-1}$ for $a, b \in F - \{0\}$. What happens in the ring $\mathbb{Z}_3$?

20. Let $(R, +, \cdot)$ be a ring with unity containing no zero-divisors. If $a, b \in R$ and $a \cdot b = 1$, show that $b \cdot a = 1$.

21. Let $(D, +, \cdot)$ be an integral domain, $a \in D$. For $n \in \mathbb{N}$, define

$$1a = a, \qquad 2a = a + a, \qquad \text{and} \qquad na = (n-1)a + a$$

Let 1 be the unity of D. D is said to be of **characteristic** 0 if $n1 \neq 0$ for all $n \in \mathbb{N}$. D is said to be of **finite characteristic** if there exists some $n \in \mathbb{N}$ such that $n1 = 0$.
a. Show that if p is a prime, then $(\mathbb{Z}_p, \oplus, \odot)$ has characteristic p.
b. Prove that if $(D, +, \cdot)$ is an integral domain of finite characteristic, then there is a smallest positive integer p such that $p1 = 0$. (p is called the **characteristic of** D.)
c. Prove that if D is of finite characteristic p, then $pa = 0$ for all $a \in D$.
d. Show that if D is of finite characteristic p, then p is prime.

22. Let us define a "ding-a-ring" as a ring $(R, \oplus, \odot)$ in which the reverse distributive laws also hold; that is, suppose

$$a \oplus (b \odot c) = (a \oplus b) \odot (a \oplus c)$$

and

$$(a \odot b) \oplus c = (a \oplus c) \odot (b \oplus c) \qquad \text{for all} \quad a, b, c \in R$$

a. Show that if D is an integral domain, then D is a "ding-a-ring" if and only if $D = \{0\}$.
b. Is $(\mathbb{Z}_2, \oplus, \odot)$ or $(\mathbb{Z}_4, \oplus, \odot)$ a "ding-a-ring"?

23. An element b in a ring $(R, +, \cdot)$ is termed **idempotent** if $b^2 = b$.
a. Determine all idempotent elements in the rings $(\mathbb{Z}_8, \oplus, \odot)$ and $(\mathscr{P}(X), \triangle, \cap)$.
b. Prove that an integral domain contains at most two idempotent elements.

24. When is the ring $(\mathscr{P}(X), \triangle, \cap)$ an integral domain?

25. In any integral domain, prove that the identity $a^2 = 1$ has $a = 1$ and $a = -1$ as its only solutions.

26. Let $M_2(\mathbb{Z}_3)$ consist of all 2×2 matrices with entries in $(\mathbb{Z}_3, \oplus, \odot)$. Define matrix addition and multiplication in terms of these modular operations. Show that $(M_2(\mathbb{Z}_3), +, \times)$ is a ring. Is this ring an integral domain?

14

Subrings and Ideals

In our study of groups, we made extensive use of subgroups to provide information on the structure of groups. We saw the close relationship between normal subgroups and group homomorphisms. Since a ring is also a group with respect to its first operation, subgroups certainly have a role to play in the study of rings. It also seems reasonable to expect that there exist subsets in a ring analogous to subgroups in a group. These subsets are called *subrings*. Special types of subrings, called *ideals*, are analogous to normal subgroups. In this chapter we will investigate both subrings and ideals and see how far we can extend our theory of groups to that of rings.

14.1

Subrings

Here we will continue the conventions we adopted in the course of development of rings in Chapter 13. We will use the notation $(R, +, \cdot)$ to refer to a ring, even though the operations on the ring may not be ordinary addition and multiplication. We will generally represent the additive identity of a ring by the symbol "0" and the unity (if it exists in the ring) by "1." Further, we will use juxtaposition and write ab for $a \cdot b$ and utilize the traditional symbols $-a$ and a^{-1} to represent inverses with respect to the ring operations.

Recall that a subset H of a group G is a subgroup of G if H itself is a group under the restriction to H of the operation on G. Motivated by this definition, we make a similar requirement for subsets of a ring.

Definition 14.1

A nonempty subset S of a ring $(R, +, \cdot)$ is called a **subring of R** if S is a ring under the restrictions of the operations $+$ and $\cdot$ to S.[1]

Before we consider examples of subrings, we will prove a result that enables us to determine easily whether or not a nonempty subset of a ring is a subring.

Theorem 14.1

A nonempty subset S of a ring $(R, +, \cdot)$ is a subring of R if and only if

    i. for all $r, s \in S$, $r + (-s) \in S$, and
    ii. for all $r, s \in S$, $r \cdot s \in S$.

Proof

First suppose that $(S, +, \cdot)$ is a subring of $(R, +, \cdot)$. Then $(S, +)$ is a subgroup of $(R, +)$, so that S is closed under group composition and the taking of inverses. Therefore part i must hold. Clearly part ii must also follow since the subring $(S, +, \cdot)$ is a ring, itself, and the second operation on a ring must be a binary operation.

Now suppose that the two conditions hold on the nonempty **subset** S of R. Condition i and Theorem 4.6 guarantee that $(S, +)$ is an abelian subgroup, while the condition ii says $\cdot$ is a binary operation on S. The facts that $\cdot$ is distributive over $+$ and $\cdot$ is associative follow from similar properties in R that are inherited by the restrictions of mappings. (See Exercise 1.)   □

---

[1] To be precise, we should require that $(S, +_S, \cdot_S)$ be a subring of $(R, +, \cdot)$.

Example 14.1

The set $2\mathbb{Z}$ of all even integers is a subring of the ring of integers with usual addition and multiplication. To show this, one simply notes that if $2m$ and $2n$ are elements of $2\mathbb{Z}$, then so is $2m - 2n = 2(m - n)$ and $(2m) \cdot (2n) = 2(2mn)$.

Example 14.2

In $\mathbb{Z}_6$, $\{0, 2, 4\}$ and $\{0, 3\}$ are both subrings. To verify this, simply construct the appropriate tables.

Example 14.3

$\{0\}$ and R are always subrings of R. $\{0\}$ is termed the **trivial** subring of R.

Observe in Examples 14.1 and 14.2 that subrings of rings with unities are not necessarily rings with unities. In the next section, we will look into this matter further.

A natural question to ask at this point is, "Based upon Theorem 14.1, I know that every subring of a ring $(R, +, \cdot)$ is a subgroup of $(R, +)$, but is the converse true; that is, if S is a subgroup of a ring R, is S a subring?" To see that the answer is no, let us consider the ring $(M_2(\mathbb{Z}), +, \times)$ of Example 13.6. It is a simple process to show that

$$\left\{ \begin{bmatrix} a & b \\ c & 0 \end{bmatrix} \middle| a, b, c \in \mathbb{Z} \right\}$$

is a subgroup of $M_2(\mathbb{Z})$. However, since

$$\begin{bmatrix} 1 & 1 \\ 1 & 0 \end{bmatrix} \times \begin{bmatrix} 1 & 1 \\ 1 & 0 \end{bmatrix} = \begin{bmatrix} 2 & 1 \\ 1 & 1 \end{bmatrix}$$

we see that the set is not closed under the second operation $\times$ and, thus, is not a subring of $M_2(\mathbb{Z})$ by Theorem 14.1.

When we studied groups, we found that certain subsets of a group, for example, the center of the group or the normalizer of some subgroup, turned out to be subgroups. A similar situation happens with rings.

Definition 14.2

> If $(R, +, \cdot)$ is a ring, then $C(R)$, **the center of** R, is the set of all elements in R which commute with all other elements for the operation $\cdot$; that is,
>
> $$C(R) = \{r \in R \mid s \cdot r = r \cdot s \text{ for all } s \in R\}$$

We leave the verification of the following theorem as Exercise 5. Compare it with Exercise 15 of Section 4.3.

Theorem 14.2

For any ring R, the center of R is a commutative subring of R.

Example 14.4   Let R be the ring $(M_2(\mathbb{Z}), +, \times)$ of matrices described in Example 13.5. To find the center of R, let

$$\begin{bmatrix} a & b \\ c & d \end{bmatrix} \in C(R)$$

What can we determine about the elements of this matrix? Since

$$\begin{bmatrix} a & b \\ c & d \end{bmatrix}$$

must commute with all elements of R, we will check the matrix equation

$$\begin{bmatrix} a & b \\ c & d \end{bmatrix} \times \begin{bmatrix} 0 & 1 \\ 0 & 0 \end{bmatrix} = \begin{bmatrix} 0 & 1 \\ 0 & 0 \end{bmatrix} \times \begin{bmatrix} a & b \\ c & d \end{bmatrix}$$

The calculation of the matrix product yields

$$\begin{bmatrix} 0 & a \\ 0 & c \end{bmatrix} = \begin{bmatrix} c & d \\ 0 & 0 \end{bmatrix}$$

so that $c = 0$ and $a = d$.

Now consider the matrix equation

$$\begin{bmatrix} a & b \\ c & d \end{bmatrix} \times \begin{bmatrix} 0 & 0 \\ 1 & 0 \end{bmatrix} = \begin{bmatrix} 0 & 0 \\ 1 & 0 \end{bmatrix} \times \begin{bmatrix} a & b \\ c & d \end{bmatrix}$$

or

$$\begin{bmatrix} b & 0 \\ d & 0 \end{bmatrix} = \begin{bmatrix} 0 & 0 \\ a & b \end{bmatrix}$$

This says that $b = 0$ and $a = d$. These two matrix conditions require that if

$$\begin{bmatrix} a & b \\ c & d \end{bmatrix} \in C(R)$$

then

$$\begin{bmatrix} a & b \\ c & d \end{bmatrix} = \begin{bmatrix} a & 0 \\ 0 & a \end{bmatrix}$$

We leave it to you to verify that any matrix of the form

$$\begin{bmatrix} a & 0 \\ 0 & a \end{bmatrix}$$

does indeed commute with any other element in R and that

$$C(R) = \left\{ \begin{bmatrix} a & 0 \\ 0 & a \end{bmatrix} \middle| a \in \mathbb{Z} \right\}$$

The similarity between the center of a ring and the center of a group is just one of the many parallels that exist between rings and groups. We list some others in Exercises 6, 10, 11 and 12.

1. Suppose $\cdot$ and $+$ are binary operations of a ring R. Prove:
   a. If $\cdot$ is associative and S is a nonempty subset of R such that S is closed under $\cdot$, then $\cdot_S$ is associative.
   b. If $\cdot$ is distributive over $+$ and S is a nonempty subset of R closed under $+$ and $\cdot$, then $\cdot_S$ is distributive over $+_S$.

2. Let $3\mathbb{Z} = \{3a \mid a \in \mathbb{Z}\}$. Prove that $(3\mathbb{Z}, +, \cdot)$ is a subring of $(\mathbb{Z}, +, \cdot)$. Is this subring also an integral domain?

3. a. Show that $\mathbb{Z}[\sqrt{2}] = \{a + b\sqrt{2} \mid a, b \in \mathbb{Z}\}$ is a subring of the field of real numbers.
   b. If $a + b\sqrt{2}$ is invertible and $(a, b) = 1$, show that $a^2 - 2b^2 = 1$ or -1.
   c. Determine the invertible elements of $\mathbb{Z}[\sqrt{2}]$.

4. Let $\mathbb{Z}[\sqrt[3]{2}] = \{a + b\sqrt[3]{3} \mid a, b \in \mathbb{Z}\}$. Is $(\mathbb{Z}[\sqrt[3]{2}], +, \cdot)$ a subring of $(\mathbb{R}, +, \cdot)$?

5. Prove that the center of a ring is a commutative subring of R.

6. Let S be a subring of a ring $(R, +, \cdot)$. Define the centralizer of S as the set of all elements $C(S) = \{r \in R \mid r \cdot s = s \cdot r \text{ for all } s \in S\}$. Prove that $C(S)$ is a subring of R containing $C(R)$.

7. Verify that the set

$$Sc_2 = \left\{ \begin{bmatrix} n & 0 \\ 0 & n \end{bmatrix} \middle| n \in \mathbb{Z} \right\}$$

is a commutative subring of $(M_2(\mathbb{Z}), +, \times)$ and that Sc_2 lies in the center of $M_2(\mathbb{Z})$.

8. Show that the set

$$G = \left\{ \begin{bmatrix} n & m \\ -m & n \end{bmatrix} \middle| m, n \in \mathbb{Z} \right\}$$

is a subring of $(M_2(\mathbb{Z}), +, \times)$.

9. Determine the relationships necessary between a, b, c so that

$$\left\{ \begin{bmatrix} a & b \\ c & 0 \end{bmatrix} \middle| a, b, c \in \mathbb{R} \right\}$$

is a subring of $(M_2(\mathbb{Z}), +, \times)$.

10. a. Show that if S_1 and S_2 are subrings of a ring $(R, +, \cdot)$, then $S_1 \cap S_2$ is a subring of $(R, +, \cdot)$.
    b. By considering the ring $\mathbb{Z}_6$, show that if S_1 and S_2 are subrings of a ring $(R, +, \cdot)$, then $S_1 \cup S_2$ is not necessarily a subring.

11. Let $(R, +, \cdot)$ be a ring and $a \in R$. Set

$$C(a) = \{r \in R \mid a \cdot r = r \cdot a\}$$

Prove that $C(R) = \bigcap_{a \in R} C(a)$.

12. Let X be a subset of a ring R. The **subring of R generated by X** is the intersection

of all subrings of R that contain X. Prove that this set is indeed a subring and that it is the smallest subring of R that contains X. What happens if $X = \emptyset$?

13. Find the smallest subring of $(\mathbb{R}, +, \cdot)$ that contains $\sqrt[3]{5}$.

14. An element a of a ring R is termed **nilpotent** if $a^n = 0$ for some $n \in \mathbb{N}$.
     a. Prove that the only nilpotent element in an integral domain is 0.
     b. If $a \in R$ is nilpotent, verify that $1 - a$ has a multiplicative inverse in R, namely, $(1 - a)^{-1} = 1 + a + a^2 + \cdots + a^{n-1}$ for some $n \in \mathbb{N}$.

15. Prove that the set of all nilpotent elements in a commutative ring R is a subring of R.

16. Prove that if S is a subgroup of $\mathbb{Z}_6$, then S is a subring of $\mathbb{Z}_6$.

17. Let $(R, +, \cdot)$ be a commutative ring. Let rad $R = \{r \in R \mid r^n = 0 \text{ for some } n \in \mathbb{Z}\}$. Prove rad R is a subring of R.

18. Prove that if S is a subring of $(\mathbb{R}, +, \cdot)$ and $1 \in S$, then $(\mathbb{Z}, +, \cdot)$ is a subring of S.

19. In Exercise 11 of Section 13.1, you were asked to show that the functions $f: \mathbb{R} \to \mathbb{R}$ form a ring under suitable definitions of function addition and multiplication. Do the continuous functions on $\mathbb{R}$ form a subring of this ring?

20. Suppose $(R, +, \cdot)$ is a ring with R a finite set. If $(S, +, \cdot)$ is a subring of R, why does $|S|$ divide $|R|$?

21. Let R be a division ring. Prove that $C(R)$, the center of R, must be a field.

14.2

Ring Homomorphisms and Ideals

In this section we will extend more of our group-theoretic ideas to rings. When we studied relationships between groups, we were interested in mappings, which preserved the operation of the group (group homomorphisms). The extension of this idea to rings (ring homomorphisms) requires that both operations be preserved.

Definition 14.3

Suppose that $(R, +, \cdot)$ and $(S, \oplus, \odot)$ are rings. Then a mapping $\theta: (R, +, \cdot) \to (S, \oplus, \odot)$ is a **ring homomorphism** if

$$\theta(a + b) = \theta(a) \oplus \theta(b) \qquad \text{and} \qquad \theta(a \cdot b) = \theta(a) \odot \theta(b)$$

for all $a, b \in R$.

If θ is a bijection, then θ is called a **ring isomorphism** and $(R, +, \cdot)$ and $(S, \oplus, \odot)$ are isomorphic as rings; we will denote this by $R \cong S$. If θ is a homomorphism, and $(R, +, \cdot) = (S, \oplus, \odot)$, we call θ an **endomorphism**; in addition, a bijective endomorphism is an **automorphism**. Thus an automorphism of a ring $(R, +, \cdot)$ is an isomorphism from R onto R.

Example 14.5

Define $\theta\colon (\mathbb{Z},\ +,\ \cdot) \to (\mathbb{Z}_n,\ \oplus,\ \odot)$ by $\theta(m) = \bar{m}$, where $\bar{m}$ denotes the equivalence class of m in $\mathbb{Z}_n$. Then θ is a ring homomorphism since

$$\theta(m + n) = \overline{m + n} = \bar{m} \oplus \bar{n} = \theta(m) \oplus \theta(n)$$

and

$$\theta(m \cdot n) = \overline{m \cdot n} = \bar{m} \odot \bar{n} = \theta(m) \odot \theta(n)$$

While θ is clearly surjective, θ is not a ring isomorphism because

$$\theta(n + 1) = \overline{n + 1} = \bar{n} \oplus \bar{1} = \bar{0} \oplus \bar{1} = \bar{1} = \theta(1)$$

that is, θ is not injective.

Example 14.6

Define $\phi\colon (\mathbb{Z},\ +,\ \cdot) \to (M_2(\mathbb{Z}),\ +,\ \times)$ by

$$\phi(n) = \begin{bmatrix} n & 0 \\ 0 & n \end{bmatrix}$$

(See Example 13.6.) Then ϕ is a ring homomorphism. To verify this, we need only note

$$\phi(n + m) = \begin{bmatrix} n + m & 0 \\ 0 & n + m \end{bmatrix} = \begin{bmatrix} n & 0 \\ 0 & n \end{bmatrix} + \begin{bmatrix} m & 0 \\ 0 & m \end{bmatrix} = \phi(n) + \phi(m)$$

and

$$\phi(n \cdot m) = \begin{bmatrix} n \cdot m & 0 \\ 0 & n \cdot m \end{bmatrix} = \begin{bmatrix} n & 0 \\ 0 & n \end{bmatrix} \times \begin{bmatrix} m & 0 \\ 0 & m \end{bmatrix} = \phi(n) \times \phi(m)$$

Example 14.7

If $(G,\ +,\ \times)$ is a ring of Gaussian integers (see Example 13.9), then $\rho\colon (G,\ +,\ \times) \to (M_2(\mathbb{Z}),\ +,\ \times)$ defined by

$$\rho(m + ni) = \begin{bmatrix} m & n \\ -n & m \end{bmatrix}$$

is a ring homomorphism. We leave the verification of this fact as Exercise 1. The matrices in $M_2(\mathbb{Z})$ of the form

$$\begin{bmatrix} n & 0 \\ 0 & n \end{bmatrix}$$

form a subring of $M_2(\mathbb{Z})$, and ρ is an isomorphism from G to this subring.

The next result follows directly from the definition of ring homomorphism. The fact that the converse is not true is the content of Exercise 2.

| Lemma 14.1 |

Suppose $\theta: (R, +, \cdot) \to (S, \oplus, \odot)$ is a ring homomorphism. Then $\theta: (R, +) \to (S, \oplus)$ is a group homomorphism.

Recall that the homomorphic image of a group turned out to be a subgroup of the image of the homomorphism. In our Examples 14.6 and 14.7, the images are the subrings Sc_2 and G in $(M_2(\mathbb{Z}), +, \times)$. (See Exercises 7 and 8 of Section 14.1.) Thus it should not surprise you to find that, in general, the homomorphic image of a ring is a subring of the range of the homomorphism.

| Theorem 14.3 |

If $\theta: (R, +, \cdot) \to (S, \oplus, \odot)$ is a ring homomorphism, then $\theta^{\to}(R)$ is a subring of S.

Proof

That $\theta^{\to}(R)$ is a subgroup of S follows from Lemma 14.1 since the image of the group $(R, +)$ is a subgroup of $(S, \oplus)$. $\theta^{\to}(R)$ is closed under the restriction of $\odot$ to $\theta^{\to}(R)$ since θ is a homomorphism. Hence $\theta(r) \odot \theta(s) = \theta(r \cdot s) \in \theta^{\to}(R)$ for all $r, s \in R$. By Theorem 14.1, $\theta^{\to}(R)$ is a subring of S. ☐

We proved in Chapter 6 that under group homomorphisms identities mapped to identities and inverses mapped to inverses. As you might expect, we can prove similar results for rings.

| Theorem 14.4 |

Suppose $\theta: (R, +, \cdot) \to (S, \oplus, \odot)$ is a ring homomorphism. The following results hold.

    i. If 0_R is the additive identity of R, then $\theta(0_R) = 0_S$, the additive identity of S.

    ii. $\theta(-a)$ is the additive inverse of $\theta(a)$ in S. In other words $\theta(-a) = -\theta(a)$.

    iii. If 1_R is the unity of R, then $\theta(1_R)$ is the unity of $\theta^{\to}(R)$.

    iv. If a has a multiplicative inverse a^{-1} in R, then $\theta(a^{-1})$ is the multiplicative inverse for $\theta(a)$ in $\theta^{\to}(R)$.

Proof

The assertions of parts i and ii follow immediately from Lemma 14.1 and the corresponding results for groups.

For part iii, $\theta(1_R) \odot \theta(a) = \theta(1_R \cdot a) = \theta(a)$. Similarly, $\theta(a) \odot \theta(1_R) = \theta(a)$. Thus $\theta(1_R)$ is a unity for $\theta^{\to}(R)$. That $\theta(1_R)$ is **the** unity for $\theta^{\to}(R)$ follows from part iii of Theorem 13.1.

Finally, to prove part iv, note that $\theta(a) \odot \theta(a^{-1}) = \theta(a \cdot a^{-1}) = \theta(1_R)$. In like manner, $\theta(a^{-1}) \odot \theta(a) = \theta(1_R)$. Therefore $\theta(a^{-1})$ is a multiplicative inverse for $\theta(a)$ in $\theta^{\to}(R)$. We leave the proof that $\theta(a^{-1})$ is the **unique** inverse of $\theta(a)$ in $\theta^{\to}(R)$ as Exercise 5. ☐

It is important to note that the preceding theorem does not say that the unity of the ring R maps to the unity of the ring S under ring homomorphisms. It simply states the unity of R maps to something that acts like an identity for multiplication in $\theta^{\rightarrow}(R)$. This element is not necessarily a unity in S. The following example illustrates this.

Example 14.8

Consider $\theta\colon (\mathbb{Z}_2, \oplus, \odot) \to (\mathbb{Z}_6, \oplus, \odot)$ defined by $\theta(\bar{0}) = 0$ and $\theta(\bar{1}) = \bar{3}$. $\bar{3}$ is the unity for the subring $\{\bar{0}, \bar{3}\}$ although $\bar{1}$, not $\bar{3}$, is the unity of $\mathbb{Z}_6$. Also, $\bar{3}$ is its own multiplicative inverse in $\theta^{\rightarrow}(\mathbb{Z}_2)$; but in $\mathbb{Z}_6$, $\bar{3}$ does not have an inverse!

Recall that if $\theta\colon R \to S$ is a mapping and $S' \subseteq S$, then $\theta^{\leftarrow}(S')$ was defined by $\theta^{\leftarrow}(S') = \{r \in R \mid \theta(r) \in S'\}$. We leave the verification of the next result to you as Exercise 7.

Theorem 14.5

If $\theta\colon (R, +, \cdot) \to (S, \oplus, \odot)$ is a ring homomorphism and S' is a subring of S, then $\theta^{\leftarrow}(S')$ is a subring of R.

Corollary 14.1

If $\theta\colon (R, +, \cdot) \to (S, \oplus, \odot)$ is a ring homomorphism and 0_S denotes the additive identitiy of S, then $\theta^{\leftarrow}(\{0_S\})$ is a subring of R.

This subring $\theta^{\leftarrow}(\{0_S\})$ is termed the **kernel of** θ and is denoted ker θ.

Example 14.9

Consider the ring homomorphism $\theta\colon \mathbb{Z} \to \mathbb{Z}_n$ of Example 14.5. We claim that ker $\theta = \{m \in \mathbb{Z} \mid \bar{m} = \bar{0}\}$ is $n\mathbb{Z}$, the set of all integer multiples of n. For if $m = nk$, then $\theta(m) = \theta(nk) = \theta(n) \odot \theta(k) = \bar{n} \odot \bar{k} = \bar{0} \odot \bar{k} = \bar{0}$. Note that if $m = t + nk$, then

$$\theta(m) = \theta(t + nk) = \theta(t) \oplus \theta(nk) = \theta(t)$$

This illustrates how the behavior of θ is determined by the subset $\{0, 1, 2, \dots, n-1\} \subseteq \mathbb{Z}$.

In our study of group homomorphisms, we discovered that the kernel of a homomorphism was a subgroup and, moreover, a normal subgroup. Since not every subgroup is a normal subgroup, not every subgroup is the kernel of a homomorphism. A similar case holds for ring homomorphisms. Consider the following theorem:

Theorem 14.6

If $\theta\colon (R, +, \cdot) \to (S, \oplus, \odot)$ is a ring homomorphism and $r \in$ ker θ, then $x \cdot r$ and $r \cdot x$ are elements of ker θ for all $x \in R$.

Proof

Since $\theta(x \cdot r) = \theta(x) \odot \theta(r) = \theta(x) \odot 0_S = 0_S$, it follows that $x \cdot r \in \ker \theta$. Similarly $r \cdot x \in \ker \theta$. $\qquad \square$

Motivated by this result, we make the following definition:

Definition 14.4

> A subring I of a ring R is an **ideal** of R if $s \cdot r \in I$ and $r \cdot s \in I$ for all $r \in R$ and $s \in I$.

In the case that R is commutative, we only need to use the postulate that $s \cdot r \in I$, since $s \cdot r = r \cdot s$. I is a **proper ideal** of R if I is an ideal of R and is also a proper subset of R.

An ideal is a subring that has an **absorption property** with respect to multiplication. The kernel of a ring homomorphism is an ideal, and all the elements of R having the same image form a coset of $\ker \theta$; that is, $\theta(r) = \theta(s)$ if and only if $r + \ker \theta = s + \ker \theta$. This follows from Theorem 6.3 and the fact (Lemma 14.1) that ring homomorphisms are group homomorphisms.

Both a ring R and the trivial subring $\{0\}$ are ideals of R. R is termed an **improper ideal of R**, while $\{0\}$ is called the **trivial ideal**. There are many other examples of nontrivial proper ideals:

Example 14.10

Consider $(5\mathbb{Z}, +, \cdot)$; we claim it is an ideal of $(\mathbb{Z}, +, \cdot)$. First, $5\mathbb{Z}$ is a subring of $\mathbb{Z}$. If $a \in 5\mathbb{Z}$, $r \in \mathbb{Z}$, then $a = 5z$ for some $z \in \mathbb{Z}$ and $r \cdot a = r(5z) = 5(rz) \in 5\mathbb{Z}$. Therefore $5\mathbb{Z}$ is an ideal of $\mathbb{Z}$.

Example 14.11

Let $R = \mathbb{Z}[x]$, the ring of Example 13.7. The subset R_1 consisting of all integers forms a subring of R. Since $2 \cdot (3 + x) \notin R_1$, R_1 is **not** an ideal.

One type of ideal that plays an extremely important role in the theory of commutative rings is described in the next definition:

Definition 14.5

> Let R be a commutative ring and $d \in R$. The intersection of all ideals of R that contain d is known as the **principal ideal generated by** d and is denoted by (d).

We saw in Exercise 10 of Section 14.1 that the intersection of two subrings of a ring is another subring. If the subrings are also ideals, then the intersection is also an ideal. The principal ideal generated by d is the smallest ideal that contains d. (See Exercise 17.)

The next two examples illustrate alternate ways to describe a principal ideal.

Example 14.12

If R is an arbitrary commutative ring with unity and $d \in R$, define $I = \{rd \mid r \in R\}$. Since

$$r_1 d - r_2 d = (r_1 - r_2)d \in I \qquad \text{and} \qquad (r_1 d)(r_2 d) = (r_1 d r_2)d \in I$$

I is a subring of R. Let $s \in R$ and $rd \in I$. Then $s(rd) = (sr)d \in I$. Since R is commutative, we can conclude that I is an ideal. Since any ideal of R that contains d must also contain I, $I = (d)$.

Example 14.13

If R is a commutative ring without unity, then the principal ideal generated by d is $\{r \cdot d + md \mid r \in R, m \in \mathbb{Z}\}$. We leave the verification that this set is an ideal to you. (See Exercise 16.)

Ideals have much the same relationship to rings and ring homomorphisms as do normal subgroups and group homomorphisms. We will investigate these relationships in Section 14.3.

Exercises 14.2

1. Prove that the mapping ρ from the Gaussian integers to $M_2(\mathbb{Z})$ defined by

$$\rho(m + ni) = \begin{bmatrix} m & n \\ -n & m \end{bmatrix}$$

is a ring homomorphism.

2. Define $\phi: \mathbb{Z} \to \mathbb{Z}$ by $\phi(a) = 2a$. Prove that ϕ is a group homomorphism of $(\mathbb{Z}, +)$ but not a ring homomorphism of $(\mathbb{Z}, +, \cdot)$.

3. Define $\theta: \mathbb{Z}[x] \to \mathbb{Z}$, where $\mathbb{Z}[x]$ is the ring of Example 13.7, by

$$\theta(a_0 + a_1 x^1 + \cdots + a_{n-1}x^{n-1} + a_n x^n) = a_0$$

Verify that θ is a ring homomorphism.

4. In Example 14.8, we investigated a ring homomorphism $\theta: \mathbb{Z}_2 \to \mathbb{Z}_6$. Let $S = \{\bar{0}, \bar{2}, \bar{4}\}$. Prove that S is a subring of $\mathbb{Z}_6$. What is $\theta^{\leftarrow}(S)$?

5. Suppose that $(R, +, \cdot)$ is a ring with unity and $a \in R$. Prove that a has at most one multiplicative inverse.

6. Prove that if $\sigma: R_1 \to R_2$ and $\tau: R_2 + R_3$ are ring homomorphisms, then $\tau \circ \sigma$ is a ring homomorphism from R_1 to R_3.

7. If $\theta: (R, +, \cdot) \to (S, \oplus, \odot)$ is a ring homomorphism and S' is a subring of S, prove $\theta^{\leftarrow}(S')$ is a subring of R (see Theorem 14.5).

8. Let R be the set of all differentiable functions $f: \mathbb{R} \to \mathbb{R}$. If addition and multiplication are defined by Exercise 11 of Section 13.1, prove that $(R, +, \cdot)$ is a

ring. Define $D: R \to R$ by $D(f(z)) = f'(z)$, the derivative of f. Is D a ring homomorphism?

9. Let $(D, +, \cdot)$ be an integral domain, and suppose $\phi: D \to D$ is a ring homomorphism. If $\phi(1) \neq 0$, prove that $\phi(1) = 1$ and that the image of each invertible element of D is again invertible.

10. Prove that if F is a field and I is an ideal of F, then $I = \{0\}$ or $I = F$.

11. Verify that a commutative ring R with unity is a field if and only if R contains no proper ideal other that $\{0\}$.

12. Suppose $(F, +, \cdot)$ is a field and $\phi: F \to F$ is a ring homomorphism. Prove that either $\phi(a) = 0$ for all $a \in F$ or ϕ is injective.

13. Show that $2\mathbb{Z}$, the set of even integers, forms an ideal of $\mathbb{Z}$. What is the group $\mathbb{Z}/2\mathbb{Z}$? Do the odd integers form an ideal of $\mathbb{Z}$? Why?

14. a. Prove that if $\theta: (R, +, \cdot) \to (S, \oplus, \odot)$ is a ring homomorphism and I is an ideal of R, then $\theta^\to(I)$ is an ideal of $\theta^\to(R)$.
    b. Give an example to show that $\theta^\to(I)$ is not necessarily ideal of S.

15. Prove that if $\theta: (R, +, \cdot) \to (S, \oplus, \odot)$ is a ring epimorphism, and J is an ideal of S, then $\theta^\gets(J)$ is an ideal of R. Recall that an epimorphism is a surjective homomorphism.

16. Let $(R, +, \cdot)$ be a commutative ring without unity and $d \in R$. Show that $(d) = \{rd + nd \,|\, r \in R, n \in \mathbb{Z}\}$.

17. If d is an element of a ring $(R, +, \cdot)$, prove that the intersection of ideals in a ring is another ideal. Verify that (d) is the smallest ideal of R containing d.

18. Suppose $(R, +, \cdot)$ is a commutative ring with unity. If $d_1, d_3 \in R$ and (d_1, d_2) is the smallest ideal of R containing d_1 and d_2, show that

$$(d_1, d_2) = \{ad_1 + bd_2 \,|\, a, b \in R\}$$

19. Let $(R, +, \cdot)$ be a ring with unity and I an ideal of R. If $1 \in I$, prove $I = R$.

20. Let I be an ideal of a ring R and J be an ideal of another ring S. Prove that

$$I \times J = \{(a, b) \,|\, a \in I, b \in J\}$$

is an ideal of $R \times S$, the external direct product of R and S. (See Exercise 12 of Section 13.1.) Determine the ideal $(2) \times (3)$ in $\mathbb{Z}_8 \times \mathbb{Z}_{12}$.

21. Let R be a commutative ring and $a_1, a_2, \ldots, a_n \in R$. Show that the set of all elements of the form $r_1 a_1 + r_2 a_2 + \ldots + r_n a_n$ for $r_1, r_2, \ldots, r_n \in R$, forms an ideal of R.

22. If $\phi: R \to R'$ is a ring homomorphism, prove that $\phi^\to(C(R)) \subseteq C(\phi^\to(R))$.

23. Let R be a ring with unity, $b \in R$, and suppose R contains the multiplicative inverse of b. Define $\phi: R \to R$ by $\phi(a) = bab^{-1}$. Verify that ϕ is an automorphism of R.

24. Suppose a is a nilpotent element of the ring R and $\phi: R \to R'$ is a ring homomorphism. Show that $\phi(a)$ is a nilpotent element of R'. Is the result true if *nilpotent* is replaced by *idempotent*?

25. We saw in the last chapter that the center of a ring is a subring. Give an example to show that the center of a ring need not be an ideal of that ring.

14.3

Factor Rings

In Chapter 8 we explored at great length the connections between factor groups and homomorphisms. We defined factor groups and saw how there exists a one-to-one correspondence between factor groups of a group G and group homomorphisms whose domain is G. In this section we wish to consider analogous results for rings. We begin by developing the proof of a basic theorem.

Let I be an ideal of a ring R. Since $(R, +)$ must be an abelian group, all subgroups are normal. Therefore I is a normal subgroup of R and $(R/I, \oplus)$ is a group, where

$$(r + I) \oplus (s + I) = (r + s) + I$$

We will make R/I into a ring by defining $\odot$ on R/I by

$$(r + I) \odot (s + I) = rs + I$$

Since the elements of R/I have many representations, it is imperative that we show that $\odot$ is well-defined. If $r_1 + I = r_2 + I$ and $s_1 + I = s_2 + I$, we must verify

$$(r_1 + I) \odot (s_1 + I) = (r_2 + I) \odot (s_2 + I)$$

We recall that $r_1 + I = r_2 + I$ if and only if $(r_2 - r_1) \in I$, which means $r_2 = r_1 + x$ for some $x \in I$. If $r_2 = r_1 + x_1$ and $s_2 = s_1 + x_2$, then

$$(r_2 + I) \odot (s_2 + I) = r_2 s_2 + I$$
$$= ((r_1 + x_1)(s_1 + x_2)) + I$$
$$= (r_1 s_1 + r_1 x_2 + x_1 s_1 + x_1 x_2) + I$$
$$= (r_1 s_1 + I) \oplus ((r_1 x_2 + x_1 s_1 + x_1 x_2) + I)$$

Since I is an ideal and $x_1, x_2 \in I$, we have $x_1 s_1 \in I$, $r_1 x_2 \in I$, and $x_1 x_2 \in I$. Therefore

$$(r_1 x_2 + x_1 s_1 + x_1 x_2) + I = 0 + I$$

and

$$(r_2 + I) \odot (s_2 + I) = (r_1 s_1 + I) \oplus (0 + I) = (r_1 + I) \odot (s_1 + I)$$

so that $\odot$ is a well-defined operation on R/I. The fact that $\odot$ is associative and distributes over $\oplus$ follows from these same properties on R. We leave the actual verification of these facts to you as Exercise 1.

We can now state our fundamental result:

Theorem 14.7

Let I be an ideal of the ring R. The mapping $\theta: R \to R/I$ defined by $\theta(r) = r + I$ is a surjective ring homomorphism with kernel I.

Proof

In light of the discussion preceding the statement of the theorem, we need only show that θ is a surjective homomorphism. This is straightforward, and we will leave its verification to you as Exercise 2. □

Definition 14.6

> The ring $(R/I, \oplus, \odot)$ defined on R/I by
>
> $$(a + I) \oplus (b + I) = (a + b) + I \qquad \text{and} \qquad (a + I) \odot (b + I) = ab + I$$
>
> is called the **factor ring** or **quotient ring of R by I**. The mapping $\theta: R \to R/I$ defined by $\theta(a) = a + I$ is called the **canonical** or **natural** or **quotient mapping** from R to R/I.

Example 14.14

In Example 14.10 we saw that $5\mathbb{Z}$ is an ideal in $\mathbb{Z}$. $\mathbb{Z}/5\mathbb{Z}$ consists of all the cosets of the form $a + 5\mathbb{Z}$ where $a = 0, 1, 2, 3,$ or 4. The Cayley tables of this ring are given in Figures 14.1a, b. We will discuss this ring again in Example 14.16.

Example 14.15

Consider $(M_2(\mathbb{Z}), +, \times)$, and let

$$R = \left\{ \begin{bmatrix} a & 0 \\ 0 & b \end{bmatrix} \middle| a, b \in \mathbb{Z} \right\}, \qquad I = \left\{ \begin{bmatrix} a & 0 \\ 0 & 0 \end{bmatrix} \middle| a \in \mathbb{Z} \right\}$$

$\oplus$	$0 + 5\mathbb{Z}$	$1 + 5\mathbb{Z}$	$2 + 5\mathbb{Z}$	$3 + 5\mathbb{Z}$	$4 + 5\mathbb{Z}$
$0 + 5\mathbb{Z}$	$0 + 5\mathbb{Z}$	$1 + 5\mathbb{Z}$	$2 + 5\mathbb{Z}$	$3 + 5\mathbb{Z}$	$4 + 5\mathbb{Z}$
$1 + 5\mathbb{Z}$	$1 + 5\mathbb{Z}$	$2 + 5\mathbb{Z}$	$3 + 5\mathbb{Z}$	$4 + 5\mathbb{Z}$	$0 + 5\mathbb{Z}$
$2 + 5\mathbb{Z}$	$2 + 5\mathbb{Z}$	$3 + 5\mathbb{Z}$	$4 + 5\mathbb{Z}$	$0 + 5\mathbb{Z}$	$1 + 5\mathbb{Z}$
$3 + 5\mathbb{Z}$	$3 + 5\mathbb{Z}$	$4 + 5\mathbb{Z}$	$0 + 5\mathbb{Z}$	$1 + 5\mathbb{Z}$	$2 + 5\mathbb{Z}$
$4 + 5\mathbb{Z}$	$4 + 5\mathbb{Z}$	$0 + 5\mathbb{Z}$	$1 + 5\mathbb{Z}$	$2 + 5\mathbb{Z}$	$3 + 5\mathbb{Z}$

(a)

$\odot$	$0 + 5\mathbb{Z}$	$1 + 5\mathbb{Z}$	$2 + 5\mathbb{Z}$	$3 + 5\mathbb{Z}$	$4 + 5\mathbb{Z}$
$0 + 5\mathbb{Z}$	$0 + 5\mathbb{Z}$	$0 + 5\mathbb{Z}$	$0 + 5\mathbb{Z}$	$0 + 5\mathbb{Z}$	$0 + 5\mathbb{Z}$
$1 + 5\mathbb{Z}$	$0 + 5\mathbb{Z}$	$1 + 5\mathbb{Z}$	$2 + 5\mathbb{Z}$	$3 + 5\mathbb{Z}$	$4 + 5\mathbb{Z}$
$2 + 5\mathbb{Z}$	$0 + 5\mathbb{Z}$	$2 + 5\mathbb{Z}$	$4 + 5\mathbb{Z}$	$1 + 5\mathbb{Z}$	$3 + 5\mathbb{Z}$
$3 + 5\mathbb{Z}$	$0 + 5\mathbb{Z}$	$3 + 5\mathbb{Z}$	$1 + 5\mathbb{Z}$	$4 + 5\mathbb{Z}$	$2 + 5\mathbb{Z}$
$4 + 5\mathbb{Z}$	$0 + 5\mathbb{Z}$	$4 + 5\mathbb{Z}$	$3 + 5\mathbb{Z}$	$2 + 5\mathbb{Z}$	$1 + 5\mathbb{Z}$

Figure 14.1

(b)

Both R and I are subrings of $M_2(\mathbb{Z})$. (See Exercise 3.) Since $I \subseteq R$, R/I is a factor ring, whose elements are of the form

$$\begin{bmatrix} 0 & 0 \\ 0 & b \end{bmatrix} + I$$

We can now state results for rings analogous to the isomorphism theorems for groups. One such result follows and we list others in Exercises 13 and 14.

Theorem 14.8

First Isomorphism Theorem

A ring $(S, \hat{+}, \hat{\cdot})$ is the homomorphic image of the ring $(R, +, \cdot)$ if and only if there is an ideal I of R such that $R/I \cong S$.

Proof

First suppose that $\sigma: R/I \to S$ is an isomorphism for some ideal I of R. Let $\theta: R \to R/I$ be the natural map defined in Theorem 14.7. Then it is straightforward to verify that $\sigma \circ h: R \to S$ is an epimorphism. (See Exercise 4.)

Conversely, suppose $\phi: R \to S$ is an epimorphism and set $I = \ker \phi$. Then I is an ideal of R. Let $\theta: R \to R/I$ be the homomorphism defined by $\theta(r) = r + I$. Define $\sigma: R/I \to S$ by $\sigma(r + I) = \phi(r)$. By the proof of Theorem 8.3, σ is a well-defined group isomorphism. We conclude by showing that σ is a ring homomorphism:

$$\sigma((r_1 + I) \odot (r_2 + I)) = \sigma(r_1 r_2 + I)$$
$$= \phi(r_1 r_2)$$
$$= \phi(r_1) \hat{\cdot} \phi(r_2)$$
$$= \sigma(r_1 + I) \hat{\cdot} \sigma(r_2 + I) \qquad \square$$

Corollary 14.2

Let $\phi: R \to S$ be a ring homomorphism. Then $R/\ker \phi \cong \phi^{\rightarrow}(R)$.

Corollary 14.3

Let $(R, +, \cdot)$ and $(S, \oplus, \odot)$ be rings and 0_R the additive identity of R. A surjective ring homomorphism $\theta: R \to S$ is an isomorphism if and only if $\ker \theta = \{0_R\}$.

Example 14.16

We saw in Example 14.5 that the mapping $\theta: \mathbb{Z} \to \mathbb{Z}_n$ defined by $\theta(m) = \bar{m}$, where $\bar{m}$ is the equivalence class of m in $\mathbb{Z}_n$, is a ring homomorphism. Clearly this is a surjective mapping. The kernel of this homomorphism is the set of all integers m such that $\bar{m} = \bar{0}$. But $\bar{m} = \bar{0}$ if and only if m is a multiple of n. Thus $\ker \theta = n\mathbb{Z} = (n)$. By Theorem 14.8, we have $\mathbb{Z}/n\mathbb{Z} \cong \mathbb{Z}_n$. For $n = 5$, the isomorphism between these rings is evident from the Cayley tables in Figure 14.1

Our results on ring homomorphisms lead us to an important discovery concerning the subring structure of an arbitrary ring with unity.

Theorem 14.9

If $(R, +, \cdot)$ is a ring with unity u, then R has a subring which is isomorphic either to $\mathbb{Z}$ or to $\mathbb{Z}_n$ for some $n \in \mathbb{N}$.

Proof

Consider the set $A = \{nu \mid n \in \mathbb{Z}\}$, where nu is defined by

$$0u = 0_R, \quad 1u = 1, \quad 2u = u + u, \quad nu = (n-1)u + u, \quad \text{and}$$
$$(-n)u = -(nu)$$

If there exists an $n \in \mathbb{N}$ such that $nu = 0_R$, the identity of R, then the Well-Ordering Property guarantees that there exists a smallest such natural number k. If $m = qk + r$ for $0 \leqslant r < k$, then

$$mu = (qk + r)u = q(ku) + ru = q(0_R) + ru = ru$$

Therefore,

$$A = \{0_R, u, 2u, \ldots, (k-1)u\}$$

The mapping $\theta \colon \mathbb{Z}_k \to \{0_R, u, \ldots, (k-1)u\}$ defined by $\theta(\bar{n}) = nu$ is an isomorphism since

$$\theta(\bar{n}_1 \oplus \bar{n}_2) = \theta(\overline{n_1 + n_2}) = (n_1 + n_2)u = n_1 u + n_2 u = \theta(\bar{n}_1) + \theta(\bar{n}_2)$$

and

$$\theta(\bar{n}_1 \odot \bar{n}_2) = \theta(\overline{n_1 \cdot n_2}) = (n_1 n_2)u = (n_1 u) \cdot (n_2 u) = \theta(\bar{n}_1) \cdot \theta(\bar{n}_2)$$

If there is no $n \in \mathbb{N}$ such that $nu = 0_R$, then the set $\{nu \mid n \in \mathbb{Z}\}$ consists of distinct elements, and there is an isomorphism $\theta \colon \mathbb{Z} \to \{nu \mid n \in \mathbb{Z}\}$ defined by $\theta(n) = nu$. The details of this verification are similar to those in the preceding paragraph and are therefore omitted. □

Definition 14.7

A ring $(R, +, \cdot)$ is said to be an **extension** of a ring $(S, \oplus, \odot)$ if R has a subring that is isomorphic to S. We also say that S is **imbedded** in R if R is an extension of S.

With the preceding definition, we may restate Theorem 14.9: Every ring with unity is an extension of either $\mathbb{Z}$ or $\mathbb{Z}_n$ for some $n \in \mathbb{N}$.

The following theorem gives an interesting result concerning extensions and imbeddings:

Theorem 14.10

Every ring may be imbedded in a ring with unity.

Proof

Let $(R, \oplus, \odot)$ be a ring, and consider $R \times \mathbb{Z} = \{(r, n) | r \in R, n \in \mathbb{Z}\}$. Define $\hat{+}$ on $R \times \mathbb{Z}$ by

$$(r_1, n_1) \hat{+} (r_2, n_2) = (r_1 \oplus r_2, n_1 + n_2)$$

and define $\hat{\cdot}$ on $R \times \mathbb{Z}$ by

$$(r_1, n_1) \hat{\cdot} (r_2, n_2) = ((r_1 \odot r_2) \oplus n_2 r_1 \oplus n_1 r_2, n_1 n_2)$$

In Exercise 6, you are asked to verify that these two operations make $(R \times \mathbb{Z}, \hat{+}, \hat{\cdot})$ a ring. There is a unity element, namely, $(0_R, 1)$ since

$$(r, n) \hat{\cdot} (0_R, 1) = ((r \odot 0_R) \oplus 1r \oplus n 0_R, n1)$$
$$= (0_R \oplus r \oplus 0_R, n)$$
$$= (r, n)$$

and similarly,

$$(0_R, 1) \hat{\cdot} (r, n) = (r, n)$$

To complete the proof, note first that $\theta: R \rightarrow R \times \mathbb{Z}$ defined by $\theta(r) = (r, 0)$ is a ring monomorphism. Moreover, $\theta^{\rightarrow}(R) = \{(r, 0) | r \in R\}$ is a subring of $R \times \mathbb{Z}$. We leave the details of the proofs of these assertions as Exercise 6. Thus $R \times \mathbb{Z}$ is an extension of R. $\square$

On the basis of the preceding result, you might assume that you need to deal only with rings with unity. One problem, however, is that the ring constructed in Theorem 14.10 can have zero-divisors even if the original ring does not. In Section 15.2, we will look at a different and most interesting imbedding, which does not introduce zero-divisors.

Exercises 14.3

1. In the proof of Theorem 14.7, verify that $\odot$ is associative and distributes over $\oplus$.

2. Show that the mapping $\theta: R \rightarrow R/I$ of Theorem 14.7 is a surjective ring homomorphism.

3. a. Prove

$$R = \left\{ \begin{bmatrix} a & 0 \\ 0 & b \end{bmatrix} \middle| a, b \in \mathbb{Z} \right\}$$

is a subring of $(M_2(\mathbb{Z}), +, \times)$.

b. Verify that

$$I = \left\{ \begin{bmatrix} a & 0 \\ 0 & 0 \end{bmatrix} \middle| ab \in \mathbb{Z} \right\}$$

is an ideal in R. Is I an ideal of $M_2(\mathbb{Z})$?

4. Show that the mapping $\sigma \circ \theta \colon R \to S$ of Theorem 14.8 is an epimorphism.

5. Let $I = (x^2 + 1)$ be the principal ideal in the ring $\mathbb{R}[x]$ of all polynomials with real coefficients; that is

$$I = \{(x^2 + 1)f(x) | f(x) \in \mathbb{R}[x]\}$$

Prove that every element of $\mathbb{R}[x]/I$ can be written in the form $(a + bx) + I$ and that $\mathbb{R}[x]/I$ is ring isomorphic to $\mathbb{C}$, the field of complex numbers.

6. a. Verify that the set $R \times \mathbb{Z}$ of Theorem 14.10 is a ring under the operations of $\hat{+}$ and $\hat{\cdot}$.
    b. If $R = 2\mathbb{Z}$, then prove that $(R \times \mathbb{Z}, \hat{+}, \hat{\cdot})$ is not an integral domain.
    c. Verify that $\theta \colon R \to R \times \mathbb{Z}$ defined by $\theta(r) = (r, 0)$ is a ring monomorphism.
    d. Prove that $\theta^\to(R) = \{(r, 0) | r \in R\}$ is a subring of $R \times \mathbb{Z}$.

7. Combine Exercise 21 of Section 13.2 and Theorem 14.9 to show that every integral domain contains a subring isomorphic either to $\mathbb{Z}$ or to $\mathbb{Z}_p$ for some prime p.

8. Find all fields with 4 or 5 elements.

9. Let A and B be ideals of the ring $(R, +, \cdot)$. Define

$$A + B = \{a + b | a \in A, b \in B\}$$

Prove that both $A + B$ and $A \cap B$ are ideals of R.

10. Let R be a commutative ring, $d \in R$, and I an ideal of R. Prove that
    i. $(d) + I$ is an ideal of R/I, and
    ii. $(d) + I$ is the principal ideal of R/I generated by $(d + I)$.

11. The set

$$Sc_2 = \left\{ \begin{bmatrix} a & 0 \\ 0 & a \end{bmatrix} \middle| a \in \mathbb{Z} \right\}$$

is a subring of $(M_2(\mathbb{Z}), +, \times)$. Is $M_2(\mathbb{Z})/Sc_2$ a ring; that is, do the cosets form a ring?

12. a. Prove that

$$S = \left\{ \begin{bmatrix} a & b \\ 0 & 0 \end{bmatrix} \middle| a, b \in \mathbb{Z} \right\}$$

is a subring of $M_2(\mathbb{Z})$, but not an ideal.

    b. Is

$$T = \left\{ \begin{bmatrix} 0 & a \\ 0 & 0 \end{bmatrix} \middle| a \in \mathbb{Z} \right\}$$

an ideal of $M_2(\mathbb{Z})$?

c. Let

$$R' = \left\{ \begin{bmatrix} a & b \\ 0 & c \end{bmatrix} \middle| a, b, c \in \mathbb{Z} \right\}$$

Prove that R' is a subring of $M_2(\mathbb{Z})$ and T is an ideal of R'.

d. Determine the factor ring R'/T.

13. Use Exercise 9 to prove that if A and B are ideals of a ring $(R, +, \cdot)$, then $(A + B)/B \cong A/(A \cap B)$. (This is the *Second Isomorphism Theorem for Rings*.)

14. Suppose I is an ideal of a ring $(R, +, \cdot)$ and J is an ideal of R contained in I. Prove that J is also an ideal of I and that I/J is an ideal of R/J. Verify that

$$\frac{R/J}{I/J} \cong R/I$$

(This is the *Third Isomorphism Theorem for Rings*.)

15. Describe the elements of the subring of $(M_2(\mathbb{Z}), +, \times)$ whose existence is guaranteed by Theorem 14.9.

16. An element a of a ring R with unity 1 is said to be a **unit** if there exists $b \in R$ such that $ab = 1$. Show that if I is a **proper** ideal of a ring with unity, then I contains no unit of R.

17. Let $\phi: \mathbb{Q} \to \mathbb{Q}$ be a ring homomorphism. Prove that either $\phi(a) = 0$ for all $a \in \mathbb{Q}$ or $\phi(a) = a$ for all $a \in \mathbb{Q}$.

18. Let I, J be ideals of a ring R, and set

$$IJ = \left\{ \sum_{i=1}^{n} a_i b_i \middle| a_i \in I, b_i \in J \text{ for some } n \in \mathbb{N} \right\}$$

Prove that IJ is an ideal of R and that $IJ \subseteq I \cap J$.

19. Let N be the set of nilpotent elements of R, a commutative ring with unity. Prove that N is an ideal of R and that R/N has no nilpotent elements other than 0.

15 Integral Domains

In the two preceding chapters, we have laid the groundwork for additional study of some important topics in ring theory. In this chapter, we will delve more deeply into the concept of integral domains. A number of examples will be developed that illustrate some important types of integral domains. Properties of ideals will be utilized to provide additional results connecting integral domains with fields. It will be shown how any integral domain can be thought of as a subring of a field.

Results on integral domains can be applied both inside and outside mathematics. The Chinese Remainder Theorem uses the theory of integral domains to guarantee that certain systems of linear congruences have a solution. In computer science, this result alone has been used to construct computer circuitry that approaches the theoretical limit for speed of addition. We will investigate both of these applications here.

Throughout this chapter, we will deal only with commutative rings. As in the last chapter, the additive identity will be denoted by 0 and the multiplicative identity or unity by 1.

15.1

Units

In this section we will examine some of the properties possessed by a ring that has cancellation laws for its second operation. In particular, we will consider the properties of integral domains. We know from the proof of Theorem 13.2 that cancellation laws hold in a commutative ring when the ring contains no zero-divisors. Therefore such a ring is an integral domain if it has a unity. Example 15.1 shows that a commutative ring may satisfy the cancellation laws without having a unity.

Example 15.1

The ring $(2\mathbb{Z}, +, \cdot)$, where $2\mathbb{Z}$ is the set of even integers, satisfies the cancellation property. (See Exercise 1.) However, the ring of even integers has no unity, and thus is not an integral domain.

The crucial difference between an integral domain and a commutative ring with cancellation laws (or, equivalently, without zero-divisors) is the unity. We know from our knowledge of the integers that $1 = 1 \cdot 1$ and $1 = (-1) \cdot (-1)$, although elements in $\mathbb{Z}$ do not, in general, possess multiplicative inverses. Do other integral domains contain similar elements? To facilitate this investigation, let us make the following definition:

Definition 15.1

> An element a of a commutative ring $(R, +, \cdot)$ with unity 1 is said to be a **unit** in R if there exists $b \in R$ such that $ab = 1$.

A unit of R is an element invertible with respect to the second ring operation. A unity is always a unit but, as with $-1 \in \mathbb{Z}$, not every unit need be a unity. One result that relates units and ideals is the following:

Theorem 15.1

If I is a proper ideal of a commutative ring R with unity, then I contains no units of R.

Proof

Recall that I is proper if $I \neq R$. If I is an ideal that contains a unit a, then there exists $b \in R$ such that $ab = 1$. Since I is an ideal, $1 = ab \in I$. But then $r = r1 \in I$ for all $r \in R$. Therefore, $R \subseteq I \subseteq R$ and $I = R$. □

This gives us one possible method for finding units in a ring with unity: Look for the largest possible ideal I; that is, look for an ideal such that any larger ideal (containing I) must be the ring itself. The elements not in the ideal I, together with the elements in I, must produce units.

Example 15.2

The even integers $2\mathbb{Z}$ is an ideal of $\mathbb{Z}$. If I is an ideal of $\mathbb{Z}$ that contains $2\mathbb{Z}$ as a proper subset, then there is an odd integer $x = 2n + 1 \in I - 2\mathbb{Z}$. Since $2n \in 2\mathbb{Z}$, $1 = 1 \cdot (2n + 1) + (-1) \cdot (2n) \in I$ and $I = \mathbb{Z}$. This shows that no proper ideal of $\mathbb{Z}$ contains $2\mathbb{Z}$ as a proper subset and that any unit of $\mathbb{Z}$ must be odd.

We know that an odd integer is a unit if and only if it is 1 or -1. Thus this method merely cuts down on the search area. To facilitate our search, let us formalize and improve upon our method.

Definition 15.2

Let $(R, +, \cdot)$ be a ring. An ideal M of R is said to be a **maximal ideal** of $M \neq R$ and given an ideal of R such that $M \subseteq I \subseteq R$, then either $M = I$ or $I = R$.

The question to ask is "Given an ideal, how do you check to see if the ideal is maximal?" We used one method in Example 15.2 to show that $2\mathbb{Z}$ is a maximal ideal of $\mathbb{Z}$. Another method can be derived from the next result.

Theorem 15.2

An ideal M of a commutative ring R is maximal in R if and only if R/M contains no proper nontrivial ideals.

Proof

Suppose M is not maximal and $M \subset I \subset R$ where I is an ideal. Let $d \in I - M$. By Exercise 10 of Section 14.3, $(d) + M = (d + M)$ is a nontrivial ideal in R/M contained in I/M. Thus $(d) + M \neq R/M$ and $(d) + M \neq M$, so $(d) + M$ is a proper ideal of R/M.

Conversely, suppose M is a maximal ideal and I/M is a proper ideal of R/M. Let $d \in I - M$. Since M is maximal, any ideal containing M and d must contain all the elements of R. But the set

$$(d) + M = \{rd + m \mid m \in M, r \in R\}$$

is an ideal of R. Thus $(d) + M = R$. But then $(d + M) = (d) + M$ is an ideal of R/M and $(d) + M/M \subseteq I/M \subseteq R/M$. Since $(d) + M = R$, this contradicts the statement that I/M is a proper ideal of R/M. The result follows. $\square$

In Example 15.2, we proved directly that $2\mathbb{Z}$ is a maximal ideal of $\mathbb{Z}$. Theorem 15.2 gives us another way to establish this result.

Example 15.3

As in Example 15.2, the even integers $2\mathbb{Z}$ form an ideal of $\mathbb{Z}$. Since $\mathbb{Z}/2\mathbb{Z} \cong \mathbb{Z}_2$ (Example 14.16) has no proper nontrivial ideals, $2\mathbb{Z}$ is a maximal ideal of $\mathbb{Z}$. Now $\mathbb{Z}/2\mathbb{Z} = \{0 + 2\mathbb{Z}, 1 + 2\mathbb{Z}\}$. In addition, we know by Theorem 15.1 that no unit of $\mathbb{Z}$ can belong to $2\mathbb{Z}$. Therefore the units of $\mathbb{Z}$ must lie in the coset $1 + 2\mathbb{Z}$ consisting of all odd numbers. We saw before that the only units of $\mathbb{Z}$ are 1 and -1. Both are odd.

Example 15.4

Consider $\mathbb{R}[x]$, the ring of all polynomials with real coefficients. The set of all polynomials with 0 as the constant term forms an ideal $I[x]$ and $\mathbb{R}[x]/I[x]$ is a factor ring with unity $1 + I[x]$. Is the ideal $I[x]$ maximal in $\mathbb{R}[x]$? Consider $p(x) + I[x]$, for $p(x) \in \mathbb{R}[x]$. If $p(x)$ has constant term p_0, then the coset $p(x) + I[x]$ can also be represented by $p_0 + I[x]$. If $p(x) \notin I[x]$, then $p_0 \neq 0$, and the coset $p_0 + I[x]$ is a unit since the element $p_0^{-1} \in \mathbb{R}$ and

$$(p_0 + I[x])(p_0^{-1} + I[x]) = p_0 p_0^{-1} + I[x] = 1 + I[x]$$

This says that every nonzero element of $\mathbb{R}[x]/I[x]$ is a unit and that $\mathbb{R}[x]/I[x]$ has no proper ideals. By Theorem 15.2, $I[x]$ is maximal in $\mathbb{R}[x]$.

In this last example, $\mathbb{R}[x]/I[x]$ is a commutative ring with unity $1 + I[x]$ in which every nonzero element is a unit, or, in other words, is invertible. But then $\mathbb{R}[x]/I[x]$ is a field. Similarly, in Example 15.3, $\mathbb{Z}/2\mathbb{Z} \cong \mathbb{Z}_2$ is also a field. Is this always the case? Is M a maximal ideal if and only if R/M is a field? The answer is yes, and the truth of this statement helps us identify maximal ideals by allowing us to check to see if each element of R/M is a unit.

Theorem 15.3

Let R be a commutative ring with unity. An ideal M of R is maximal if and only if R/M is a field.

Proof

If R/M is a field, then any nonzero element of R/M is a unit. But then R/M has no proper ideals. To see this, suppose I/M were an ideal of R/M and $a \in I - M$. Then $a + M$ is a unit, contrary to Theorem 15.1. Therefore $I = M$, and M is maximal in R.

Conversely, if M is maximal, let $a + M \in R/M$ for $a \notin M$. Then $(a) + M$ is an ideal of R that properly contains M. Since M is maximal in R, $(a) + M = R$, and there are $b \in R$, $m \in M$ such that $1 = ab + m$. Thus $(a + M)(b + M) = 1 + M$ and $a + M$ is a unit. Therefore R/M is a field. $\square$

Example 15.5

Let p be a prime in $\mathbb{Z}$. Then $(p) = p\mathbb{Z}$ and, since $\mathbb{Z}/p\mathbb{Z} \cong \mathbb{Z}_p$ is a field (Example 13.12), (p) is a maximal ideal. For $p = 3$,

$$\mathbb{Z}/3\mathbb{Z} = \{0 + 3\mathbb{Z}, 1 + 3\mathbb{Z}, 2 + 3\mathbb{Z}\}$$

Now $2 + 3\mathbb{Z}$ is a unit of $\mathbb{Z}/3\mathbb{Z}$, although 2 is not a unit in $\mathbb{Z}$. Note, however, that $-1 = 2 + 3(-1)$ **is** a unit in $\mathbb{Z}$. This shows how the units in a ring are related to the elements of a maximal ideal. Note that it is **not** true in general that every element outside a maximal ideal is a unit of R. (See Example 15.3.)

In our search for maximal ideals, we must be certain that the factor ring is a field and not just an integral domain. Consider the next definition and the discussion that follows.

Definition 15.3

> An ideal P of a commutative ring R is a **prime ideal** if $P \neq R$ and for all $a, b \in R$, $ab \in P$ implies $a \in P$ or $b \in P$.

The fact that any maximal ideal in a commutative ring with unity is also a prime ideal follows from the following argument: Suppose P is maximal, $a \notin P$, $b \notin P$. Then $a + P$ and $b + P$ are nonidentity cosets in R/P. If $ab \in P$, then

$$(a + P)(b + P) = ab + P = 0 + P$$

Since $a + P \neq P$ and $b + P \neq P$, the field R/P contains zero-divisors, contradicting Corollary 13.1. Therefore $ab \notin P$ and P is prime. We state this result as the following theorem.

Theorem 15.4

In a commutative ring with unity, every maximal ideal is a prime ideal.

The converse to this theorem is not true. To see this, consider the next example:

Example 15.6

Consider $\mathbb{Z}[x]$, the ring of polynomials with integer coefficients. We will see in Chapter 17 that $\mathbb{Z}[x]$ is an integral domain. The principal ideal (x) consisting of all polynomials whose constant term is 0 is not a maximal ideal since $\mathbb{Z}[x]/(x) \cong \mathbb{Z}$ (see Exercise 11) and $\mathbb{Z}$ is not a field. Nevertheless, (x) is a prime ideal. To see this, suppose $p(x), q(x) \in \mathbb{Z}[x]$, but $p(x) \notin (x)$ and $q(x) \notin (x)$. This says that the constant terms p_0 and q_0 of $p(x)$ and $q(x)$, respectively, are nonzero. The polynomial $p(x)q(x)$ has constant term $p_0 q_0$, which is not zero since $\mathbb{Z}$ is an integral domain. Therefore, $p(x)q(x) \notin (x)$, and (x) is a prime ideal.

In Example 15.6, we noted that $\mathbb{Z}[x]/(x) \cong \mathbb{Z}$ and $\mathbb{Z}$ is an integral domain. In fact, any quotient ring of a commutative ring with unity factored by a prime ideal is an integral domain. This property characterizes prime ideals:

| Theorem 15.5 |

An ideal I of a commutative ring R with unity is prime if and only if R/I is a nontrivial integral domain.

Proof

If I is prime, we know R/I is a nontrivial commutative ring. We need to show that there is a unity and that there are no zero-divisors.

If 1 is the unity of R, then $1 + R$ is the unity of R/I since

$$(a + I)(1 + I) = a1 + I = a + I$$

If $(a + I)(b + I) = I = ab + I$, then $ab \in I$. This means $a \in I$ or $b \in I$ since I is prime. Thus $a + I = I$ or $b + I = I$, and R/I has no zero-divisors.

Conversely, if I is not prime, then there exist $a, b \in R$ such that $ab \in I$ and $a \notin I$, $b \notin I$. Therefore $(a + I)(b + I) = ab + I = I$. Since R/I has zero-divisors, it is not an integral domain. □

Compare the relationship between prime ideals and integral domains given in this result with the relationship between maximal ideals and fields given in Theorem 15.3.

The idea behind the definition of prime ideal comes from the concept of prime integers. If $p \in \mathbb{Z}$ is not a prime, then p is a composite and $p = ab$, $a \neq \pm 1$, $b \neq \pm 1$. Then $a \notin (p)$ and $b \notin (p)$, but $ab \in (p)$, so (p) is not a prime ideal.

If p is a prime, it is easy to see that $(p) = p\mathbb{Z}$ is a prime ideal since $\mathbb{Z}/p\mathbb{Z} \cong \mathbb{Z}_p$, which is an integral domain. But $\mathbb{Z}_p$ is a **finite** integral domain. By Theorem 13.3, $\mathbb{Z}_p \cong \mathbb{Z}/p\mathbb{Z}$ is a field and, therefore, $p\mathbb{Z} = (p)$ is a maximal ideal. Since any maximal ideal of $\mathbb{Z}$ is prime by Theorem 15.4, the result follows.

Prime ideals are quite useful in the study of polynomials. For example, in $\mathbb{R}[x]$ consider $x^2 + 1$. Is $(x^2 + 1)$ prime? Is it maximal? Some of the elements of $\mathbb{R}[x]/(x^2 + 1)$ are obvious: $r + (x^2 + 1)$ for $r \in \mathbb{R}$, and $x + (x^2 + 1)$, and, of course, $a + bx + (x^2 + 1)$ for $a, b \in \mathbb{R}$. With a bit of algebra, you can see that any polynomial $p(x) \in \mathbb{R}[x]$ may be written as $p_0 + p_1 x + q(x)(x^2 + 1)$. (See Exercise 12.)

We know $\mathbb{R}[x]/(x^2 + 1)$ is a commutative ring with $1 + (x^2 + 1)$ being the unity. Let $p(x), q(x) \notin (x^2 + 1)$. Then

$$p(x) = p_0 + p_1 x + r_1(x)(x^2 + 1)$$

with p_0 and p_1 not both zero, and

$$q(x) = q_0 + q_1 x + r_2(x)(x^2 + 1)$$

with q_0 and q_1 not both zero.

Now,

$$p(x)q(x) = p_0 q_0 + (p_0 q_1 + p_1 q_0)x + (p_1 q_1)x^2 + r_3(x)(x^2 + 1)$$

$$= (p_0 q_0 - p_1 q_1) + (p_0 q_1 + p_1 q_0)x + p_1 q_1(x^2 + 1) + r_3(x)(x^2 + 1)$$

$$= (p_0 q_0 - p_1 q_1) + (p_0 q_1 + p_1 q_0)x + r_4(x)(x^2 + 1)$$

If $p(x)q(x) \in (x^2 + 1)$, then both

$$p_0 q_0 = p_1 q_1 \qquad \text{and} \qquad p_0 q_1 = -p_1 q_0$$

Thus,

$$p_0^2 q_0 = p_0 p_1 q_1 = -p_1^2 q_0$$

so that

$$(p_0^2 + p_1^2) q_0 = 0$$

Similarly,

$$p_1^2 q_1 = p_1 p_0 q_0 = p_0^2 q_1 \qquad \text{and} \qquad (p_0^2 + p_1^2) q_1 = 0$$

Since $p_0^2 + p_1^2 \neq 0$, we conclude $q_0 = q_1 = 0$, contrary to the choice of $q(x)$. This shows that $p(x)q(x) \notin (x^2 + 1)$ and $(x^2 + 1)$ is a prime ideal; hence, $\mathbb{R}[x]/(x^2 + 1)$ is an integral domain.

To show $(x^2 + 1)$ is a maximal ideal, we will show $\mathbb{R}[x]/(x^2 + 1)$ is a field. To do this, set $I = (x^2 + 1)$. The preceding calculations show that

$$(a + bx + I)(c + dx + I) = (ac - bd) + (ad + bc)x + I$$

To compute $(a + bx + I)^{-1}$, we must solve

$$ac - bd = 1 \qquad \text{and} \qquad ad + bc = 0$$

for c and d. As long as $a^2 + b^2 \neq 0$ (which means that $a + bx + I \neq I$), this system has the solution

$$c = \frac{a}{a^2 + b^2}, \qquad d = \frac{-b}{a^2 + b^2}$$

(See Exercise 13.) Thus

$$(a + bx + I)^{-1} = \frac{a}{a^2 + b^2} + \frac{-b}{a^2 + b^2} x + I$$

If you followed this discussion, you might sense that this formula looks familiar. It should—it is similar to the formula for the inverse of the complex number $a + bi$ where $i = \sqrt{-1}$. In fact, it can be shown that $\mathbb{R}[x]/(x^2 + 1)$ is isomorphic to $\mathbb{C}$. The construction of our ring $\mathbb{R}[x]/(x^2 + 1)$ might lead us to conclude that complex numbers are not so imaginary after all!

The next chapter contains a good deal of information about special types of integral domains. In Chapter 17, we will investigate polynomial rings in greater detail. In Chapter 20, we will mimic the preceding construction to form other fields that are factor rings of integral domains. Next, however, we consider the way the integers are contained in the field of rational numbers.

1. Prove that if $(R, +, \cdot)$ is a ring with cancellation laws, then every subring of R has cancellation laws.

2. Let R be a ring with unity. An element $a \in R$ is said to have a **right inverse** $b \in R$ if $ab = 1$. Similarly, $c \in R$ is a **left inverse** if $ca = 1$. Prove that if a has both a right inverse b and a left inverse c, then $b = c$.

3. Suppose M is an ideal in a ring R in which $1 \neq 0$. If R/M is a division ring, show that M is a maximal ideal of R.

4. Let d be a positive integer, not a perfect square. Set

$$\mathbb{Z}[\sqrt{d}] = \{a + b\sqrt{d} \,|\, a, b \in \mathbb{Z}\}$$

   a. Prove that $\mathbb{Z}[\sqrt{d}]$, under the operations of real number addition and multiplication, is a commutative ring with unity.
   b. If $(a, b) = 1$, then show that $a + b\sqrt{d}$ is a unit if and only if $a^2 - b^2 d = 1$ or -1.
   c. Find all units in $\mathbb{Z}[\sqrt{d}]$.
   d. Verify that $\mathbb{Z}[\sqrt{d}]$ is an integral domain.

5. By Exercise 4, we know that $\mathbb{Z}[\sqrt{2}] = \{a + b\sqrt{2} \,|\, a, b \in \mathbb{Z}\}$ is an integral domain.
   a. By examining the quotient ring $\mathbb{Z}[\sqrt{2}]/(\sqrt{2})$, determine whether the ideal $(\sqrt{2})$ is a prime and/or a maximal ideal.
   b. Show that $(x^2 - 2)$ is a prime ideal in $\mathbb{Z}[x]$ by showing that $\mathbb{Z}[x]/(x^2 - 2)$ is isomorphic to $\mathbb{Z}[\sqrt{2}]$.
   c. Use the preceding to develop an argument that $\sqrt{2}$ is **not** a rational number.

6. Follow the pattern of Exercise 5, and show that $\sqrt{3}$ is an irrational number.

7. a. Prove the Correspondence Theorem for Rings: If $\phi\colon R \to R'$ is a ring homomorphism with $I = \ker \phi$, then there is a one-to-one correspondence between the set of subrings of R containing I and subrings of Im ϕ. (Compare with Theorem 8.6.)
   b. Suppose M is a maximal ideal of R, a commutative ring with unity. Apply part a to the natural mapping $\phi\colon R \to R/M$ defined by $\phi(r) = r + M$ to show that R/M has no proper ideals.

8. Prove that the set of all polynomials in $\mathbb{Z}[x]$ with zero constant term forms an ideal. Is this ideal principal?

9. Let R be a ring. An ideal I in R is termed **primary** if $ab \in I$ for $a, b \in R$ implies that either $a \in I$ or $b^n \in I$ for some positive integer n.
   a. Verify that every prime ideal is primary.
   b. Prove that I is primary if and only if each zero-divisor of R is nilpotent.
   c. Show that (4) is a primary ideal of $\mathbb{Z}$ but not a prime ideal.

10. Suppose $\phi\colon R \to R$ is a ring homomorphism. Suppose $a \in R$ is a unit. Show it is not necessarily true that $\phi(a)$ is a unit of R. What conditions on R and ϕ guarantee that $\phi(a)$ is a unit?

11. Verify $\mathbb{Z}[x]/(x) \cong \mathbb{Z}$.

12. Use mathematical induction to show that if $p(x)$ is a polynomial of degree $\geqslant 2$, then $p(x) = a + bx + q(x)(x^2 + 1)$.

13. Prove that if a and b are not both equal to zero and $ax - by = 1$ and $bx + ay = 0$, then

$$x = \frac{a}{a^2 + b^2} \quad \text{and} \quad y = \frac{-b}{a^2 + b^2}$$

14. Prove that I is a prime ideal in a commutative ring R with unity if and only if $R - I$ is closed under the second ring operation.

15. If I is an ideal in a commutative ring R with unity, then set $Rad(I) = \cap P$, where $\cap P$ is the intersection of all prime ideals P of R containing I. $Rad(I)$ is an ideal called the **nilradical** of I. Verify that in $\mathbb{Z}$, $Rad(\{0\}) = \{0\}$, $Rad((16)) = (2)$ and $Rad((48)) = (2) \cap (3) = (6)$.

16. If R is a commutative ring with unity, let $J(R) = \cap M$ where $\cap M$ is the intersection of all maximal ideals M of R. $J(R)$ is termed the **Jacobson radical** of R. Prove $J(\mathbb{Z}) = \{0\}$. What is $J(2\mathbb{Z})$?

17. Is Theorem 15.2 also true for noncommutative rings?

18. Suppose that $\phi: R \to R'$ is a ring epimorphism and P is a prime ideal of R containing $\ker(\phi)$.
    a. Prove that $\phi^{\to}(P)$ is prime ideal of R'.
    b. If P' is a prime ideal of R', prove that $\phi^{\leftarrow}(P')$ is a prime ideal of R.

19. Show that the requirement in Theorem 15.3 that R possesses a unity is essential by finding a maximal ideal M in $2\mathbb{Z}$ such that $2\mathbb{Z}/M$ is not a field.

20. Let X be a nonempty set and $A \subseteq X$. Prove that $\mathscr{P}(A)$ is an ideal in the ring $(\mathscr{P}(X), \triangle, \cap)$. Find a maximal ideal in this ring.

15.2

Field of Quotients

In Section 14.3, we discussed how any ring can be thought of as a subring of a ring with unity. We expressed this result by proving "every ring is imbeddable in a ring with unity." Recall Definition 14.6 where we defined *imbeddable*: A ring $(R, +, \cdot)$ is **imbeddable** in a ring $(S, \oplus, \odot)$ if there is an injective ring homomorphism from R into S or, equivalently, if $(R, +, \cdot)$ is isomorphic to a subring of $(S, \oplus, \odot)$. However, as we demonstrated in that section, the ring we constructed in Theorem 14.10 had zero-divisors. Must this be the case? Or can we imbed a commutative ring without zero-divisors in a field? These questions are answered by Theorem 15.6, which is one of the most useful and important imbedding theorems. As a special case, it will allow us to construct the rational numbers from the ring of integers.

Theorem 15.6

Every integral domain is imbeddable in a field.

The proof of this theorem involves a construction and several definitions. For this reason we will prove several lemmas first. For simplicity, we will use $(D, +, \cdot)$ to denote our integral domain and juxtaposition for the second operation in D. Set

$$Q^* = \{(a, b) \,|\, a, b \in D, b \neq 0\}$$

In following this proof, it may help if you think of D as the integers and Q as the rational numbers.

Lemma 15.1

The relation $\sim$ on Q^* defined by $(a, b) \sim (c, d)$ if $ad = bc$ is an equivalence relation.

Proof

Since $ab = ba$, we have $(a, b) \sim (a, b)$; and so $\sim$ is reflexive. If $ad = bc$, then $cb = da$; so $(a, b) \sim (c, d)$ implies $(c, d) \sim (a, b)$. Therefore $\sim$ is symmetric.

To show that $\sim$ is transitive requires some tricky manipulations. Assume that $(a, b) \sim (c, d)$ and $(c, d) \sim (g, h)$. Then $ad = bc$ and $ch = dg$. Utilizing these facts, we obtain $(ad)h = (bc)h$ and $(ch)b = (dg)b$. Knowing that $\cdot$ is commutative and associative, we have $(bc)h = (ah)d = (bg)d$. Using the equality of the last two elements, and the fact that $d \neq 0$, coupled with the cancellation laws in our integral domain, we obtain $ah = bg$ or $(a, b) \sim (g, h)$. $\square$

By Q we will denote the set of all $\sim$ equivalence classes on Q^*, and by a/b we will denote the equivalence class of which (a, b) is an element.

By a/b we **do not mean** "a divided by b," since there is not necessarily an operation of division. "a/b" is a set of elements, an equivalence class: For example, if $D = \mathbb{Z}$, then

$$1/2 = \{(c, d) \,|\, (c, d) \sim (1, 2)\}$$
$$= \{(c, d) \,|\, 2c = d, d \neq 0\}$$
$$= \{(c, 2c) \,|\, c \neq 0\}$$
$$= \{(1, 2), (2, 4), (3, 6), \ldots, (-1, -2), (-2, -4), (-3, -6), \ldots\}$$

We say that $a/b = c/d$ if the two equivalence classes are equal as sets; that is, $a/b = c/d$ if $(a, b) \sim (c, d)$. For our example above, $1/2 = 2/4$.

Definition 15.4

We define the operations $\oplus$ and $\otimes$ on Q by

$$a/b \oplus c/d = (ad + bc)/bd \qquad \text{and} \qquad (a/b) \otimes (c/d) = ac/bd$$

Recall that whenever you define operations on a set, especially a set of equivalence classes, you must be sure that the definitions are consistent with the basic properties of the set. In this case, the basic properties include an equivalence relation.

Lemma 15.2

$\oplus$ and $\otimes$ are well defined.

Proof

Let $(a, b) \sim (s, t)$ and $(c, d) \sim (x, y)$; then $at = bs$ and $cy = dx$. To show $a/b \oplus c/d = s/t \oplus x/y$, we must show that

$$(ty)((ad) + (bc)) = (bd)((sy) + (tx))$$

But

$$(ty)((ad) + (bc)) = ((dy)(at)) + ((bt)(cy))$$

$$= ((dy)(bs)) + ((bt)(dx))$$

$$= ((bd)(sy)) + ((bd)(tx))$$

$$= (bd)((sy) + (tx))$$

Thus $\oplus$ is well defined.

We leave the verification that $\otimes$ is well defined as Exercise 1.   □

Theorem 15.7

If $(D, +, \cdot)$ is an integral domain, and Q, $\oplus$, and $\otimes$ are defined as in Lemma 15.2, then $(Q, \oplus, \otimes)$ is a field. (We will call this field the **field of quotients of** D.)

Proof

We will require you to verify that both operations are commutative and associative and that $\otimes$ distributes over $\oplus$. (See Exercise 2.) We will complete the proof that Q is a field in a number of steps.

First, if $a/b \in Q$, then

$$a/b \oplus 0/1 = ((a1) + (0b))/b1 = (a + 0)/b = a/b$$

Therefore $0/1$ is an identity for $\oplus$.

To show that $(Q, \oplus)$ is an abelian group, we need only show that every $a/b \in Q$ has an additive inverse. Since

$$a/b \oplus (-a)/b = ((ab) + (b(-a))/bb$$

$$= ((ab) - (ab))/bb$$

$$= 0/bb = 0/b = 0/1$$

due to the nature of the equivalence relation on Q, we have

$$(-a)/b = -(a/b)$$

Therefore, $(Q, \oplus)$ is an abelian group.

Now let us consider $Q - \{0/1\}$. If $a/b \in Q$, then

$$a/b \oplus 1/1 = a1/b1 = a/b$$

so that $1/1$ is a unity for any element of Q. If $a/b \in Q - \{0/1\}$, then $b/a \in Q - \{0/1\}$ and

$$a/b \otimes b/a = ab/ba = ab/ab = 1/1$$

by the definition of the equivalence relation. Thus $(Q - \{0/1\}, \otimes)$ is an abelian group and $(Q, \oplus, \otimes)$ is a field. □

Proof of Theorem 15.6

In light of Lemmas 15.1 and 15.2 and Theorem 15.7, we now need to show that $(D, +, \cdot)$ is imbeddable in $(Q, \oplus, \otimes)$. Define $h: D \to Q$ by $h(a) = a/1$. We must show that h is a homomorphism and is injective (that is, h is a ring monomorphism). Now,

$$h(a + b) = (a + b)/1 = ((a1) + (1b))/(1 \cdot 1) = a/1 \oplus b/1 = h(a) \oplus h(b)$$

Similarly,

$$h(a \cdot b) = ab/1 = ab/(1 \cdot 1) = a/1 \otimes b/1 = h(a) \otimes h(b)$$

Therefore h is a homomorphism.

If $h(a) = h(b)$, then $a/1 = b/1$ or $a1 = b1$ and $a = b$. Therefore h is injective. This completes the proof of Theorem 15.6. □

If we identify a with $h(a)$, we can think of D as a subring of Q. In order to give an alternate characterization of the field of quotients of an integral domain, we need the next result.

Theorem 15.8

No proper subfield of Q contains $\{a/1 \mid a \in D\}$.

Proof

If F is a subfield of Q that contains $\{a/1 \mid a \in D\}$, then for $a \neq 0$, $1/a \in F$, and thus for $a, b \in D$, $b \neq 0$,

$$a/1 \otimes 1/b = a/b \in F$$

Therefore $Q \subseteq F$ and $F = Q$. □

Corollary 15.1

If an integral domain $(D, +, \cdot)$ is imbeddable in a field $(F, \hat{+}, \hat{\times})$, then there is a subfield of F isomorphic to the field of quotients of D.

Proof

If $h: D \to F$ is a monomorphism and Q is the field of quotients of R, define $h^*: Q \to F$ by

$$h^*(a/b) = h(a) \hat{\times} (h(b))^{-1}$$

We will let you verify that h^* is a monomorphism. (See Exercise 6.) □

For an integral domain D, the following statements can be derived as consequences of our results:

1. If D is a subring of a field F, then the field of quotients of D is also a

subring of F; in fact, the field of quotients of D is (isomorphic to) the smallest subfield of F that contains D.
2. The field of quotients of the field F is (isomorphic to) the field F itself.

Figure 15.1 helps to symbolize this discussion.

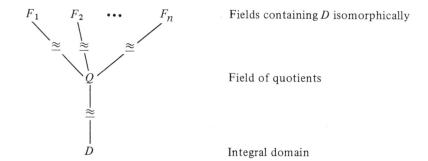

Figure 15.1

You should note that the field of quotients of the ring of integers $\mathbb{Z}$ is none other than the rational number system $\mathbb{Q}$. Thus, we have a characterization of the rational number system as the smallest field (isomorphically) containing the integers. Throughout the remainder of this text, we will write the integer a and its corresponding rational number $a/1$ interchangeably.

We have carried out our construction of the field of quotients for an integral domain D. Actually, the construction works for any commutative ring with no zero-divisors; that is, we do not really need to postulate the existence of a unity in our ring. We consider this in Exercise 4.

Historically, the construction of the rational numbers in terms of integers was part of a movement to axiomatize mathematics, which flourished in the latter half of the nineteenth century. It should be clear to you how the complex number system can be defined in terms of real numbers. In 1872, Richard Dedekind used the concept of **Dedekind cut** to define the real number system in terms of rational numbers. The first effort towards giving a logical foundation of the rational numbers had been made earlier by Martin Ohm (1792–1872), a professor in Berlin and brother of the famed physicist. Later, in 1860, K. Weierstrauss (1815–1897) used a method employing ordered pairs to define the rationals. This study was taken over by G. Peano, who, in 1889, developed the approach we have considered in this section. G. Peano completed this effort the same year by listing the basic axioms for natural numbers.

In Chapter 17, we will have occasion again to consider fields of quotients as they relate to rings of polynomials and special types of integral domains.

Exercises 15.2

1. Show that $\otimes$, as defined in Definition 15.4, is a well-defined operation on Q.

2. Complete the proof of Theorem 15.7 by showing that $\oplus$ and $\otimes$ are associative and commutative operations and that $\otimes$ distributes over $\oplus$.

3. Show that if $(D, +, \cdot)$ is an integral domain and $b \in R$, $b \neq 0$, then $0/b$ is an identity for $\oplus$ in Q. Doesn't this contradict the fact that $(Q, \oplus)$ is a group and that the additive identity in a group is unique?

4. If $(R, +, \cdot)$ is simply a commutative ring with no zero-divisors, show that R is, nevertheless, still imbeddable in a field. In this case describe the additive and multiplicative identities of the field in terms of equivalence classes. What is the field of quotients of the ring of even integers?

5. Show that if $(R, +, \cdot)$ has zero-divisors, then R is not imbeddable in a field by our process outlined in this section, or by any other method.

6. Verify that the mapping h^* of Corollary 15.1 is an injective homomorphism.

7. Supply in detail the proof that the field of quotients of a field is isomorphic to the field itself.

8. Describe the field of quotients of the Gaussian integers of Example 13.9.

9. Do the quaternions of Hamilton (see Example 13.13) have a field of quotients? Why?

10. Prove that if F is a field of characteristic 0 (see Exercise 21 of Section 13.2), then F contains a subfield isomorphic to $\mathbb{Q}$.

11. Verify that if $(D, +, \cdot)$ and $(D', +', \cdot')$ are isomorphic integral domains, then their corresponding fields of quotients are isomorphic.

12. Let $R^* = \mathbb{N} \times \mathbb{N} = \{(a, b) \mid a, b \in \mathbb{N}\}$. Define $\approx$ on R^* by $(a, b) \approx (c, d)$ if $a + d = b + c$.
    a. Prove that $\approx$ is an equivalence relation on R^*.
    b. Let $a - b$ denote the equivalence class of (a, b) and R the set of these equivalence classes. Define $\oplus$ and $\odot$ on R^* by

    $$(a - b) \oplus (c - d) = (a + c) - (b + d) \qquad \text{and}$$

    $$(a - b) \odot (c - d) = (ac + bd) - (ad + bc)$$

    c. Verify that $\oplus$ and $\odot$ are well defined on R.
    d. Prove that $(R, \oplus, \odot)$ is a commutative ring with unity.
    e. Verify that $(R, \oplus, \odot)$ is an integral domain.
    f. What are the equivalence classes of $(2, 1)$, $(7, 3)$, and $(1, 4)$?
    g. If $a > b$, then $a = b + h$, $h > 0.$, and $(a, b) = (b + h, b)$. If $a < b$, then $b = a + h$, $h > 0$, and $(a, b) = (a, a + h)$. Prove that the mapping $\phi \colon \mathbb{Z} \to R$ defined by

    $$\phi(h) = \begin{cases} (1 + h, 1), & h > 0 \\ (h, 1 + h), & h > 0 \end{cases}$$

    is an isomorphism.

13. What is the field of quotients of $\mathbb{Z}[\sqrt{2}]$. (See Exercise 3 of Section 14.1.)

14. Extend the order relation of $\mathbb{Z}$ to $\mathbb{Q}$ in the following manner: Define

    $$(a/b) \trianglelefteq (c/d) \qquad \text{if} \qquad ad < bc$$

    a. Show that $\trianglelefteq$ is a well-defined relation.
    b. Prove that $a < b$ if and only if $a/1 \trianglelefteq b/1$.
    c. Let $\mathbb{Q}^+ = \{a/b \in \mathbb{Q} \mid a \cdot b > 0\}$. Show that $\mathbb{Q}^+$ is closed under addition and multiplication.

d. Show that $\trianglelefteq$ satisfies the Rule of Trichotomy: if $a/b \in \mathbb{Q}$, then either
$$a/b \in \mathbb{Q}^+, \qquad -a/b \in \mathbb{Q}^+, \qquad \text{or} \qquad a/b = 0/1 = 0$$

e. Prove that $\mathbb{Q}$ satisfies the **Archimedean Property**: if $a/b, c/d \in \mathbb{Q}^+$, then there exists some $n \in \mathbb{N}$ such that $(c/d) \trianglelefteq n(a/b)$.

15. Suppose D is an integral domain and P a prime ideal of D. Set S equal to the set of all quotients of the form a/b where $a, b \in D$ and $b \notin P$. Prove that S is a subring of $Q(D)$, the field of quotients of D.

16. Let R be a ring. Prove that the fact that R is isomorphic to a subring of a field is a necessary and sufficient condition for R being an integral domain.

17. Suppose that Q_1 and Q_2 are both fields of quotients of an integral domain D. Show that there exists an isomorphism from Q_1 to Q_2 that fixes all the elements of D.

15.3

The Chinese Remainder Theorem

In an integral domain, the existence of cancellation laws permits the solution of a large variety of problems. The fact that not every nonzero element in an integral domain D need be a unit means, however, that not every equation of form $ax = b$ for $a, b \in D$ has a solution in D. For example, $2x = 3$ has no solution in $\mathbb{Z}$. The purpose of this section is to show that, despite this deficiency, there is a certain specialized type of problem that can always be solved in an integral domain. The proof of the existence of these solutions and the method of solution itself are of special interest to both mathematicians and computer scientists. In fact, in Section 15.4, we will apply these results to circuit design. To begin our discussion, however, consider the following problem:

> A gang of 17 Chinese bandits captured a caravan of the emperor. Among the treasures, they obtained a quantity of solid gold eggs. Upon dividing the eggs into equal portions, they found three eggs remaining which they agreed they ought to give their cook, Foo Yun. But six of the bandits were killed in a brawl, and now when the total amount of eggs was divided equally among them, there were four eggs left over, which they planned to give to Foo Yun. In an attack that followed, only six of the bandits, the eggs, and the cook were saved. This time an equal distribution left a remainder of five eggs for the cook. Becoming weary of his masters' stinginess, Foo Yun poisoned the stew the next night, so that all the eggs became his! How many eggs did Foo Yun obtain?

While this problem may appear silly, the mathematics involved is not. Stated in other terms, we need to find a number x such that when we divide x by 17, we get remainder 3; when we divide x by 11, we get remainder 4; and when we divide x by 6, we get remainder 5. If we let n_1, n_2, and n_3 denote the respective integral quotients, we can write the problem as a system of equations:

$$\begin{cases} x = 17n_1 + 3 \\ x = 11n_2 + 4 \\ x = 6n_3 + 5 \end{cases}$$

In terms of congruences, we have

$$\begin{cases} x \equiv 3 \ (\text{mod } 17) \\ x \equiv 4 \ (\text{mod } 11) \\ x \equiv 5 \ (\text{mod } 6) \end{cases}$$

The result, which tells us that there is a solution to this system, and, in fact, an infinite number of solutions, is known as the Chinese Remainder Theorem. We will prove this result and then apply it to solve our problem.

In an earlier chapter, we defined the product of groups. To solve the preceding problem, we need the concept of the product of ring first encountered in the Exercise 12 of Section 13.1.

Definition 15.5

If $(R, \mp, \bar{\cdot})$ and $(S, \hat{+}, \hat{\cdot})$ are rings, then define

$$(a_1, b_1) + (a_2, b_2) = (a_1 \mp a_2, b_1 \hat{+} b_2)$$

and

$$(a_1, b_1) \cdot (a_2, b_2) = (a_1 \bar{\cdot} a_2, b_1 \hat{\cdot} b_2)$$

for $a_1, a_2 \in R$ and $b_1, b_2 \in S$.

Theorem 15.9

If $(R, \mp, \bar{\cdot})$ and $(S, \hat{+}, \hat{\cdot})$ are rings, then $(R \times S, +, \cdot)$ is a ring, termed the **external direct product of R and S.**

Proof

The operations $+, \cdot$ in Definition 15.5 are well defined. Also, since $(R, \mp)$, $(S, \hat{+})$ are abelian groups, $(R \times S, +)$ is an abelian group by Exercise 15 of Section 3.2. The establishment of the other ring properties requires tedious verification. We will examine the left distributive property and leave the others as Exercise 1.

$$\begin{aligned}
(a_1, b_1) \cdot ((a_2, b_2) + (a_3, b_3)) &= (a_1, b_1) \cdot (a_2 \mp a_3, b_2 \hat{+} b_3) \\
&= ((a_1 \bar{\cdot} (a_2 \mp a_3), b_1 \hat{+} (b_2 \hat{+} b_3)) \\
&= ((a_1 \bar{\cdot} a_2) \mp (a_1 \bar{\cdot} a_3), (b_1 \hat{\cdot} b_2) \hat{+} (b_1 \hat{\cdot} b_3)) \\
&= (a_1 \bar{\cdot} a_2, b_1 \hat{\cdot} b_2) + (a_1 \bar{\cdot} a_3, b_1 \hat{\cdot} b_3) \\
&= [(a_1, b_1) \cdot (a_2, b_2)] + [(a_1, b_1) \cdot (a_3, b_3)]
\end{aligned}$$

$\square$

Although many of the results we will discuss in this section can be extended to arbitrary rings, our concentration will be on the integers.

Recall Theorem 1.4, which says that if (a, b) is the greatest common divisor of integers $a, b \in \mathbb{Z}$, then there exist $s, t \in \mathbb{Z}$ such that $(a, b) = sa + tb$. (We will obtain the proof of this theorem as a consequence of a more general result in Chapter 17.) If $(a, b) = 1$, then $1 = sa + tb$ for suitable $s, t \in \mathbb{Z}$. We will utilize these ideas in the proof of our main result.

We begin first with a lemma.

Lemma 15.3

If $n = ab$, $(a, b) = 1$, then $\mathbb{Z}_n$ is a ring that is isomorphic to $\mathbb{Z}_a \times \mathbb{Z}_b$.

Proof

In $\mathbb{Z}_a \times \mathbb{Z}_b$, the element $(1, 0)$ has order a, while the element $(0, 1)$ has order b. By Lemma 10.2, $(1, 1) = (1, 0) + (0, 1)$ has order $ab = n$ in the **group** $\mathbb{Z}_a \times \mathbb{Z}_b$. Thus $\mathbb{Z}_a \times \mathbb{Z}_b$ is cyclic. Since all cyclic groups of order n are isomorphic, we know $\mathbb{Z}_n$ is a **group** isomorphic to $\mathbb{Z}_a \times \mathbb{Z}_b$. In fact, the mapping $\phi \colon \mathbb{Z}_n \to \mathbb{Z}_a \times \mathbb{Z}_b$ defined by $\phi(k) = (k, k)$ is a group isomorphism. (See Exercise 2.) Since $\mathbb{Z}_n \cong \mathbb{Z}/n\mathbb{Z}$, we can describe ϕ in another way by defining $\phi \colon \mathbb{Z}/n\mathbb{Z} \to (\mathbb{Z}/a\mathbb{Z}) \times (\mathbb{Z}/b\mathbb{Z})$ through the formula

$$\phi(k + n\mathbb{Z}) = (k + a\mathbb{Z}, k + b\mathbb{Z})$$

To complete our proof, we show that ϕ preserves ring multiplication; that is,

$$\phi((k_1 + n\mathbb{Z}) \odot (k_2 + n\mathbb{Z})) = \phi(k_1 + n\mathbb{Z}) \cdot \phi(k_2 + n\mathbb{Z})$$

This is straightforward, and we leave it as Exercise 4.   □

If we have that $n = p_1^{e_1} p_2^{e_1}$, where p_1 and p_2 are distinct primes, then Lemma 15.3 says that $\mathbb{Z}_n$ is ring isomorphic to $\mathbb{Z}_{p_1^{e_1}} \times \mathbb{Z}_{p_2^{e_2}}$. Induction applied to $n = p_1^{e_1} p_2^{e_2} \cdots p_k^{e_k}$, where the p_i's are distinct primes, yields the following analog of Theorem 10.4:

Theorem 15.10

If $n = p_1^{e_1} p_2^{e_2} \cdots p_k^{e_k}$, where the p_i's are distinct primes, then the ring $\mathbb{Z}_n$ is isomorphic to

$$\mathbb{Z}_{p_1^{e_1}} \times \mathbb{Z}_{p_2^{e_2}} \times \cdots \times \mathbb{Z}_{p_k^{e_k}}$$

As a consequence of Lemma 15.3, we know the mapping $\phi \colon \mathbb{Z}_n \to \mathbb{Z}_a \times \mathbb{Z}_b$ given by $\phi(k) = (k, k)$ is surjective. The proof does not indicate, however, how you would find an element $k \in \mathbb{Z}_n$ that maps onto a given element $(c, d) \in \mathbb{Z}_a \times \mathbb{Z}_b$. If you can do this, then the problem of finding a solution to Foo Yun's problem is much easier.

Therefore, we want to give a direct and constructive argument to show that ϕ is surjective. Toward that end, let

$$(c + a\mathbb{Z}, d + a\mathbb{Z}) \in (\mathbb{Z}/a\mathbb{Z}) \times (\mathbb{Z}/b\mathbb{Z})$$

Now we can write $1 = sa + tb$ for $s, t \in \mathbb{Z}$ since $(a, b) = 1$. Thus

$$c - d = (c - d)1 = (c - d)sa + (c - d)tb = s'a + t'b$$

Then $c - s'a = d + t'b = x$ satisfies

$$\phi(x) = (x + a\mathbb{Z}, x + b\mathbb{Z})$$
$$= ((c - s'a) + a\mathbb{Z}, (c + t'b) + b\mathbb{Z})$$
$$= (c + a\mathbb{Z}, d + b\mathbb{Z})$$

Let us consider how to go about determining x for a specific value of (c, d).

| Example 15.7 |

Consider $\mathbb{Z}_{15} \cong \mathbb{Z}_3 \times \mathbb{Z}_5$. What element x in $\mathbb{Z}_{15}$ maps onto the element $(2, 4) \in \mathbb{Z}_3 \times \mathbb{Z}_5$? If

$$\phi(x + 15\mathbb{Z}) = (x + 3\mathbb{Z}, x + 5\mathbb{Z})$$

what are the integers m, n such that $x = 2 + 3m$ and $x = 4 + 5n$? This is equivalent to solving the congruences

$$x \equiv 2 \ (\text{mod } 3) \qquad \text{and} \qquad x \equiv 4 \ (\text{mod } 5)$$

Using the first equation in the second congruence, we have

$$x = 2 + 3m \equiv 4 \ (\text{mod } 5) \qquad \text{or} \qquad 3m \equiv 2 \ (\text{mod } 5)$$

Since $2 \cdot 3 \equiv 1 \ (\text{mod } 5)$, 3 has a multiplicative inverse in $\mathbb{Z}_5$, namely, $3^{-1} = 2$. Therefore

$$m \equiv 2(3m) \equiv 2(2) \equiv 4 \ (\text{mod } 5)$$

and $m = 4 + 5k$ for some k. Substituting this value in the first equation, we have

$$x = 2 + 3m = 2 + 3(4 + 5k) = 14 + 15k$$

Therefore, $x \equiv 14 \ (\text{mod } 15)$ and

$$\phi(14 + 15\mathbb{Z}) = (14 + 3\mathbb{Z}, 14 + 5\mathbb{Z})$$
$$= (2 + 12 + 3\mathbb{Z}, 4 + 10 + 5\mathbb{Z})$$
$$= (2 + 3\mathbb{Z}, 4 + 5\mathbb{Z})$$

Key to the solution of this example is the fact that 3 is a unit of $\mathbb{Z}_5$. If $(a, b) = 1$, then a is always a unit in $\mathbb{Z}_b$. (See Exercise 3.) As long as the orders of the various cyclic summands, or equivalently, the moduli of the various congruences, are relatively prime in pairs, the method outlined in Example 15.7 will work.

Since the elements of $\mathbb{Z}_{15}$ are equivalence classes, the values $x = 14 + 15 = 29$ or $x = 14 + 15 \cdot 3 = 59$ would also work equally well. There is only one class of $\mathbb{Z}_{15}$ that maps onto $(2, 4)$, but there are many ways to represent that class.

Theorem 15.10 can be stated in an alternate version using the notation of linear congruences rather than that of ring products.

Corollary 15.2

The Chinese Remainder Theorem

Let $\{m_i\}_{i=1}^{k}$ be a set of k integers such that $(m_i, m_j) = 1$ for all $i \neq j$. Then the system of congruences

$$\{x \equiv a_i (\mathrm{mod}\ m_i)\}_{i=1}^{k}$$

where $a_i \in \mathbb{Z}$, has a unique integral solution modulo $n = m_1 m_2 \cdots m_k$.

In essence, the corollary asserts that exactly one class in $\mathbb{Z}_n$ maps onto the element $(a_1, a_2, \ldots, a_k)$ in $\mathbb{Z}_{m_1} \times \mathbb{Z}_{m_2} \times \cdots \times \mathbb{Z}_{m_k}$.

Problems involving simultaneous linear congruences are found in the early Chinese writings of Sun-Tse and were known as early as the first century A.D. Brahmagupta also wrote on these types of problems in the seventh century.

Let us use the congruence form of the Chinese Remainder Theorem to solve our original problem. We want to find x such that

$$\begin{cases} x \equiv 3\ (\mathrm{mod}\ 17), \\ x \equiv 4\ (\mathrm{mod}\ 11),\quad \text{and} \\ x \equiv 5\ (\mathrm{mod}\ 6) \end{cases}$$

From the first equation, $x = 3 + 17r$. Substituting this value into the second equation in the system, we obtain

$$3 + 17r \equiv 4\ (\mathrm{mod}\ 11) \qquad \text{or} \qquad 6r \equiv 17r \equiv 1\ (\mathrm{mod}\ 11)$$

Since $2 \cdot 6 = 12 \equiv 1\ (\mathrm{mod}\ 11)$, we can multiply both sides of our equation by 2 to obtain

$$r \equiv 12r \equiv 2(6r) \equiv 2(1) \equiv 2\ (\mathrm{mod}\ 11)$$

Then $r = 2 + 11s$. But then

$$x = 3 + 17(r) = 3 + 17(2 + 11s) = 37 + 187s$$

This value for x in the third equation of the system gives us

$$37 + 187s \equiv 5\ (\mathrm{mod}\ 6)$$

or,

$$1 + 1s \equiv 5\ (\mathrm{mod}\ 6)\quad \text{and}\quad s \equiv 4\ (\mathrm{mod}\ 6)$$

We conclude that $x = 37 + 187(s) = 37 + 187(4 + 6t) = 785 + 1122t$. Since

$1122t = 17 \cdot 11 \cdot 6$, we see that $x \equiv 785 \pmod{1122}$. Therefore Foo Yun obtained at least 785 eggs, and the next possible number would be $785 + 1122 = 1907$ eggs.

Modifications to the Chinese Remainder Theorem permit the solution of some systems where the congruences are of the form $ax \equiv b \pmod{m}$. We give some examples in the exercises.

Exercises 15.3

1. Complete the proof of Theorem 15.9.

2. Verify the assertion of Lemma 15.3 that $\phi(k) = (k, \; k)$ defines a group isomorphism.

3. a. Prove that a is a unit in $(\mathbb{Z}_n, \oplus, \odot)$ if and only if $(a, n) = 1$.
   b. If $(a, n) = 1$ and $sa + tn = 1$, show that $ax \equiv b \pmod{n}$ is equivalent to $x \equiv sb \pmod{n}$.

4. Verify that ϕ of Lemma 15.3 satisfies

$$\phi((k_1 + n\mathbb{Z}) \odot (k_2 + n\mathbb{Z})) = \phi(k_1 + n\mathbb{Z}) \cdot \phi(k_2 + n\mathbb{Z})$$

5. Supply the inductive proof of Theorem 15.10.

6. Show that every linear congruence of the form $ax \equiv b$ (modulo n), where $(a, n) = 1$, can be written in the form $x \equiv c$ (modulo n).

7. Verify that a linear congruence of the form $ax \equiv b$ (modulo n) has a solution in $\mathbb{Z}$ if and only if the greatest common divisor of a and n divides b. If this is the case, then how many distinct solutions are there less than or equal to n?

8. Suppose $a, b, c \in \mathbb{Z}$. Prove that the equation $ax + by = c$ has a solution (x, y) in $\mathbb{Z} \times \mathbb{Z}$ if and only if $(a, b) | c$.

9. Solve where possible; if there is no solution, state the reason for your answer.

   a. $\begin{cases} x \equiv 1 \pmod{7} \\ x \equiv 3 \pmod{5} \\ x \equiv 2 \pmod{8} \end{cases}$

   b. $\begin{cases} x \equiv 1 \pmod{2} \\ x \equiv 3 \pmod{4} \\ x \equiv 9 \pmod{11} \end{cases}$

   c. $\begin{cases} 3x \equiv 4 \pmod{5} \\ x \equiv 3 \pmod{4} \\ 5x \equiv 3 \pmod{9} \\ 3x \equiv 2 \pmod{7} \end{cases}$

   d. $\begin{cases} 3x \equiv 1 \pmod{5} \\ 2x \equiv 3 \pmod{7} \\ x \equiv 3 \pmod{4} \end{cases}$

   e. $\begin{cases} 2x \equiv 3 \pmod{4} \\ x \equiv 3 \pmod{5} \\ 3x \equiv 5 \pmod{7} \end{cases}$

10. a. If $a = da_1$, $b = db_1$, and $n = dm$, show that $ax \equiv b \pmod{n}$ is equivalent to $a_1 x \equiv b_1 \pmod{m}$.
    b. Solve

$$\begin{cases} 2x \equiv 4 \pmod{6} \\ 2x \equiv 3 \pmod{5} \\ x \equiv 4 \pmod{11} \end{cases}$$

11. Let

$$\begin{cases} x \equiv a_1 \ (\text{mod } m_1) \\ x \equiv a_2 \ (\text{mod } m_2) \\ \quad \vdots \\ x \equiv a_k \ (\text{mod } m_k) \end{cases}$$

be a system of linear congruences in which $i \neq j$ implies $(m_1, \ m_j) = 1$. Set $M = m_1 m_2 \cdots m_k$, and suppose b_i is the solution of $(M/m_i)x \equiv 1 \ (\text{mod } m_i)$. Verify that

$$x = a_1 b_1 \frac{M}{m_1} + a_2 b_2 \frac{M}{m_2} + \cdots + a_k b_k \frac{M}{m_k}$$

is a solution of the original system.

12. Prove that if $(p, q) = 1$ and $(r, q) = 1$, then $(pr, q) = 1$.

13. Seventeen Chinese cousins appeared before the mighty Emperor Ming in order to obtain settlement on their uncle's estate consisting of many rice paddies. Upon equal division of the land parcels, there remained three paddies that Ming acquired for himself, in the traditional manner. At their celebration banquet (at Toe Main's American Pizza Palace), seven of the brothers became ill and died. Ming did his thing again with the original estate and had five parcels remaining for himself. On their homeward trip, three cousins perished in rickshaw accidents. Redistribution of their uncle's estate found an equal apportionment of the land. Upset with the whole affair, Ming declared that the remaining Chinese cousins forfeit all to the emperor. What is the smallest number of rice paddies that Ming could obtain from the estate of the well-known rice merchant, Uncle Bien?

*14. Write a computer program to implement the Chinese Remainder Theorem.

15.4

The Construction of Fast Adders

Adders are central components of modern high-speed digital computers. The addition in these machines is usually performed modulo n for some large value of n. For example, usual binary addition has n equal to some power of 2. In some microcomputers, the value of n is 2^{32}.

$$\begin{array}{rcl} 0111 & \leftrightarrow & 7 \\ 1101 & \leftrightarrow & 13 \\ \hline 10100 & \leftrightarrow & 20 \end{array}$$

Figure 15.2

Consider the examples of ordinary binary addition in Figures 15.2 and 15.3. In performing this addition, one works from right to left summing the two digits and carrying digits to the next column when necessary. Note that because of the carry digit. the sum of two 4-bit binary numbers can be a 5-bit number. This sequence or **serial** addition of two binary numbers of length 2^m requires the same number 2^m of additions or time steps. In our example, $4 = 2^2$ steps were required.

$$\begin{array}{rccccc} & a_4 & a_3 & a_2 & a_1 \\ + & b_4 & b_3 & b_2 & b_1 \\ \hline c_5 & c_4 & c_3 & c_2 & c_1 \end{array}$$

Figure 15.3

A number of questions arise from this example. Is it possible to design circuitry so that the speed or time of addition is faster than ordinary binary addition? If so, just *how fast* can one add binary numbers on a computer?

Finally, if some lower bound on the speed of addition is found theoretically, can one actually design practical circuitry that meets or, at least, approaches this theoretical bound?

In 1965, S. Winograd of Bell Labs published a fundamental article wherein he discussed these basic ideas.[1] In looking into this material, we will see that the basic questions posed in the preceding paragraph involve some familiar topics in abstract algebra, namely, the Chinese Remainder Theorem and the subgroup structure of the groups $\mathbb{Z}_n$.

Adders are switching networks composed of elementary devices that perform addition. In Figure 15.4, we show an example of a 4-input device. Four input wires carry voltages that we will represent by 1 or 0, and the device can be considered as a switching function from the group

$$(\mathbb{Z}_2)^4 \cong \mathbb{Z}_2 \oplus \mathbb{Z}_2 \oplus \mathbb{Z}_2 \oplus \mathbb{Z}_2$$

to the group $\mathbb{Z}_2$. While various r-input devices are utilized, most devices usually have at most $r = 4$ inputs for reasons of economy, speed, and reliability. These basic adder components can be connected together to form multiple-level devices. For example, the circuit depicted in Figure 15.5 represents a 3-level network utilizing five 2-input devices that yields a 6-variable switching circuit.

A second example is given in Figure 15.6. It is a 9-variable, 2-level switching circuit using 3-input devices.

If we assume that each adder device uses one unit of time, then the first

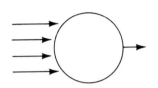

Figure 15.4

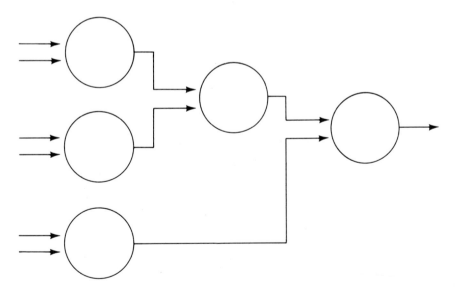

Figure 15.5

---

[1]S. Winograd, "On the time required to perform addition," *J. of the Assoc. for Computer Machinery*, Vol. 12, No. 2, pp. 277–285.

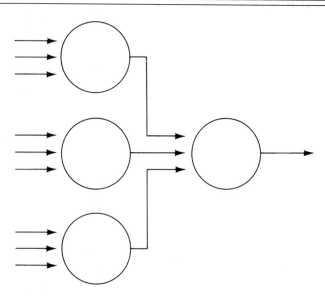

Figure 15.6

example represents an adder circuit that requires three units of time, while the second requires only two.

The key to designing high-speed addition circuitry is not minimizing the number of devices used but, instead, the number of levels necessary to complete the addition. Recall our description of sequential or serial addition of $a_4a_3a_2a_1 = 0111$ and $b_4b_3b_2b_1 = 1101$. This binary addition scheme for adding two 2^m-digit numbers required 2^m steps or time units. Figure 15.7 is an example of **parallel addition** of these same numbers. This 4-bit parallel adder is composed of 2-input adders and 3-input logic devices. The adders in the top row simultaneously compute the **sum** c_i and **carry** d_i. Each one can be thought of as a combination of two single-output adders. This parallel adder requires only three times units. In general, parallel adders of 2^m-digit numbers require only $m + 1$ stages or time units.

A parallel adder that can combine two 2^m-bit numbers in $m + 1$ time units is certainly an improvement over a slower serial adder that requires 2^m time units. (We are emphasizing speed over simplicity of construction.) Can we do better? Suppose we have an adder network made up of r-input, single-output devices that represents an m-variable switching circuit. How many time units does it take to operate? Note that a 1-level network can handle r inputs, while a 2-level network can accept $r \cdot r = r^2$ inputs. (Look at our second example wherein our 2-level network of 3-input devices accepted $9 = 3^2$ inputs.) Then a 3-level network will take $r \cdot r^2 = r^3$ variables, and so on. Since we have m variables to start with, our network must have at least m-input wires. If t levels are required, we therefore conclude that $r^t \geqslant m$ or $t \geqslant \log_r m$.

For example, if we have $m = 15$ and $r = 2$, our result shows $2^t \geqslant 15$ or $t \geqslant \log_2 15$. The smallest such integer t is 4. If $m = 15$ and $r = 3$, then $3^t > 15$

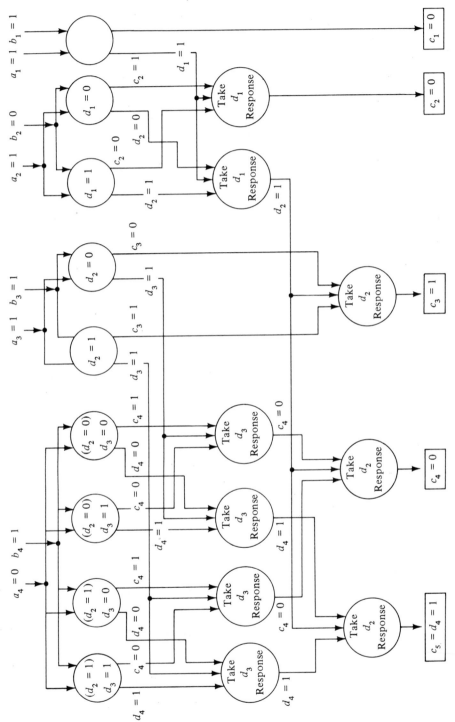

Figure 15.7

allows the value $t = 3$. Thus a network utilizing 3-input devices is conceivably "faster" than a network using 2-input devices for $m = 15$. For $m = 4$, however, both types of devices require $t = 2$ levels.

Since the values of t must be integers, we can use the ceiling function to express the result we have just established. If $x \in \mathbb{R}$, the **ceiling of** x, denoted $\lceil x \rceil$, is defined to be the smallest integer greater than or equal to x. Thus $\lceil 2 \rceil = 2$ and $\lceil \sqrt[3]{10} \rceil = 3$. The preceding discussion can be stated as follows.

Theorem 15.11

An m-variable switching circuit utilizing r-input devices requires at least $\lceil \log_r m \rceil$ levels.

In other words, the minimum time of computation is at least $\lceil \log_r m \rceil$ time units. In order to apply this result to the construction of fast adders, recall from our initial examples that both serial and parallel binary addition of two 4-bit binary involves all of the 8-bits. The value of c_5 in the sum involves a switching function of the variables a_4, a_3, a_2, a_1, b_4, b_3, b_2, and b_1. Therefore, any binary addition of two k-digit binary numbers requires a switching circuit to compute with $2k$ variables just to obtain the left-most digit in the sum. We have therefore obtained the following corollary.

Corollary 15.3

Binary addition of two k-digit binary numbers requires at least $\lceil \log_r 2k \rceil$ time units.

Recall we said that addition in a computer is performed modulo n for some large value of n. To add two numbers modulo n, you must first represent them in binary form. Since there are 2^k binary numbers of length k, we must have $2^k \geqslant n$ if all n numbers are to be written in length k or less. Thus $k \geqslant \lceil \log_2 n \rceil$. Combining this observation with our last result yields another consequence.

Corollary 15.4

Adding two numbers modulo n using binary addition requires at least $\lceil \log_r 2\lceil \log_2 n \rceil \rceil$ time units.

Say, for example, that $n = 2^{35}$. Then, for $r = 3$, the minimum bound is $\lceil \log_3 2\lceil \log_2 2^{35} \rceil \rceil = \lceil \log_3 2(35) \rceil = 4$ time units. Therefore, this says that if you wish to add these numbers in less than 4 time units, you must represent your numbers in some form other than usual binary form. Can this be done?

To answer this question requires some group theory and the Chinese Remainder Theorem. We will show that by utilizing a process intuitively allied to parallel processing and by carefully choosing our modulus n, we can beat the theoretical lower bound for binary addition. We will consider only the mathematical aspect of this problem and leave the discussion of the practical construction of such an adder to Winograd's paper.

0 → 000	6 → 001	12 → 002	18 → 003	24 → 004
1 → 111	7 → 112	13 → 113	19 → 114	25 → 110
2 → 022	8 → 023	14 → 024	20 → 020	26 → 021
3 → 103	9 → 104	15 → 100	21 → 101	27 → 102
4 → 014	10 → 010	16 → 011	22 → 012	28 → 013
5 → 120	11 → 121	17 → 122	23 → 123	29 → 124

Figure 15.8

We know by the Chinese Remainder Theorem that if $n = p_1^{e_1} p_2^{e_2} \cdots p_k^{e_k}$, where the p_i's are distinct primes, then

$$\mathbb{Z}_n \cong \mathbb{Z}_{p_1^{e_1}} \oplus \mathbb{Z}_{p_2^{e_2}} \oplus \cdots \oplus \mathbb{Z}_{p_k^{e_k}}$$

For example $\mathbb{Z}_{30} \cong \mathbb{Z}_2 \oplus \mathbb{Z}_3 \oplus \mathbb{Z}_5$. The isomorphism can be viewed as taking $k \in \mathbb{Z}_{30}$ and calculating $(k \pmod 2), k \pmod 3), k \pmod 5))$. Thus

$$10 \to (10 \pmod 2), 10 \pmod 3), 10 \pmod 5)) = (0, 1, 0)$$

which, for brevity, we write as 010. The complete table of this correspondence is given in Figure 15.8.

The idea behind Winograd's results is to take two given numbers in $\mathbb{Z}_n$, say 11 and 14, represent them as 121 and 024, and simultaneously add each component to obtain 145 or equivalently, 110. This answer is then decoded as 25. The speed of this addition in $\mathbb{Z}_n$, not counting the time for coding and decoding the numbers according to the Chinese Remainder Theorem, depends upon the time of addition for groups of the form $\mathbb{Z}_{p_i^{e_i}}$.

In order to determine this time, we return to the underlying group structure of $\mathbb{Z}_n$. A nonidentity element u in a group G is termed **ubiquitous** in G if it is contained in every nontrivial subgroup. For example, since the only nontrivial subgroups of $\mathbb{Z}_4$ are $\langle 2 \rangle$ and $\mathbb{Z}_4$ itself, 2 is a ubiquitous element of $\mathbb{Z}_4$. On the other hand, $\mathbb{Z}_6$ has subgroups $\langle 2 \rangle = \{0, 2, 4\}$ and $\langle 3 \rangle = \{0, 3\}$. Since $\langle 2 \rangle \cap \langle 3 \rangle = \{0\}$, $\mathbb{Z}_6$ has no ubiquitous elements.

Why are we concerned with ubiquitous elements? The answer is contained in Winograd's major result. In order to state it, we assume that the elements of $\mathbb{Z}_n$ are first represented in a one-to-one fashion as m-bit binary numbers $(a_1, a_2, \ldots, a_m) = a_1 a_2 \cdots a_m$ and that $0 \in \mathbb{Z}_n$ is represented by $(0, 0, \ldots, 0) = 00 \cdots 0$. Suppose, furthermore, that we have constructed a modulo n adder that works with these m-bit representations. Winograd's result has the following form:

Theorem 15.12

Suppose $\mathbb{Z}_n$ has a ubiquitous element and is represented as m-bit binary numbers. Then at least one output of any modulo n adder using these binary numbers depends upon at least $2 \lceil \log_2 n \rceil$ distinct inputs.

Proof

Suppose $u \in \mathbb{Z}_n$ is ubiquitous and has binary representation as $u_1 u_2 \cdots u_m$. Since $u \neq 0$ and our binary representation is injective, at least one $u_i \neq 0$. The key to proving this result lies in showing that the output of the ith component of the adder, that is, the value of $(x + y)_i$ for $x, y \in \mathbb{Z}_n$, requires at least $\lceil \log_2 n \rceil$ distinct inputs of each addend in the sum.

By way of contradiction, assume not. Suppose $x \in \mathbb{Z}_n$ has representation $x_1 x_2 \cdots x_m$ and only $t < \lceil \log_2 n \rceil$ components of the form x_j are necessary to compute $(x + y)_i$ for $y \in \mathbb{Z}_n$. Then $2^t < n < 2^m$, so that there must exist $x' \in \mathbb{Z}_n$, $x' \neq x$, having the same identical components as x in these t locations. But then $(x + y)_i = (x' + y)_i$ for all $y' \in \mathbb{Z}_n$. This says

$$(w)_i = (x' + (w - x'))_i = (x + (w - x'))_i = ((x - x') + w)_i$$

for all $w \in \mathbb{Z}_n$. For $w = 0$, this says

$$(x - x')_i = ((x - x') + 0)_i = 0_i = 0$$

Setting $w = x - x'$, we obtain

$$0 = (x - x')_i = ((x - x') + (x - x'))_i = 2(x - x')_i$$

An inductive argument (see Exercise 14) shows that $z_i = 0$ for all $z \in \langle x - x' \rangle$.

But this is a contradiction, since $u \in \langle x - x' \rangle$ by definition and $u_i \neq 0$. Our result follows. $\qquad\square$

Winograd's result says if $\mathbb{Z}_n$ has a ubiquitous element, then any modulo n adder is a m-variable switching circuit for $m \geqslant 2 \lceil \log_2 n \rceil$. Combining this result with Theorem 15.11, we obtain the following corollary.

Corollary 15.5

If $\mathbb{Z}_n$ has a ubiquitous element, then adding two numbers modulo n in binary form with any type of addition requires at least $\lceil \log_r 2 \lceil \log_2 n \rceil \rceil$ time units.

We see, therefore, that if we hope to beat this lower bound, we must find a $\mathbb{Z}_n$ with **no** ubiquitous elements. Even in this case, however, we can still apply Winograd's Theorem. If $H \leqslant \mathbb{Z}_n$ and H has a ubiquitous element, then the sum of two elements in H is the same as the sum of these two elements in $\mathbb{Z}_n$. The time for addition in $\mathbb{Z}_n$ is bounded below by the time for addition in H.

Corollary 15.6

Let H be a subgroup of $\mathbb{Z}_n$ of largest order such that H contains a ubiquitous element. Then adding two numbers modulo n in binary form requires at least $\lceil \log_r 2 \lceil \log_2 |H| \rceil \rceil$ time units.

This last corollary tells us to choose a value of n where the orders of the subgroups H containing ubiquitous elements are as small as possible. By the Chinese Remainder Theorem, if $n = p_1^{e_1} p_2^{e_2} \cdots p_k^{e_k}$, then

$$\mathbb{Z}_n \cong \mathbb{Z}_{p_1^{e_1}} \oplus \mathbb{Z}_{p_2^{e_2}} \oplus \cdots \oplus \mathbb{Z}_{p_k^{e_k}}$$

Our next result tells us that we need to focus our attention on the direct summands.

Theorem 15.13

$\mathbb{Z}_n$ has a ubiquitous element if and only if n is a prime power.

Proof

Suppose first that n is not a prime power and there exist distinct prime divisors p and q of n. By Cayley's Theorem, there exist $x, y \in \mathbb{Z}_n$ of orders p and q, respectively. If $a \in \langle x \rangle \cap \langle y \rangle$, then $o(a)$ divides both p and q. Thus $o(a) = 1$, and a is not ubiquitous.

Now suppose $n = p^m$, where p is a prime. By Theorem 4.7, any subgroup of the cyclic group $\mathbb{Z}_n$ is also cyclic. In multiplicative notation, if $\mathbb{Z}_n = \langle a \rangle$, then any proper subgroup H is of the form $\langle a^t \rangle$ for t, a divisor of $n = p^m$. Since $t \neq p^m$, t must divide p^{m-1}. If $p^{m-1} = ts$, then $a^{p^{m-1}} = (a^t)^s \in \langle a^t \rangle = H$. Thus $a^{p^{m-1}}$ is a ubiquitous element of $\mathbb{Z}_n$. □

Corollary 15.7

If $\mathbb{Z}_n \cong \mathbb{Z}_{p_1^{e_1}} \oplus \mathbb{Z}_{p_2^{e_2}} \oplus \cdots \oplus \mathbb{Z}_{p_k^{e_k}}$, then the order of the largest subgroup of $\mathbb{Z}_n$ possessing a ubiquitous element $= \max \{ p_j^{e_j} \}$.

Proof

It suffices to note that subgroups of $\mathbb{Z}_n$ isomorphic to $\mathbb{Z}_{p_j^{e_j}}$ each contain a ubiquitous element, and that any subgroup H of $\mathbb{Z}_n$ whose order is divisible by two distinct primes cannot contain a ubiquitous element. □

We now have a way of choosing n in order that $\mathbb{Z}_n$ have only "small" subgroups containing ubiquitous elements. Choose a value of n so that its prime-power factors are small in comparison to the size of n. For example,

$$n = 2^5 \cdot 3^3 \cdot 5^2 \cdot 7 \cdot 11 \cdot 13 \cdot 17 \cdot 19 \cdot 23 \cdot 29 \cdot 31 > 1.4 \times 10^{14} > 2^{32}$$

and the size of the largest prime-power factor is $2^5 = 32$. Then Corollary 15.6 says that the theoretical lower bound for addition modulo n is

$$\lceil \log_r 2 \lceil \log_2 2^5 \rceil \rceil = \lceil \log_r 10 \rceil$$

For $r = 2$, this value is 4, while for $r = 3$, the value is 3. For binary addition modulo n, these values are 7 and 5. Even greater differences in value are obtainable for larger values of n.

Assuming that these lower bounds for addition can be approached in actuality, you must wonder why modern computers still utilize ordinary binary addition circuitry. The answer is that these computers are designed to do much more than add. Among the other requirements is the ability to compare two numbers and decide which is larger. Circuitry constructed according to Winograd's scheme for maximizing the speed of addition is much slower than ordinary binary circuitry for performing comparisons.

Therefore, the design of circuitry is a "give and take" proposition. In order to improve some aspect, one often causes problems in another area. Winograd's work is still important and not only from a theoretical point of view. Society is currently experiencing a large growth in the area of computer circuitry dedicated to some specific function. In designing hardware required for the performance of rapid addition (and multiplication) only, Winograd's work was a pioneering effort.

Exercises 15.4

1. Use the 4-bit parallel adder pictured in Figure 15.7 to add 1010 and 1011.

2. What is the theoretical lower limit on the number of time units necessary for the computation of a 16-variable switching circuit utilizing 3-input devices?

3. Draw a diagram to perform ordinary binary addition of 4-bit binary numbers.

4. Use the standard isomorphism of the Chinese Remainder Theorem to code the numbers 147 and 251 in Z_{315} as 3-tuples in $Z_9 \oplus Z_5 \oplus Z_7$. Add these numbers both in Z_{315} and in $Z_9 \oplus Z_5 \oplus Z_7$. Do the answers coincide?

5. Show that bounds for "fast" multiplications in a digital computer depend upon bounds for addition.

6. If p, q are distinct primes, suppose that $x \in Z_{p^m q^n}$ has isomorphic image $(1, 0)$ in $Z_{p^m} \oplus Z_{q^n}$, while $y \in Z_{p^m q^n}$ has image $(0, 1)$ in $Z_{p^m} \oplus Z_{q^n}$. Prove that the elements $(a, b) \in Z_{p^m} \oplus Z_{q^n}$ has preimage $ax + by \in Z_{p^m q^n}$. Find x and y for $p^m = 25$, $q^n = 16$. What element in Z_{400} has image $(21, 15) \in Z_{25} \oplus Z_{16}$?

*7. Use the results of Exercise 6 to write a computer program to take elements in $Z_{25} \oplus Z_{16}$ and find the corresponding element in Z_{400}.

8. Find all ubiquitous elements in Z_8, Z_9, Z_{16}, Z_{25}, and Z_{27}.

9. The group of motions of a square can be described by the presentation $G = \langle a, b \mid a^4 = b^2 = 1, bab = a^3 \rangle$.
   a. List all elements of G.
   b. Show that G contains no ubiquitous elements.
   c. Why doesn't this answer to part b contradict Theorem 15.13?

10. Prove that if G is a group and $a \neq 1$ is an element of G contained in all nontrivial **cyclic** subgroups of G, then a is a ubiquitous element of G.

11. Suppose Q is a group defined by the presentation

    $$Q = \langle a, b \mid a^4 = b^4 = 1, a^2 = b^2, b^{-1}ab = a^3 \rangle$$

    a. Verify that $|Q| = 8$.
    b. Show that Q contains a ubiquitous element.

12. Give an example of a one-to-one binary representation $f: Z_{16} \rightarrow B(4)$ taking 0 to $(0, 0, 0, 0)$ such that f is not a homomorphism.

13. Prove that Z_{p^n} has $p - 1$ distinct ubiquitous elements.

14. Complete the proof of Theorem 15.12 to show that $z_i = 0$ for all $z \in \langle x - x' \rangle$.

15. In Theorem 15.13 we used multiplicative notation to prove that if $Z_{p^m} = \langle a \rangle$, then $a^{p^{m-1}}$ is ubiquitous. Reformulate this result in additive notation, and prove that p^{m-1} is a ubiquitous element in $Z_{p^m} = \{0, 1, \ldots, p^{m-1}\}$.

16

Unique Factorization

Domains

In the preceding chapter, we investigated integral domains. The fact that integral domains possess cancellation properties allowed us to solve certain equations without the need to require the existence of multiplicative inverses. But there are other benefits to be gained in working with integral domains (and particularly, the integers), and these are equally useful. For example, we often have need to factor a polynomial such as $x^2 - 5x + 6$ in an effort to solve equations like $x^2 - 5x + 6 = 0$. With great confidence we write

$$x^2 - 5x + 6 = (x - 3)(x - 2) = 0$$

and conclude that $x = 3$ and $x = 2$ are roots (zeros) of this polynomial equation. Furthermore, we state that these two values are the *only* solutions.

What gives us the right to make such statements? Are we not assuming that $\mathbb{Z}[x]$ contains no zero-divisors? Why do we know that there is no other way to factor this equation; that is, why is our factorization unique? In this chapter we will consider integral domains in which unique factorization occurs. In the next chapter, we will apply the results of this investigation to rings of polynomials.

16.1

The Division Algorithm

Often when we try to factor an integer, say 4129, or a polynomial, say $x^2 - 5x + 6$, we are unaware of the exact nature of the factors. What we usually do is choose numbers or polynomials that may be factors and divide them into the number or polynomial in question to see if the remainder is zero. If it *is* zero, we then conclude that we have a factor. In this section, we will provide some justification for this approach and look at associated terminology.

Throughout this chapter, D will denote an integral domain with identity 0 and unity 1.

Definition 16.1

Let $(D, +, \cdot)$ be an integral domain and $a, b \in D$. If $a \neq 0$, we say that a **divides** b, denoted $a \mid b$, if there exists some $c \in D$ such that $b = ac$; otherwise, we write $a \nmid b$. An element $v \in D$ is called a **unit** of D if $v \mid 1$, where 1 is the unity of D. Two elements $a, b \in D$ are termed **associates** (and a is said to be an **associate** of b) if there is some unit v in D such that $a = bv$. Let $p \neq 0$ be a nonunit of D. If $p \mid ab$ implies $p \mid a$ or $p \mid b$, then p is said to be **prime**. If $p = ab$ implies that either a or b is a unit, then p is termed **irreducible**.

You might note that, when we defined **prime integer** in Chapter 1, we gave a definition that is the same as that for *irreducible* in Definition 16.1. Are there two different types of prime integers? The answer is no; the concepts of *prime* and *irreducible* agree for the ring of integers. To see this, suppose first that $p \in \mathbb{Z}$, $p = ab$, and p is *prime* as in Definition 16.1. Then $p \mid a$ or $p \mid b$. If $p \mid a$, then $a = px$ and $p = ab = (px)b = p(xb)$. By cancellation in $\mathbb{Z}$, $xb = 1$ and $b \mid 1$ so b is a unit. A similar argument holds if $p \mid b$. In either case, p is irreducible. Conversely, if $p \in \mathbb{Z}$ is *irreducible* as above and $p \mid ab$, suppose $p \nmid a$. Then $(p, a) = 1$, and by Theorem 1.5 there exist $s, t \in \mathbb{Z}$ such that $1 = sp + ta$. Then $b = spb + tab$. The right side of this equation is divisible by p; therefore $p \mid b$ and p is *prime*.

In the discussion that follows, we will encounter many rings. In some, the ideas of *prime* and *irreducible* elements agree. In others, they do not. One such example is given in Exercise 8.

We allow you to verify that the definition of **unit** (Definition 16.1) is equivalent to the one we gave in the last chapter. (See Exercise 1.)

Example 16.1

Let $D = \{a + b\sqrt{5} \mid a, b \in \mathbb{Z}\}$. Then D is an integral domain with respect to the operations of real number addition and multiplication. What are the units? Note that $(a - b\sqrt{5})(a + b\sqrt{5}) = (a^2 - 5b^2)$. Thus if $(a + b\sqrt{5})(c + d\sqrt{5}) = 1$, multiplying both sides by $a - b\sqrt{5}$, we obtain $(a^2 - 5b^2)(c + d\sqrt{5}) = a - b\sqrt{5}$. If we choose a and b to be relatively prime, then we leave it to you to show (in Exercise 2) that $a^2 - 5b^2 = \pm 1$ and $c + d\sqrt{5} = \pm(a - b\sqrt{5})$. Therefore, the set of units includes all numbers of the form $a + b\sqrt{5}$ where $a^2 - 5b^2 = \pm 1$. For example, $2 + \sqrt{5}$ is a unit.

Lemma 16.1

Let $(D, +, \cdot)$ be an integral domain. Then, if $a, b \in D - \{0\}$,

   i. $a \mid b$ and $b \mid a$ if and only if a and b are associates;
   ii. $a \mid 0$;
   iii. v is a unit implies that v^{-1} is a unit; and
   iv. $a \mid b$ and $b \mid c$ for $c \in D$ imply $a \mid c$.

Proof

   i. If $a \mid b$ and $b \mid a$, then $b = ad_1$ and $a = bd_2$ for d_1 and d_2 in D. But then

$$a1 = a = bd_2 = (ad_1)d_2 = a(d_1 d_2)$$

By cancellation, $1 = d_1 d_2$ and d_1 is a unit in D. Thus a and b are associates. Conversely, if $a = bv$, where v is a unit, then $b \mid a$ and there exists $v_0 \in D$ such that $vv_0 = 1$. But then

$$av_0 = (bv)v_0 = b(vv_0) = b1 = b$$

so that $a \mid b$.
   ii. Clearly $0 = a0$ for all $a \in D$.
   iii. If $v \mid 1$, then $1 = vv_0$. Therefore $v_0 = v^{-1}$ is also a unit.
   iv. $a \mid b$ and $b \mid c$ imply that $b = ad_1$ and $c = bd_2$, so that $c = bd_2 = (ad_1)d_2 = a(d_1 d_2)$. Thus $a \mid c$. $\square$

Note how similar these results are to those we know are true for the integers. Consequently, we define the following:

Definition 16.2

Let $(D, +, \cdot)$ be an integral domain and $a, b \in D - \{0\}$. An element $d \in D$ is said to be a **greatest common divisor** or **gcd** of a and b (denoted (a, b)) if

   i. $d \mid a$ and $d \mid b$,
   ii. whenever $c \mid a$ and $c \mid b$, then $c \mid d$.

Two elements a and b are said to be **relatively prime** if $(a, b) = 1$.

Although we have defined what we mean by the greatest common divisor of two elements in D, we have in no way shown that such objects ever exist. Indeed, in any field, the concept of greatest common divisor is meaningless, since any nonzero real number is the gcd of any two others. (See Exercise 3.) In some integral domains, there exist elements a and b where (a, b) does not exist. Exercise 8 provides an example of such a domain.

How does this definition of greatest common divisor compare with that of greatest common divisor of two integers given in Chapter 1? If you check, you will see that if $a, b \in \mathbb{Z}$, we previously required (a, b) to be positive. We now amend our earlier definition to this one, inasmuch as not every ring has an order connected with it. The statements $(6, -10) = 2$ and $(6, -10) = -2$ are now both true.

What is also true is that a gcd of two elements in an integral domain is unique up to associates; that is, if $d_1 = (a, b)$ and $d_2 = (a, b)$, then $d_2 = d_1 u$ where u is a unit. (See Exercise 6.)

In the next section we will consider a type of integral domain in which the gcd of two elements always exists. Here, however, we wish to consider divisors themselves and methods relating to discovering divisors of a given element. For the ring of integers, the key result in determining divisors is the Division Algorithm. You recall that we stated this result in Chapter 1 but deferred the proof to a later time. Well, the time is now!

Theorem 16.1

The Division Algorithm

Let a and $b > 0$ be integers. Then there exist unique integers $q, r \in \mathbb{Z}$ such that $a = bq + r$ and $0 \leqslant r < b$.

Proof

The most straightforward method of proof here is not elegant but does demonstrate the reason for naming this result The Division Algorithm. Recall that an algorithm is a process employed to achieve an end. For that reason, we might describe the procedure used in the example following the Chinese Remainder Theorem as the Chinese Remainder Algorithm.

In the case where $a > b > 0$, the intuitive idea is that we add b to itself q times such that $qb \leqslant a$, but $(q + 1)b > a$. Then we let $r = a - qb$. In algorithmic terms, let $r_1 = a - b$. Then $a = b(1) + r_1$. If $r_1 < b$, we are done. If not, then $r_2 = r_1 - b \geqslant 0$ and

$$a = b(1) + r_1 = b(1) + (b + r_2) = b(2) + r_2$$

Again, if $r_2 < b$, we are done; otherwise, we proceed with $r_3 = r_2 - b \geqslant 0$ and continue this process. Since $r_1 > r_2 > r_3 > \dots$, this algorithm terminates, and the result follows.

Let us give an alternate proof of this theorem. If $a = 0$, then let $q = r = 0$. Assume $a \neq 0$, and let $M = \{m \in \mathbb{Z}^+ \cup \{0\} \mid m = bs - a \text{ for some } s \in \mathbb{Z}\}$.

Now $M \neq \emptyset$, for if $a \geqslant 0$, we can let $s = a$; and if $a < 0$, we can let $s = 0$.

If $0 \in M$, then there exists a unique $q \in \mathbb{Z}$ such that $a = bq$, and we can let $r = 0$. Thus, we may assume that $0 \notin M$. By the Well-Ordering Property, there exists a least element $r \in \mathbb{Z}^+$ such that $r = bq - a$ for some $q \in \mathbb{Z}$.

Suppose $r \geq b$ and $r = b + r_0$ where $r > r_0 \geq 0$. Then

$$a = bq + r = bq + (b + r_0) = b(q + 1) + r_0$$

so that $r_0 \in M$. This contradicts the choice of r. Therefore, $r < b$.

To show the uniqueness of q and r, let us assume $a = bq_1 + r_1$ where $0 \leq r_1 < b$. Then $r_1 \in M$; hence $r \leq r_1$. Let $r_1 = r + x$, $x \geq 0$. Then $a = bq + r = bq_1 + r + x$, so that $x = b(q - q_1)$. Since $x \leq r_1 < b$, we must have $x = 0$, $q = q_1$, and therefore $r = r_1$. This completes the proof of the theorem. $\square$

It would be nice to prove this theorem true in every integral domain. Sadly enough, a proof of such a result is not possible, since the mere statement of this theorem requires an order on the integral domain. This is a problem? Yes, not all integral domains are orderable. To see this, just note that if $\mathbb{C}$ were orderable, then z^2 should be nonnegative for all $z \in \mathbb{C}$. Since $i^2 = -1 < 0$, $\mathbb{C}$ is not orderable.

The Division Algorithm for $\mathbb{Z}$ is one of the most useful results that we have thus far discussed. We have utilized it numerous times in many of the previous chapters. It was especially helpful in proving results on finite groups. The key portion of the theorem is the part that tells us that $r < b$. In Section 16.2, we will resort to the Division Algorithm to prove that every ideal in $\mathbb{Z}$ is principal, that is, generated by a single element.

Exercises 16.1

1. Let $(D, +, \cdot)$ be an integral domain and $a \in D$. Prove that a is a unit in D if and only if a is invertible with respect to the second operation of D.

2. a. Show that $D = \{a + b\sqrt{5} | a, b \in \mathbb{Z}\}$ is an integral domain under the operation of real number addition and multiplication.

   b. Verify that if $a + b\sqrt{5}$ is a unit in D, then $a^2 - 5b^2 = 1$ or -1.

3. Show that if $a, b, c \in \mathbb{R} - \{0\}$, then c is a greatest common divisor of a and b.

4. Suppose $(D, +, \cdot)$ is an integral domain with $p, q \in D$. Prove that if p and q are primes such that $p | q$, then p and q are associates.

5. Prove that if $(D, +, \cdot)$ is an integral domain, then any element $p \in D$ that is prime is also irreducible.

6. Show that the gcd of two elements in an integral domain is unique up to associates; that is, if $d_1 = (a, b)$ and $d_2 = (a, b)$, then $d_2 = d_1 u$, where u is a unit.

7. Use arguments similar to those in the text to show that $d = \{a + b\sqrt{7} | a, b \in \mathbb{Z}\}$ is an integral domain. Determine the units in D.

8. a. Show that $D = \{a + b\sqrt{-5} | a, b \in \mathbb{Z}\}$ is an integral domain under the operations of real number addition and multiplication.

   b. Use the fact that $9 = 3 \cdot 3 = (2 + \sqrt{-5})(2 - \sqrt{-5})$ to show that the gcd of 9 and $6 + 3\sqrt{-5}$ does not exist in D.

   c. Use part b to show that 3 is irreducible but not prime in D.

9. Let $(D, +, \cdot)$ be an integral domain. For $a, b \in D$, define $a \mathbin{\#} b$ if a and b are associates. Prove $\#$ is an equivalence relation on D. What are the equivalence classes for this relation on $\mathbb{Z}$?

10. In any integral domain, verify that $(a, b) = (-a, -b)$.

11. For what values of $q \in \mathbb{Z}$ is $D = \{a + b\sqrt{q} \mid a, b \in \mathbb{Z}\}$ an integral domain with respect to the operations of real number addition and multiplication?

12. Let $(D, +, \cdot)$ be an integral domain. Prove or disprove each of the following:
    a. If $a \mid b$ and $a \mid c$, then $a \mid (b + c)$.
    b. If $a \mid (b + c)$, then $a \mid b$ and $a \mid c$.
    c. If a and b are units, then they are associates.
    d. If $a \mid bc$, $a \mid b$ and $a \mid c$, then a is not irreducible.

13. Explain the reason for not *defining* division by zero. (See Definition 16.1.)

14. Prove that the Division Algorithm (Theorem 16.1) is still true if we replace $b \in \mathbb{Z}^+$ with $b \neq 0$ in the statement of the theorem.

15. *Definition:* An element $a \neq 0$ of an integral domain is a **composite** if there exist irreducible elements $p_1, p_2, \ldots, p_k$ such that $a = p_1 p_2 \cdots p_k$. Prove or disprove that an element $a \neq 0$ of an integral domain is either a unit, irreducible, or composite. If false, what additional hypotheses are needed to obtain a true statement?

16. Find all associates of $2 + 3i$ in $\mathbb{Z}[i]$ and in $\mathbb{C}$.

17. Suppose D is an integral domain and $a, b \in D$. Verify that a and b are associates if and only if $(a) = (b)$.

18. Prove that $a + bi$ is prime in $\mathbb{Z}[i]$ if and only if $a - bi$ is prime in $\mathbb{Z}[i]$.

*19. Write a computer program that will determine whether an integer is a prime, composite, or unit.

*20. Write a computer program to implement the algorithm discussed in the second paragraph of the proof of Theorem 16.1.

16.2

Principal Ideal Domains

Recall that we are investigating divisibility especially as it relates to factoring and the search for greatest common divisors. The existence of greatest common divisors also relates to ideas about irreducible factorizations and solutions to polynomial equations. A fruitful method of inquiry about divisibility involves the study of a special type of integral domain.

In Definition 14.5, we defined principal ideal. Let us restate this in light of Example 14.12.

Definition 16.3

An ideal I in an integral domain $(D, +, \cdot)$ is a **principal ideal** if $I = (a) = \{ra \mid r \in D\}$ for some $a \in I$.

Principal ideals can be thought of as the ring equivalent of cyclic normal subgroups. Not every ideal is principal. For example, $\mathbb{Z}[x]$, the ring of polynomials with integer coefficients, is an integral domain. In this ring, however, the intersection of all ideals that contain both 2 and x is an ideal, but this ideal is not principal. (See Exercise 1.) In this chapter, we wish to avoid rings such as $\mathbb{Z}[x]$, which contain nonprincipal ideals. The rings that interest us are those in which greatest common divisors exist. We will see that these two concepts are related.

Definition 16.4

An integral domain $(D, +, \cdot)$ is said to be a **principal ideal domain** or PID if every ideal I in D is principal.

Since our motivation for this section is the investigation of properties of the integers, we should show that the integers themselves form a principal ideal domain.

Theorem 16.2

$\mathbb{Z}$ is a principal ideal domain.

Proof

Let I be an ideal in $\mathbb{Z}$. If I is the ideal containing only 0, then clearly $I = (0)$. We may assume that $I \neq (0)$. Set $I^+ = \{a \in I \mid a > 0\}$. Then $I^+ \neq \varnothing$. (Why?) By our assumption and the Well-Ordering Property for $\mathbb{Z}^+$, there exists a least element $d \in I^+$. Let $a \in I$; then by the Division Algorithm for $\mathbb{Z}$, there exist $q, r \in \mathbb{Z}$, with $0 \leqslant r < d$, such that $a = dq + r$. Since d is minimal in I^+ and $r = a - qd \in I$, we are forced to conclude that $r = 0$. But then $a = qd \in (d)$. Since the choice of a was arbitrary, $I \subseteq (d)$. But $(d) \subseteq I$. Therefore, $I = (d)$ is principal. $\square$

We will now prove that principal ideal domains do provide us with a class of rings in which greatest common divisors exist.

Theorem 16.3

Let $(D, +, \cdot)$ be a PID. If a and b are nonzero elements of D, then (a, b) exists and is expressible as $(a, b) = sa + tb$ for some $s, t \in D$. (In other words, (a, b) can be expressed as a **linear combination** of a and b.)

Proof

Let $I = \{ya + zb \mid y, z \in D\}$. Since

$$(y_1 a + z_1 b) - (y_2 a + z_2 b) = (y_1 - y_2)a + (z_1 - z_2)b$$

and $\qquad r(y_1 a + z_1 b) = (ry_1)a + (rz_1)b$

I is an ideal of D. Since D is a PID, there exists $d \in D$ such that $I = (d)$, and so

$d = sa + tb$ for some $s, t \in D$. The remainder of the proof consists of showing that d is indeed a gcd of a and b.

Since $a = 1a + 0b$, we have $a \in I$. Thus $a = rd$ and $d \mid a$. Similarly $d \mid b$. Now suppose that $c \mid a$ and $c \mid b$. Then $c \mid sa$ and $c \mid tb$, implying that $c \mid (sa + tb) = d$. The result follows.   □

Since $\mathbb{Z}$ is a PID, Theorem 16.3 not only proves Theorem 1.5, which guarantees that any two integers have a gcd, but the proof of this theorem also provides a characterization of this element.

Corollary 16.1

Let $a, b \in \mathbb{Z}$ and d be the smallest positive integer such that $d = sa + tb$ for some $s, t \in \mathbb{Z}$. Then $d = (a, b)$.

Note that this corollary yields Theorem 1.5.
Theorem 16.3 and Corollary 16.1 are existence results. They assert that a gcd of two integers exists. Their proofs, however, do not show *how* to find the gcd of two integers. To demonstrate it requires the following lemma:

Lemma 16.2

Let $a, b \in \mathbb{Z}$, $b > 0$. If $a = bq + r$, $0 \leqslant r < b$, then $(a, b) = (b, r)$.

Proof

Let $d = (a, b)$. Then $a = c_1 d$, $b = c_2 d$, and $r = a - bq = (c_1 - c_2 q)d$. This says $d \mid (b, r)$.

On the other hand, if $t = (b, r)$, then $b = tu_1$, $r = tu_2$, and $a = t(u_1 q + u_2)$. Thus, $t \mid a$. Since $t \mid b$, we have $t \mid (a, b)$. Thus $(a, b) = d = t = (b, r)$.   □

The repeated use of this lemma and the Division Algorithm for $\mathbb{Z}$ to find the gcd of two integers is a process known as the **Euclidean Algorithm**.

Theorem 16.4

Euclidean Algorithm

Let $a, b \in \mathbb{Z}$, $b \neq 0$. Suppose

$$a = q_1 b + r_1, \qquad 0 \leqslant r_1 < |b|$$
$$b = q_2 r_1 + r_2, \qquad 0 \leqslant r_2 < r_1$$
$$r_1 = q_3 r_2 + r_3, \qquad 0 \leqslant r_3 < r_2$$
$$\vdots$$
$$r_{k-1} = q_{k+1} r_k + r_{k+1}, \quad 0 \leqslant r_{k+1} < r_k$$

If $r_k \neq 0$ and $r_{k+1} = 0$, then $r_k = (a, b)$.

Example 16.2

Find the gcd of 1,118 and 91. We solve using the Euclidean Algorithm: By long division, we find

$$1,118 = 91(12) + 26 \qquad r_1 = 26 < 91 = b$$
$$91 = 26(3) + 13 \qquad r_2 = 13 < 26 = r_1$$
$$26 = 13(2) + 0 \qquad r_3 = 0 < 13 = r_2$$

By repeated use of the lemma, we know that

$$13 = (26, 13) = (91, 26) = (1,118, 91)$$

This procedure always terminates since the remainders obtained by this process get progressively smaller.

Example 16.3

Find $(94, 74)$.

$$94 = 74(1) + 20 \qquad (94, 74) = (74, 20)$$
$$74 = 20(3) + 14 \qquad (74, 20) = (20, 14)$$
$$20 = 14(1) + 6 \qquad (20, 14) = (14, 6)$$
$$14 = 6(2) + 2 \qquad (14, 6) = (6, 2)$$
$$6 = 2(3) + 0 \qquad (6, 2) = 2$$

Therefore, the greatest common divisor of 94 and 74 is 2.

Theorem 16.3 also asserts that the gcd of $a, b \in \mathbb{Z}$ can also be expressed as an integral linear combination of a and b. To do so, we need only reverse the process of the Euclidean Algorithm.

Example 16.4

Express $(94, 74)$ in terms of 94 and 74. By Example 16.3, we have

$$2 = 14 - 6(2)$$
$$= 14 - (20 - 14(1))(2)$$
$$= 14(3) - 20(2)$$
$$= (74 - 20(3))(3) - 20(2)$$
$$= 74(3) - 20(11)$$
$$= 74(3) - (94 - 74(1))(11)$$
$$= (74)(14) + (94)(-11)$$

The Euclidean Algorithm can also be used to find inverses of elements in modular rings of the form $\mathbb{Z}_n$. For example, consider the ring $\mathbb{Z}_{144}$. The class $\overline{91}$ is an element of this ring. Is this element a unit? By the Euclidean Algorithm,

$$144 = 91(1) + 53$$
$$91 = 53(1) + 38$$
$$53 = 38(1) + 15$$
$$38 = 15(2) + 8$$
$$15 = 8(1) + 7$$
$$8 = 7(1) + 1$$
$$7 = 7(1) + 0$$

Thus $(144, 91) = 1$, and 91 is relatively prime to 144. Reversing this process, we obtain

$$1 = 144(-12) + 91(19)$$

In $\mathbb{Z}_{144}$, this equation becomes

$$\overline{1} = \overline{144}(-\overline{12}) + \overline{91}(\overline{19}) = \overline{0}(-\overline{12}) + \overline{91}(\overline{19}) = (\overline{91})(\overline{19})$$

We conclude that $\overline{91}$ is a unit and $\overline{91}^{-1} = \overline{19}$ in $\mathbb{Z}_{144}$.

There are other useful results that come from Theorem 16.3:

Corollary 16.2

Let $(D, +, \cdot)$ be a PID and $a, b, c, p \in D - \{0\}$. Then

    i. $a \mid bc$ and $(a, b) = 1$ imply $a \mid c$, and
    ii. if p is irreducible, then p is prime.

Proof

    i. $(a, b) = 1$ implies $sa + tb = 1$ for $s, t \in D$. Thus $sac + tbc = c$. Since $a \mid bc$, we have $a \mid sac$ and $a \mid tbc$. Thus $a \mid (sac + tbc) = c$.
    ii. Suppose $p \mid ab$ and p do not divide a. Then $(p, a) = 1$. By part i, $p \mid b$. $\square$

Note that the Euclidean Algorithm is a consequence of the Division Algorithm, not the result that $\mathbb{Z}$ is a PID. But since $\mathbb{Z}$ is a PID, results about arbitrary PIDs are also true for $\mathbb{Z}$. It is, therefore, of interest to see what results in $\mathbb{Z}$ can be extended to results that are true for any PID. We will give an important example of such an extension in the next section, but it requires a preliminary result.

Consider (24), the principal ideal in $\mathbb{Z}$ generated by 24. We have

$$(24) \subset (12) \subset (6) \subset (3) \subset (1) = \mathbb{Z}$$

We will show this is always the case in a principal ideal domain; that is, every ascending sequence of principal ideals eventually stops or stabilizes.

Lemma 16.3

Let $I_1 \subseteq I_2 \subseteq I_3 \subseteq \cdots \subseteq I_k \subseteq I_{k+1} \subseteq \cdots$ be an ascending sequence of ideals in a principal ideal domain $(D, +, \cdot)$. Then there exists $m \in \mathbb{N}$ such that $I_m = I_{m+i}$ for all $i \in \mathbb{N}$.

Proof

Let

$$I = \bigcup_{i=1}^{\infty} I_i$$

We leave it as an exercise to show that I is an ideal of D. (See Exercise 12.) Since D is a PID, there exists some $d \in I$ such that $I = (d)$. Since $d \in I$, d must be an element of some I_m. But then $I_m \subseteq I = (d) \subseteq I_m$ implies $I_m = I$ and $I_m \subseteq I_{m+i} \subseteq I = I_m$, so that $I_m = I_{m+i}$ for all $I \in \mathbb{N}$.   □

In the next section, we will see how this result relates to the question of whether a ring element can be factored as a product of irreducible elements.

Exercises 16.2

1. Show that the set $I = \{2f(x) + xg(x) | f(x), g(x) \in \mathbb{Z}[x]\}$ is an ideal in $\mathbb{Z}[x]$. Prove that this ideal is not principal.

2. Prove that every ideal in $2\mathbb{Z}$, the ring of even integers, is principal. Is $2\mathbb{Z}$ a PID?

3. Let $I = \{(2m, 3n) | m, n \in \mathbb{Z}\}$. Verify that I is an ideal in $\mathbb{Z} \times \mathbb{Z}$. Is this ideal principal?

4. Prove that every field is a PID.

5. Show that if $a, b \in D$, an integral domain, and $d = (a, b)$ with $a = da_1$ and $b = db_1$, then a_1 and b_1 are relatively prime.

6. Use the Euclidean Algorithm to find (a, b) and integers s and t such that $(a, b) = sa + tb$ if $(a, b) =$
   a. $(15, 115)$ b. $(256, 312)$ c. $(42, 128)$
   d. $(39, 236)$ e. $(811, 322)$ f. $(751, 347)$

7. Find the inverse (if it exists) of:
   a. $\overline{17}$ in $\mathbb{Z}_{35}$ b. $\overline{42}$ in $\mathbb{Z}_{153}$ c. $\overline{15}$ in $\mathbb{Z}_{16}$

8. Prove that the gcd of 2 and 4 fails to exist in $2\mathbb{Z}$.

9. Use Exercise 14 of Section 14.2 to prove that $\mathbb{Z}_n$ is a PID for all $n \geqslant 1$.

10. If $(D, +, \cdot)$ is a PID and $U = (a)$, $V = (b)$ are ideals of D, then prove that $U + V = (d)$ where $d = (a, b)$ is the gcd of a and b. $U + V$ was shown to be an ideal in Exercise 9 of Section 14.3.

11. A commutative ring $(R, +, \cdot)$ with unity is said to be a **Noetherian ring**[1] if it has the following property: If

---

[1] Emmy Noether (1882–1935) was possibly the greatest woman in the history of mathematics. Although universally recognized as a brilliant algebraist specializing in the theory of rings, she suffered the fate of being a woman in a man's world and was never given a professorship at the University of Göttingen. The original position she obtained there had no duties and no pay.

$$I_1 \subseteq I_2 \subseteq I_3 \subseteq \cdots \subseteq I_n \subseteq I_{n+1} \subseteq \cdots$$

is any ascending sequence of ideals of R, then there exists $m \in \mathbb{N}$ such that $I_m = I_{m+i}$ for all $i \in \mathbb{N}$. (Lemma 16.3 states that every PID is a Noetherian ring.) Prove that if $(R, +, \cdot)$ is a Noetherian ring, then so, too, is any subring or quotient ring of R.

12. Complete the proof of Lemma 16.3 by showing that if

$$I_1 \subseteq I_2 \subseteq \cdots \subseteq I_k \subseteq I_{k+1} \subseteq \cdots$$

is an ascending chain of ideals in a PID, then

$$I = \bigcup_{i=k}^{\infty} I_k$$

is an ideal.

13. Use induction to extend part ii of Corollary 16.2 to obtain: if p is irreducible and $p \mid a_1 a_2 \cdots a_n$, then $p \mid a_i$ for some $i = 1, 2, \ldots, n$.

14. Let D be a PID. The **least common multiple** or **lcm** of a and b in $D - \{0\}$ is an element $m \in D$ such that
    i. $a \mid m$ and $b \mid m$, and
    ii. $a \mid c$ and $b \mid c$ imply $m \mid c$.
    The lcm of a and b is denoted $[a, b]$. Prove that if $m = [a, b]$ and $d = (a, b)$, then $md = ab$. Why does the lcm of two nonzero elements of D always exist?

15. Let D be an integral domain, and suppose $a, b \in D$.
    a. Prove $(a) \subseteq (b)$ if and only if $b \mid a$.
    b. D is said to satisfy the **descending chain condition** if there does not exist any collection of ideals $\{I_i\}_{i=0}^{\infty}$ such that $I_0 \supseteq I_1 \supseteq I_2 \supseteq \cdots$. Does $\mathbb{Z}$ satisfy the descending chain condition?

16. Prove that the homomorphic image of any principal ideal domain is a commutative ring in which every ideal is principal. Is the homomorphic image of a principal ideal domain another principal ideal domain?

17. Suppose D is an integral domain. Prove that $p \in D$ is prime if and only if (p) is a nonzero prime ideal.

18. Let D be an integral domain, $a_1, a_2, \ldots, a_n \in D$. Verify that the gcd of $a_1, a_2, \ldots, a_n$ exists and is expressible in the form $r_1 a_1 + r_2 a_2 + \cdots + r_n a_n$, for $r_1, r_2, \ldots, r_n \in D$, if and only if the ideal generated by $a_1, a_2, \ldots, a_n$ is principal.

19. If D_1 and D_2 are principal ideal domains, is $D_1 \times D_2$ also a principal ideal domain?

20. Find the gcd of $3 - 5i$ and $4 + 6i$ in the principal ideal domain $\mathbb{Z}[i]$.

*21. Write a computer program to implement the Euclidean Algorithm.

16.3

Unique Factorization Domains

We are now ready to verify unique factorization for integers. This result will be considered in the light of a more general setting that will enable us to make similar statements about other rings, including the ring of polynomials, which we will consider in the next chapter. Since we are aware of our goal, let us start with the type of structure we want.

Definition 16.5

An integral domain $(D, +, \cdot)$ is called a **unique factorization domain** or **UFD** if:

i. every $a \in D$, $a \neq 0$, is either a unit in D or $a = p_1 p_2 \cdots p_n$, where each p_i is irreducible in D; and

ii. whenever $a = p_1 p_2 \cdots p_n = q_1 q_2 \cdots q_m$, where the p_i's and q_i's are irreducible elements in D, then $n = m$ and there exists $\sigma \in S_n$ such that $p_{\sigma(i)}$ is an associate of q_i for $i = 1, 2, \ldots, n$.

In other words, every nonzero element of D is either a unit or has an **irreducible factorization** unique up to associates and a reordering of the terms. Note that since every nonidentity element of a field is a unit, every field is a UFD. In addition, the very definition of UFD forces every irreducible element to be prime. (See Exercise 4.) Therefore, one can also use the expression **prime factorization** in dealing with a UFD.

In actuality, what we have done by making this definition is to state a theorem which, when applied to the integers, is called the Fundamental Theorem of Arithmetic. We first encountered this in Chapter 1. The proof of this theorem is the content of the following result.

Theorem 16.5

Every PID is a UFD.

Proof

Let $(D, +, \cdot)$ be a PID and $a \neq 0$ be a nonunit in D. By way of contradiction, let us assume that a has no irreducible factorization. Then a itself is not irreducible in D; so $a = a_1 b_1$ for some $a_1, b_1 \in D$, neither of which is a unit. Now irreducible factorizations for a_1 and b_1 would combine to yield an irreducible factorization for a, contrary to assumption. Therefore, either a_1 or b_1, say a_1, has no such factorization. Repetition of this argument yields $a_1 = a_2 b_2$, where a_2 or b_2, say a_2, has no such factorization, and so on. Hence, we get an infinite properly ascending sequence $(a) \subset (a_1) \subset (a_2) \subset (a_3) \subset \cdots$ of principal ideals in D. By Lemma 16.3, this yields a contradiction.

Therefore, $a = p_1 p_2 \cdots p_n$, for p_i irreducible in D. Now suppose that a has another factorization $a = q_1 q_2 \cdots q_m$, where each q_j is also irreducible in D. Then $p_1 p_2 \cdots p_n = q_1 q_2 \cdots q_m$. Since $p_1 \mid (p_1 p_2 \cdots p_n)$, we have $p_1 \mid (q_1 q_2 \cdots q_m)$. But then p_1 must divide some q_j (Exercise 13 of Section 16.2). Since p_1 and q_j are both irreducible, p_1 and q_j must be associates (Exercise 4 of

Section 16.1). Therefore for some unit, v_1, $q_j = p_1 v_1$ and

$$p_1 p_2 \cdots p_n = (p_1 v_1) q_1 q_2 \cdots q_{j-1} q_{j+1} \cdots q_m$$
$$= p_1 (v_1 q_1 q_2 \cdots q_{j-1} q_{j+1} \cdots q_m)$$

Cancelling p_1, we conclude

$$p_2 \cdots p_n = v_1 q_1 \cdots q_{j-1} q_{j+1} \cdots q_m$$

Repeat this argument; after n cancellations, the left side becomes 1. Therefore, $m \leqslant n$. If we had started on the right side and repeated this argument, we would have concluded that $n \leqslant m$. Thus $m = n$, and the correspondence of each p_i that we cancel with its associate q_j yields an element in S_n. □

An immediate consequence is the statement of this result for the ring of integers. We first saw this result in Chapter 1.

Corollary 16.3

Fundamental Theorem of Arithmetic

For any natural number $n > 1$, there exist positive primes $p_1, p_2, \ldots, p_m$ and natural numbers $e_1, e_2, \ldots, e_m$ such that

$$n = p_1^{e_1} p_2^{e_2} \cdots p_m^{e_m}$$

where the factorization is unique up to a rearrangement of the prime powers.

In the last section, we looked at principal ideal domains in search of rings in which a greatest common divisor of two elements always exists. Since every PID is a UFD, we should ask whether a gcd of any two elements in a UFD exists. The answer is yes.

Corollary 16.4

If D is a UFD and $a, b \in D$, then (a, b) exists.

Proof

Suppose $a = p_1^{e_1} p_2^{e_2} \cdots p_k^{e_k}$ and $b = q_1^{t_1} q_2^{t_2} \cdots q_m^{t_m}$, where the p_i's and q_j's are irreducible in D and $e_i > 0$, $f_j > 0$. By relabeling the irreducible factors, if necessary, and grouping together associates, we can write

$$a = p_1^{r_1} p_2^{r_2} \cdots p_w^{r_w} \qquad \text{and} \qquad b = v p_1^{s_1} p_2^{s_2} \cdots p_w^{s_w}$$

where v is a unit in D and $r_i \geqslant 0$, $s_j \geqslant 0$. (To get the same elements to appear in each factorization, we are using the fact $p^0 = 1$.) We claim

$$d = p_1^{n_1} p_2^{n_2} \cdots p_w^{n_w},$$

where n_i is the minimum of r_i and s_i, is the gcd of a and b. We leave the proof of this assertion to you as Exercise 3. □

We pointed out earlier that, although a gcd of two elements in a PID or a UFD always exists, a gcd may not be unique. For example, in $\mathbb{Z}$ both 2 and -2 satisfy our definition of gcd for $a = 14$ and $b = 18$. In an arbitrary integral domain, if d_1 and d_2 are both gcd's of $a, b \in D$, then $d_1 \mid d_2$ and $d_2 \mid d_1$, so that d_1 and d_2 are associates by Lemma 16.2. Therefore we can write d_2 as $d_1 v$ where v is a unit in D. It is for this reason that we speak of "a" gcd of two elements rather than "the" gcd.

We have now seen a good many types of rings. The set of UFDs contains the set of PIDs, while the PIDs are a special type of integral domain. There are many other types of rings, some of which we discuss in the exercises. For our purposes, however, it is most important to understand the workings of the types of rings already considered and to develop some feeling for the ways they are related.

Not every integral domain is a unique factorization domain. For example,

$$\mathbb{Z}[\sqrt{-5}] = \{a + b\sqrt{-5} \mid a, b \in \mathbb{Z}\}$$

is an integral domain in which $2 \cdot 3 = (1 + \sqrt{-5})(1 - \sqrt{-5})$. Now 2, 3, $1 + \sqrt{-5}$, $1 - \sqrt{-5}$ are irreducible elements in $\mathbb{Z}[\sqrt{-5}]$, and 2 is not an associate of either $1 + \sqrt{-5}$ or $1 - \sqrt{-5}$. (See Exercise 7.) Thus $\mathbb{Z}[\sqrt{-5}]$ is not a UFD, and also it is not a PID. Moreover, there are unique factorization domains that are not principal ideal domains. We will treat polynomial rings in the next chapter, wherein we will classify them according to properties. We will show at that time that $\mathbb{Z}[x]$, the ring of polynomials with integer coefficients, is a unique factorization domain that is **not** a principal ideal domain.

Incidentally, it was the fact that not every integral domain is a UFD which prompted Kummer to devise *ideals* in order to restore unique factorization to domains in which it did not exist. He investigated integral domains of the form $\mathbb{Z}[\sqrt{n}]$ for $n \in \mathbb{Z}$ and determined much about their structure.

Exercises 16.3

1. Suppose that $\mathbb{Z}[i]$ is the ring of Gaussian integers.
   a. Show that $a + bi \in \mathbb{Z}[i]$ is irreducible if $a^2 + b^2$ is a prime in $\mathbb{Z}$.
   b. Find an irreducible factorization of 25 in $\mathbb{Z}$ and in $\mathbb{Z}[i]$.
   c. If $z = a + bi$ is prime in $\mathbb{Z}[i]$, prove that there exists a prime integer p such that $p = zz'$ for some $z' \in \mathbb{Z}[i]$.

2. Prove that 2 is not irreducible in $\mathbb{Z}[i]$, the ring of Gaussian integers, but that 3 is irreducible. Find all associates of 3.

3. Suppose D is a UFD and $a, b \in D$.
   a. Suppose $a = p_1^{e_1} p_2^{e_2}$ and $b = p_1^{f_1} p_2^{f_2}$, where p_1, p_2 are irreducible in D and $e_i \geqslant 0$, $f_j \geqslant 0$ for $i, j = 1, 2$. Set $d = p_1^{n_1} p_2^{n_2}$ where $n_i = $ minimum $\{e_i, f_i\}$. Prove that $d = (a, b)$.
   b. Use induction and part a to complete the proof of Corollary 16.4.

4. Let D be a UFD and $a, b \in D$. Prove the following:
   a. If p is irreducible in D and $p \mid ab$, then $p \mid a$ or $p \mid b$.
   b. If p is irreducible and $p \mid a^n$ for $n > 0$, then $p \mid a$.

5. a. Prove that if D is an integral domain and p is irreducible in D, then vp is irreducible for any unit $v \in D$.
   b. Does the fact that $6 = (2)(3) = (-2)(-3)$ destroy unique factorization in $\mathbb{Z}$?

6. Prove that if D is a UFD, then the intersection of any two principal ideals in D is a principal ideal. Use Exercise 8 of Section 16.1 to show that $(3) \cap (2 + \sqrt{-5})$ is not a principal ideal in $\mathbb{Z}[2 + \sqrt{-5}]$. Conclude that $\mathbb{Z}[2 + \sqrt{-5}]$ is not a UFD.

7. Verify that $2, 3, 1 + \sqrt{-5}$, and $1 - \sqrt{-5}$ are irreducible in the ring $\mathbb{Z}[\sqrt{-5}]$ and that 2 is not an associate of either $1 + \sqrt{-5}$ or $1 - \sqrt{-5}$.

8. An integral domain D is termed a **euclidean domain** if there exists a function $N: D \to \mathbb{Z}^+ \cup \{0\}$ such that: (i) $N(a) \leqslant N(ab)$ for all $a, b \in D - \{0\}$, and (ii) for any $a \in D$, $b \in D - \{0\}$, there exist $q, r \in D$ such that $a = bq + r$ where $r = 0$ or $N(r) < N(b)$.
   a. Why is $(\mathbb{Z}, +, \cdot)$ a euclidean domain?
   b. If D is a euclidean domain, prove that $N(a) = N(1)$ if and only if a is a unit.
   c. Is every field a euclidean domain? Why?
   d. Prove that every euclidean domain is a PID.
   e. Define $N: \mathbb{Z}[i] \to \mathbb{Z}^+ \cup \{0\}$ by $N(a + bi) = a^2 + b^2$. Prove that the Gaussian integers form a euclidean domain.

9. Verify that if D is a UFD and $a, b \in D$, then a lcm of a and b exists. (See Exercise 14 of Section 16.2.)

10. Prove that if D is a UFD and $(a) + (b) = (d)$ for $a, d \in D$, then d is a gcd of a and b.

11. Under what conditions on an integral domain D is a gcd of two elements unique?

12. Prove that $\mathbb{Z}_n$ is a UFD. What are the primes in $\mathbb{Z}_5$, $\mathbb{Z}_9$, and $\mathbb{Z}_{12}$?

*13. Write a computer program that will write any integer as a product of powers of its prime divisors.

17 | Polynomials

Among all the structures of mathematics, the one that is probably the most universally recognized is the polynomial. In high school, students learn to add, subtract, multiply, divide, and factor polynomials. In elementary calculus, polynomials provide examples to illustrate differentiation and integration. Because of this familiarity, we have used the set of polynomials in Example 13.6 in our study of rings.

In this chapter, we will classify polynomials and investigate what properties of the ring of coefficients extend to the ring of polynomials themselves. We will also consider polynomials over specific rings, such as the integers, the rationals, and the reals. In order to give an algebraic basis to our identification of polynomials, we will adopt an approach that may seem somewhat unfamiliar. The purpose is to avoid accepting as valid those properties of polynomials which have not yet been verified. We urge you to bear with us and promise to return you to the more standard notation, little the worse for wear, but, hopefully, with a better understanding of the underlying concepts.

17.1

Polynomials as Sequences

Throughout this chapter, we will use R to denote a ring with additive identity 0. When R has a unity element, we will indicate it by the symbol 1. Recall that a **sequence** $(a_0, a_1, a_2, a_3, \ldots)$ of elements in R can be thought of as an injective function on $\mathbb{N}$ or on $\mathbb{N} \cup \{0\}$. For example, the mapping $f: \mathbb{N} \cup \{0\} \to \mathbb{Q}$ defined by $f(n) = 1/(n+1)$ has image

$$\{f(0),\ f(1),\ f(2),\ f(3),\ \ldots\} = \left\{1, \frac{1}{2}, \frac{1}{3}, \frac{1}{4}, \ldots\right\}$$

and can be alternately described by

$$\left(1, \frac{1}{2}, \frac{1}{3}, \frac{1}{4}, \ldots, \frac{1}{n+1}, \ldots\right)$$

We wish to identify polynomials whose coefficients lie in a ring R with certain sequences of elements of R.

Definition 17.1

Let $(R, +, \cdot)$ be a ring. A **polynomial** is a sequence $f = (a_0, a_1, a_2, \ldots)$ of elements of R with the property that only a finite number of such elements are not equal to 0. The set of all such sequences is termed the set of **polynomials in one variable over** R and is denoted $R[x]$.

The notation $R[x]$ is used for a number of reasons: to remind us that the ring of coefficients is R, to differentiate between this ring R and the associated ring of polynomials, and to indicate that x is simply the polynomial place-holder.

While this approach to polynomials may appear quite foreign and, perhaps, formidable, let us consider a few examples that relate this definition to more recognizable concepts.

Example 17.1

The polynomials $3x^2 + 4x - 5$ in conventional notation can be written in sequence form as $(-5, 4, 3, 0, 0, \ldots) \in \mathbb{Z}[x]$. This sequence is the mapping $f: \mathbb{N} \cup \{0\} \to \mathbb{Z}$ defined by

$$f(0) = -5, \quad f(1) = 4, \quad f(2) = 3, \quad \text{and} \quad f(i) = 0 \quad \text{for } i > 2$$

Example 17.2

The polynomial $(0, -1, 0, \frac{1}{2}, 4, 0, 0, \ldots) \in \mathbb{R}[x]$ represents the mapping $g: \mathbb{N} \cup \{0\} \to \mathbb{R}$ defined by

$$g(0) = 0, \quad g(1) = -1, \quad g(2) = 0, \quad g(3) = \tfrac{1}{2}, \quad g(4) = 4, \quad \text{and}$$

$$g(i) = 0 \quad \text{for } i > 4$$

In traditional notation, g is the polynomial $4x^4 + (\tfrac{1}{2})x^3 - x$.

In light of our definition, we will reconsider polynomials in $R[x]$ as mappings $f: \mathbb{N} \cup \{0\} \to R$ with certain restrictions on the images. Recall that two mappings are equal if and only if they have the same domain, image, and rule of correspondence. Therefore two polynomials $f = (a_0, a_1, a_2, \ldots)$ and $g = (b_0, b_1, b_2, \ldots)$ are **equal** if and only if $a_i = b_i$ for all $i = 0, 1, 2, \ldots$.

The next definition shows how the concept of *polynomial degree* relates to our sequence notation.

Definition 17.2

If $f = (a_0, a_1, a_2, \ldots) \in R[x]$, $a_n \neq 0$, but $a_k = 0$ for all $k > n$, then a_n is termed the **leading coefficient of** f and n the **degree of** f, denoted $\deg(f)$. If r has unity 1, then a polynomial whose leading coefficient is 1 is known as a **monic polynomial**. The leading coefficient and the degree of the polynomial $\mathbf{0} = (0, 0, 0, \ldots)$ are not defined. $\mathbf{0}$ is known as the **zero polynomial**.

In $\mathbb{Z}[x]$, the polynomial $f = (1, 1, 0, 0, 0, \ldots)$ is a monic polynomial whose degree is 1. $(0, 2, -3, 4, 0, 0, 0, \ldots)$ is a polynomial with degree 3 and leading coefficient 4, while $(2, 0, 0, \ldots)$ has degree 0 and leading coefficient 2.

From the definition of a polynomial, we see that every nonzero polynomial has both a leading coefficient and a degree. If the degree of $f = (a_0, a_1, a_2, \ldots)$ is n, then it is often easier to express f as $(a_0, a_1, a_2, \ldots, a_n)$. As is customary, we term polynomials of degree 1 **linear** and degree 2 **quadratic**. The zero polynomial and all polynomials of degree 0 are called **constant polynomials**.

The next two definitions will allow us to show that $R[x]$ is a ring under certain operations.

Definition 17.3

Let R be a ring and $f = (a_0, a_1, \ldots, a_m)$ and $g = (b_0, b_1, \ldots, b_n)$ be elements of $R[x]$. Define

$$f \oplus g = (a_0 + b_0, a_1 + b_1, \ldots, a_i + b_i, \ldots)$$

and

$$f \odot g = (c_0, c_1, \ldots, c_k, \ldots)$$

where

$$c_k = \sum_{i+j=k} a_i b_j$$

$$= \sum_{i=0}^{k} a_i b_{k-i}$$

$$= a_0 b_k + a_1 b_{k-1} + \cdots + a_{k-1} b_1 + a_k b_0$$

These definitions are really nothing more than a restatement of the well-known formulas for addition and multiplication of polynomials couched in our algebraic format.

According to Definition 17.3,

$$(1, 2, -3) \oplus (0, 1, 2) = (1, 3, -1)$$

and

$$(1, 2, -3) \odot (0, 1, 2) = (0, 1, 4, 1, -6)$$

This is the same as we would obtain in a more familiar fashion. Since $(1, 2, -3)$ represents $-3x^2 + 2x + 1$ and $(0, 1, 2)$ represents $2x^2 + x$, we see that

$$(-3x^2 + 2x + 1) \oplus (2x^2 + x) = -x^2 + 3x + 1$$

and

$$(-3x^2 + 2x + 1) \odot (2x^2 + x) = -6x^4 - 3x^3 + 4x^3 + 2x^2 + 2x^2 + x$$

$$= -6x^4 + x^3 + 4x^2 + x$$

which we write as $(0, 1, 4, 1, -6, 0, 0, 0, \ldots)$ or $(0, 1, 4, 1, -6)$.

From Definition 17.3, it should be clear that $\oplus$ and $\odot$ are binary operations on $R[x]$. The next result establishes that $R[x]$ is a ring and describes the connection between R and $R[x]$.

Theorem 17.1

Let R be a ring and f, g be nonzero polynomials in $R[x]$. Then:

    i. $(R[x], \oplus, \odot)$ is a ring.
    ii. The constant polynomials form a subring of $R[x]$ isomorphic to R.
    iii. $\deg(f \oplus g) \leqslant \max(\deg(f), \deg(g))$.
    iv. $\deg(f \odot g) \leqslant (\deg(f) + \deg(g))$.

Proof

The proof of part i is a straightforward exercise in applying the basic definitions to show that $(R[x], \oplus)$ is an abelian group in which the identity is the zero polynomial and the additive inverse of $f = (a_0, a_1, a_2, \ldots)$ is $-f = (-a_0, -a_1, -a_2, \ldots)$. (See Exercise 2.)

To show that $\odot$ is associative on $R[x]$, let $f = (a_0, a_1, a_2, \ldots)$, $g = (b_0, b_1, b_2, \ldots)$, and $h = (c_0, c_1, c_2, \ldots)$ be arbitrary elements of $R[x]$. Then

$$f \odot g = (d_0, d_1, d_2, \ldots) \qquad \text{where} \quad d_k = \sum_{i+j=k} a_i b_j$$

Therefore

$$(f \odot g) \odot h = (r_0, r_1, r_2, \ldots)$$

with

$$r_n = \sum_{k+m=n} d_k c_m$$

$$= \sum_{k+m=n} \left(\sum_{i+j=k} a_i b_j \right) c_m$$

$$= \sum_{(i+j)+m=n} (a_i b_j) c_m$$

But

$$g \odot h = (s_0, s_1, s_2, \ldots) \qquad \text{where} \quad s_p = \sum_{j+m=p} b_j c_m$$

so that

$$f \odot (g \odot h) = (t_0, t_1, t_2, \ldots)$$

with

$$t_n = \sum_{i+p=n} a_i s_p$$

$$= \sum_{i+p=n} a_i \left(\sum_{j+m=p} b_j c_m \right)$$

$$= \sum_{i+(j+m)=n} a_i (b_j c_m)$$

by the distributive property of R. Then $f \odot (g \odot h) = (f \odot g) \odot h$ by the associativity of integer addition and the associativity of multiplication on R.

To show that $\odot$ distributes over $\oplus$, we note that if

$$f \odot g = (u_0, u_1, u_2, \ldots) \qquad \text{where} \quad u_k = \sum_{i+j=k} a_i b_j$$

and

$$f \odot h = (v_0, v_1, v_2, \ldots) \qquad \text{where} \quad v_k = \sum_{i+j=k} a_i c_j$$

then

$$(f \odot g) \oplus (f \odot h) = (u_0 + v_0, u_1 + v_1, u_2 + v_2, \ldots)$$

where

$$u_k + v_k = \sum_{i+j=k} a_i b_j + \sum_{i+j=k} a_i c_j$$

$$= \sum_{i+j=k} a_i (b_j + c_j)$$

by the distributive property in R. But $g \oplus h = (b_0 + c_0, b_1 + c_1, \ldots)$, so

$$f \odot (g \oplus h) = (w_0, w_1, w_2, \ldots) \qquad \text{where} \quad w_k = \sum_{i+j=k} a_i (b_j + c_j)$$

Therefore the left distributive law holds. We leave the proof of the right distributive law to you. (See Exercise 4.)

In part ii, to show that R is isomorphic to a subring of $R[x]$, define $\theta: R \to R[x]$ by $\theta(a) = (a, 0, 0, 0, \ldots)$ for $a \in R$. It follows from the definition of equality of polynomials that θ is injective. Moreover, if $a, b \in R$, then

$$\theta(a + b) = (a + b, 0, 0, 0, \ldots) = (a, 0, 0, 0, \ldots) \oplus (b, 0, 0, 0, \ldots)$$
$$= \theta(a) \oplus \theta(b)$$

and

$$\theta(ab) = (ab, 0, 0, 0, \ldots) = (a, 0, 0, 0, \ldots) \odot (b, 0, 0, 0, \ldots)$$
$$= \theta(a) \odot \theta(b)$$

Thus, θ is a ring homomorphism whose image is the set of all constant polynomials of $R[x]$, and θ is a ring isomorphism between R and the set of constant polynomials (which is then a subring of $R[x]$).

To prove part iii, suppose that f has degree m and g has degree n. Then, if $m < n$,

$$f \oplus g = (a_0 + b_0, a_1 + b_1, a_2 + b_2, \ldots, a_m + b_m, b_{m+1}, \ldots, b_n)$$

so that $\deg(f \oplus g) = \max(\deg(f), \deg(g))$. A similar argument can be given if $m > n$. If $m = n$, and $b_m = -a_n$, then

$$f \oplus g = (a_0 + b_0, a_1 + b_1, \ldots, a_{m-1} + b_{m-1}, 0)$$

so that $\deg(f \oplus g) < \max(\deg(f), \deg(g))$. Hence,

$$\deg(f \oplus g) \leqslant \max(\deg(f), \deg(g))$$

To prove part iv, again suppose that f has degree m and g has degree n. Then

$$f \odot g = (u_0, u_1, u_2, \ldots) \qquad \text{where} \quad u_k = \sum_{i+j=k} a_i b_j$$

Since $a_i = 0$ for all $i > m$ and $b_j = 0$ for all $j > n$, we see that $u_{m+n} = a_m b_n$ and $u_k = 0$ for all $k > m + n$. If $u_{m+n} \neq 0$, then $\deg(f \odot g) = \deg(f) \oplus \deg(g)$. If a_m and b_n are zero-divisors of R, it could happen that $a_m b_n = 0$. In that case,

$$\deg(f \odot g) < (\deg(f) \oplus \deg(g)) \qquad\qquad \square$$

It may be helpful for you to consider three arbitrary polynomials (say, of degree two, three, and four) and verify the results of the preceding theorem for these examples. Bear in mind, however, that specific cases do not prove general results, unless all possibilities are considered.

Using the fact that R is isomorphic to a subring of $R[x]$, we will make the identification $a = (a, 0, 0, \ldots)$; in other words, we will replace the constant polynomials by the constants themselves.

Similarly, let us denote the polynomial $(0, a, 0, \ldots)$ by ax. Then we leave it to you to show that the set of such polynomials is an additive subgroup of $R[x]$. (See Exercise 6.) Is this set a subring? The answer is no. If

$$ax = (0, a, 0, 0, \ldots) \qquad \text{and} \qquad bx = (0, b, 0, 0, \ldots)$$

then $ax \odot bx$ is the polynomial $(0, 0, ab, 0, 0, \ldots)$, which is not abx. We denote this last polynomial $ax \odot bx$ by abx^2. This idea can be generalized. Denote the polynomial with the ring element a in the ith position (starting our count at 0) and 0 in all other positions by ax^i. The verification of $ax^i \odot bx^j = abx^{i+j}$ is left to you in Exercise 7.

Now things should begin to look familiar. Let $f = (a_0, a_1, \ldots, a_n)$ be any polynomial of degree n. Then the identifications of the preceding paragraph yield

$$f = a_0 \oplus a_1 x \oplus a_2 x^2 \oplus \cdots \oplus a_n x^n$$

Inasmuch as R is isomorphic to a subring of $R[x]$, we will simply write

$$f = a_0 + a_1 x + a_2 x^2 + \cdots + a_n x^n, \quad f \oplus g = f + g, \quad \text{and} \quad f \odot g = fg$$

keeping in mind that we have actually extended $+$ and $\cdot$ (isomorphically) to a larger domain.

If R is a ring with unity, then we can define $x = (0, 1, 0, 0, \ldots)$ and

$$ax = a \odot x = (a, 0, 0, 0, \ldots) \odot (0, 1, 0, 0, \ldots)$$

Not all rings, however, have a unity (or multiplicative identity), so our first definition of ax is more general. Nevertheless, $R[x]$ now looks more familiar, as we promised, and the meaning of x has been made explicit: the symbol x is merely a positional device, not necessarily an element of $R[x]$.

Exercises 17.1

1. Verify that $\oplus$ and $\odot$ are binary operations on $R[x]$.

2. Show that $(R[x], \oplus)$ is an abelian group in which the identity is the polynomial **0** and the additive inverse of $f = (a_0, a_1, a_2, \ldots)$ is $-f = (-a_0, -a_1, -a_2, \ldots)$.

3. Verify the fact that polynomial multiplication is distributive using arbitrary polynomials of degree two, three, and four.

4. Prove that $\odot$ is right distributive over $\oplus$. (See Theorem 17.1.)

5. Find a ring R and two polynomials $p(x)$ and $q(x)$ in $R[x]$ such that

$$\deg(p(x) \odot q(x)) < (\deg p(x) + \deg q(x))$$

6. Verify that all polynomials

$$ax^{n-1} = (0, 0, \ldots, a, \ldots)$$
$$\uparrow$$
$$n\text{th position}$$

for $a \in R$, form an additive subgroup of $R[x]$.

7. Prove $ax^i \odot bx^j = abx^{i+j}$. (Be careful! Do not rashly assume $x^i \odot x^j = x^{i+j}$.)

8. Polynomial addition is defined termwise: if $p(x) = (a_0, a_1, a_2, \ldots)$ and $q(x) = (b_0, b_1, \ldots)$, then

$$p(x) + q(x) = (a_0 + b_0, a_1 + b_1, a_2 + b_2, \ldots)$$

Why do we not define polynomial multiplication termwise?

9. Use Definition 17.1 to compute the following polynomials in $\mathbb{Z}[x]$:
   a. $(3 + 4x - x^2) + (1 + x^2 + x^3)$
   b. $(3 + 4x - x^2)(1 + x^2 + x^3)$
   c. $(3x^2 - x^3)(2x - x^4)$

10. Let $R = (\mathbb{Z}_6, \oplus, \odot)$ and $f(x) = \bar{2}x^2 + \bar{1}$, $g(x) = \bar{4}x^2 + \bar{5}x + \bar{3}$, and $h(x) = \bar{3}x^3 + \bar{1}$. Compute $f(x) + g(x)$ and $f(x)h(x)$. Compare your results with Theorem 17.1.

11. Let R be a ring, and define $\phi: R \to R[x]$ by $\phi(a) = ax$. Is ϕ a ring homomorphism?

12. If R is a ring and $I = \{ax \mid a \in R\}$, then show that I is **not** an ideal in $R[x]$.

*13. Write a computer program to accept a finite sequence $(a_0, a_1, a_2, \dots, a_{10})$ of integers and output a polynomial $a_0 + a_1 x + a_2 x^2 + \cdots + a_{10}x^{10}$.

*14. Write a computer program to add and multiply two polynomials of degree at most 10. Assume the polynomials are given as finite sequences.

17.2

Properties of Polynomial Rings

Since the constant polynomials form a subring of $R[x]$, which is isomorphic to R, it would be valuable to see which properties of R continue to hold in $R[x]$. This is the purpose of our first result.

Theorem 17.2

If a ring R

    i. is commutative, then $R[x]$ is commutative;
    ii. has a unity, then $R[x]$ has a unity;
    iii. is an integral domain, then $R[x]$ is an integral domain.

Proof

    i. If R is commutative and $f = a_0 + a_1 x + \cdots + a_m x^m$ and $g = b_0 + b_1 x + \cdots + b_n x^n$ are any two elements of $R[x]$, then

$$fg = c_0 + c_1 x + \cdots + c_{m+n}x^{m+n} \qquad \text{where} \quad c_k = \sum_{i+j=k} a_i b_j$$

Since

$$gf = d_0 + d_1 x + \cdots + d_{m+n}x^{m+n} \qquad \text{where} \quad d_k = \sum_{i+j=k} b_i a_j$$

the commutativity of R implies that $fg = gf$.

    ii. If R has unity 1, then the polynomial $1 = (1, 0, 0, \dots)$ has the property that $f \cdot 1 = 1 \cdot f = f$ for any nonzero polynomial f in $R[x]$. To see this, suppose $f = a_0 + a_1 x + \cdots + a_m x^m$. Then the product $f \cdot 1 = c_0 + c_1 x + \cdots + c_m x^m$, where

$$c_k = a_0 0 + a_1 0 + \cdots + a_{k-1}0 + a_k 1 = a_k$$

The proof of the assertion $1 \cdot f = f$ follows in a similar way.

iii. If R is an integral domain, then the proof of Theorem 17.1 shows
$\deg(fg) = \deg(f) + \deg(g)$ where f and g are any nonzero elements
of $R[x]$, since R contains no zero-divisors. But then the product of
two nonzero polynomials is a polynomial with nonzero constant
term and degree at least 0. Therefore, the product is a nonzero
element of $R[x]$. (Recall that the zero polynomial in $R[x]$ has no
degree.)   □

From the proposition in Theorem 17.2, one might guess that if R is a
PID, then $R[x]$ is also a PID. This is not true, however, since the ideal I on
$\mathbb{Z}[x]$ generated by x and 2 cannot be generated by a single polynomial. To
see this, suppose $I = (d(x))$ is the smallest ideal in $R[x]$ containing both 2 and
x. Then $2 \in (d(x))$ implies that $d(x)$ divides 2, so that $d(x)$ is a constant
polynomial. On the other hand, $x \in (d(x))$ implies $d(x)$ divides x, which forces
$d(x) = 1$. But then $I = \mathbb{Z}[x]$, and this contradicts the fact that the set of all
polynomials with an even constant term is an ideal containing both 2 and x.
Therefore I cannot be a principal ideal and $\mathbb{Z}[x]$ is not a PID.

The question remains, "Under what conditions on R is $R[x]$ a PID?" In
order to give an answer, we need the next result, which is important in and of
itself. Its assertion for certain polynomial rings is quite similar to that of the
Division Algorithm for $\mathbb{Z}$ (Theorem 16.1).

| Theorem 17.3 |

Polynomial Division Algorithm

Let R be an integral domain, $f \in R[x], g \in R[x]$, and the leading coefficient of
g be a unit in R. Then there exist unique $q, r \in R[x]$ such that $f = gq + r$ where
$r = 0$ or $\deg(r) < \deg(g)$.

Proof

Note that the restriction on g forces $g \neq 0$. First we wish to show the existence
of q and r. We prove their uniqueness last. Clearly, if $\deg(g) = 0$, the result
holds; for then $g = b_0 \in R$, and we can let $q = b_0^{-1}f$ and $r = 0$. Likewise, if
$\deg(g) > \deg(f)$, then we can set $q = 0$ and $r = f$.

So, without loss of generality, let us assume $0 < \deg(g) < \deg(f)$. We
proceed by induction on $n = \deg(f)$ and assume the result true for all
polynomials of degree less than n. (This is an application of the Second
Principle of Mathematical Induction, discussed in Exercise 10 of Section 1.4.)
Since the preceding paragraph handles the case in which $\deg(f) = 0$, we need
only show that the result holds for $\deg(f) = n$.

Set $m = \deg(g)$, a_n equal to the leading coefficient of f, and b_m equal to
the leading coefficient of g. The polynomial $a_n b_m^{-1}g$ has degree m, but has the
same leading coefficient as f. Hence, $a_n b_m^{-1}x^{n-m}g$ is a polynomial with the
same degree and leading coefficient as f, so that the polynomial $f_1 = f - a_n b_m^{-1}x^{n-m}g$ is a polynomial of degree less than n. By the induction
assumption, there exist $q_1, r_1 \in R[x]$ such that $f_1 = gq_1 + r_1$ where $r_1 = 0$ or

$\deg(r_1) < \deg(g)$. Therefore

$$f - a_n b_m^{-1} x^{n-m} g = g q_1 + r_1 \qquad \text{and} \qquad f = (a_n b_m x^{n-m} + q_1)g + r_1$$

If we set $q = a_n b_m^{-1} x^{n-m} + q_1$ and $r = r_1$, then q and r have the desired properties. By the Second Principle of Mathematical Induction, we conclude such q and r exist for $f \in R[x]$.

To show the uniqueness of q and r, suppose that $f = qg + r = q_0 g + r_0$, where either $r = 0$ or $\deg(r) < \deg(g)$ and either $r_0 = 0$ or $\deg(r_0) < \deg(g)$. If $q \neq q_0$, then $(q - q_0)g = r_0 - r$. By Theorem 17.1, the polynomial on the left has degree greater than or equal to $m = \deg(g)$, while the polynomial on the right has degree less than m. This contradiction forces $q = q_0$, and so $r = r_0$. $\square$

Since any field F is an integral domain and, hence, every leading coefficient of a nonzero polynomial $f(x) \in F[x]$ is a unit, we are now able to give a partial answer to the question we posed earlier.

Theorem 17.4

If F is a field, then $F[x]$ is a PID.

Proof

Let A be any nonzero ideal in $F[x]$. Let $d \in A$ be an element of A of least degree. (Such an element exists by the Well-Ordering Property.) Let f be any element of A. By Theorem 17.3, there exist $q, r \in F[x]$ such that $f = dq + r$, and $r = 0$ or $\deg(r) < \deg(d)$. Since $f, d \in A$, $r = f - dq$, and A is an ideal, we have $r \in A$. By the choice of d, r must be 0. Therefore, $f \in (d)$ and $A = (d)$. $\square$

By this result, together with Theorem 16.3, we know that any two polynomials in $F[x]$ have a gcd. The process that we used in Example 16.2 to compute the gcd of two integers, the Euclidean Algorithm, works also in $F[x]$. You recall that the key to the Euclidean Algorithm was repeated division using the Division Algorithm for $\mathbb{Z}$. Since Theorem 17.3 holds, we can extend our earlier methods. Problems of this type are given in Exercise 12.

Theorem 17.3 is one of the most useful theorems we will encounter in our study of rings. With the aid of a little notation, we now give consequences of this result, which connects the ideas of polynomials and equations.

Definition 17.4

Let $f = a_0 + a_1 x + \cdots + a_n x^n \in R[x]$ and $d \in R$. Then the element $a_0 + a_1 d + \cdots + a_n d^n$ is said to be the **value obtained by evaluating f at d** and is usually symbolized by $f(d)$. In keeping with this notation, f is often written as $f(x)$. We say d is a **root of** f or $f(x)$ if $f(d) = 0$.

Definition 17.4 allows us to treat a polynomial as both an element of an algebraic structure and as a function from R to R.

Definition 17.5

> A nonzero polynomial $f(x) \in R[x]$ of degree n is **irreducible over** R if whenever there exist polynomials $s(x)$, $t(x) \in R[x]$ with the property that $f(x) = s(x)t(x)$, then either $s(x)$ or $t(x)$ is a unit in $R[x]$; otherwise, $f(x)$ is said to be **reducible over** R. In that case, $s(x)$ and $t(x)$ are said to be **factors** of $f(x)$, and $f(x)$ is said to be **factored**.

If F is a field and $f(x)$ is a reducible polynomial of degree n in $F[x]$, then $f(x) = s(x)t(x)$, with $0 < \deg(s(x)) < n$ and $0 < \deg(t(x)) < n$. (See Exercise 15.) Recall that if an element $p = ab$ for a, b in an integral domain D, is irreducible, then a or b must be a unit in D. If F is a field, the units in $F[x]$ are precisely the nonzero constant polynomials. Note the similarity between the concepts of *irreducible in* D and *irreducible in* $F[x]$ when F is a field. If $f(x)$ is irreducible in $F[x]$ and $f(x) = s(x)t(x)$, then $\deg(s(x)) = 0$ or $\deg(t(x)) = 0$. In either case, one of the factors of f has degree 0 and is, therefore, a unit in $F[x]$.

Note that, by our definition, $2x^2 + 4 = 2(x^2 + 2)$ is reducible in $\mathbb{Z}[x]$, but irreducible in $\mathbb{Q}[x]$. The polynomials $x^2 - 3x + 2$ is reducible over $\mathbb{Z}$ since it can be factored by $(x - 2)(x - 1)$. The polynomial $x^2 - 2$ is irreducible over $\mathbb{Q}$ since $\sqrt{2}$ is irrational; however, $x^2 - 2$ **is** reducible over $\mathbb{R}$ and can be written as $(x - \sqrt{2})(x + \sqrt{2})$.

From these examples it is clear that one must specify the ring over which one is working when speaking of polynomial reducibility.

The next two results, like Theorem 17.4, are really corollaries of Theorem 17.3 but will be called theorems because of their importance.

Theorem 17.5

Remainder Theorem

If R is an integral domain and $f(x) \in R[x]$, then the remainder obtained by dividing $f(x)$ by $g(x) = x - a$, where $a \in R$, is $f(a)$.

Proof

By Theorem 17.3, $f(x) = (x - a)q(x) + r(x)$ where either $r(x) = 0$ or $\deg(r(x)) < \deg(x - a)$. Therefore, $r(x)$ must be a constant polynomial. But $f(a) = (a - a)q(a) + r(a) = r(a)$, so that $r(x) = r(a) = f(a)$ is that constant. □

Theorem 17.6

Factor Theorem

Let R be an integral domain and $f(x) \in R[x]$. If $a \in R$, then a is a root of $f(x)$ if and only if $x - a$ is a factor of $f(x)$.

Proof

If $x - a$ is a factor of $f(x)$, then in Theorem 17.3, $f(x) = q(x)(x - a) + 0$, so that $r(x) = 0$. But, by the Remainder Theorem, $r(x) = f(a)$. Hence $f(a) = 0$, and a is a root of $f(x)$.

Conversely, if a is a root of $f(x)$, then $f(a) = 0$. Again by the Remainder Theorem,

$$f(x) = q(x)(x - a) + f(a) = q(x)(x - a)$$

and $x - a$ is a factor of $f(x)$.   □

If a polynomial $f(x) \in R[x]$ can be written in the form $f(x) = (x - a)^m g(x)$, where $m \geqslant 1$, and $x - a$ is not a factor of $g(x)$, then the element $a \in R$ is said to be a **root of multiplicity** m in $f(x)$. For example,

$$f(x) = x^4 - 4x^3 + 5x^2 - 4x + 4 = (x^2 - 4x + 4)(x^2 + 1)$$
$$= (x - 2)^2(x^2 + 1)$$

is a polynomial in $\mathbb{Z}[x]$ with a root of multiplicity two.

You probably have encountered our last two theorems previously in your mathematics courses. They are quite useful in graphing functions and form the theoretical basis for the method of **synthetic division**. This technique aids in root-finding. But when have you found *enough* roots? Most students rely on the next result.

Theorem 17.7

Let R be an integral domain and $f(x) \in R[x]$. If the degree of $f(x)$ is n, then $f(x)$ can have at most n distinct roots.

Proof

If $n = 0$, then $f(x)$ is a constant and, thus, has no roots. We now use induction on n. If $n = 1$, then $f(x)$ is a linear polynomial $ax + b$, where $a, b \in R$ and $a \neq 0$. If f has no roots, the result is true. If f has two roots, say x_1 and x_2, then $ax_1 + b = ax_2 + b = 0$. But then $ax_1 = ax_2$, and cancellation in the integral domain R forces $x_1 = x_2$. Thus f has at most one root.

Assume the result is true for polynomials of degree k and that $n = k + 1$. If a is a root of $f(x)$, then we know by the Factor Theorem that $f(x) = (x - a)q(x)$ where $\deg(q(x)) = k$. Suppose b is a root of $q(x)$, $b \in R$, $b \neq a$. Then $f(b) = (b - a)q(b) = (b - a)0 = 0$, and b is also a root of $f(x)$. Similarly, any root $c \in R$, $c \neq a$, of $f(x)$ must be a root of $q(x)$, since R is an integral domain. By our induction hypothesis, $q(x)$ has at most k distinct roots. Thus, $f(x)$ can have at most $k + 1$ distinct roots.   □

It is interesting to note that Theorem 17.7 does not hold for all rings R. The polynomial $x^2 - x$ has four distinct roots in $\mathbb{Z}_6$. (See Exercise 10.) Also, the theorem does not say that every polynomial of degree n has n distinct roots. For example, a polynomial of degree zero has no roots; a polynomial of degree one whose leading coefficient is not a unit may not have a root; a polynomial can have multiple roots. In $\mathbb{Z}_6$, the polynomial $\bar{3}x + \bar{5}$ has no root although the polynomial $\bar{3}x + \bar{3}$ does. The polynomial $x^2 + 2x + 1 \in \mathbb{Z}[x]$ does not have two distinct roots. We saw in our previous example, $f(x) = x^4 - 4x^3 + 5x^2 - 4x + 4$, has but one root in $\mathbb{Z}[x]$.

We have seen that if F is a field, then $F[x]$ is a PID. In Theorem 16.5, we proved every PID is a UFD; therefore, if F is a field, then $F[x]$ is a UFD. A question that might arise at this point is, "If R is a UFD, is $R[x]$ also a UFD?" The answer is yes, but to prove the desired conclusion, we need some additional concepts and related results.

Definition 17.6

> Let $f(x) = a_n x^n + a_{n-1} x^{n-1} + \cdots + a_1 x + a_0$, $n \geqslant 1$, be an element of $R[x]$ where R is a UFD. The **content** of $f(x)$ is the greatest common divisor of its coefficients $a_n, \ldots, a_1, a_0$. $f(x)$ is said to be **primitive** if its content is 1.

For example, the content of $f(x) = 2x^2 + 8 \in \mathbb{Z}[x]$ is 2, while $g(x) = x^2 + 3x + 6 \in \mathbb{Z}[x]$ is a primitive polynomial.

Lemma 17.1

Suppose R is a UFD and $p(x)$, $q(x) \in R[x]$. Then $p(x)q(x)$ is a primitive polynomial if and only if both $p(x)$ and $q(x)$ are primitive polynomials.

Proof

Let

$$f(x) = \sum_{i=0}^{n} a_i x^i \qquad \text{and} \qquad g(x) = \sum_{j=0}^{m} b_j x^j$$

be primitive polynomials in $R[x]$. If

$$h(x) = f(x)g(x) = \sum_{k=0}^{n+m} c_k x^k$$

is not primitive, then there exists some prime integer p that divides all the c_k's. Since $f(x)$ and $g(x)$ are primitive, there exists at least one a_i and one b_j not divisible by p. Let r and s be the least indices such that a_r and b_s are not divisible by p; then the elements $a_0, \ldots, a_{r-1}, b_0, \ldots, b_{s-1}$ are all divisible by p. Consider

$$c_{r+s} = a_{r+s}b_0 + a_{r+s-1}b_1 + \cdots + a_{r+1}b_{s-1} + a_r b_s + \cdots + a_0 b_{r+s}$$

Then

$$a_r b_s = c_{r+s} - (a_{r+s}b_0 + \cdots + a_{r+1}b_{s-1}) - (a_{r-1}b_{s+1} + \cdots + a_0 b_{r+s})$$

By assumption, $p \mid c_{r+s}$ and so p divides every term on the right side of the equation. Therefore $p \mid a_r b_s$; but then $p \mid a_r$ or $p \mid b_s$ by Corollary 16.2, contrary to our choice of a_r and b_s. Thus $h(x)$ must be primitive.

Conversely, suppose $h(x) = f(x)g(x)$ where the content of $f(x)$ is $c \neq 1$. Then the definition of polynomial multiplication implies that c divides every coefficient of $h(x)$. Therefore, c divides the content of $h(x)$, and $h(x)$ is not primitive.   □

Lemma 17.2

If R is a UFD and $p(x)$ is a nonzero element of $R[x]$, then there exist a scalar $a \in R$ and a primitive polynomial $q(x) \in R[x]$ such that $p(x) = aq(x)$. The scalar and primitive polynomial are unique up to associates.

Proof

Let $p(x) = a_n x^n + a_{n-1} x^{n-1} + \cdots + a_1 x + a_0$, and a be the greatest common divisor of the set of coefficients $\{a_n, a_{n-1}, \ldots, a_1, a_0\}$. Set $a_k = ab_k$. If

$$q(x) = b_n x^n + b_{n-1} x^{n-1} + \cdots + b_1 x + b_0$$

then we see $p(x) = aq(x)$ where $q(x)$ is primitive.

To show uniqueness, suppose $p(x) = a_0 q_0(x)$ where $a_0 \in R$, and $q_0(x)$ is a primitive polynomial in $R[x]$. Then a and a_0 are both greatest common divisors in R of $p(x) \in R[x]$. But the greatest common divisor is unique up to associates. Thus $a = a_0 u$ where u is a unit in R. Since $a_0 q_0(x) = a p_0(x) = (a_0 u)p_0(x)$ implies $q_0(x) = u p_0(x)$, we see that $q_0(x)$ and $p_0(x)$ are associates.   □

Theorem 17.8

If R is a UFD and $f(x) \in R[x]$, then either $f(x)$ is irreducible in $R[x]$ or $f(x)$ can be written as a product of irreducible elements of $R[x]$.

Proof

For $f(x) \in R[x]$, there exists by Lemma 17.2 an element $a_0 \in R$ such that $f(x) = a_0 p(x)$, where $p(x)$ is a primitive polynomial in $R[x]$. If $\deg(p(x)) = 0$, then $f(x)$ is a constant polynomial, and $f(x) = a_0 \in R$. Since R is a UFD, $f(x) = a_0$ can be written as a product of irreducible elements in R and, hence, in $R[x]$.

Without loss of generality, we can assume $\deg(p(x)) \geq 1$. Let us proceed inductively on the degree of $p(x)$, assuming the result true for all polynomials of degree less than the degree of $p(x)$. If $p(x)$ is irreducible, we are done; if not, then $p(x) = q_1(x)q_2(x)$, where neither $q_1(x)$ nor $q_2(x)$ is a unit. If $\deg(q_1(x)) = 0$, then $q_1(x)$ is a constant polynomial, contradicting the fact that $p(x)$ is primitive. A similar statement can be made for $q_2(x)$. Thus

$$0 < \deg(q_1(x)) < \deg(p(x)) \quad \text{and} \quad 0 < \deg(q_2(x)) < \deg(p(x))$$

Moreover, both $q_1(x)$ and $q_2(x)$ are primitive by Lemma 17.1. By our induction hypothesis, both $q_1(x)$ and $q_2(x)$ can be factored as a product of primitive irreducible polynomials. If $a_0 = a_1 a_2 \cdots a_n$ is a factorization of a_0 as a product of irreducible elements of R and $q_1(x)q_2(x) = q'_1(x)q'_2(x) \cdots q'_k(x)$ is the factorization of $p(x) = q_1(x)q_2(x)$ as a product of irreducible primitive polynomials in $R[x]$, then

$$p(x) = a_1 a_2 \cdots a_n q'_1(x) q'_2(x) \cdots q'_k(x)$$

is the desired factorization.   □

In order to show that the factorization obtained in Theorem 17.8 is unique up to multiplication by associates and, hence, that $R[x]$ is a UFD, we must investigate the relationship between a unique factorization domain R and its field of quotients.

Let R be a UFD and Q its field of quotients as developed in Section 15.2. Every polynomial in $R[x]$ can be identified with the corresponding polynomial in $Q[x]$. For example, $f(x) = 2x^2 - 3 \in \mathbb{Z}[x]$ can be identified with $f(x) = \frac{2}{1}x^2 - \frac{6}{2} \in \mathbb{Q}[x]$. Any polynomial reducible in $R[x]$ is still reducible in $Q[x]$; therefore, if $f(x) \in R[x]$ is irreducible as a polynomial in $Q[x]$, then $f(x)$ is already irreducible in $R[x]$. In the case of primitive polynomials, the converse is true.

Lemma 17.3

If R is a UFD, Q its field of quotients, and $p(x)$ a primitive polynomial in $R[x]$, then $p(x)$ is irreducible in $R[x]$ if and only if $p(x)$ is irreducible in $Q[x]$.

Proof

In light of the comments preceding this result, we need only consider the case in which $p(x) \in R[x]$ is reducible in $Q[x]$; that is, $p(x) = f(x)g(x)$ where $f(x), g(x) \in Q[x]$ with $\deg(f(x)) \geq 1$ and $\deg(g(x)) \geq 1$. We will prove that $p(x)$ is reducible in $R[x]$.

Suppose

$$f(x) = \sum_{i=0}^{n} \left(\frac{a_i}{b_i} \right) x^i \quad \text{and} \quad g(x) = \sum_{j=0}^{m} \left(\frac{c_j}{d_j} \right) x^j$$

Set

$$b = b_0 b_1 \cdots b_n, \qquad d = d_0 d_1 \cdots d_m,$$

$$f_1(x) = bf(x) = \sum_{i=0}^{n} \hat{a}_i x^i, \quad \text{and} \quad g_1(x) = dg(x) = \sum_{j=0}^{m} \hat{c}_j x^j$$

Then both $f_1(x)$ and $g_1(x)$ have their coefficients lying in R and $f_1(x)g_1(x) = bdp(x)$.

By Lemma 17.2, there exist $a, c \in R$ such that $f_1(x) = af_2(x)$ and $g_1(x) = cg_2(x)$ with $f_2(x)$ and $g_2(x)$ primitive in $R[x]$. Then $acf_2(x)g_2(x) = bdp(x)$. By Lemma 17.1, $f_2(x)g_2(x)$ is a primitive polynomial, as is $p(x)$. The uniqueness guaranteed in Lemma 17.2 forces ac and bd to be associates; that is, $ac = bdv$ where v is a unit in R. Then $p(x) = vf_2(x)g_2(x)$ follows by cancellation in the integral domain $R[x]$, and $p(x)$ is reducible in $R[x]$.   □

The key to proving uniqueness of the irreducible factors in factorizations of $f(x) \in R[x]$, when R is a UFD, is the observation that irreducible factorizations in $R[x]$ are also irreducible factorizations in $Q[x]$, where Q is the field of quotients of R. But $Q[x]$ is a UFD, since $Q[x]$ is a PID by Theorem 17.4 and every PID is a UFD by Theorem 16.5.

Theorem 17.9

If R is a UFD, then $R[x]$ is a UFD.

Proof

Let $f(x) \in R[x]$ be a nonunit. Then $\deg(f(x)) \geqslant 1$ and, by Lemma 17.1, $f(x) = ap(x)$ where $a \in R$ and $p(x)$ is primitive in $R[x]$. It suffices to show that $p(x)$ has a unique decomposition as a product of irreducible polynomials in $R[x]$. Since $p(x) \in Q[x]$ and $Q[x]$ is a UFD, there exist unique polynomials $f_1(x), \ldots, f_m(x) \in Q[x]$ such that $p(x) = f_1(x) \cdots f_m(x)$. Set $f_i(x) = (q_i(x)/b_i$ where $b_i \in R$ and $q_i(x) \in R[x]$, and then set $q_i(x) = a_i p_i(x)$ where $a_i \in R$ and $p_i(x)$ is primitive in $R[x]$. We have

$$p(x) = \frac{a_1 a_2 \cdots a_m}{b_1 b_2 \cdots b_m} p_1(x) p_2(x) \cdots p_m(x)$$

Since both $p(x)$ and $p_1(x)p_2(x) \cdots p_m(x)$ are primitive, an argument similar to that used in Lemma 17.3 shows that $a_1 a_2 \cdots a_m = (b_1 b_2 \cdots b_m)v$ where v is a unit in R. Therefore the factors of $p(x)$, and thus of $f(x)$, are unique up to associates. ☐

Knowing that a polynomial $f(x)$ *can* be factored uniquely is not the same as knowing *how* to factor $f(x)$. In the next two sections, we investigate methods that aid in factoring a polynomial or determining if the polynomial is irreducible.

Exercises 17.2

1. Prove that if F is a field, then the only units of $F[x]$ are the nonzero constant polynomials.

2. Prove or disprove: If R is a ring, the only units of $R[x]$ are the constant polynomials corresponding to the units of R.

3. The fact that a polynomial $f(x) \in R[x]$ is defined to be **prime** in $R[x]$ whenever $f(x)$ divides $g(x)h(x)$ implies that either $f(x)$ divides $g(x)$ or $f(x)$ divides $h(x)$.
   a. Prove or disprove: Every irreducible polynomial is prime.
   b. Prove or disprove: Every prime polynomial is irreducible.

4. Suppose $f(x) \in \mathbb{Z}[x]$ and a is an element of $\mathbb{Z}$ such that $f(a)$ is a prime in $\mathbb{Z}$. Does it follow that $f(x)$ is irreducible in $\mathbb{Z}[x]$?

5. Show that if $f(x), g(x) \in \mathbb{Z}[x]$, then f and g are associates in $\mathbb{Z}[x]$ if and only if $f(x) = g(x)$ or $f(x) = -g(x)$.

6. Prove or disprove: For an arbitrary ring R, two polynomials in $R[x]$ are associates if and only if they generate the same ideal.

7. Use the Remainder Theorem and the Factor Theorem (Theorems 17.5 and 17.6)

to determine which of the following are roots of the polynomial $x^3 - 2x + 1$ in $\mathbb{Z}[x]$:

$$3, \quad 1, \quad 0, \quad -2, \quad 2$$

8. a. Utilize the knowledge obtained in Exercise 7 to determine the irreducible factors of $x^3 - 2x + 1$ in $\mathbb{Z}[x]$.

   b. What are the irreducible factors of $x^3 - 2x + 1$ in $\mathbb{Q}[x]$, $\mathbb{R}[x]$, $\mathbb{C}[x]$?

9. Use the Well-Ordering Property rather than induction to prove the Polynomial Division Algorithm.

10. Verify that the polynomial $x^2 - x$ has four distinct roots in $\mathbb{Z}_6$.

11. Let R be a UFD, $a \in R$, and $f(x) \in R[x]$. How is the content of $af(x)$ related to the content of f?

12. Find the greatest common divisor of the following pairs of polynomials:

   a. $3x^3 + x - 4,\ x^3 - 1$ b. $x^4 - x^3 + x^2 + 2,\ x^3 - x + 1$

   c. $x^4 - 16,\ x^3 - 6x + 4$ d. $x^4 - 3x^2 - 4,\ x^3 - 2x^2 + 2x - 2$

13. Let D be an integral domain, and let $D[x, y] = (D[x])\,[y]$ be the set of polynomials in y with coefficients in $D[x]$. Verify that $D[x, y] = D[y, x]$ and that $D[x, y]$ is an integral domain. If F is a field, is $F[x, y]$ a PID?

14. Do there exist $f(x), g(x) \in \mathbb{Q}[x]$ such that $1 = f(x)(x^2 - 2) + g(x)(x^2 - 3x + 2)$?

15. Prove that if R is a field and $f(x)$ is a reducible polynomial in $R[x]$ of degree n, then $f(x) = s(x)t(x)$, where $s(x)$ and $t(x)$ are polynomials in $R[x]$ of positive degree. Is this result true if R is an integral domain?

*16. Modify Exercise 21 of Section 16.2 to determine the content of an arbitrary polynomial in $\mathbb{Z}[x]$.

*17. Write a computer program to divide one polynomial in $\mathbb{Z}[x]$ by another, where both polynomials are input as finite sequences.

*18. Use the computer program in the previous exercise to find the gcd of $f(x), g(x) \in \mathbb{Z}[x]$. Each polynomial may be given by a finite sequence of integers.

17.3
Reducibility

In this section we will consider $R[x]$, where R is a specific ring or field, and attempt to find methods of determining whether a given polynomial in $R[x]$ is irreducible. This problem is certainly not an easy one to solve, and we will be content to indicate some of the methods available. If the theory of equations intrigues you, then consult the numerous references on this subject for a more detailed discussion.

   Let us first consider the case in which $R = \mathbb{Z}$. If $f(x) \in \mathbb{Z}[x]$, then what determines whether or not $f(x)$ has a root in $\mathbb{Z}$ or in $\mathbb{Q}$? We answer this question in the next result:

Theorem 17.10   Let $f(x) = a_n x^n + a_{n-1}x^{n-1} + \cdots + a_1 x + a_0 \in \mathbb{Z}[x]$. If r/s is a rational root of $f(x)$ and $(r, s) = 1$, then $r \mid a_0$ and $s \mid a_n$.

Proof

By our assumptions,

$$0 = f\left(\frac{r}{s}\right) = a_n\left(\frac{r}{s}\right)^n + a_{n-1}\left(\frac{r}{s}\right)^{n-1} + \cdots + a_1\left(\frac{r}{s}\right) + a_0$$

By multiplying by s^n, we get

$$0 = a_n r^n + a_{n-1} r^{n-1} s + \cdots + a_1 r s^{n-1} + a_0 s^n$$

Thus

$$-a_0 s^n = r(a_n r^{n-1} + \cdots + a_1 s^{n-1})$$

If $r = 1$, then $r \mid a_0$. Otherwise, $r \neq 1$ and $(r, s) = 1$ implies $(r, s^n) = 1$ for any $n \geq 1$. (See Exercise 1.) From the preceding equation, we see $r \mid (-a_0 s^n)$. By Corollary 16.2, we have $r \mid a_0$.

Taking our original equation and multiplying both sides by r^n, we get

$$-a_n r^n = s(a_{n-1} r^{n-1} + \cdots + a_0 s^{n-1})$$

By an argument similar to that used in the previous paragraph, we have $s \mid a_n$. The result follows. $\qquad\square$

This theorem gives us a way to check if a polynomial in $\mathbb{Z}[x]$ has a rational root and, so, a linear factor. We establish in Example 17.4, however, that showing a polynomial in $\mathbb{Z}[x]$ has no rational roots does **not** guarantee that it is irreducible over $\mathbb{Q}$.

Example 17.3

Consider $f(x) = 2x^3 - x^2 + 4x - 2 \in \mathbb{Z}[x]$. By Theorem 17.10, if $f(x)$ has a rational root r/s in reduced form; that is, if $(r, s) = 1$, then $r = \pm 1$ or ± 2 and $s = \pm 1$ or ± 2. The possible candidates for rational roots are $r/s = \pm 1$, ± 2, $\pm\frac{1}{2}$. A check of these shows that only $\frac{1}{2}$ is a root. Thus

$$f(x) = (2x^2 + 4)(x - \tfrac{1}{2}) = 2(x - \tfrac{1}{2})(x^2 + 2)$$

Example 17.4

Let $g(x) = x^4 + 2x^2 + 1$. Then $g(x)$ has no rational roots since the only possibilities are ± 1 and neither is a root. Yet $g(x)$ is not irreducible over $\mathbb{Q}$ since $g(x) = (x^2 + 1)(x^2 + 1)$.

From this last example, we see that it would be advantageous to have some way of determining whether or not a given polynomial in $\mathbb{Z}[x]$ is irreducible over $\mathbb{Z}$, not just that it has no linear factors in $\mathbb{Z}[x]$. Such a method does exist, but first we will restate Lemma 17.3 as it applies to $\mathbb{Z}$ and $\mathbb{Q}$. This classical result is known as **Gauss' Theorem**. It asserts that if a polynomial with integer coefficients cannot be factored in $\mathbb{Z}[x]$, then it cannot be factored in $\mathbb{Q}[x]$.

Theorem 17.11 | **Gauss' Theorem**

If $f(x) \in \mathbb{Z}[x]$ and $f(x)$ can be factored as a product of two polynomials with rational coefficients, then it is possible to factor $f(x)$ as a product of two polynomials with integer coefficients.

While this result is interesting, it is not immediately applicable, for we still have not determined whether a given polynomial is irreducible. The next result, however, does yield a useful tool to aid in making such a determination.

Theorem 17.12 | **Eisenstein's Criterion**[1]

Let $f(x) = a_n x^n + \cdots + a_1 x + a_0 \in \mathbb{Z}[x]$. If there exists some prime p in $\mathbb{Z}$ such that $p \nmid a_n$, $p \mid a_i$, $i = 0, 1, \ldots, n-1$, and $p^2 \nmid a_0$, then $f(x)$ is irreducible over $\mathbb{Q}$.

Proof

Suppose the conditions of the hypothesis hold and $f(x)$ is reducible over $\mathbb{Q}$. By Gauss' Theorem, we can assume $f(x)$ is reducible over $\mathbb{Z}$ and

$$f(x) = (c_0 + c_1 x + \cdots + c_r x^r)(d_0 + d_1 x + \cdots + d_s x^s)$$

where the c_i's and d_j's are integers and $r < n$ and $s < n$. Then $a_0 = c_0 d_0$ and, since $p \mid a_0$ but $p^2 \nmid a_0$, we have either $p \mid c_0$ or $p \mid d_0$, but not both. For the sake of definiteness, assume $p \mid c_0$ and $p \nmid d_0$. Since $p \nmid a_n$, we know p does not divide all the c_i's. Let j be the least index such that $p \nmid c_j$. Then

$$a_j = c_0 d_j + c_1 d_{j-1} + \cdots + c_{j-1} d_1 + c_j d_0$$

and

$$c_j d_0 = a_j - c_0 d_j - \cdots - c_{j-1} d_1$$

By assumption, p divides the expression on the right side of the equation. Therefore, $p \mid c_j d_0$. But then Corollary 16.2 implies that $p \mid c_j$ or $p \mid d_0$, contrary to our assumptions. This contradiction yields the desired result. $\qquad\square$

Eisenstein's Criterion has numerous applications, as shown in the following examples:

---

[1] "There have been only three epoch-making mathematicians, Archimedes, Newton, and Eisenstein." This remark (certainly an exaggeration) has been attributed to Gauss speaking about his much admired friend Ferdinand Gotthold Eisenstein (1823–1852). The Eisenstein conjecture, unverified at present time, is that numbers of the form $2^2 + 1$, $2^{2^2} + 1$, and so on, are primes.

Example 17.5

Let $f(x) = x^{201} - 6x^{107} + 21$. Then $f(x)$ is irreducible over $\mathbb{Q}$ by Eisenstein's Criterion with $p = 3$.

Example 17.6

Let $g(x) = x^2 - 4x + 9$. Eisenstein's Criterion cannot be immediately applied to $g(x)$; but if we let

$$h(x) = g(x+1) = (x+1)^2 - 4(x+1) + 9 = x^2 - 2x + 6$$

we see that $h(x)$ **is** irreducible over $\mathbb{Q}$ by Eisenstein's Criterion. We leave it to you to show $g(x)$ is irreducible if and only if $g(x+1)$ is irreducible. (See Exercise 2.) Note we could have simply used Theorem 17.10 to show $g(x)$ has no rational roots and then argued that if a rational quadratic polynomial is reducible over $\mathbb{Q}$, then its factors must be linear.

Example 17.7

Let

$$f(x) = \frac{x^p - 1}{x - 1} = x^{p-1} + x^{p-2} + \cdots + x + 1$$

where p is a prime integer. Again the Eisenstein Criterion cannot be immediately applied, but the substitution we used in the last example yields

$$f(x+1) = \frac{(x+1)^p - 1}{(x+1) - 1}$$

$$= \frac{x^p + px^{p-1} + \cdots + px + 1 - 1}{x}$$

$$= x^{p-1} + px^{p-2} + \cdots + p$$

where the omitted terms are all multiples of p. Now Eisenstein's Criterion shows that $f(x+1)$, and hence $f(x)$, is irreducible over $\mathbb{Q}$.

Example 17.8

Let $h(x) = x^2 + 4$. Then Eisenstein's Criterion is not applicable since the only acceptable prime(s) are $+2$ and -2, but $(-2)^2 | 4$ and $2^2 | 4$. Therefore no decision can be reached by this test. If we try the approach of Examples 17.6 and 17.7 and consider $h(x+1)$, we find $h(x+1) = x^2 + 2x + 5$; again Eisenstein's Criterion does not apply. Note, however, $h(x) = x^2 + 4$ is irreducible over $\mathbb{Q}$.

Example 17.9

Let $g(x) = x^4 - 4 \in \mathbb{Z}[x]$. As in Example 17.8, the hypothesis of Eisenstein's Criterion is not satisfied, and no conclusion is therefore available utilizing this result. In this case, however, $x^4 - 4 = (x^2 + 2)(x^2 - 2)$ is reducible over $\mathbb{Q}$.

While Eisenstein's Criterion is not a universal catch-all, it is useful in many situations. You should realize that, while all of these examples refer to polynomials in $\mathbb{Z}[x]$, Eisenstein's Criterion can be applied to polynomials in $\mathbb{Q}[x]$. To do this, simply multiply the polynomial by the least common multiple (Exercise 14 of Section 16.2) of the denominators of the coefficients to obtain an equivalent polynomial in $\mathbb{Z}[x]$.

For example, if $f(x) = x^2 + \frac{1}{2}x - \frac{2}{3}$, we multiply by 6 to obtain $f_1(x) = 6x^2 + 3x - 4$. Since

$$f_1(x+1) = 6(x+1)^2 + 3(x+1) - 4 = 6x^2 + 15x + 5$$

we can apply Eisenstein's Criterion with $p = 5$ to conclude that $f(x)$ is irreducible.

We now turn to consideration of polynomials with real and complex coefficients. The theory of polynomial reducibility in $\mathbb{R}[x]$ and $\mathbb{C}[x]$ is intimately entwined. Many results concerning the nature of $\mathbb{C}$ can be applied to derive results about $\mathbb{R}[x]$. The following is a typical example:

Theorem 17.13

The mapping $\zeta\colon \mathbb{C} \to \mathbb{C}$ defined by $\zeta(a+bi) = a - bi$ is an automorphism of $\mathbb{C}$.

Proof

ζ is onto, because if $a + bi \in \mathbb{C}$, then $\zeta(a - bi) = a + bi$. ζ is injective, because if $a - bi = c - di$, then $(a-c) + (b-d)i = 0 + 0i$; therefore $a - c = 0$, or $a = c$, and $b - d = 0$, or $b = d$. Now, to show that ζ is a homomorphism, we must verify that ζ preserves addition and multiplication.

$$\zeta((a+bi) + (c+di)) = \zeta((a+c) + (b+d)i)$$
$$= (a+c) - (b+d)i$$
$$= a + c - bi - di$$
$$= (a - bi) + (c - di)$$
$$= \zeta(a+bi) + \zeta(c+di)$$

We will leave the verification of the fact that $\zeta(zw) = \zeta(z)\zeta(w)$ as Exercise 9. □

If $z = a + bi$, then $\zeta(z) = a - bi$ is termed the **complex conjugate** of $a + bi$. Sometimes the conjugate of $z = a + bi$ is denoted by $\bar{z} = a - bi$.

In order to apply the result to $\mathbb{R}[x]$, we need a theorem from that part of mathematics known as analysis. When we discussed the concept of *field of quotients* and applied the construction to the integers to obtain the field of rationals, we mentioned that partial motivation for this process was the desire to be able to solve a larger number of equations. For example, while $2x + 3 = 0$ has no solution in $\mathbb{Z}$, it does have a solution in $\mathbb{Q}$. The fact that $x^2 - 2 = 0$ has no solution in $\mathbb{Q}$ led, in part, to the study of the reals, $\mathbb{R}$. Even with the reals, however, there are polynomial equations with real coefficients that have no real solution. The equation $2x^2 = -5$ is one such example.

The complex number system does contain the solutions to this equation, namely, $x = i\sqrt{5/2}$ and $x = -i\sqrt{5/2}$. Conceivably, this process might go on ad infinitum, with each new field giving rise to new polynomial equations whose solution requires still larger fields. This is not the case, however.

Theorem 17.14

The Fundamental Theorem of Algebra

Any equation of the form $a_0 + a_1x + \cdots + a_nx^n = 0$, where $n \geqslant 1$, $a_i \in \mathbb{C}$, and $a_n \neq 0$, has a solution in $\mathbb{C}$.

The Fundamental Theorem of Algebra, like so many other "fundamental theorems," has a proof that is anything but fundamentally simple. Gauss developed a number of different proofs for the celebrated result, but all involve some concept of analysis or topology. Proofs of the Fundamental Theorem of Algebra can be found in almost all introductory texts on complex analysis.[2] From an aesthetical point of view, however, the name "Fundamental Theorem of Algebra" is probably preferable to "Difficult Theorem of Complex Analysis."

Let us now consider the problem of factoring a polynomial in $\mathbb{R}[x]$. Suppose $f(x) \in \mathbb{R}[x]$ can be written as

$$f(x) = a_nx^n + a_{n-1}x^{n-1} + \cdots + a_1x + a_0$$

where $a_i \in \mathbb{R}$ and $n > 1$. By the Fundamental Theorem of Algebra, $f(x)$ has a complex root $z = a + bi$. (We do not rule out the possibility that $b = 0$ and $a \in \mathbb{R}$.) If we apply the automorphism ζ of Theorem 17.13 to the equation

$$0 = f(z) = a_nz^n + a_{n-1}z^{n-1} + \cdots + a_1z + a_0$$

we obtain

$$0 = f(\bar{z}) = a_n\bar{z}^n + a_{n-1}\bar{z}^{n-1} + \cdots + a_1\bar{z} + a_0$$

(See Exercise 11.) Therefore $\bar{z} = a - bi$ is another root of $f(x)$ and

$$(x - z)(x - \bar{z}) = x^2 - 2ax + (a^2 + b^2)$$

is a factor of $f(x)$; moreover, this factor is in $\mathbb{R}[x]$. In this way, we can pair up every complex nonreal root of $f(x)$ with its conjugate to obtain a real quadratic factor of $f(x)$. Thus, the only irreducible polynomials in $\mathbb{R}[x]$ must be either linear or quadratic. Since $\mathbb{R}[x]$ is a UFD, we have the following:

Theorem 17.15

Let $f(x)$ be a nonconstant real polynomial. Then $f(x)$ can be factored

---

[2] See, for example, Jerrold E. Marsden, *Basic Complex Analysis*, San Francisco: W. F. Freeman and Company, 1973, p. 124.

uniquely as a product of a real number by a product of monic linear and/or quadratic real polynomials.

The following examples illustrate this result.

Example 17.10

Let $f(x) = x^3 - 7x^2 + 17x - 15$. By Theorem 17.10 we find that 3 is a root. Thus $f(x) = (x - 3)(x^2 - 4x + 5)$. The roots of $x^2 - 4x + 5$ are $2 - i$ and $2 + i$; so that

$$f(x) = (x - 3)(x^2 - 4x + 5)$$

is the complete factorization of $f(x)$ over $\mathbb{R}$, and

$$f(x) = (x - 3)(x - 2 + i)(x - 2 - i)$$

is the complete factorization over $\mathbb{C}$.

Example 17.11

Let $h(x) = 2x^4 + 3x^2 - 5$. This is a quadratic in x^2. We factor $h(x)$ as

$$(2x^2 + 5)(x^2 - 1) = 2(x^2 + \tfrac{5}{2})(x - 1)(x + 1)$$

over $\mathbb{R}$ and

$$2\left(x + \sqrt{\frac{5}{2}}\,i\right)\left(x - \sqrt{\frac{5}{2}}\,i\right)(x - 1)(x + 1)$$

over $\mathbb{C}$.

While the Fundamental Theorem of Algebra guarantees the existence of at least one root of $f(x) \in \mathbb{C}[x]$ when $\deg f \geqslant 1$, it does not indicate how to find the root. Numerous algorithms for locating roots of f have been devised. You can find many of them in any introductory text in numerical analysis. What we want to do here is discuss a method for finding all solutions to a special type of polynomial equation, namely, one of the form $x^n + a = 0$ for $a \in \mathbb{C}$. To do so, requires us to investigate an alternate form for expressing complex numbers.

If $a + bi$ is a complex number, then $a + bi$ can be graphically represented as the point (a, b) in the Cartesian plane. (See Figure 17.1.) If $r = \sqrt{a^2 + b^2}$, then $\cos \theta = a/r$ and $\sin \theta = b/r$ so that

$$a + bi = r\left(\frac{a}{r} + i\frac{b}{r}\right) = r(\cos \theta + i \sin \theta)$$

The next result uses this trigonometric representation of complex numbers.

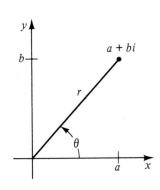

Figure 17.1

Theorem 17.16	DeMoivre's Theorem[3]

If n is any integer and $z = r(\cos\theta + i\sin\theta)$, then $z^n = r^n(\cos(n\theta) + i\sin(n\theta))$.

Proof

The proof is divided into three cases: $n = 0$, n positive, and n negative.

i. If $n = 0$, then $z^0 = 1 = |z|^0(\cos 0 + i\sin 0)$.

ii. If $n = 1$, the conclusion is obvious. Using induction, we will assume that the statement is valid for $n = k$ and show that it is valid for $n = k + 1$.

$$z^{k+1} = z \cdot z^k$$

$$= r(\cos\theta + i\sin\theta) \cdot r^k(\cos(k\theta) + i\sin(k\theta))$$

$$= r^{k+1}[\cos\theta\cos(k\theta) - \sin\theta\sin(k\theta)$$

$$+ i(\sin\theta\cos(k\theta) + \cos\theta\sin(k\theta))]$$

$$= r^{k+1}[\cos((k+1)\theta) + i\sin((k+1)\theta)]$$

iii. The case for n, a negative integer, is proved by utilizing the preceding statement and fact that $z^{-n} = (z^n)^{-1}$. We will leave the actual verification to Exercise 10. □

Let us apply DeMoivre's Theorem to find all the roots to $x^4 + 2 = 0$. If $x = r(\cos\theta + i\sin\theta)$, we have

$$x^4 = r^4(\cos 4\theta + i\sin 4\theta) = -2 = 2(-1 + 0i)$$

$$= 2\left(\cos\left(\frac{\pi}{2} + 2\pi k\right) + i\sin\left(\frac{\pi}{2} + 2\pi k\right)\right)$$

where k is an integer. Comparing both sides of this equation, we see that

$$r^4 = 2 \quad \text{and} \quad 4\theta = \frac{\pi}{2} + 2\pi k$$

For $k = 0, 1, 2, 3$, we obtain

$$r = \sqrt[4]{2} \quad \text{and} \quad \theta = \frac{\pi}{8}, \frac{5\pi}{8}, \frac{9\pi}{8}, \text{and } \frac{13\pi}{8}$$

---

[3] Abraham DeMoivre (1667–1754), a French Huguenot, settled in England after the revocation of the Edict of Nantes. He was forced to make his living as a private teacher of mathematics after he failed to obtain a position at a university. Despite this lack of success, his mathematical ability was highly respected by his peers as evidenced by the fact that Newton, in his twilight years, often referred questioners to him.

The solutions to our equation are

$$x_1 = \sqrt[4]{2}\left(\cos\frac{\pi}{8} + i\sin\frac{\pi}{8}\right)$$

$$x_2 = \sqrt[4]{2}\left(\cos\frac{5\pi}{8} + i\sin\frac{5\pi}{8}\right)$$

$$x_3 = \sqrt[4]{2}\left(\cos\frac{9\pi}{8} + i\sin\frac{9\pi}{8}\right)$$

$$x_4 = \sqrt[4]{2}\left(\cos\frac{13\pi}{8} + i\sin\frac{13\pi}{8}\right)$$

While DeMoivre's Theorem is extremely useful in finding solutions of equations having the form $x^n - a = 0$ (or, finding the nth roots of a), it does not apply to a general polynomial equation of degree n.

In Section 17.4, we will investigate some classical methods for finding the roots of all third-degree and fourth-degree polynomials in $\mathbb{R}[x]$.

Exercises 17.3

1. If $r, s \in \mathbb{Z}$ and $(r, s) = 1$, then $(r, s^n) = 1$ for all $n \geqslant 1$.

2. Show that $g(x) \in \mathbb{Z}$ is irreducible if and only if $g(x + 1)$ is irreducible.

3. Prove that if $f(x)$ is a monic polynomial with integer coefficients, then any rational root of $f(x)$ is already an integer.

4. Suppose $f(x) = a_n x^n + \cdots + a_1 x + a_0 \in \mathbb{Z}[x]$ and $n \in \mathbb{N}$. Let

   $$\bar{f}(x) = \bar{a}_n x^n + \cdots + \bar{a}_1 x + \bar{a}_0$$

   where $\bar{a}_i$ is the congruence class in $\mathbb{Z}_n$ of a_i in $\mathbb{Z}$. For example, the polynomial $f(x) = 41x^2 + 9x - 2$ becomes $\bar{f}(x) = \bar{2}x^2 + \bar{1}$ in $\mathbb{Z}_3$.

   a. Show that when $f(x)$ is reducible in $\mathbb{Z}[x]$, then $\bar{f}(x)$ is reducible in $\mathbb{Z}_n[x]$. Conclude that to show $f(x)$ is irreducible over $\mathbb{Z}$, it suffices to show $\bar{f}(x)$ is irreducible over $\mathbb{Z}_n$.

   b. Use part a and the fact that a polynomial of degree two or three is reducible if and only if it has a linear factor and, hence, a root to show that $x^3 + 280x^2 - 191x + 1127$ is irreducible over $\mathbb{Z}$.

   c. By considering $f(x) = x^2 + 1$ and $n = 2$, show that the fact that $f(x)$ is irreducible over $\mathbb{Z}$ does not necessarily imply that $\bar{f}(x)$ is irreducible over $\mathbb{Z}_n$.

   d. Show that $x^4 + 3x + 1$ is irreducible over $\mathbb{Q}$. Be careful!

5. If $ax^2 + bx + c \in \mathbb{C}[x]$, verify that the roots of this polynomial are

   $$\frac{-b + \sqrt{b^2 - 4ac}}{2a} \quad \text{and} \quad \frac{-b - \sqrt{b^2 - 4ac}}{2a}$$

6. In the preceding problem, the expression $b^2 - 4ac$ under the radical is called the **discriminant**. Verify the following facts about $ax^2 + bx + c \in \mathbb{R}[x]$.

   a. If $b^2 - 4ac > 0$, the roots are real and distinct.

    b. If $b^2 - 4ac = 0$, the roots are real and equal.

    c. If $b^2 - 4ac < 0$, the roots are nonreal complex conjugates.

7. Factor each of the following polynomials over $\mathbb{Q}$.

    a. $x^3 + 1$ b. $6x^3 - 7x^2 - x + 2$

    c. $x^5 - 16$ d. $x^4 + 6x^3 + x^2 - 24x - 20$

    e. $x^3 + 6x - 3$ f. $x^4 - x^3 + 6x^2 - 9x + 3$

    g. $x^8 + 1$ h. $2x^4 + 13x^2 + 20$

    i. $x^4 - x^2 + 4x - 4$ j. $x^5 - 9x^3 + 12x^2 - 6x + 15$

8. Let $z \in \mathbb{C}$ and $\bar{z}$ denote its complex conjugate. Show that:

    a. $z + \bar{z} \in \mathbb{R}$ b. $z \cdot \bar{z} \in \mathbb{R}$

    c. $(z^{-1}) = (\bar{z})^{-1}$ d. $z = \bar{z}$ if and only if $z \in \mathbb{R}$

9. Prove that if z, $w \in \mathbb{C}$, then $\zeta(zw) = \zeta(z)\zeta(w)$ where ζ denotes complex conjugation.

10. Prove that if $z = r(\cos \theta + i \sin \theta)$ and $n > 0$ then $z^{-n} = r^{-n}(\cos \theta - i \sin \theta)$.

11. Suppose $f(x) \in \mathbb{R}[x]$ and $f(z) = 0$ for some $z \in \mathbb{C}$. Prove $f(\bar{z}) = 0$.

12. Verify that if $z = a + bi$, then $(x - z)(x - \bar{z}) = x^2 - 2ax + (a^2 + b^2)$.

13. Let $f(x) \in \mathbb{R}[x]$, and suppose $\deg(f)$ is an odd integer. Prove that f has a real root.

14. Write the following complex numbers in trigonometric form:

    a. $(1, -2)$ b. $(2, 1)$ c. $(-3, -2)$ d. $(0, -3)$ e. $(12, 1)$

    f. $(3, 4)$ g. $(-4, 3)$ h. $(3, -4)$ i. $(8, -6)$ j. $(4, 2)$

15. Use DeMoivre's Theorem to find all distinct solutions to each of the following:

    a. $x^3 - 2 = 0$ b. $x^3 + 2 = 0$

    c. $x^4 - (2 - i) = 0$ d. $ix^4 - 2 = 0$

    e. $x^5 + 32 = 0$ f. $x^6 + 64 = 0$

    g. $3x^3 - 81 = 0$ h. $8x^3 + 27 = 0$

    i. $x^{10} + 1 = 0$ j. $x^{12} - 1 = 0$

16. Prove that the set of all solutions to $x^n = 1$ forms a cyclic group under multiplication.

17. Let x_0, x_1, x_2 be distinct real numbers and y_0, y_1, y_2 any elements of $\mathbb{R}$. Set

$$f_0(x) = \frac{(x - x_1)(x - x_2)}{(x_0 - x_1)(x_0 - x_2)}$$

$$f_1(x) = \frac{(x - x_0)(x - x_2)}{(x_1 - x_0)(x_1 - x_2)}$$

$$f_2(x) = \frac{(x - x_0)(x - x_1)}{(x_2 - x_0)(x_2 - x_1)}$$

    Prove that if $f(x) = y_0 f_0(x) + y_1 f_1(x) + y_2 f_2(x)$, then $f(x_i) = y_i$ for $i = 0, 1, 2$.

18. Use the Fundamental Theorem of Algebra to show if $f(x) \in \mathbb{C}[x]$ has degree $n \geqslant 1$ and leading coefficient a_n, then there exist $z_1, z_2, \ldots, z_n \in \mathbb{C}$, not necessarily distinct, such that

$$f(x) = a_n(x - z_1)(x - z_2) \cdots (x - z_n)$$

*19. Write a computer program that uses Theorem 17.10 to determine the possible rational roots of $f(x) \in \mathbb{Z}[x]$.

*20. Compose a computer program to implement Eisenstein's Criteria. Input a polynomial in $\mathbb{Z}[x]$ as an array or string of integers.

17.4

Solution of Cubic and Quartic Equations

In order to factor real polynomials, it is usually necessary to determine one or more of the roots, so that one can apply the techniques of the quadratic formula and the preceding discussion. The problem is in finding these roots. With quadratic polynomials, one can use the quadratic formula to determine the roots (Exercise 5 of Section 17.3). But what about cubic and quartic (fourth degree) polynomials and those of higher degree?

The theory of equations and the corresponding problem of factoring polynomials have kept mathematicians busy for hundreds of years. In the sixteenth century, mathematicians used to challenge each other with equations in much the same way that modern man occupies his time with long distance chess matches.

In the early 1500's, two mathematicians by the names of Ferro (1465–1525) and Tartaglia (1506–1559) happened upon a method for solving cubic equations. They made the mistake of showing their secret to G. Cardano (1501–1576), who published it in 1545. Although Cardano gave the others credit, the method of solution has come to be known as the **Cardano solution**. The method involves reducing a general cubic equation to a monic cubic with no second degree term. We will first investigate the method for solving a special polynomial of this type and then analyze the general case.

Lemma 17.4

Let $f(x) = x^3 + 3px + q \in \mathbb{R}[x]$. Set

$$\omega = -\frac{1}{2} + i\frac{\sqrt{3}}{2}, \quad A = \frac{-q + \sqrt{q^2 + 4p^3}}{2}, \quad \text{and } B = \frac{-q - \sqrt{q^2 + 4p^3}}{2}$$

Then the roots of $f(x)$ are $\sqrt[3]{A} + \sqrt[3]{B}$, $\omega\sqrt[3]{A} + \omega^2\sqrt[3]{B}$, and $\omega^2\sqrt[3]{A} + \omega\sqrt[3]{B}$.

Proof

The result may be seen by direct computation; however, we will derive the formulas. Set $x = y + z$. Upon substitution, the equation $f(x) = 0$ becomes

$$y^3 + z^3 + q + 3(p + yz)(y + z) = 0$$

Choose z so that $p + yz = 0$; that is, $z = -p/y$. Then

$$f(y + z) = y^3 - \frac{p^3}{y^3} + q,$$

so that $f(x) = 0$ if and only if

$$y^3 - \frac{p^3}{y^3} + q = 0$$

This equation is equivalent to

$$y^6 + qy^3 - p^3 = 0$$

a quadratic equation in y^3. Therefore $f(x) = 0$ if and only if

$$y^3 = \frac{-q + \sqrt{q^2 + 4p^3}}{2}$$

Suppose

$$y^3 = A = \frac{-q + \sqrt{q^2 + 4p^3}}{2}$$

Then

$$z^3 = \frac{-p^3}{y^3} = B = \frac{-q + \sqrt{q^2 + 4p^3}}{2}$$

(See Exercise 1.) By DeMoivre's Theorem, we have $y = \sqrt[3]{A}, \omega\sqrt[3]{A},$ or $\omega^2\sqrt[3]{A}$ and $z = \sqrt[3]{B}, \omega\sqrt[3]{B},$ or $\omega^2\sqrt[3]{B}$, where

$$\omega = \cos\left(\frac{2\pi}{3}\right) + i\sin\left(\frac{2\pi}{3}\right) = -\frac{1}{2} + i\frac{\sqrt{3}}{2}$$

From the nine possible choices for $x = y + z$, we see that only the following three values satisfy $yz = -p$:

$$x_1 = \sqrt[3]{A} + \sqrt[3]{B}, \quad x_2 = \omega\sqrt[3]{A} + \omega^2\sqrt[3]{B} \quad \text{and} \quad x_3 = \omega^2\sqrt[3]{A} + \omega\sqrt[3]{B}$$

$\square$

Example 17.12

Consider $f(x) = x^3 + 6x + 2$. Then $p = \frac{6}{3} = 2$ and $q = 2$, so that

$$A = -\frac{2}{2} + \frac{1}{2}\sqrt{4 + 4\cdot 2^3} = -1 + 3 = 2 \quad \text{and} \quad B = -1 - \sqrt{9} = -4$$

Therefore,

$$x_1 = \sqrt[3]{2} - \sqrt[3]{4}$$

$$x_2 = \left(-\frac{1}{2} - i\frac{\sqrt{3}}{2}\right)\sqrt[3]{2} + \left(-\frac{1}{2} - i\frac{\sqrt{3}}{2}\right)\sqrt[3]{4}$$

$$= -\frac{1}{2}\left(\sqrt[3]{2} + \sqrt[3]{4}\right) + \frac{i}{2}\left(\sqrt{3}\sqrt[3]{2} - \sqrt{3}\sqrt[3]{4}\right)$$

$$x_3 = \left(-\frac{1}{2} - i\frac{\sqrt{3}}{2}\right)\sqrt[3]{2} + \left(-\frac{1}{2} + i\frac{\sqrt{3}}{2}\right)\sqrt[3]{4}$$

$$= -\frac{1}{2}\left(\sqrt[3]{2} + \sqrt[3]{4}\right) - \frac{i}{2}\left(\sqrt{3}\sqrt[3]{2} - \sqrt{3}\sqrt[3]{4}\right)$$

are the three roots of our equation.

Lemma 17.4 is essentially the method of Ferro and Tartaglia. Cardano's contribution, besides publishing their method and accrediting it to them, was to note that an elementary substitution solves the general case. This extension is sufficient reason to justify the association of his name with the general method.

Theorem 17.17

Let $f(x) = x^3 + bx^2 + cx + d \in \mathbb{R}[x]$, and let $g(t) = f(t - b/3) = t^3 + 3pt + q$. Let A, B, and ω have the values given in Lemma 17.4. The roots of $f(x)$ are then $x_1 = \sqrt[3]{A} + \sqrt[3]{B} - (b/3)$, $x_2 = \omega\sqrt[3]{A} + \omega^2\sqrt[3]{B} - (b/3)$, and $x_3 = \omega^2\sqrt[3]{A} + \omega\sqrt[3]{B} - (b/3)$.

Proof

Straightforward calculation shows $g(t) = f(t - b/3) = t^3 + 3pt + q$ where

$$p = \frac{c}{3} - \frac{b^2}{9} \quad \text{and} \quad q = \frac{2b^3}{27} - \frac{bc}{3} + d$$

By Lemma 17.4, the roots of $f(t - b/3) = g(t)$ are $t_1 = \sqrt[3]{A} + \sqrt[3]{B}$, $t_2 = \omega\sqrt[3]{A} + \omega^2\sqrt[3]{B}$, and $t_3 = \omega^2\sqrt[3]{A} + \omega\sqrt[3]{B}$. Therefore, the roots of $f(x)$ are

$$x_1 = t_1 - \frac{b}{3}, \quad x_2 = t_2 - \frac{b}{3}, \quad \text{and} \quad x_3 = t_3 - \frac{b}{3} \qquad \square$$

Example 17.13

Suppose $f(x) = x^3 + 9x^2 + 33x + 47$. Then $b = 9$ and $f(t - b/3) = f(t - 3) = t^3 + 6t + 3$. But this is the cubic that we solved in Example 17.12. The roots of the polynomial are, therefore, $x_1 - 3$, $x_2 - 3$, and $x_3 - 3$, where x_1, x_2, x_3 are the three roots of $f(x) = x^3 + 6x + 3$, which we found earlier.

Although L. Ferraro (1522–1565), a student of Cardano, is credited with the discovery of the solution of the general quartic equation, we now describe a method due to Euler. First, as in the case of the cubic, we will consider a monic quartic polynomial with a missing term.

Let $f(x) = x^4 + px^2 + qx + r \in \mathbb{C}[x]$. Assume $f(x)$ can be factored as a product of two quadratic polynomials, and write

$$x^4 + px^2 + qx + r = (x^2 + 2k_1 x + m)(x^2 + 2k_2 x + n)$$

$$= x^4 + 2(k_1 + k_2)x^3 + (m + n + 4k_1k_2)x^2$$
$$+ 2(nk_1 + mk_2) + mn$$

By equating coefficients, the absence of the x^3 term in $f(x)$ forces $k_1 = -k_2$, and so

$$m + n - 4k_1^2 = p, \quad 2k_1(n - m) = q, \quad \text{and} \quad mn = r$$

Setting $m + n = p + 4k_1^2$ and $n - m = q/2k_1$, we have

$$m = \frac{p}{2} + 2k_1^2 - \frac{q}{4k_1} \quad \text{and} \quad n = \frac{p}{2} + 2k_1^2 + \frac{q}{4k_1}$$

(See Exercise 3.) The equation $r = mn$ implies

$$64k_1^6 + 32pk_1^4 + (4p^2 - 16r)k_1^2 - q^2 = 0$$

But this is a cubic equation in k_1^2. By Cardano's method we can solve for k_1^2 and, hence, for m, n, and k_1 in terms of p, q, and r. Therefore the roots of $f(x)$ will be the roots of $x^2 + 2k_1x + m$ and $x^2 - 2k_1x + n$.

| Example 17.14 |

Let $f(x) = x^4 - 2x^2 - 8x - 3$. Then $p = -2$, $q = -8$, and $r = -3$, and

$$64k_1^6 - 64k_1^4 + 64k_1^2 - 64 = 0 \quad \text{or} \quad k_1^6 - k_1^4 + k_1^2 - 1 = 0$$

Immediately we see that $k_1^2 = 1$ is a solution. Letting $k_1 = 1$, we find $m = 3$ and $n = -1$, so that we need only solve the equations

$$x^2 + 2x + 3 = 0 \quad \text{and} \quad x^2 - 2x - 1 = 0$$

to see that

$$x = \frac{-2 \pm \sqrt{4 + 4(3)}}{2} = -1 \pm i\sqrt{2} \quad \text{and}$$

$$x = \frac{2 \pm \sqrt{4 + 4}}{2} = 1 \pm \sqrt{2}$$

are the desired solutions for $f(x) = 0$.

In order to handle the general fourth degree polynomial $f(x) = x^4 + ax^3 + bx^2 + cx + d$, we make the substitution $x = t - a/4$. Then

$$f(x) = f\left(t - \frac{a}{4}\right) = t^4 + pt^2 + qt + r = g(t)$$

where $g(t)$ is of the form we have just discussed. The roots of $f(x)$ are then $x_i = t_i - a/4$, with t_i being the roots of $g(t)$. We obtain the following theorem.

Theorem 17.18

Let $f(x) = x^4 + ax^3 + bx^2 + cx + d \in \mathbb{R}[x]$. The roots of $f(x)$ are $x_i = t_i - a/4$, for $i = 1, 2, 3, 4$, where the number t_i is one of the four roots of the equation $g(t) = f(t - a/4)$.

The next question is that of finding some way to solve the general polynomial equation of degree five. A number of mathematicians, including Abel, published what they thought were solutions for this equation. Abel found his mistake, however, and decided that there is no general solution; that is, there is no algorithm or method for finding a root of an arbitrary polynomial of degree five or higher. While Abel was not the first to come to this conclusion,[4] his proof was the first to contain all the necessary reasoning. Abel did his work before he reached the age of twenty, yet the quality and importance of his work was not recognized until 1830, one year after he died from tuberculosis. It is important to realize that Abel did not prove it was impossible to solve any polynomial equation of degree five or greater by suitable substitutions; what he proved was that *not every* polynomial equation could be solved by the same method. In Chapter 22, we will see how such a proof is done, but, in order to understand this discussion, we must investigate a most important algebraic structure, namely, the vector space.

Exercises 17.4

1. Verify the assertion in Lemma 17.4 that

$$z^3 = \frac{-q + \sqrt{q^2 + 4p^3}}{2}$$

   and that x_1, x_2, and x_3 are the only values for $x = y + z$ that satisfy $yz = -p$.

2. Find the prime factorization of the following polynomials in $\mathbb{Q}[x]$.
   a. $2x^5 - 25x^3 + 5$ b. $x^3 + 26x^2 - 52x + 13$
   c. $2x^3 - 3x^2 - 3x - 2$ d. $x^5 + x^4 - 2x^3 - x^2 - x + 2$
   e. $x^3 - 3x - 52$ f. $x^3 + 9x^2 + 18x - 28$

3. Show that if $m + n = p + 4k_1^2$ and $n - m = 9/2k_1$, then

$$m = \frac{p}{2} + 2k_1^2 - \frac{9}{4k_1}, \quad n = \frac{p}{2} + 2k_1^2 + \frac{q}{4k_1}, \quad \text{and} \quad r = mn$$

   imply $64k_1^6 + 32pk_1^4 + (4p^2 - 16r)k_1^2 - q^2 = 0$.

4. Verify directly that $\{-1 + i\sqrt{2}, \ -1 - i\sqrt{2}, \ 1 + \sqrt{2}, \ 1 - \sqrt{2}\}$ is the set of solutions of $x^4 - 2x^2 - 8x - 3 = 0$.

5. Find the prime factorization of the following polynomials in $\mathbb{Q}[x]$, in $\mathbb{R}[x]$, and in $\mathbb{C}[x]$.
   a. $x^3 + 6x^2 - 3x - 148$ b. $x^4 - 2x^2 + 8x - 3$
   c. $x^4 + 3x^2 + 4$ d. $x^4 - 10x^2 + 32x - 7$

---

[4] Paolo Ruffini (1765–1833) had offered several proofs of this fact, the first as early as 1799.

e. $8x^3 + 36x^2 + 102x - 61$ f. $x^5 + 32$

g. $x^4 - 10x^2 + 32x - 7$ h. $x^4 + x^3 + x^2 + x + 1$

6. What happens in Example 17.14 if we choose $k_1 = -1$?

7. Verify Theorem 17.18.

*8. Write a computer program that accepts as input a cubic or quartic polynomial with real coefficients and returns the roots of the polynomial.

18

Vector Spaces

In this and the next chapter, we will present an introductory look at an algebraic structure that is one of the most widely applied in all of mathematics. Linear algebra is the study of vector spaces and vector space homomorphisms, commonly represented by matrices. Linear algebra, together with its alter-ego matrix theory, appears in many diverse areas. Matrices are a unifying force in mathematics since they tie together analysis, geometry, and algebra. In our brief examination of vector spaces, you will note that, although the structure is algebraic, much of the intuition comes from geometric considerations and a good deal of the methodology is analytic in nature.

In this chapter, we examine the basic definitions and properties of vector spaces. We also look at mappings that preserve the vector space structure. These mappings are vector space homomorphisms, but they are more commonly known as *linear transformations*. They are major tools not only in the study of vector spaces but also in numerous applications. In the next chapter, we will see how matrices can be used to represent these linear transformations, as well as how they can be applied to solve systems of linear equations and to develop group codes.

Definition of Vector Space

If you have ever taken a calculus course, then you probably have a knowledge of vectors in two- and three-dimensional space and the properties these vectors possess. The usefulness of these directed magnitudes motivates their study in greater detail and this analysis, in turn, requires a generalization of the terms **vector** and **vector space**. The purpose of this section is to present the formal definition of *vector space*.

In order to make this generalization, let us first introduce the properties of three-dimensional vectors from calculus. To simplify our notation, we will write our vectors in a manner similar to ordered triples, which are normally used to represent points in $\mathbb{R}^3$. If $\mathbf{v} = [v_1,\, v_2,\, v_3]$ and $\mathbf{w} = [w_1,\, w_2,\, w_3]$ are vectors in $\mathbb{R}^3$, we define

$$\mathbf{v} + \mathbf{w} = [v_1 + w_1,\, v_2 + w_2,\, v_3 + w_3]$$

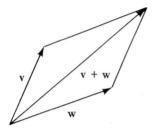

Figure 18.1

This method of adding vectors is often referred to as the **parallelogram law** because of its pictorial representation. (See Figure 18.1.)

The fact that the vectors in $\mathbb{R}^3$, together with the operation of vector addition just defined, form an abelian group is easily verified: For $\mathbf{v}, \mathbf{w}, \mathbf{u} \in \mathbb{R}^3$, we have

    i. $\mathbf{v} + \mathbf{w} = \mathbf{w} + \mathbf{v}$,
    ii. $\mathbf{v} + (\mathbf{w} + \mathbf{u}) = (\mathbf{v} + \mathbf{w}) + \mathbf{u}$,
    iii. $\mathbf{v} + \mathbf{0} = \mathbf{0} + \mathbf{v}$ where $\mathbf{0} = [0,\, 0,\, 0]$, and
    iv. $\mathbf{v} + (-\mathbf{v}) = (-\mathbf{v}) + \mathbf{v} = \mathbf{0}$, where $-\mathbf{v} = [-v_1,\, -v_2,\, -v_3]$.

There is also an operation of **scalar multiplication**[1], which is defined as followed:

If $a \in \mathbb{R}$ and $\mathbf{v} = [v_1,\, v_2,\, v_3]$, then $a\mathbf{v} = [av_1,\, av_2,\, av_3]$

Scalar multiplication is **not** an operation on $\mathbb{R}^3$; rather, it is a mapping from $\mathbb{R} \times \mathbb{R}^3$ to $\mathbb{R}^3$, that is, a way of combining a real number with a vector in $\mathbb{R}^3$ to obtain another vector in $\mathbb{R}^3$. It is easily verified that scalar multiplication satisfies the following properties:

If $a, b \in \mathbb{R}$, and $\mathbf{v}, \mathbf{w} \in \mathbb{R}^3$, then

$$(a + b)\mathbf{v} = a\mathbf{v} + b\mathbf{v}$$

$$(ab)\mathbf{v} = a(b\mathbf{v})$$

$$a(\mathbf{v} + \mathbf{w}) = a\mathbf{v} + b\mathbf{w}$$

$$1\mathbf{w} = \mathbf{w}$$

---

[1]Note that this operation of scalar multiplication is symbolized by juxtaposition.

It can be shown that these properties also hold for vectors in $\mathbb{R}^n$, for $n \geqslant 4$, if similar definitions are made for vector addition and scalar multiplication.

To enable us to study vectors in $\mathbb{R}^2$, $\mathbb{R}^3$, and $\mathbb{R}^n$, simultaneously, we will generalize these concepts. Let us utilize F to denote an arbitrary field with identity 0 and unity 1. You may want to review the definition and properties of fields in Chapter 13.

Definition 18.1

> A **vector space** V over a field F is an abelian group $(V, \oplus)$, together with a mapping $\circ: F \times V \to V$, satisfying the following properties for all a, $b \in F$ and all $\mathbf{v}$, $\mathbf{w} \in V$:
>
>   i. $(a + b) \circ \mathbf{v} = (a \circ \mathbf{v}) + (b \circ \mathbf{v})$,
> ii. $a \circ (\mathbf{v} \oplus \mathbf{w}) = (a \circ \mathbf{v}) \oplus (b \circ \mathbf{w})$,
> iii. $(a \cdot b) \circ \mathbf{v} = a \circ (b \circ \mathbf{v})$, and
> iv. $1 \circ \mathbf{v} = \mathbf{v}$ where 1 is the unity of F.

We will call V a **vector space over** F or an F**-vector space** and denote it by $V(F)$. The elements of V are termed **vectors**. We will be consistent with our geometric notation and write these elements as lower-case letters set in boldface type. The elements of the field F are called **scalars** and will be denoted by lower-case letters. The mapping $\circ: F \times V \to V$ is termed **scalar multiplication**.

To simplify the notation, we will use juxtaposition to denote not only the multiplication operation $\cdot$ on F but also the operation of scalar multiplication. This should not prove confusing since we have been careful to distinguish between scalars and vectors. Therefore, ab is $a \cdot b$ and $a\mathbf{v}$ is $a \circ \mathbf{v}$.

Example 18.1

Let $(E, +, \cdot)$ be a field and $(F, +, \cdot)$ a subfield. Then $(E, +)$ is a vector space over $(F, +, \cdot)$ since $(E, +)$ is an abelian group. In this way, $\mathbb{R}$ can be thought of as a vector space over $\mathbb{Q}$ or rational vector space, while $\mathbb{C}$ can be considered to be a vector space over $\mathbb{R}$ or $\mathbb{Q}$. The justification of these assertions, as well as those that follow in Examples 8.2–8.6, is straightforward and is left to you as Exercises 1–6, respectively.

Example 18.2

Let $(F, +, \cdot)$ be a field. The ring of polynomials $F[x]$ is a vector space over F, where we identify the constant polynomials in $F[x]$ with the scalars in F.

Example 18.3

If $(F, +, \cdot)$ is a field, set $M_2(F)$ equal to the set of all 2×2 matrices with entries in F. Define $\oplus$ on $M_2(F)$ by

$$\begin{bmatrix} a & b \\ c & d \end{bmatrix} \oplus \begin{bmatrix} e & f \\ g & h \end{bmatrix} = \begin{bmatrix} a+e & b+f \\ c+g & d+h \end{bmatrix}$$

and scalar multiplication on $M_2(F)$ by

$$r\begin{bmatrix} a & b \\ c & d \end{bmatrix} = \begin{bmatrix} ra & rb \\ rc & rd \end{bmatrix}$$

Then $M_2(F)$ is an F-vector space.

Example 18.4

If $(F, +, \cdot)$ is a field and $\mathbf{v} = [v_1, v_2, \ldots, v_n]$, $\mathbf{w} = [w_1, w_2, \ldots, w_n] \in F^n$, we define

$$\mathbf{v} \oplus \mathbf{w} = [v_1 + w_1, v_2 + w_2, \ldots, v_n + w_n]$$

For $a \in F$, we set

$$a\mathbf{v} = [av_1, av_2, \ldots, av_n]$$

Then F^n is a vector space over F and the elements of F^n are termed **row vectors**. For $F = \mathbb{R}$ and $n = 3$, we have the example of three-dimensional vectors used to motivate our definition.

Example 18.5

Let $(F, +, \cdot)$ be a field. Set

$$F_n = \left\{ \begin{bmatrix} v_1 \\ v_2 \\ \vdots \\ v_n \end{bmatrix} \;\middle|\; v_i \in F \right\}$$

If

$$\mathbf{v} = \begin{bmatrix} v_1 \\ v_2 \\ \vdots \\ v_n \end{bmatrix}, \quad \mathbf{w} = \begin{bmatrix} w_1 \\ w_2 \\ \vdots \\ w_n \end{bmatrix} \in F_n, \quad \text{and} \quad a \in F$$

define

$$\mathbf{v} \oplus \mathbf{w} = \begin{bmatrix} v_1 + w_1 \\ v_2 + w_2 \\ \vdots \\ v_n + w_n \end{bmatrix} \quad \text{and} \quad a\mathbf{v} = \begin{bmatrix} av_1 \\ av_2 \\ \vdots \\ av_n \end{bmatrix}$$

Then F_n is a vector space over F whose elements are called **column vectors**.

Example 18.6

Let $C[0, 1]$ be the set of all real-valued continuous mappings with domain $[0, 1]$. If $f, g \in C[0, 1]$, we define

$$(f + g)(x) = f(x) + g(x)$$

Recall from calculus that the sum of two continuous functions is also continuous. With this result, it is not difficult to show that $(C[0, 1], +)$ is an abelian group. For $r \in \mathbb{R}$, we define rf to be the mapping $(rf)(x) = r(f(x))$. Then rf is also continuous and $C[0, 1]$ is a vector space over $\mathbb{R}$.

There is often a tendency to identify the additive identity of the field of scalars with the additive identity of the vector space. For example, in $\mathbb{R}^3$, students sometimes write 0 for $\mathbf{0}$. This is misleading and leads to confusion since $\mathbf{0} = [0, 0, 0]$ is a vector. Thus $0\mathbf{v}$ and $0 + 1$ make sense while $\mathbf{0}\mathbf{v}$ and $\mathbf{0} + 1$ do not. Care should be taken to avoid confusing these identities, although these elements are related to one another. This relationship is expressed in the following theorem:

Theorem 18.1

Let V be a vector space over a field F. Then the following hold:

   i. $0\mathbf{v} = \mathbf{0}$ for all $v \in V$;
   ii. $(-1)\mathbf{v} = -\mathbf{v}$ for all $v \in V$;
   iii. $a\mathbf{0} = \mathbf{0}$ for all $a \in F$; and
   iv. if $a\mathbf{v} = \mathbf{0}$, then either $a = 0$ or $\mathbf{v} = \mathbf{0}$.

Proof

In the following, let $\mathbf{v} \in V$, $a \in F$.

   i. $0\mathbf{v} \oplus \mathbf{v} = 0\mathbf{v} \oplus 1\mathbf{v} = (0 + 1)\mathbf{v} = 1\mathbf{v} = \mathbf{v} = \mathbf{0} \oplus \mathbf{v}$. Since $(V, \oplus)$ is a group, we have $0\mathbf{v} = \mathbf{0}$ by cancellation.
   ii. Note that $\mathbf{v} \oplus (-1)\mathbf{v} = 1\mathbf{v} \oplus (-1)\mathbf{v} = (1 + (-1))\mathbf{v} = 0\mathbf{v} = \mathbf{0}$ by part i. Therefore, $(-1)\mathbf{v} = -\mathbf{v}$.
   iii. Since $a\mathbf{0} \oplus a\mathbf{0} = a(\mathbf{0} \oplus \mathbf{0}) = a\mathbf{0} = a\mathbf{0} \oplus \mathbf{0}$, we conclude $a\mathbf{0} = \mathbf{0}$.
   iv. Suppose $a\mathbf{v} = \mathbf{0}$ and $a \neq 0$. By part iii, $\mathbf{0} = (a^{-1})\mathbf{0}$. Thus

$$\mathbf{0} = (a^{-1})\mathbf{0} = a^{-1}(a\mathbf{v}) = (a^{-1}a)\mathbf{v} = 1\mathbf{v} = \mathbf{v}. \qquad \square$$

The preceding result allows us to define another operation on an arbitrary vector space V over a field F. That operation is one of **vector subtraction** and is symbolized by $\ominus$. We define $\ominus$ by

$$\mathbf{v} \ominus \mathbf{w} = \mathbf{v} \oplus (-1)\mathbf{w}$$

Then Theorem 18.1 says $\mathbf{v} \ominus \mathbf{w} = \mathbf{v} \oplus (-\mathbf{w})$ and $\mathbf{v} \ominus \mathbf{v} = \mathbf{0}$ for all $\mathbf{v}, \mathbf{w} \in V$.

Exercises 18.1

1. Verify that a field $(F, +, \cdot)$ is a vector space over any of its subfields.

2. If $(F, +, \cdot)$ is a field, prove that $F[x]$ is a vector space over F.

3. If $(F, +, \cdot)$ is a field, verify the assertion in Example 18.3 that $M_2(F)$ is an F-vector space.

4. Prove that F^n is a vector space over the field F.

5. If $(F, +, \cdot)$ is a field, show that F_n is an F-vector space.

6. Prove that the set of all real-valued continuous functions with domain $[0, 1]$ constitute a real vector space under the operations defined in Example 18.6.

7. In $\mathbb{R}^4$, compute:
   a. $[1, 0, 1, 2] \ominus [2, 1, 4, 3]$
   b. $2[0, 0, 7, 3] \oplus 3[1, 0, 2, 7]$
   c. $7[1, 0, 0, 1] \oplus 8[\sqrt{2}, 0, 2, 3]$
   d. $(-4)[1, 0, -1, 3] \ominus (-2)[-1, 2, 0, 4]$
   e. $2[\pi, \sqrt{2}, 1, 3] \oplus (-1)[\pi/2, 2\sqrt{2}, 4, 7]$

8. Let V be an F-vector space, $a, b \in F$, and $\mathbf{v} \in V$. Prove if $\mathbf{v} \neq \mathbf{0}$, then $a\mathbf{v} = b\mathbf{v}$ implies $a = b$.

9. If V is the set of all real-valued differentiable functions of the form $y = f(x)$ such that $(dy/dx) + y = 0$, then define addition and scalar multiplication on V as in Example 18.6. Prove that V is a real vector space.

10. Since $(\mathbb{Z}_5, \oplus, \odot)$ is a field, we know that $(\mathbb{Z}_5)^2 = \mathbb{Z}_5 \times \mathbb{Z}_5$ is a vector space over $\mathbb{Z}_5$. Let vector addition be defined as usual, and define "vector multiplication" by

$$(\bar{a}, \bar{b}) \cdot (\bar{c}, \bar{d}) = (\overline{ac}, \overline{bd})$$

Show that $\mathbb{Z}_5$ is *not* a field under these operations.

11. Consider $\mathbb{Z}_4 \times \mathbb{Z}_4$, the direct product of the abelian group $(\mathbb{Z}_4, \oplus)$ with itself. Why is this direct product not a vector space over $\mathbb{Z}_4$? Which properties of a vector space fail to hold? Is this group a vector space over $\mathbb{Z}_2$?

12. Suppose F is a field and an operation of addition $\hat{+}$ is defined on F^2 by

$$(a, b) \hat{+} (c, d) = (b, d)$$

If scalar multiplication is defined by $r(a, b) = ra$, what properties of a vector space fail to hold for $(F^2, \hat{+})$?

13. Let F be a field, and let vector addition be defined on F^3 as in Example 18.4. If scalar multiplication is defined on F^3 by

$$a[x, y, z] = [0, az, ay]$$

what properties of a vector space no longer hold for F^3?

14. a. Let V be the set of $(x, y, z) \in \mathbb{R}^3$ satisfying the equation $2x + y - z = 0$. Show that V is a real vector space.
    b. Let V be the set of all $(x, y, z) \in \mathbb{R}^3$ that satisfy $2x + y - z = 1$. Is V a vector space over $\mathbb{R}$? Why?

15. Prove that if V is a vector space over F, $a \in F$, $a \neq 0$, and $\mathbf{u}, \mathbf{v} \in V$, then $a\mathbf{u} = a\mathbf{v}$ implies $\mathbf{u} = \mathbf{v}$.

16. Let $V = \mathbb{R}^+ = \{x \in \mathbb{R} \mid x > 0\}$. Define addition on V by $x \oplus y = xy$, that is, normal multiplication, and define scalar multiplication by $r \circ x = x^r$, that is, exponentiation. Verify that V, so defined, is a real vector space.

18.2

Subspaces and Spanning

In our investigation of groups and rings, we considered subgroups and subrings, that is, certain subsets that were themselves groups and rings. Knowledge of the subgroup structure of a group or the subring structure of the ring often provided a good deal of information about the group or ring itself. It seems reasonable that we should obtain similar results if we investigate subspaces of vector spaces. This is the major objective of this section.

Basic to the definition of a vector space V is the requirement that V be an abelian group with respect to the operation of vector addition. Any definition of vector subspace should coincide with our definition of subgroup. In addition, since a vector space possesses a scalar multiplication, our definition should also require the subset to be closed under scalar multiplication. These observations lead us to make the following definition and to formulate subsequent results.

Definition 18.2

Let $(V, \oplus)$ be a vector space over a field $(F, +, \cdot)$ with scalar product $\circ \colon F \times V \to V$. A nonempty subset S of V is a **subspace of** V if $(S, +_S)$ is a vector space over the field F with scalar product $\circ_{F \times S} \colon F \times S \to S$.

Stripped of notation, this definition says that in order for a subset S of a vector space V to be a subspace, vector sums and scalar multiples in S must be the same as they would be in V; moreover, S and V must have the same field of scalars.

Theorem 18.2

Let S be a nonempty subset of a vector space V over a field F. Then S is a subspace of V if and only if

    i. $\mathbf{u} \ominus \mathbf{v} \in S$ for all $\mathbf{u}, \mathbf{v} \in S$, and
    ii. $a\mathbf{u} \in S$ for all $a \in F$ and $\mathbf{u} \in S$

Proof

Clearly, if S is a subspace of V, then the two conditions must hold. Conversely, in Theorem 4.6, we saw that condition i was a necessary and sufficient condition for $(S, \oplus_S)$ to be a subgroup of $(V, \oplus)$. Condition ii insures that $\circ_{F \times S}$ preserves the scalar multiplication. The four properties of scalar multiplication in Definition 18.1 are then inherited by S.   □

We can combine these two conditions into one result. We leave the proof to you in Exercise 1.

Corollary 18.1

A nonempty subset S of a vector space V over a field F is a subspace of V if and only if $a\mathbf{v}_1 \oplus b\mathbf{v}_2 \in S$ for all $\mathbf{v}_1, \mathbf{v}_2 \in S$ and $a, b \in F$.

The verification of the assertions made in the following examples is left to you in the exercises.

Example 18.7

If V is an F-vector space, then both $\{\mathbf{0}\}$ and V, itself, are subspaces of V. (See Exercise 2.)

Example 18.8

Let F be a field, n a nonnegative integer, and $F^n[x]$ be the set consisting of the zero polynomial and all polynomials $p(x) \in F[x]$ such that $\deg p(x) \leqslant n$. Then $F^n[x]$ is a subspace of $F[x]$. (See Exercise 3.)

Example 18.9

Let F be a field and $p(x) \in F[x]$. Then $S = \{ap(x) \mid a \in F\}$ is a subspace of $F[x]$. (See Exercise 4.)

Example 18.10

The set of all real-valued differentiable functions on the interval $[0, 1]$ is a subspace of the set of all real-valued continuous functions on that interval. (See Exercise 5.)

Example 18.11

Let F be a field, $i \in \{1, 2, \ldots, n\}$, and $H^i = \{[x_1, x_2, \ldots, x_n] \in F^n \mid x_i = 0\}$. Then H^i is a subspace of F^n. (See Exercise 6.)

Example 18.12

If V is a vector space over a field F and $\mathbf{v} \in V$, then $\{a\mathbf{v} \mid a \in F\}$ is a subspace of V. (See Exercise 7.)

If A and B are subspaces of a vector space V, then it is a straightforward matter to verify that $A \cap B$ is also a subspace of V. (See Exercise 8.) However, this process is not the only way to combine known subspaces to obtain new ones. Consider the next result.

Theorem 18.3

Let V be a vector space over a field F with A and B subspaces of V. Then

$$A + B = \{\mathbf{a} \oplus \mathbf{b} \mid \mathbf{a} \in A, \mathbf{b} \in B\}$$

is a subspace of V.

Proof

Let $\mathbf{a}_1 \oplus \mathbf{b}_1$ and $\mathbf{a}_2 \oplus \mathbf{b}_2$ be elements of $A + B$ with $\mathbf{a}_1, \mathbf{a}_2 \in A$ and $\mathbf{b}_1, \mathbf{b}_2 \in B$. Since A and B are subspaces of V, Theorem 18.2 guarantees that $\mathbf{a} \ominus \mathbf{a}_2 \in A$

and $\mathbf{b}_1 \ominus \mathbf{b}_2 \in B$, so that

$$(\mathbf{a}_1 \oplus \mathbf{b}_1) \ominus (\mathbf{a}_2 \oplus \mathbf{b}_2) = (\mathbf{a}_1 \oplus \mathbf{b}_1) \oplus - (\mathbf{a}_2 \oplus \mathbf{b}_2)$$
$$= (\mathbf{a}_1 \oplus \mathbf{b}_1) \oplus ((-\mathbf{a}_2) \oplus (-\mathbf{b}_2))$$
$$= (\mathbf{a}_1 \oplus (-\mathbf{a}_2)) \oplus (\mathbf{b}_1 \oplus (-\mathbf{b}_2))$$
$$= (\mathbf{a}_1 \ominus \mathbf{a}_2) \oplus (\mathbf{b}_1 \ominus \mathbf{b}_2) \in A + B$$

Similarly, if $r \in F$ and $\mathbf{a} \oplus \mathbf{b} \in A + B$, then $r\mathbf{a} \in A$ and $r\mathbf{b} \in B$, so that $r(\mathbf{a} \oplus \mathbf{b}) = r\mathbf{a} \oplus r\mathbf{b} \in A + B$. □

The notation $A + B$, rather than $A \cup B$, is used to denote this subspace because $A \cup B$ has a different meaning. In general, $A \cup B$ fails to be a subspace of V. Note that $A + B \neq A \cup B$ unless $A \subseteq B$ or $B \subseteq A$. (See Exercise 9.) If F is a finite field of scalars and A and B contain but a finite number of vectors, then the number of vectors in the space $A + B$ is finite. (See Exercise 10.)

One important question concerning subspaces of the form $A \oplus B$ is whether or not there is more than one way to write an element $\mathbf{w} \in A + B$ as a sum of two vectors, one from A and one from B.

| Theorem 18.4 |

If A and B are subspaces of a vector space V, then each element of $A + B$ has a unique representation $\mathbf{a} \oplus \mathbf{b}$ for $\mathbf{a} \in A$, $\mathbf{b} \in B$ if and only if $A \cap B = \{\mathbf{0}\}$.

Proof

First suppose $A \cap B \neq \{\mathbf{0}\}$ and let $\mathbf{w} \in A \cap B$, $\mathbf{w} \neq \mathbf{0}$. Then for any $\mathbf{a} \in A$, we have $\mathbf{a} = \mathbf{a} \oplus \mathbf{0}$ and also $\mathbf{a} = (\mathbf{a} \ominus \mathbf{w}) \oplus \mathbf{w} \in A + B$. Therefore, $\mathbf{a}$ cannot be written in a unique way.

Conversely, if $A \cap B = \{\mathbf{0}\}$ and $\mathbf{a}_1 \oplus \mathbf{b}_1 = \mathbf{a}_2 \oplus \mathbf{b}_2$, where $\mathbf{a}_1, \mathbf{a}_2 \in A$ and $\mathbf{b}_1, \mathbf{b}_2 \in B$. Then

$$\mathbf{a}_1 \ominus \mathbf{a}_2 = \mathbf{b}_2 \ominus \mathbf{b}_1 \in A \cap B$$

which implies $\mathbf{a}_1 \ominus \mathbf{a}_2 = \mathbf{0}$ and $\mathbf{b}_2 \ominus \mathbf{b}_1 = \mathbf{0}$. Therefore, $\mathbf{a}_1 = \mathbf{a}_2$ and $\mathbf{b}_1 = \mathbf{b}_2$. The result follows. □

Theorems 18.3 and 18.4 motivate two very important questions concerning vector spaces:

1. How many vectors do you need so that every other vector is a sum of scalar multiples of these vectors?
2. If such a set exists, then under what conditions is a representation of a vector as a sum of scalar multiples of these vectors unique?

These two questions are answered through the study of spanning sets and bases. We will discuss spanning sets in this section and bases in the next.

Definition 18.3

Let $\{v_1, v_2, \ldots, v_n\}$ be a set of vectors contained in the F-vector space V. If $c_1, c_2, \ldots, c_n \in F$, then the vector $c_1v_1 \oplus c_2v_2 \oplus \cdots \oplus c_nv_n$ is called a **linear combination of the vectors** $v_1, v_2, \ldots, v_n$. The set of all linear combinations

$$a_1v_1 \oplus a_2v_2 \oplus \cdots \oplus a_nv_n \qquad \text{for} \quad a_i \in F$$

is denoted by the symbol $\langle v_1, v_2, \ldots, v_n \rangle$ and is known as the **span** of $\{v_1, v_2, \ldots, v_n\}$. Often the vector $v = c_1v_1 \oplus c_2v_2 \oplus \cdots \oplus c_nv_n$ is written in the form

$$\sum_{i=1}^{n} c_iv_i$$

Example 18.13

If $p(x) = c_nx^n + \cdots + c_1x + c_0 \in F[x]$, where F is a field, then $p(x)$ is a linear combination of $x^n, \ldots, x^1$, and $x^0 = 1$.

Example 18.14

If $v = [x_1, x_2, \ldots, x_n] \in \mathbb{R}^n$, then v can be written as a linear combination of the vectors in the set

$$\{[1, 0, 0, \ldots, 0], [0, 1, 0, \ldots, 0], \ldots, [0, 0, 0, \ldots, 1]\}$$

For example, $[x_1, x_2, x_3] = x_1[1, 0, 0] \oplus x_2[0, 1, 0] \oplus x_3[0, 0, 1]$.

If v is a vector in the F-vector space V and $a \in F$, then $u = av$ is a linear combination of v. The zero vector 0 is a linear combination of any nonempty, finite set of vectors.

Definition 18.4

A nonempty set of vectors S of a vector space V over a field F is said to be a **spanning set for** V (or **to span** V) if every vector in V can be written as a linear combination of a finite number of vectors in S.

Equivalently, if $S = \{v_1, v_2, \ldots, v_n\}$, then S is a spanning set for V if every $v \in V$ can be expressed in the form $v = a_1v_1 \oplus a_2v_2 \oplus \cdots \oplus a_nv_n$ for some $a_1, a_2, \ldots, a_n \in F$.

Example 18.15

If V is an F-vector space and $v \in V$, then clearly v spans the subspace $\{av \mid a \in F\}$. The set of vectors in V is a spanning set for V, itself.

Example 18.16

If F is a field, then the set $\{x^0, x^1, x^2, \ldots, x^n, \ldots\}$ spans $F[x]$.

Example 18.17

The set $\{[1, 0, 0, \dots, 0], [0, 1, 0, \dots, 0], \dots, [0, 0, 0, \dots, 1]\}$ spans F^n since

$$[x_1, x_2, \dots, x_n] = x_1[1, 0, 0, \dots, 0] \oplus x_2[0, 1, 0, \dots, 0]$$
$$\oplus \cdots \oplus x_n[0, 0, 0, \dots, 1]$$

In talking about a vector space V, it is extremely important to know the field of scalars. The set of complex numbers $\mathbb{C}$ is both a real vector space and a complex vector space. By our identification of $\mathbb{R}$ as a subfield of $\mathbb{C}$, one can show that every complex vector space is a real vector space. (See Exercise 12.) The converse is not true, however; every real vector space is not necessarily a complex vector space. For example, although $\mathbb{R}[x]$ is a real vector space, $(2 + i)(x + x^2)$ is not an element of $\mathbb{R}[x]$.

Another point needs to be made along these lines. Suppose V is an F-vector space, F' a proper subfield of F, and $S \subseteq V$. If S is an F'-vector space under the restriction of the vector addition in V to S, then it is true that S is a subspace of the F'-vector space V. However, it is wrong to assume that S is a subspace of the F-vector space V. To see this, let $V = \mathbb{R}^2$ and $S = \mathbb{Q}^2$. Then S is a $\mathbb{Q}$-vector space, and $\mathbb{R}^2$ is a $\mathbb{Q}$-vector space; but S is not a real vector subspace of V since $\sqrt{2}[1, 2] \notin S$.

A fundamental relationship between spanning sets and subspaces is suggested by the next result.

Theorem 18.5

Let $S = \{v_1, v_2, \dots, v_n\}$ be a nonempty subset of an F-vector space V. Then $S = \langle v_1, v_2, \dots, v_n \rangle$ is a subspace of V.

Proof

Let

$$v = \sum_{i=1}^{n} a_i v_i \qquad \text{and} \qquad w = \sum_{i=1}^{n} b_i v_i$$

be elements of S. Then

$$v \ominus w = \sum_{i=1}^{n} a_i v_i \ominus \sum_{i=1}^{n} b_i v_i = \sum_{i=1}^{n} (a_i - b_i) v_i$$

Therefore, $v \ominus w$ is an element of S.

Similarly, if $a \in F$, then

$$av = a \sum_{i=1}^{n} a_i v_i = \sum_{i=1}^{n} (aa_i) v_i$$

is also an element of S.

To apply Theorem 18.2 and conclude that S is a subspace of V, we need only check that S is nonempty. But $\mathbf{0}$ is clearly a trivial linear combination of the elements of S. Thus S is nonempty and, so, a subspace of V. $\qquad\square$

By Example 18.16, $F[x]$ is a vector space spanned by an infinite number of vectors. It can be shown that $F[x]$ *cannot* be spanned by any finite set of polynomials. (See Exercise 20.) Other vector spaces, such as F^n, *can* be spanned by a finite set of vectors. This distinction motivates the following definition:

Definition 18.5

> An F-vector space V is termed **finite-dimensional** (and said to have **finite dimension**) if there exists some finite spanning set for V. If no such finite spanning set exists, V is termed **infinite dimensional**.

In our treatment of vector spaces, we will concentrate almost exclusively on finite-dimensional vector spaces. You should not feel shortchanged, however, for the theory of finite-dimensional vector spaces is rich in structure and applications. Moreover, many of the definitions and ideas of finite-dimensional vector spaces can be extended to infinite-dimensional vector spaces. We will indicate a number of these extensions in subsequent exercise sets and leave others to more advanced courses in algebraic structures and real variable theory.

As an example of some immediate results concerning finite-dimensional vector spaces, we list the following three theorems. The proofs are left to Exercises 17, 18, and 19, respectively.

Theorem 18.6

If A and B are subsets of the F-vector space V such that $\{v_1, v_2, \ldots, v_n\}$ spans A and $\{w_1, w_2, \ldots, w_m\}$ spans B, then

$$\{v_1, v_2, \ldots, v_n, w_1, w_2, \ldots, w_m\}$$

spans $A + B$.

Theorem 18.7

If $\{v_1, v_2, \ldots, v_n\}$ spans the F-vector space V and $v \in V$, then $\{v_1, v_2, \ldots, v_n, v\}$ spans V.

Theorem 18.8

If S is a subset of the F-vector space V and $\{v_1, v_2, \ldots, v_n\} \subseteq S$, then $\langle v_1, v_2, \ldots, v_n \rangle$ is contained in the span of S.

In the next section, we will investigate the concept of dimension in more detail.

Exercises 18.2

1. Prove Corollary 18.1: A nonempty subset S of a vector space V over a field F is a subspace of V if and only if $a\mathbf{v}_1 \oplus b\mathbf{v}_2 \in S$ for all $\mathbf{v}_1, \mathbf{v}_2 \in S$ and $a, b \in F$.

2. Verify that both V and $\{\mathbf{0}\}$ are subspaces of a vector space V.

3. Prove the assertion in Example 18.8 that $F^n[x]$ is a subspace of $F[x]$. Is the set of all polynomials with degree greater than n a subspace of $F[x]$?

4. Prove that if $p(x) \in F[x]$, where F is a field, then $\{ap(x) | a \in F\}$ is a subspace of $F[x]$.

5. Show that the set of all real-valued differentiable mappings on $[0, 1]$ is a subspace of the set of all real-valued continuous mappings on $[0, 1]$.

6. Verify that if $1 \leqslant i \leqslant n$ and $H^i = \{[x_1, x_2, \ldots, x_n] | x_i = 0\}$ is a subspace of F^n.

7. If V is an F-vector space and $\mathbf{v} \in V$, prove that $\{a\mathbf{v} | a \in F\}$ is a subspace of V. How is this exercise related to Exercise 4?

8. Prove that if A and B are both subspaces of the F-vector space V, then $A \cap B$ is also a subspace.

9. If A and B are subspaces of the F-vector space V, prove that $A \cup B$ is a subspace of V if and only if $A \subseteq B$ or $B \subseteq A$.

10. Suppose A and B are subspaces of the F-vector space V, and V has but a finite number of vectors. Verify

$$|A + B| = |A| + |B| - |A \cap B|$$

11. Prove that if A and B are both subspaces of the F-vector space V, then $A + B = A$ if and only if B is a subspace of A.

12. Verify that if V is a vector space over F and F' is a subfield of F, then V is a vector space over F'.

13. Prove that $\{[1, 0, 0], [0, 1, 0], [0, 0, 1]\}$ spans $\mathbb{R}^3$.

14. Given that $\{[1, 0, -1], [0, 2, 1], [-1, 3, 1]\}$ spans $\mathbb{R}^3$, determine how to express each of the following as a linear combination of these vectors:
    a. $[2, -1, 3]$ b. $[7, 4, 9]$
    c. $[3, -4, 5]$ d. $[1, 1, 0]$

15. Construct an example to show that if the set of vectors S spans the F-vector space V and the set T also spans V, then it is possible that $S \cap T = \emptyset$.

16. Determine a finite spanning set for $M_2(\mathbb{R})$, the space of all 2×2 real matrices.

17. Verify Theorem 18.6: If A and B are subsets of the F-vector space V such that $\{\mathbf{v}_1, \mathbf{v}_2, \ldots, \mathbf{v}_n\}$ spans A and $\{\mathbf{w}_1, \mathbf{w}_2, \ldots, \mathbf{w}_m\}$ spans B, then

$$\{\mathbf{v}_1, \mathbf{v}_2, \ldots, \mathbf{v}_n, \mathbf{w}_1, \mathbf{w}_2, \ldots, \mathbf{w}_m\}$$

spans $A + B$.

18. Prove Theorem 18.7: If $\{\mathbf{v}_1, \mathbf{v}_2, \ldots, \mathbf{v}_n\}$ spans the F-vector space V and $\mathbf{v} \in V$, then $\{\mathbf{v}_1, \mathbf{v}_2, \ldots, \mathbf{v}_n, v\}$ spans V.

19. Verify Theorem 18.8: If S is a subset of the F-vector space V and the set $\{\mathbf{v}_1, \mathbf{v}_2, \ldots, \mathbf{v}_n\} \subseteq S$, then $\langle \mathbf{v}_1, \mathbf{v}_2, \ldots, \mathbf{v}_n \rangle$ is contained in the span of S.

20. Prove that $F[x]$ cannot be spanned by a finite number of polynomials in $F[x]$.

18.3

Linear Independence and Bases

If V is an F-vector space, it is often possible to obtain several representations for a given vector in terms of the elements of a spanning set for V. This does not always appeal to mathematicians since conciseness and uniqueness of expression are important considerations. To aid in determining uniqueness of vector representations, we will first study how the zero vector $\mathbf{0}$ can be written in terms of the elements of a spanning set. We then generalize these concepts.

If $\mathbf{v}_1, \mathbf{v}_2, \ldots, \mathbf{v}_n$ are vectors in an F-vector space, then $\mathbf{0}$ can always be written as a linear combination of these vectors:

$$\mathbf{0} = 0\mathbf{v}_1 \oplus 0\mathbf{v}_2 \oplus \cdots \oplus 0\mathbf{v}_n$$

If there is another representation for $\mathbf{0}$ in terms of these vectors, that is, if this way of writing $\mathbf{0}$ is not unique, then

$$\mathbf{0} = a_1\mathbf{v}_1 \oplus a_2\mathbf{v}_2 \oplus \cdots \oplus a_n\mathbf{v}_n$$

for $a_i \in F$, where $a_i \neq 0$ for some $i = 1, 2, \ldots, n$. We can assume without loss of generality that $a_1 \neq 0$. Then

$$a_1\mathbf{v}_1 = (-a_2)\mathbf{v}_2 \oplus (-a_3)\mathbf{v}_3 \oplus \cdots \oplus (-a_n)\mathbf{v}_n$$

Since F is a field, $a_1^{-1} \in F$ and therefore

$$\mathbf{v}_1 = (-a_1^{-1}a_2)\mathbf{v}_2 \oplus (-a_1^{-1}a_3)\mathbf{v}_3 \oplus \cdots \oplus (-a_1^{-1}a_n)\mathbf{v}_n$$

The preceding discussion shows that if the representation is not unique, then one of the vectors is a linear combination of the others. The converse is also true: if one of the vectors $\mathbf{v}_1, \mathbf{v}_2, \ldots, \mathbf{v}_n$ is a linear combination of the others, then the zero vector does not have a unique representation as a linear combination of all these vectors. (See Exercise 1.) We therefore make the following definition:

Definition 18.6

A set of vectors $\{\mathbf{v}_1, \mathbf{v}_2, \ldots, \mathbf{v}_n\}$ of the F-vector space V is a **linearly independent set** (and termed **linearly independent** or, simply, **independent**) if

$$a_1\mathbf{v}_1 \oplus a_2\mathbf{v}_2 \oplus \cdots \oplus a_n\mathbf{v}_n = \mathbf{0} \qquad \text{implies} \qquad a_1 = a_2 = \cdots = a_n = 0$$

If $\{\mathbf{v}_1, \mathbf{v}_2, \ldots, \mathbf{v}_n\}$ is not linear independent, then the set is a **linearly dependent set** (and the vectors are called **linearly dependent** or, simply, **dependent**.) In this case, there is a set of scalars $a_1, a_2, \ldots, a_n$ such that $a_j \neq 0$ for some j and

$$a_1\mathbf{v}_1 \oplus a_2\mathbf{v}_2 \oplus \cdots \oplus a_n\mathbf{v}_n = \mathbf{0}$$

Intuitively, a set of vectors is linearly independent if the zero vector can be expressed as a linear combination of these vectors in one and only one way. Similarly a set of vectors is linearly dependent if the zero vector can be expressed in more than one way as a linear combination of these vectors.

Example 18.18

If F is a field, then the set of vectors

$$\{[1, 0, 0, \ldots, 0], [0, 1, 0, \ldots, 0], \ldots, [0, 0, 0, \ldots, 1]\}$$

is a linearly independent subset of F^n, since if

$$a_1[1, 0, 0, \ldots, 0] \oplus a_2[0, 0, 0, \ldots, 1] \oplus \cdots \oplus a_n[0, 0, 0, \ldots, 1]$$
$$= [0, 0, 0, \ldots, 0]$$

then

$$[a_1, a_2, \ldots, a_n] = [0, 0, \ldots, 0]$$

By our definition of F^n, we have $a_1 = a_2 = \cdots = a_n$.

Example 18.19

The set $\{[1, 1, 0], [1, 2, 1], [1, 0, -1]\}$ is a linearly dependent subset of $\mathbb{R}^3$. To see this, we set

$$a[1, 1, 0] \oplus b[1, 2, 1] \oplus c[1, 0, -1] = [0, 0, 0]$$

Solving the system of equations

$$a + b + c = 0$$
$$a + 2b \phantom{{}+c} = 0$$
$$\phantom{a + {}}b - c = 0$$

we see that $b = c$ and $a = -2c$. Hence if we let $c = 1$, then

$$(-2)[1, 1, 0] \oplus (1)[1, 2, 1] \oplus (1)[1, 0, -1]$$
$$= [-2, -2, 0] \oplus [1, 2, 1] \oplus [1, 0, -1]$$
$$= [-2 + 1 + 1, -2 + 2, 1 - 1]$$
$$= [0, 0, 0]$$

Thus the vectors $\{[1, 1, 0], [1, 2, 1], [1, 0, -1]\}$ are dependent.

The choice of $c = 1$ in this example was arbitrary. Any nonzero value of c would work just as well in showing that the set is dependent.

The definitions of linear dependence and independence have many consequences. We list some of them in the next result.

Theorem 18.9

Let V be an F-vector space, and let $\{v_1, v_2, \ldots, v_n\}$ be any subset of V. The following hold:

    i. $\{v_1, v_2\}$ is linearly dependent if and only if $v_1 = cv_2$ or $v_2 = cv_1$ for some $c \in F$.

    ii. If $0 \in \{v_1, v_2, \ldots, v_n\}$, then $\{v_1, v_2, \ldots, v_n\}$ is linearly dependent.

    iii. If $\{v_1, v_2, \ldots, v_n\}$ is dependent and $v \in \langle v_1, v_2, \ldots, v_n \rangle$, then v does not have a unique representation as a linear combination of the vectors $v_1, v_2, \ldots, v_n$.

    iv. If $\{v_1, v_2, \ldots, v_n\}$ has a linearly dependent subset, then $\{v_1, v_2, \ldots, v_n\}$ is dependent.

    v. If $v \in \langle v_1, v_2, \ldots, v_n \rangle$, then $\{v, v_1, v_2, \ldots, v_n\}$ is linearly dependent.

    vi. If $\{v_1, v_2, \ldots, v_n\}$ is linearly independent and $v \notin \langle v_1, v_2, \ldots, v_n \rangle$, then $\{v, v_1, v_2, \ldots, v_n\}$ is linearly independent.

Proof

We will prove parts i and iii and leave the proofs of the other parts as Exercises 3–6.

To prove part i, let us first suppose that $\{v_1, v_2\}$ is dependent. Then we can write $0 = av_1 \oplus bv_2$ for $a, b \in F$. If $a = 0$, then $0 = bv_2$ and v_2 must equal 0 since a and b cannot both equal 0. But then $v_2 = 0v_1$, and part i is true. Therefore, we can assume that $a \neq 0$. Then

$$v_1 = (a^{-1})(-b)v_2 = (-a^{-1}b)v_2$$

Conversely, suppose $v_1 = cv_2$ for some $c \in F$. Then $0 = (1)v_1 \oplus (-c)v_2$, and $\{v_1, v_2\}$ is a dependent set.

For part iii, suppose

$$a_1 v_1 \oplus a_2 v_2 \oplus \cdots \oplus a_n v_n = 0$$

where $a_i \neq 0$, and

$$v = b_1 v_1 \oplus b_2 v_2 \oplus \cdots \oplus b_n v_n$$

Then

$$v = v \oplus 0$$
$$= (b_1 v_1 \oplus b_2 v_2 \oplus \cdots \oplus b_n v_n) \oplus (a_1 v_1 \oplus a_2 v_2 \oplus \cdots \oplus a_n v_n)$$
$$= (a_1 + b_1)v_1 \oplus (a_2 + b_2)v_2 \oplus \cdots \oplus (a_n + b_n)v_n$$

Since $a_i \neq 0$, we know $b_i \neq b_i + a_i$, and we have found two different representations of v as an element of $\langle v_1, v_2, \ldots, v_n \rangle$. $\qquad\square$

We now combine the concepts of linear independence and spanning sets in the next definition.

Definition 18.7

Let V be a finite dimensional F-vector space. Then a set of vectors $\{v_1, v_2, \ldots, v_n\}$ in V is a **basis of V** if:

i. $\{v_1, v_2, \ldots, v_n\}$ spans V, and
ii. $\{v_1, v_2, \ldots, v_n\}$ is linearly independent.

Example 18.20

The set

$$\{[1, 0, 0, \ldots, 0], [0, 1, 0, \ldots, 0], \ldots, [0, 0, 0, \ldots, 1]\}$$

is a basis for F^n, termed the **usual** or **unit basis of F^n**.

Example 18.21

The set $\{1 = x^0, x^1, x^2, \ldots, x^n\}$ is a basis for $F^n[x]$. (See Exercise 7.)

It is possible to define the concept of a basis for an infinite-dimensional vector space similarly. (See Exercise 14.) Many of the results on bases of finite-dimensional vector spaces can be extended to the infinite-dimensional case.

One interesting and useful property of bases is the fact that they provide uniqueness of expression for not only the zero vector, but for every vector in the vector space.

Theorem 18.10

Let $S = \{v_1, v_2, \ldots, v_n\}$ be a basis of the F-vector space V. Then every vector of V has a unique representation as a linear combination of the vectors in S.

Proof

Let $v \in V$. Since S spans V, there exist $a_1, a_2, \ldots, a_n \in F$ such that $v = a_1 v_1 \oplus a_2 v_2 \oplus \cdots \oplus a_n v_n$. If there also exist scalars $b_1, b_2, \ldots, b_n \in F$ such that $v = b_1 v_1 \oplus b_2 v_2 \oplus \cdots \oplus b_n v_n$. Then

$$a_1 v_1 \oplus a_2 v_2 \oplus \cdots \oplus a_n v_n = b_1 v_1 \oplus b_2 v_2 \oplus \cdots \oplus b_n v_n$$

implies

$$(a_1 - b_1)v_1 \oplus (a_2 - b_2)v_2 \oplus \cdots \oplus (a_n - b_n)v_n = 0$$

The independence of S forces

$$a_1 - b_1 = a_2 - b_2 = \cdots = a_n - b_n = 0$$

and $a_1 = b_1, a_2 = b_2, \ldots, a_n = b_n$. Therefore, there is exactly one representation of v as an element of the span of S. $\square$

The preceding definition enables us to define the coordinates of a vector with respect to an arbitrary basis.

Definition 18.8

Let $\{\mathbf{v}_1, \mathbf{v}_2, \ldots, \mathbf{v}_n\}$ be a basis of V and $\mathbf{v} \in V$. If

$$\mathbf{v} = a_1 \mathbf{v}_1 \oplus a_2 \mathbf{v}_2 \oplus \cdots \oplus a_n \mathbf{v}_n$$

then the elements of the n-tuple $(a_1, a_2, \ldots, a_n)$ are termed **the coordinates of v with respect to the basis** $\{\mathbf{v}_1, \mathbf{v}_2, \ldots, \mathbf{v}_n\}$.

If $V = F^n$ and $\{\mathbf{v}_1, \mathbf{v}_2, \ldots, \mathbf{v}_n\}$ is the usual or unit basis, then the coordinates of $\mathbf{v} = [a_1, a_2, \ldots, a_n]$ with respect to the usual basis is simply $(a_1, a_2, \ldots, a_n)$. This identification is not always true if we alter the basis. To see this, look at the next example.

Example 18.22

Let $S = \{[1, 1], [1, -1]\}$. Then S is a basis of $\mathbb{R}^2$. Consider the vector $\mathbf{v} = [3, 5]$. We can write

$$[3, 5] = 4[1, 1] \oplus (-1)[1, -1]$$

Therefore the coordinate 2-tuple of $[3, 5]$ with respect to the basis S is $(4, -1)$ and the corresponding coordinates are 4 and -1.

The usefulness of coordinates will be studied in Section 19.2.

Another fundamental property of vector spaces is that every vector space has a basis. The verification of this result in the infinite-dimensional case involves concepts beyond the scope of this text. We will prove this result, however, in the finite-dimensional case.

Theorem 18.11

Every finite-dimensional vector space $V \neq \{\mathbf{0}\}$ has a basis.

Proof

Let $\{\mathbf{v}_1, \mathbf{v}_2, \ldots, \mathbf{v}_n\}$ span V. We will assume that $\mathbf{v}_i \neq \mathbf{0}$ for each $i = 1, 2, \ldots, n$. If $V = \langle \mathbf{v}_1 \rangle$, then $\{\mathbf{v}_1\}$ is a basis for V. If $V \neq \langle \mathbf{v}_1 \rangle$, consider $\mathbf{v}_2$. If $\mathbf{v}_2 \in \langle \mathbf{v}_1 \rangle$, discard $\mathbf{v}_2$. If $\mathbf{v}_2 \notin \langle \mathbf{v}_1 \rangle$, then Theorem 18.9 (vi) asserts that $\{\mathbf{v}_1, \mathbf{v}_2\}$ is an independent set. If $V = \langle \mathbf{v}_1, \mathbf{v}_2 \rangle$, then we are done; if not, consider $\mathbf{v}_3$. As before, if $\mathbf{v}_3 \in \langle \mathbf{v}_1, \mathbf{v}_2 \rangle$, discard it; otherwise we can again apply Theorem 18.9 (vi) to conclude that $\{\mathbf{v}_1, \mathbf{v}_2, \mathbf{v}_3\}$ is an independent set. We can then proceed by induction, and eventually consider each element in the spanning set, either keeping or discarding it. The set that remains will be a basis of V. $\square$

$V = \{\mathbf{0}\}$ is excluded in Theorem 18.11 because the set $\{\mathbf{0}\}$ is linearly dependent. We assumed the spanning set used in the proof contained only nonzero vectors inasmuch as any spanning set containing $\mathbf{0}$ is automatically dependent. (See Exercise 3.)

The proof of Theorem 18.11 contains an algorithm for obtaining a basis of a vector space V from a spanning set of nonzero vectors $\{v_1, v_2, \ldots, v_n\}$:

For $i = 2, 3, \ldots, n - 1$, if $v_i \in \langle v_1, v_2, \ldots, v_{i-1} \rangle$, discard v_i from the spanning set; otherwise, keep v_i in the spanning set. If $i \neq n$, proceed to the next value of i.

It would be helpful to know if there is a fixed number of vectors in any basis of a vector space V, and even more helpful if we knew how many linearly independent vectors were necessary to guarantee that such a set forms a basis. The following theorem gives us a method for answering both of these questions:

| Theorem 18.12 |

Steinitz Replacement Theorem[2]

If $S = \{v_1, v_2, \ldots, v_n\}$ is a subset of the F-vector space V and $\{w_1, w_2, \ldots, w_k\}$ is a linearly independent subset of $W = \langle v_1, v_2, \ldots, v_n \rangle$, the span of S, then $k \leqslant n$ and there exist k of the v_i's (say, $v_1, v_2, \ldots, v_k$) such that

$$W = \langle w_1, w_2, \ldots, w_k, v_{k+1}, \ldots, v_n \rangle$$

Proof

We will use induction on the value of k. If $k = 1$, then, since $w_1 \neq 0$, we have

$$w_1 = a_1 v_1 \oplus a_2 v_2 \oplus \cdots \oplus a_n v_n$$

and $a_i \neq 0$ for some i. Without loss of generality, we can assume that $a_1 \neq 0$. Solving for v_1, we have

$$v_1 = a_1^{-1}(w_1 \ominus (a_2 v_2 \oplus \cdots \oplus a_n v_n))$$

But then $v_1 \in \langle w_1, v_2, \ldots, v_n \rangle$ and $\langle v_1, v_2, \ldots, v_n \rangle \subseteq \langle w_1, v_2, \ldots, v_n \rangle$. Since $w_1 \in W$ implies $\langle w_1, v_2, \ldots, v_n \rangle \subseteq \langle v_1, v_2, \ldots, v_n \rangle$ by Theorem 18.8, we have

$$\langle v_1, v_2, \ldots, v_n \rangle = \langle w_1, v_2, \ldots, v_2 \rangle$$

and the result is true for $k = 1$.

Let us now assume the result true for $k = j$ and prove it true for $k = j + 1$. By our induction hypothesis, we have

$$\langle w_1, w_2, \ldots, w_j, v_{j+1}, \ldots, v_n \rangle = \langle v_1, v_2, \ldots, v_n \rangle$$

Since $w_{j+1} \in \langle v_1, v_2, \ldots, v_n \rangle = \langle w_1, \ldots, w_j, v_{j+1}, \ldots, v_n \rangle$, there exist scalars b_i, not all zero, such that

$$w_{j+1} = b_1 w_1 \oplus \cdots \oplus b_j w_j \oplus b_{j+1} v_{j+1} \oplus \cdots \oplus b_n v_n$$

---

[2] E. Steinitz (1871–1928) was a German mathematician whose fundamental contributions to the theory of fields gave a strong impulse to the abstraction of algebra.

Since the set $\{w_1, w_2, \ldots, w_k\}$ is linearly independent, it follows that $b_i \neq 0$ for some $j + 1 \leqslant i \leqslant n$. We can assume (by suitable renumbering) that $b_{j+1} \neq 0$. Solving for v_{j+1}, we have

$$v_{j+1} = b_{j+1}^{-1}(w_{j+1} \ominus (b_1 w_1 \oplus \cdots \oplus b_j w_j \oplus b_{j+2} v_{j+2} \oplus \cdots \oplus b_n v_n))$$

Therefore $v_{j+1} \in \langle w_1, \ldots, w_j, w_{j+1}, v_{j+2}, \ldots, v_n \rangle$ and, as before,

$$\langle w_1, \ldots, w_j, w_{j+1}, v_{j+2}, \ldots, v_n \rangle = \langle v_1, v_2, \ldots, v_n \rangle$$

(See Exercise 11.)

Therefore by induction, we may replace k of the v_i's by the linearly independent vectors $w_1, w_2, \ldots, w_k$ and have a set which still spans $W = \langle v_1, v_2, \ldots, v_n \rangle$. $\square$

The following results are really corollaries of the Steinitz Replacement Theorem, but we feel they are important enough to be considered as theorems in their own right.

Theorem 18.13

If an F-vector space V has a basis of n elements, then every spanning set has at least n elements.

Proof

If $\{v_1, v_2, \ldots, v_n\}$ is a basis and $\{w_1, w_2, \ldots, w_k\}$ spans V, then the Steinitz Replacement Theorem says that n of the w_i's may be replaced by the v_j's. This is only possible if $k \geqslant n$. $\square$

Theorem 18.14

If an F-vector space V has a basis of n elements, then every basis has n elements.

Proof

If $\{v_1, v_2, \ldots, v_n\}$ is a basis and $\{w_1, w_2, \ldots, w_k\}$ is another basis, then the last result implies that $k \leqslant n$ and $n \leqslant k$. We conclude that $n = k$. $\square$

Theorem 18.15

If $\{v_1, v_2, \ldots, v_n\}$ is a basis for the F-vector space V and $\{w_1, w_2, \ldots, w_k\}$ spans V, then $n = k$ if and only if $\{w_1, w_2, \ldots, w_k\}$ is a basis.

Proof

If $\{w_1, w_2, \ldots, w_k\}$ is a basis, then by the preceding theorem $k = n$. Suppose $\{w_1, w_2, \ldots, w_k\}$ is not a basis, then $k \geqslant n$ and $\{w_1, w_2, \ldots, w_k\}$ is a linearly dependent set. Then one of the w_i's is a linear combination of the rest. (See Exercise 12.) We will assume without loss of generality that w_k is a linear combination of $\{w_1, w_2, \ldots, w_{k-1}\}$. Then

$$\langle w_1, w_2, \ldots, w_{k-1} \rangle = \langle w_1, w_2, \ldots, w_k \rangle$$

Therefore $k - 1 \geqslant n$ by Theorem 18.11 and $k > n$. $\square$

Theorem 18.14 shows that there is a unique nonnegative integer associated with every finite-dimensional vector space. This integer is the number of elements in a basis.

Definition 18.9

If V is an F-vector space and V has a basis possessing n elements of V, then n is termed the **dimension of V over F**, and we write $n = \dim_F(V)$.[3] If $V = \{\mathbf{0}\}$, we will set $\dim_F(V) = 0$.

It is important to point out that the dimension of a vector space V is fundamentally related to its field of scalars. The complex number system is a field and, as such, is a complex vector space. As a complex vector space, its dimension is one. (Any nonzero complex number serves as a basis.) As a real vector space, the complex number system has dimension two (1 and i form a basis). Thus, it is important to keep in mind the field over which one is working.

While the next result is another consequence of the Steinitz Replacement Theorem, we give a proof that demonstrates a practical method for extending or completing an independent set to a basis.

Theorem 18.16

If $\{\mathbf{v}_1, \mathbf{v}_2, \ldots, \mathbf{v}_k\}$ is a linearly independent subset of an F-vector space V of dimension n, then there exist $\mathbf{w}_1, \mathbf{w}_2, \ldots, \mathbf{w}_{n-k} \in V$ such that

$$\{\mathbf{v}_1, \ldots, \mathbf{v}_k, \mathbf{w}_1, \ldots, \mathbf{w}_{n-k}\}$$

is a basis of V.

Proof

Let $V_0 = \langle \mathbf{v}_1, \mathbf{v}_2, \ldots, \mathbf{v}_k \rangle$. If $V_0 = V$, then $n = k$ and the result holds. If $V_0 \neq V$, let $\mathbf{w}_1 \in V - V_0$. By Theorem 18.9 (vi), $\{\mathbf{v}_1, \ldots, \mathbf{v}_k, \mathbf{w}_1\}$ is linearly independent. Consider $V_1 = \langle \mathbf{v}_1, \ldots, \mathbf{v}_k, \mathbf{w}_1 \rangle$. If $V_1 = V$, we are done.

If $V_1 \neq V$, then we proceed by induction. Assume $\{\mathbf{v}_1, \ldots, \mathbf{v}_k, \mathbf{w}_1, \ldots, \mathbf{w}_m\}$ is independent. If $V_m = \langle \mathbf{v}_1, \ldots, \mathbf{v}_k, \mathbf{w}_1, \ldots, \mathbf{w}_m \rangle = V$, then we are done. If $V_m \neq V$, then let $\mathbf{w}_{m+1} \in V - V_m$. Again $\{\mathbf{v}_1, \ldots, \mathbf{v}_k, \mathbf{w}_1, \ldots, \mathbf{w}_m, \mathbf{w}_{m+1}\}$ is independent by Theorem 18.9 (vi). Continuing in this way, we eventually obtain a set $\{\mathbf{v}_1, \ldots, \mathbf{v}_k, \mathbf{w}_1, \ldots, \mathbf{w}_{n-k}\}$ of linearly independent vectors, which must be a basis of V. $\qquad\square$

Theorem 18.16 says that every linearly independent subset of a finite-dimensional vector space can be extended or completed to a basis. The

---

[3]Where there is no danger of confusion, we will simply write "$\dim(V)$" or "$\dim V$" for "$\dim_F(V)$".

following is a straightforward consequence of this theorem. The proof of this result is left to you as Exercise 13.

Corollary 18.2

Let V be an F-vector space of dimension n. Then:

    i. Every linearly independent subset of n elements of V is a basis.

    ii. If $\{\mathbf{w}_1, \mathbf{w}_2, \ldots, \mathbf{w}_k\}$ is a subset of V and $k > n$, $\{\mathbf{w}_1, \mathbf{w}_2, \ldots, \mathbf{w}_k\}$ is linearly dependent.

We conclude this section by relating the idea of dimensionality to the concept of a sum of subspaces. This is useful if we are given a basis of a subspace A and a basis of a subspace B and asked to find a basis of $A + B$ or $A \cap B$.

Theorem 18.17

Let V be a finite dimensional F-vector space, and let A and B be subspaces of V. Then

$$\dim_F(A + B) = \dim_F(A) + \dim_F(B) - \dim_F(A \cap B)$$

Proof

Let $\{\mathbf{v}_1, \mathbf{v}_2, \ldots, \mathbf{v}_k\}$ be a basis of $A \cap B$. By Theorem 18.16, we know there exists $\{\mathbf{w}_1, \mathbf{w}_2, \ldots, \mathbf{w}_m\} \subseteq V$ such that $V_1 = \{\mathbf{v}_1, \ldots, \mathbf{v}_k, \mathbf{w}_1, \ldots, \mathbf{w}_m\}$ is a basis of A, and there exists $\{\mathbf{u}_1, \mathbf{u}_2, \ldots, \mathbf{u}_n\} \subseteq V$ such that $V_2 = \{\mathbf{v}_1, \ldots, \mathbf{v}_k, \mathbf{u}_1, \ldots, \mathbf{u}_n\}$ is a basis of B. The assertion will be proved if we can show that

$$V_3 = \{\mathbf{v}_1, \ldots, \mathbf{v}_k, \mathbf{w}_1, \ldots, \mathbf{w}_m, \mathbf{u}_1, \ldots, \mathbf{u}_n\}$$

is a basis of $A + B$.

We first show that V_3 is a spanning set for $A + B$. Let $\mathbf{v} \in A + B$; then $\mathbf{v} = \mathbf{w} \oplus \mathbf{u}$ where $\mathbf{w} \in A$ and $\mathbf{u} \in B$. Since V_3 is a basis of A, there exist scalars $a_i \in F$ such that

$$\mathbf{w} = a_1\mathbf{v}_1 \oplus \cdots \oplus a_k\mathbf{v}_k \oplus a_{k+1}\mathbf{w}_1 \oplus \cdots \oplus a_{k+m}\mathbf{w}_m$$

Similarly, there exist scalars $b_i \in F$ such that

$$\mathbf{u} = b_1\mathbf{v}_1 \oplus \cdots \oplus b_k\mathbf{v}_k \oplus b_{k+1}\mathbf{u}_1 \oplus \cdots \oplus b_{k+n}\mathbf{u}_n$$

But then

$$\mathbf{v} = \mathbf{w} \oplus \mathbf{u}$$

$$= (a_1 + b_1)\mathbf{v}_1 \oplus \cdots \oplus (a_k + b_k)\mathbf{v}_k \oplus a_{k+1}\mathbf{w}_1$$

$$\oplus \cdots \oplus a_{k+m}\mathbf{w}_m \oplus b_{k+1}\mathbf{u}_1 \oplus \cdots \oplus b_{k+n}\mathbf{u}_n$$

so $\mathbf{v}$ lies in the span of V_3. Since the choice of $\mathbf{v}$ was arbitrary, V_3 spans V.

To show that V_3 is independent, suppose there exist scalars $a_i \in F$ such that

$$\mathbf{0} = a_1\mathbf{v}_1 \oplus \cdots \oplus a_k\mathbf{v}_k \oplus a_{k+1}\mathbf{w}_1 \oplus \cdots \oplus a_{k+m}\mathbf{w}_m$$

$$\oplus a_{k+m+1}\mathbf{u}_1 \oplus \cdots \oplus a_{k+m+n}\mathbf{u}_n$$

Then the vector

$$\mathbf{v} = a_1\mathbf{v}_1 \oplus \cdots \oplus a_k\mathbf{v}_k \oplus a_{k+1}\mathbf{w}_1 \oplus \cdots \oplus a_{k+m}\mathbf{w}_m$$

$$= (-a_{k+m+1})\mathbf{u}_1 \oplus \cdots \oplus (-a_{k+m+n})\mathbf{u}_n$$

is contained in both A and B. Thus $\mathbf{v} \in A \cap B$. Since $\{\mathbf{v}_1, \mathbf{v}_2, \ldots, \mathbf{v}_k\}$ is a basis of $A \cap B$ and, hence, provides a unique way of expressing $\mathbf{v}$, we have

$$\mathbf{v} = b_1\mathbf{v}_1 \oplus b_2\mathbf{v}_2 \oplus \cdots \oplus b_k\mathbf{v}_k$$

So

$$\mathbf{0} = \mathbf{v} \ominus \mathbf{v} = (a_1 - b_1)\mathbf{v}_1 \oplus \cdots \oplus (a_k - b_k)\mathbf{v}_k \oplus a_{k+1}\mathbf{w}_1 \oplus \cdots \oplus a_{k+m}\mathbf{w}_m$$

But $V_1 = \{\mathbf{v}_1, \ldots, \mathbf{v}_k, \mathbf{w}_1, \ldots, \mathbf{w}_m\}$ is linearly independent; so we have

$$a_1 = b_1, a_2 = b_2, \ldots, a_k = b_k, \quad \text{and} \quad a_{k+1} = a_{k+2} = \cdots = a_{k+m} = 0$$

Since $V_2 = \{\mathbf{v}_1, \ldots, \mathbf{v}_k, \mathbf{u}_1, \ldots, \mathbf{u}_n\}$ is independent, we see from our original equation that

$$a_1 = \cdots = a_k = a_{k+m+1} = \cdots = a_{k+m+n} = 0$$

We conclude that V_3 is independent.   □

1. Let $\{\mathbf{v}_1, \mathbf{v}_2, \ldots, \mathbf{v}_n\}$ be a subset of the F-vector space V. Prove that if one of the vectors $\mathbf{v}_1, \mathbf{v}_2, \ldots, \mathbf{v}_n$ is a linear combination of the others, then the zero vector does not have a unique representation as a linear combination of these vectors.

2. Which of the following sets are independent?
    a. $\{[1, 0, 2]\}$
    b. $\{[1, 1], [2, 3]\}$
    c. $\{[2, 1], [-7, 2], [4, 6]\}$
    d. $\{[1, 2, 3], [1, 0, 2], [2, 3, 1]\}$
    e. $\{[1, 0, 2, 1], [2, 4, 3, 0], [1, -4, 3, 3]\}$

3. Prove Theorem 18.9 (ii): If an F-vector space V contains a subset $S = \{\mathbf{v}_1, \mathbf{v}_2, \ldots, \mathbf{v}_n\}$ and $\mathbf{0} \in S$, then S is dependent.

4. Prove Theorem 18.9 (iv): If $S = \{\mathbf{v}_1, \mathbf{v}_2, \ldots, \mathbf{v}_n\}$ has a linearly dependent subset, then show S itself is dependent.

5. Prove Theorem 18.9 (v): If $\{\mathbf{v}_1, \mathbf{v}_2, \ldots, \mathbf{v}_n\}$ is a subset of the F-vector space V and $\mathbf{v} \in V$, then $\mathbf{v} \in \langle \mathbf{v}_1, \mathbf{v}_2, \ldots, \mathbf{v}_n \rangle$ implies $\{\mathbf{v}, \mathbf{v}_1, \mathbf{v}_2, \ldots, \mathbf{v}_n\}$ is dependent.

6. Prove Theorem 18.9 (vi): If $\{\mathbf{v}_1, \mathbf{v}_2, \ldots, \mathbf{v}_n\}$ is independent and $\mathbf{v} \notin \langle \mathbf{v}_1, \mathbf{v}_2, \ldots, \mathbf{v}_n \rangle$, then $\{\mathbf{v}, \mathbf{v}_1, \mathbf{v}_2, \ldots, \mathbf{v}_n\}$ is independent.

7. Verify that $\{1 = x^0, x^1, x^2, \ldots, x^n\}$ is a basis of $F^n[x]$, the set of polynomials of degree at most n.

8. Let

    $$W = \langle [0, 1, 3, -1], [1, 0, -1, 2], [2, 0, 0, -1], [1, 2, -1, 1], [1, 3, 0, 1] \rangle$$

    Use the method shown in the proof of Theorem 18.11 to find a basis for W. What is the dimension of W?

9. Let $W = \langle [0, 1, -1], [1, -1, 1], [2, 0, -1] \rangle$; then $\{[1, 2, -3], [0, -3, 4]\}$ is a linearly independent subset of W. Use the Steinitz Replacement Theorem to replace two of the vectors in the given spanning set by the linearly independent vectors. (Be careful!) What is dim W?

10. If $\{[1, 2, -3], [0, -3, 4]\}$ is a linearly independent subset of $\mathbb{R}^3$, find a vector that can be added to this set to form a basis for $\mathbb{R}^3$.

11. Complete the proof of Theorem 18.12 by showing that

$$\langle \mathbf{w}_1, \ldots, \mathbf{w}_j, \mathbf{w}_{j+1}, \mathbf{v}_{j+2}, \ldots, \mathbf{v}_n \rangle = \langle \mathbf{v}_1, \mathbf{v}_2, \ldots, \mathbf{v}_n \rangle$$

12. Suppose $\{\mathbf{v}_1, \mathbf{v}_2, \ldots, \mathbf{v}_n\}$ is a linearly dependent subset of an F-vector space V. Show that at least one of the vectors in the set can be written as a linear combination of those vectors that precede it; that is, prove there exists $\mathbf{v}_i$ such that $\mathbf{v}_i \in \langle \mathbf{v}_1, \mathbf{v}_2, \ldots, \mathbf{v}_{i-1} \rangle$.

13. Prove Corollary 18.2: Let V be an F-vector space of dimension n. Then:
    a. Every linearly independent subset of n elements of V is a basis.
    b. If $\{\mathbf{w}_1, \mathbf{w}_2, \ldots, \mathbf{w}_k\}$ is a subset of V and $k > n$, then $\{\mathbf{w}_1, \mathbf{w}_2, \ldots, \mathbf{w}_k\}$ is linearly dependent.

14. Let S be a subset of V, where V is an infinite-dimensional F-vector space. S is said to be a **spanning set for** V if for every vector $\mathbf{v} \in V$, there exists a finite subset of S whose span contains $\mathbf{v}$. S is termed **linearly independent** if every finite subset of S is **linearly independent**. A **basis** for V is a linearly independent spanning set for V.
    a. Check that these definitions are consistent with the corresponding definitions given for finite-dimensional vector spaces.
    b. Show that $\{x^0, x^1, \ldots, x^n, \ldots\}$ is a basis of $F[x]$.

15. Let $A = \langle [1, 0, 2], [2, 1, 0] \rangle$ and $B = \langle [1, 1, 0], [1, 0, 1] \rangle$ be subspaces of $\mathbb{R}^3$. Find
    a. A basis of $A \cap B$
    b. $\dim_F(A + B)$
    c. A basis of $A + B$

16. Let $V = M_3(\mathbb{R})$, the set of all 3×3 real matrices.
    a. Define addition and scalar multiplication in order to make V a vector space.
    b. Verify that

$$W_1 = \left\{ \begin{bmatrix} a & 0 & a \\ 0 & b & 0 \\ a & 0 & b \end{bmatrix} \middle| a, b \in \mathbb{R} \right\} \quad \text{and}$$

$$W_2 = \left\{ \begin{bmatrix} x & 0 & y \\ 0 & y & 0 \\ z & 0 & x \end{bmatrix} \middle| x, y, z \in \mathbb{R} \right\}$$

are subspaces of V.
    c. Find $\dim(W_1 + W_2)$.

18.4

Linear Transformations

When we investigated groups and rings, we eventually reached a point where we analyzed mappings on these structures. We termed these structure-preserving mappings *homomorphisms*. Now as we study vector spaces and desire a means to compare properties on two spaces, we encounter mappings between vector spaces. The name "vector space homomorphism" is not commonly used; instead, the term *linear transformation* is utilized to describe the identical concept.

Definition 18.10

> Let $(V, \oplus)$ and $(V', \oplus')$ be two vector spaces over a common field of scalars F. A mapping $\tau: V \to V'$ is termed a **linear transformation** or **linear operator** if
>
> $$\tau(\mathbf{v}_1 \oplus \mathbf{v}_2) = \tau(\mathbf{v}_1) \oplus' \tau(\mathbf{v}_2) \qquad \text{and} \qquad \tau(c\mathbf{v}_1) = c\tau(\mathbf{v}_1)$$
>
> for all $\mathbf{v}_1, \mathbf{v}_2 \in V$ and $c \in F$.

From Definition 18.10, you can see that τ must be a group homomorphism from the abelian group $(V, \oplus)$ to the abelian group $(V', \oplus')$ and that τ must also preserve scalar multiplication. In addition, we know $\tau(\mathbf{0}) = \mathbf{0}$ since a group homomorphism takes an identity to an identity.

Example 18.23

Let $V = V' = \mathbb{Q}^2$, and define $\tau([x_1, x_2]) = [x_1 + x_2, x_1 - x_2]$. Then

$$\tau([a_1, a_2] \oplus [b_1, b_2]) = \tau([a_1 + b_1, a_2 + b_2])$$
$$= [(a_1 + b_1) + (a_2 + b_2), (a_1 + b_1) - (a_2 + b_2)]$$
$$= [(a_1 + b_1) + (a_2 + b_2), (a_1 - a_2) + (b_1 - b_2)]$$
$$= [a_1 + a_2, a_1 - a_2] \oplus [b_1 + b_2, b_1 - b_2]$$
$$= \tau([a_1, a_2]) \oplus \tau([b_1, b_2])$$

Also, for $c \in F$,

$$\tau(c[a_1, a_2]) = \tau([ca_1, ca_2])$$
$$= [ca_1 + ca_2, ca_1 - ca_2]$$
$$= c[a_1 + a_2, a_1 - a_2]$$
$$= c\tau([a_1, a_2])$$

Thus, τ is a linear transformation.

Example 18.24

Let $V = F_2$, $V' = F_3$ where F is a field. Define $\tau: V \to V'$ by

$$\tau\left(\begin{bmatrix} x_1 \\ x_2 \end{bmatrix}\right) = \begin{bmatrix} x_1 \\ -x_2 \\ x_1 + x_2 \end{bmatrix}$$

Then

$$\tau\left(\begin{bmatrix} x_1 \\ x_2 \end{bmatrix} \oplus \begin{bmatrix} y_1 \\ y_2 \end{bmatrix}\right) = \tau\left(\begin{bmatrix} x_1 + y_1 \\ x_2 + y_2 \end{bmatrix}\right)$$

$$= \begin{bmatrix} x_1 + y_1 \\ -x_2 - y_2 \\ (x_1 + y_1) + (x_2 + y_2) \end{bmatrix}$$

$$= \begin{bmatrix} x_1 \\ -x_2 \\ x_1 + x_2 \end{bmatrix} \oplus \begin{bmatrix} y_1 \\ -y_1 \\ y_1 + y_2 \end{bmatrix}$$

$$= \tau\left(\begin{bmatrix} x_1 \\ x_2 \end{bmatrix}\right) \oplus \tau\left(\begin{bmatrix} y_1 \\ y_2 \end{bmatrix}\right)$$

$$\tau\left(c\begin{bmatrix} x_1 \\ x_2 \end{bmatrix}\right) = \tau\left(\begin{bmatrix} cx_1 \\ cx_2 \end{bmatrix}\right) = \begin{bmatrix} cx_1 \\ -cx_2 \\ cx_1 + cx_2 \end{bmatrix}$$

$$= c\begin{bmatrix} x_1 \\ -x_2 \\ x_1 + x_2 \end{bmatrix} = c\tau\left(\begin{bmatrix} x_1 \\ x_2 \end{bmatrix}\right)$$

Therefore, τ is a linear transformation.

Example 18.25

Let $V = V' = \mathbb{R}[x]$ and $\tau: \mathbb{R}[x] \to \mathbb{R}[x]$ be the mapping $\tau(f) = df/dx$, the derivative of f with respect to x. In calculus, it is shown that

$$\frac{d}{dx}(f + g) = \frac{df}{dx} + \frac{df}{dx} \qquad \text{and} \qquad \frac{d}{dx}(rf) = r\frac{df}{dx}$$

for all $f, g \in \mathbb{R}[x]$ and $r \in \mathbb{R}$. Thus the differentiation operator can be viewed as a linear transformation.

Example 18.26

Let $\tau: \mathbb{R}^2 \to \mathbb{R}^2$ be the mapping defined by $\tau([x_1, x_2]) = [x_1 + x_2, x_2 + 1]$. Then τ is *not* a linear transformation, since

$$\tau([1, 0] + [1, 0]) = \tau([2, 0]) = [2, 1]$$

while

$$\tau([1, 0]) \oplus \tau([1, 0]) = [1, 1] \oplus [1, 1] = [2, 2]$$

The fact that any finite-dimensional vector space V over a field F has a basis means that in order to define a linear transformation τ on V, it is sufficient to know the images of a basis for V. For example, suppose $\{v_1, v_2, \ldots, v_n\}$ is a basis for V and $\tau(v_1) = w_i$ for $i = 1, \ldots, n$. For $v \in V$ we know there exist $a_1, a_2, \ldots, a_n \in F$ such that

$$v = \sum_{i=1}^{n} a_i v_i$$

Since τ is a linear transformation,

$$\tau(v) = \tau \left(\sum_{i=1}^{n} a_i v_i \right) = \sum_{i=1}^{n} a_i \tau(v_i) = \sum_{i=1}^{n} a_i w_i$$

Example 18.27

Let $\tau \colon \mathbb{R}^3 \to \mathbb{R}^2$ be a linear transformation with

$$\tau([2, 1, 0]) = [1, 3], \qquad \tau([-1, 0, 1]) = [0, 2],$$

and

$$\tau([1, 0, -2]) = [-1, 1]$$

Since

$$[3, 1, 0] = [2, 1, 0] \ominus 2[-1, 0, 1] \ominus [1, 0, -2]$$

we have

$$\tau([3, 1, 0]) = \tau([2, 1, 0]) \ominus 2\tau([-1, 0, -1]) \ominus \tau([1, 0, -2])$$
$$= [1, 3] \ominus 2[0, 2] \ominus [-1, 1]$$
$$= [2, -2]$$

Example 18.28

The determination of the image of a vector under a linear transformation is easier if the usual basis is used to define the transformation. For example, if $\tau \colon \mathbb{R}^2 \to \mathbb{R}^3$ is defined by

$$\tau([1, 0]) = [2, 3, -1] \qquad \text{and} \qquad \tau([0, 1]) = [4, 0, 5]$$

then we have

$$\tau([3, -2]) = 3\tau([1, 0]) \ominus 2\tau([0, 1])$$
$$= 3[2, 3, -1] \ominus 2[4, 0, 5]$$
$$= [-2, 9, -13]$$

Linear transformations are one of the most important tools in mathematics. Vector spaces have sufficient structures so that their homomorphisms can be exceedingly complicated. In the next chapter we will show how matrices can be used to simplify representations of such mapping. A detailed study of matrices is more appropriately part of a course in linear algebra. For now, we restrict ourselves to an investigation of some basic properties of linear transformations.

Theorem 18.18

If $\tau: V \to V'$ is a linear transformation between F-vector spaces, then $\tau^{\to}(W)$ is a subspace of V' for each subspace W of V, and $\tau^{\gets}(W')$ is a subspace of V for each subspace W' of V'.

Proof

If W is a subspace of V, then for $v_1, v_2 \in W$

$$a\tau(\mathbf{v}_1) + b\tau(\mathbf{v}_2) = \tau(a\mathbf{v}_1 + a\mathbf{v}_2) \in \tau^{\to}(W)$$

since $a\mathbf{v}_1 + b\mathbf{v}_2 \in W$. Thus $\tau^{\to}(W)$ is a subspace of V by Corollary 18.1.

If W' is a subspace of V', then for $\mathbf{v}_1, \mathbf{v}_2 \in \tau^{\gets}(W')$ and $a, b \in F$,

$$a\tau(\mathbf{v}_1) + b\tau(\mathbf{v}_2) = \tau(a\mathbf{v}_1 + b\mathbf{v}_2) \in W'$$

Therefore, $a\mathbf{v}_1 + b\mathbf{v}_2 \in \tau^{\gets}(W')$ and $\tau^{\gets}(W')$ is a subspace of V. □

Recall that our previous considerations of homomorphisms involved the ideas of kernels and isomorphisms. Our current investigation uses similar concepts.

Definition 18.11

Let $\tau: V \to V'$ be a linear transformation of F-vector spaces. The **null space of** τ, denoted N_τ, is the kernel of τ, that is, the set of all vectors $\mathbf{v} \in V$ such that $\tau(\mathbf{v}) = \mathbf{0}'$. The **rank space of** τ, denoted $\tau^{\to}(V)$ or simply $\tau(V)$, is the image of τ, that is, the set of all vectors of V' of the form $\tau(\mathbf{v})$ for some $\mathbf{v} \in V$. $\dim(N_\tau)$ is the **nullity of** τ and $\dim(\tau^{\to}(V))$ is the **rank of** τ.

Theorem 18.19

Let $\tau: V \to V'$ be a linear transformation of F-vector spaces. If V is finite-dimensional, then $\dim(V) = \dim(N_\tau) + \dim(\tau(V))$.

Proof

Let $\{\mathbf{u}_1, \mathbf{u}_2, \ldots, \mathbf{u}_m\}$ be a basis of N_τ. By Theorem 18.16 of the preceding section, we know we can extend $\{\mathbf{u}_1, \mathbf{u}_2, \ldots, \mathbf{u}_m\}$ to a basis

$$\{\mathbf{u}_1, \mathbf{u}_2, \ldots, \mathbf{u}_m, \mathbf{v}_1, \mathbf{v}_2, \ldots, \mathbf{v}_n\}$$

of V. Our result will hold if we can show that $\{\tau(\mathbf{v}_1), \tau(\mathbf{v}_2), \ldots, \tau(\mathbf{v}_n)\}$ is a basis of $\tau(V)$.

Let $\mathbf{v}' \in \tau(V)$; then there exists $\mathbf{v} \in V$ such that $\tau(\mathbf{v}) = \mathbf{v}'$. Since

$$\{\mathbf{u}_1, \mathbf{u}_2, \ldots, \mathbf{u}_m, \mathbf{v}_1, \mathbf{v}_2, \ldots, \mathbf{v}_n\}$$

is a basis of V, there exist scalars a_i, a_j, $i = 1, \ldots, m$, $j = 1, \ldots, n$ such that

$$\mathbf{v} = \sum_{i=1}^{m} a_i \mathbf{u}_i \oplus \sum_{j=1}^{n} a_j \mathbf{v}_j$$

Then

$$\mathbf{v}' = \tau(\mathbf{v}) = \tau \left(\sum_{i=1}^{m} a_i \mathbf{u}_i \oplus \sum_{j=1}^{n} a_j \mathbf{v}_j \right)$$

$$= \sum_{i=1}^{m} a_i \tau(\mathbf{u}_i) \oplus' \sum_{j=1}^{n} a_j \tau(\mathbf{v}_j)$$

$$= \sum_{i=1}^{m} a_i \mathbf{0}' \oplus' \sum_{j=1}^{n} a_j \tau(\mathbf{v}_j)$$

$$= \sum_{j=1}^{n} a_j \tau(\mathbf{v}_j)$$

Hence, $\{\tau(\mathbf{v}_1), \tau(\mathbf{v}_2), \ldots, \tau(\mathbf{v}_n)\}$ spans $\tau(V)$.

To complete our proof, we need only show that $\{\tau(\mathbf{v}_1), \tau(\mathbf{v}_2), \ldots, \tau(\mathbf{v}_n)\}$ is a linearly independent subset of V'. So suppose

$$\mathbf{0}' = \sum_{j=1}^{n} b_j \tau(\mathbf{v}_j)$$

Then

$$\mathbf{0}' = \tau \left(\sum_{j=1}^{n} b_j \mathbf{v}_j \right)$$

so that

$$\sum_{j=1}^{n} b_j \mathbf{v}_j \in N_\tau$$

But then there exists $a_i \in F$, $i = 1, \ldots, m$, such that

$$\sum_{j=1}^{n} b_j \mathbf{v}_j = \sum_{i=1}^{m} a_i \mathbf{u}_i$$

since $\{\mathbf{u}_1, \mathbf{u}_2, \ldots, \mathbf{u}_m\}$ is a basis for N_τ. This implies

$$\sum_{i=1}^{m} (-a)\mathbf{u}_i \oplus \sum_{j=1}^{n} b_j \mathbf{v}_j = \mathbf{0}$$

Then $a_i = b_j = 0$ for all i and j because $\{\mathbf{u}_1, \ldots, \mathbf{u}_m, \mathbf{v}_1, \ldots, \mathbf{v}_n\}$ is a basis of V and, hence, linearly independent. We conclude, since all $b_j = 0$, that $\{\tau(\mathbf{v}_1), \tau(\mathbf{v}_2), \ldots, \tau(\mathbf{v}_n)\}$ is linearly independent in V'. □

Let V be a vector space over F and W a subspace of V. Then W is a subgroup of the abelian group V, so that V/W is an abelian factor group of V. If we define scalar multiplication of V/W by $a(\mathbf{v} \oplus W) = a\mathbf{v} \oplus W$, then V/W can be shown to be an F-vector space termed the **factor space** or **quotient** space of V by W. (See Exercise 4.) Let us apply the theorem we have just proved to this construction of V/W.

Corollary 18.3

Let W be a subspace of the F-vector space V. Then, if $\dim(V)$ is finite, $\dim(V) = \dim(W) + \dim(V/W)$.

Proof

Let $\tau: V \to V/W$ be the natural map defined by $\tau(\mathbf{v}) = \mathbf{v} \oplus W$. Then τ is an epimorphism with kernel W. Substituting $W = N_\tau$ and $V/W = \tau(V)$ in Theorem 18.19 yields the desired result. $\square$

To utilize the results of Corollary 18.3, we make the following definitions:

Definition 18.12

Let $\tau: V \to V'$ be a linear transformation of F-vector spaces. If τ is a bijection, τ is said to be an **isomorphism** between the vector spaces V and V', and we write $V \cong V'$. In this case, τ is invertible and termed **nonsingular**. Similarly, if τ is not a bijection, τ is termed **singular**.

Example 18.29

Let F be a field. The mapping $\tau: F^n \to F_n$ defined by

$$\tau([x_1, x_2, \ldots, x_n]) = \begin{bmatrix} x_1 \\ x_2 \\ \vdots \\ x_n \end{bmatrix}$$

is an isomorphism. (See Exercise 5.)

If σ is a nonsingular linear transformation on the F-vector space V, then σ^{-1} is not only a mapping on V but a linear transformation as well. (See Exercise 20.)

The importance of the vector spaces of the form F^n and F_n is shown by the next result.

Theorem 18.20

Let V be an F-vector space and $\dim(V) = n$. Then V is isomorphic to F^n (and, hence, to F_n).

Proof

Let $\{\mathbf{v}_1, \mathbf{v}_2, \ldots, \mathbf{v}_n\}$ be a basis of V. Then every element $\mathbf{v} \in V$ can be uniquely expressed as

$$v = \sum_{i=1}^{n} a_i v_i \qquad \text{for} \quad a_i \in F$$

Define $\tau: V \to F_n$ by $\tau(v) = [a_1, a_2, \ldots, a_n]$. We will show that τ is the desired isomorphism.

Since every element of V is uniquely expressible as a linear combination of the v_i's, τ is both well defined and injective. Let $[b_1, b_2, \ldots, b_n] \in F^n$. Setting

$$w = \sum_{j=1}^{n} b_i v_i$$

we see $\tau(w) = [b_1, b_2, \ldots, b_n]$; thus τ is also a surjection. It remains to show that τ is a homomorphism of vector addition and scalar multiplication. Let

$$v = \sum_{i=1}^{n} a_i v_i \qquad \text{and} \qquad w = \sum_{i=1}^{n} b_i v_i$$

Then $v \oplus w = \sum_{i=1}^{n} (a_i + b_i) v_i$, so that

$$\begin{aligned}
\tau(v \oplus w) &= [a_1 + b_1, a_2 + b_2, \ldots, a_n + b_n] \\
&= [a_1, a_2, \ldots, a_n] \oplus [b_1, b_2, \ldots, b_n] \\
&= \tau(v) \oplus \tau(w)
\end{aligned}$$

Similarly,

$$\begin{aligned}
\tau(cv) &= \tau\left(c \sum_{i=1}^{n} a_i v_i \right) = \tau\left(\sum_{i=1}^{n} c a_i v_i \right) \\
&= [c a_1, c a_2, \ldots, c a_n] \\
&= c[a_1, a_2, \ldots, a_n] = c\tau(v)
\end{aligned}$$

Therefore τ is a linear transformation.   □

The usefulness of this theorem lies in the fact that whenever we wish to investigate a finite-dimensional vector space V over a field F, we can simply consider F^n or F_n where $n = \dim V$. In Section 19.1, we will discuss a convenient representation of linear transformations of these vector spaces: matrices. Thus, the identification of a vector space as F^n or F_n allows us to view all linear transformations on finite-dimensional vector spaces in a simple form.

We conclude this section by proving a very interesting result; namely, any injective linear transformation on a finite-dimensional vector space is surjective, and conversely.

Theorem 18.21   Let $\tau: V \to V$ be a linear transformation of an F-vector space V of dimension *n*. Then

> i. τ is surjective if τ is injective, and
> ii. τ is injective if τ is surjective.

Proof

We will prove i and leave ii as Exercise 6. Suppose τ is injective. Then $N_\tau = \{\mathbf{0}\}$ and $\dim(N_\tau) = 0$. By Theorem 18.19, we have $\dim(\tau(V)) = \dim(V)$. Then any basis of $\tau(V)$ is already a basis of V, since we know that any basis of $\tau(V)$ can be completed to a basis of V and $\dim(V) = \dim(\tau(V))$ implies that any basis of $\tau(V)$ has the same number of elements as a basis of V. Hence, $\tau(V) = V$, and τ is a surjection. ☐

Theorem 18.21 gives us a way to analyze the mapping underlying a linear transformation by studying its action on a basis. We leave the proof of this result to you as Exercises 11 and 12.

Corollary 18.4

A linear transformation is an isomorphism if and only if it takes a basis to a basis.

We conclude this chapter with one last remark. If $\tau\colon V \to V'$ and $\sigma\colon V' \to V''$ are linear transformations of F-vector spaces, then $\sigma\tau\colon V \to V''$ is not only a mapping but a linear transformation as well. (See Exercise 10.) We will utilize this result in Chapter 19.

Exercises 18.4

1. Determine which of the following are linear transformations from $\mathbb{R}^4$ to $\mathbb{R}^4$.
   a. $\tau_1([x, y, z, w]) = [x - y + z, y - z, w + 2z, w + z - x]$
   b. $\tau_2([x, y, z, w]) = [xy - z, y + z, z - y, z + y]$
   c. $\tau_3([x, y, z, w]) = [\log x, e^{xy}, z + w, 0]$
   d. $\tau_4([x, y, z, w]) = [x^2, y^3, x + y, z - w]$
   e. $\tau_5([x, y, z, w]) = [x - w + y + 5z, y - 2z + 6w, z, w]$

2. Prove that if $\tau\colon V \to V'$ is a linear transformation of F-vector spaces and $c \in F$, then $c\tau\colon V \to V'$ defined by $(c\tau)(\mathbf{v}) = c(\tau(\mathbf{v}))$, for $\mathbf{v} \in V$, is another transformation.

3. For the mappings of Exercise 1, give a formula for
   a. $\tau_2 + \tau_5$ b. $4\tau_5$

4. If V is a vector space over F and W is a subspace of V, define $a(\mathbf{v} \oplus W) = a\mathbf{v} \oplus W$. Prove that, with this definition of scalar multiplication, V/W is an F-vector space.

5. Let F be a field. Show that $\tau\colon F^n \to F_n$ defined by

$$\tau([x_1, x_2, \ldots, x_n]) = \begin{bmatrix} x_1 \\ x_2 \\ \vdots \\ x_n \end{bmatrix}$$

is an isomorphism.

6. Prove Theorem 18.21 (ii): Let $\tau: V \to V$ be a linear transformation on a finite dimensional vector space V. Prove that if τ is a surjection, then τ is an injection.

7. Let V be the vector space of all continuous functions and $I: V \to V$ be defined by

$$I(f) = \int f(t) \, dt \qquad \text{where} \quad f \in V$$

Show that I is a linear transformation on V. What is N_I?

8. Let $\tau: \mathbb{R}^2 \to \mathbb{R}^3$ be defined by $\tau([x_1, x_2]) = [x_1, x_1 + x_2, -x_2]$.
   a. Compute N_τ.
   b. Determine $\dim(\tau^\to(\mathbb{R}^2))$.

9. Let $\tau: \mathbb{R}^3 \to \mathbb{R}^2$ be defined by $\tau([x_1, x_2, x_3]) = [x_2 - x_1, x_3]$.
   a. What is N_τ?
   b. Determine $\dim(\tau^\to(\mathbb{R}^3))$.

10. Let V, V', V'' be F-vector spaces and $\tau: V \to V'$ and $\sigma: V' \to V''$ be linear transformations. Prove that $\sigma \circ \tau: V \to V''$ is a linear transformation.

11. Let $\tau: V \to V'$ be an isomorphism of vector spaces. Show if $\{v_1, v_2, \ldots, v_n\}$ is a basis of V, then $\{\tau(v_1), \tau(v_2), \ldots, \tau(v_n)\}$ is a basis of V'.

12. Suppose $\tau: V \to V'$ is a linear transformation and $\{v_1, v_2, \ldots, v_n\}$ is a basis of V. If $\{\tau(v_1), \tau(v_2), \ldots, \tau(v_n)\}$ is a basis of V', prove that τ is an isomorphism.

13. Let $V = \mathbb{R}[x]$ and D be the differentiation operator on V. What are N_D and $D^\to(V)$? Do the results of Theorem 18.2 hold for V?

14. Suppose $\tau: V \to V'$ is a linear transformation and that V is a finite-dimensional vector space. Show that $\tau^\to(V)$ is finite-dimensional.

15. Find the kernel, nullity, and rank of each linear transformation in Exercise 1.

16. If $\tau: \mathbb{R}^3 \to \mathbb{R}^3$ is a linear transformation defined by

$$\tau([2, 1, -1]) = [1, 0, 3], \qquad \tau([1, 2, 0]) = [-1, 1, 1]$$

and

$$\tau([0, 3, -1]) = [2, 1, 2]$$

find the image of each of the following:
   a. $[3, 6, -2]$ b. $[3, 3, -1]$
   c. $[1, 2, -1]$ d. $[4, 2, -2]$
   e. $[1, 5, -1]$ f. $[1, -1, 1]$

17. For the transformation in Exercise 16, find the kernel of τ and the nullity of τ. What is the rank of τ?

18. Let V and V' be F-vector spaces. Set

$$V \times V' = \{(v, v') \mid v \in V, v' \in V'\}$$

Define vector addition on $V \times V'$ by $(v_1, v_1') + (v_2, v_2') = (v_1 \oplus v_2, v_1' \oplus' v_2')$ and scalar multiplication on $V \times V'$ by $a(v, v') = (av, av')$.
   a. Prove that $V \times V'$, with these definitions of vector addition and scalar multiplication, forms a vector space.
   b. Show if A and B are subspaces of the finite-dimensional F-vector space V and $A \cap B = \{0\}$, then $A + B \cong A \times B$.

19. Find the kernel, nullity, and rank of each of the following linear transformations:

a. $\tau_1: \mathbb{R}^3 \to \mathbb{R}^2$ defined by $\tau_1([x, y, z]) = [x - y, y - z]$

b. $\tau_2: \mathbb{R}^2 \to \mathbb{R}^3$ defined by $\tau_2([x, y]) = [2x - 3y, 2x + y, y - 3x]$

c. $\tau_3: \mathbb{R}^2 \to \mathbb{R}^2$ defined by $\tau_3([x, y]) = [x, -y]$

d. $\tau_4: \mathbb{R}_3 \to \mathbb{R}_3$ defined by $\tau_4 \left(\begin{bmatrix} x \\ y \\ z \end{bmatrix} \right) = \begin{bmatrix} x + z \\ y + z \\ -x + 3y + 2z \end{bmatrix}$

e. $\tau_5: \mathbb{R}_3 \to \mathbb{R}_2$ defined by $\tau_5 \left(\begin{bmatrix} x \\ y \\ z \end{bmatrix} \right) = \begin{bmatrix} 2x - 4y - 2z \\ -x + 2y + z \end{bmatrix}$

20. If $\tau: V \to V$ is a nonsingular linear transformation, verify that $\tau^{-1}: V \to V$ is also a linear transformation.

21. For the transformations of Exercise 19, compute a formula for each of the following:

a. $\tau_1 \tau_2$ b. $\tau_2 \tau_1$ c. $\tau_2 \tau_3 \tau_1$ d. $\tau_5 \tau_4$

19

Linear Transformations

and Matrices

We have seen that vector spaces have an underlying group structure. This motivated us to consider vector spaces in a manner similar to groups. The concept of scalar multiplication adds another dimension to this study, however, especially as it relates to vector space homomorphisms or linear transformations. In order to simplify our consideration of linear transformations, we turn to matrices.

Matrices are the primary tool in linear algebra. A matrix is usually much easier to work with than is an arbitrary linear transformation. In Section 19.1, we show how to represent a linear transformation on a finite-dimensional vector space by a matrix. We then apply matrices in a variety of settings in the remainder of this chapter to investigate how bases of a vector space are related, to solve systems of linear equations, and to develop group codes.

Matrix Representations of Linear Transformations

Throughout this text, we have utilized the set of 2×2 matrices as an example of various algebraic structures. In this section we shall generalize that concept and show how we can view a matrix as a linear transformation between finite-dimensional vector spaces and, conversely, a linear transformation between finite-dimensional vector spaces as a matrix. Let us begin with the basic definition of a matrix:

Definition 19.1

> An $m \times n$ **matrix over a ring** R is a rectangular array consisting of m rows and n columns of elements of R. (See Figure 19.1). $R^{m \times n}$ is the set of all such $m \times n$ matrices with entries in R.

$$[a_{ij}] = \begin{bmatrix} a_{11} & a_{12} & \cdots & a_{1n} \\ a_{21} & a_{22} & \cdots & a_{2n} \\ \vdots & \vdots & \ddots & \vdots \\ a_{m1} & a_{m2} & \cdots & a_{mn} \end{bmatrix}$$

Figure 19.1

Example 19.1

The following are examples of matrices:

$$\begin{bmatrix} 1 & 3 \\ 2 & \frac{1}{2} \end{bmatrix} \in \mathbb{Q}^{2 \times 2} \qquad \begin{bmatrix} 1 & 2 & 3 \\ 0 & 1 & 4 \end{bmatrix} \in \mathbb{Z}^{2 \times 3}$$

$$[1 \quad \pi \quad 3] \in \mathbb{R}^{1 \times 3} \qquad \begin{bmatrix} i \\ -3 \\ 2+i \end{bmatrix} \in \mathbb{C}^{3 \times 1}$$

$$\begin{bmatrix} 2 & 1 \\ 3 & 2 \\ 0 & -1 \end{bmatrix} \in \mathbb{R}^{3 \times 2} \qquad \begin{bmatrix} 0 & 0 & 0 \\ 0 & 0 & 0 \\ 0 & 0 & 0 \end{bmatrix} \in \mathbb{Z}^{3 \times 3}$$

Let F be a field. Our initial objective is to develop a way of representing every linear transformation from F_n to F_m by a matrix in $F^{m \times n}$. Inasmuch as every finite-dimensional F-vector space V is isomorphic to F_n for $n = \dim(V)$, this will provide a means to represent any linear transformation between finite-dimensional F-vector spaces as a matrix.

Let us begin with a concrete example. Recall that the usual or unit base of F_n is the set of vectors

$$\left\{ \begin{bmatrix} 1 \\ 0 \\ 0 \\ \vdots \\ 0 \end{bmatrix}, \begin{bmatrix} 0 \\ 1 \\ 0 \\ \vdots \\ 0 \end{bmatrix}, \dots, \begin{bmatrix} 0 \\ 0 \\ 0 \\ \vdots \\ 1 \end{bmatrix} \right\}$$

Example 19.2

Define $\tau: \mathbb{Q}_2 \to \mathbb{Q}_2$ by

$$\tau\left(\begin{bmatrix} x_1 \\ x_2 \end{bmatrix} \right) = \begin{bmatrix} x_1 \\ x_1 + x_2 \end{bmatrix}$$

Then

$$\tau\left(\begin{bmatrix} 1 \\ 0 \end{bmatrix} \right) = \begin{bmatrix} 1 \\ 1 \end{bmatrix} \quad \text{and} \quad \tau\left(\begin{bmatrix} 0 \\ 1 \end{bmatrix} \right) = \begin{bmatrix} 0 \\ 1 \end{bmatrix}$$

We define

$$M(\tau) = \begin{bmatrix} 1 & 0 \\ 1 & 1 \end{bmatrix}$$

as **the matrix of τ with respect to the usual basis of** $\mathbb{Q}_2$. Note that the columns of $M(\tau)$ are the images of the vectors

$$\begin{bmatrix} 1 \\ 0 \end{bmatrix} \quad \text{and} \quad \begin{bmatrix} 0 \\ 1 \end{bmatrix}$$

under the actions of τ.

Example 19.3

Let us look at another example. If $\sigma: \mathbb{Q}_2 \to \mathbb{Q}_2$ is defined by

$$\sigma\left(\begin{bmatrix} x_1 \\ x_2 \end{bmatrix} \right) = \begin{bmatrix} x_1 + x_2 \\ 2x_1 \end{bmatrix}$$

then

$$\sigma\left(\begin{bmatrix} 1 \\ 0 \end{bmatrix} \right) = \begin{bmatrix} 1 \\ 2 \end{bmatrix} \quad \text{and} \quad \sigma\left(\begin{bmatrix} 0 \\ 1 \end{bmatrix} \right) = \begin{bmatrix} 1 \\ 0 \end{bmatrix}$$

We set

$$M(\sigma) = \begin{bmatrix} 1 & 1 \\ 2 & 0 \end{bmatrix}$$

We now make explicit the idea behind these examples in a more general setting.

Definition 19.2

Let F be a field, and let $\tau: F_n \to F_m$ be a linear transformation. Let $\{\mathbf{e}_1^n, \mathbf{e}_2^n, \ldots, \mathbf{e}_n^n\}$ denote the unit column vectors of F_n. Suppose

$$\tau(\mathbf{e}_1^n) = \begin{bmatrix} a_{11} \\ a_{21} \\ \vdots \\ a_{m1} \end{bmatrix}, \quad \tau(\mathbf{e}_2^n) = \begin{bmatrix} a_{12} \\ a_{22} \\ \vdots \\ a_{m2} \end{bmatrix}, \quad \ldots, \quad \tau(\mathbf{e}_n^n) = \begin{bmatrix} a_{1n} \\ a_{2n} \\ \vdots \\ a_{mn} \end{bmatrix}$$

Define

$$M(\tau) = [a_{ij}] = \begin{bmatrix} a_{11} & a_{12} & \cdots & a_{1n} \\ a_{21} & a_{22} & \cdots & a_{2n} \\ \vdots & \vdots & \ddots & \vdots \\ a_{m1} & a_{m2} & \cdots & a_{mn} \end{bmatrix}$$

to be **the matrix of τ with respect to the usual basis in F_n and F_m.**

From Definition 19.2, we see that to obtain the matrix representation of a linear transformation τ from F_n to F_m we simply take the elements in F_m of the form

$$\tau\left(\begin{bmatrix} 1 \\ 0 \\ 0 \\ \vdots \\ 0 \end{bmatrix}\right), \quad \tau\left(\begin{bmatrix} 0 \\ 1 \\ 0 \\ \vdots \\ 0 \end{bmatrix}\right), \quad \ldots, \quad \tau\left(\begin{bmatrix} 0 \\ 0 \\ 0 \\ \vdots \\ 1 \end{bmatrix}\right)$$

and set them beside one another in the proper order. The resulting array is the desired matrix. Later we will see that it is possible to have another matrix representation of τ if we choose bases of F_n and F_m other than the usual bases.

In the last chapter, we saw that if σ and τ are two linear transformations from the F-vector space V to the F-vector space V', then $\sigma + \tau$ is also a linear transformation from V to V'. (Recall that $(\sigma + \tau)(\mathbf{v}) = (\sigma)(\mathbf{v}) \oplus (\tau)(\mathbf{v})$.) In order to have our matrix representations of linear transformations consistent with transformation addition, we now need to define matrix *addition* so that $M(\sigma + \tau) = M(\sigma) + M(\tau)$. To aid our formation of this definition, let us consider the following example:

Example 19.4

Let σ and τ be linear transformations on $\mathbb{Q}_2$ as described in Examples 19.2 and 19.3. Then

$$(\sigma + \tau)\left(\begin{bmatrix} x_1 \\ x_2 \end{bmatrix}\right) = \begin{bmatrix} x_1 \\ x_1 + x_2 \end{bmatrix} + \begin{bmatrix} x_1 + x_2 \\ 2x_1 \end{bmatrix} = \begin{bmatrix} 2x_1 + x_2 \\ 3x_1 + x_2 \end{bmatrix}$$

Since

$$(\sigma + \tau)\left(\begin{bmatrix} 1 \\ 0 \end{bmatrix}\right) = \begin{bmatrix} 2 \\ 3 \end{bmatrix} \quad \text{and} \quad (\sigma + \tau)\left(\begin{bmatrix} 0 \\ 1 \end{bmatrix}\right) = \begin{bmatrix} 1 \\ 1 \end{bmatrix}$$

we have

$$M(\sigma + \tau) = \begin{bmatrix} 2 & 1 \\ 3 & 1 \end{bmatrix}$$

Inasmuch as we want $M(\sigma + \tau) = M(\sigma) + M(\tau)$, that is

$$\begin{bmatrix} 2 & 1 \\ 3 & 1 \end{bmatrix} = \begin{bmatrix} 1 & 1 \\ 2 & 0 \end{bmatrix} + \begin{bmatrix} 1 & 0 \\ 1 & 1 \end{bmatrix}$$

it seems reasonable to define *addition* of matrices as follows:

Definition 19.3

Let $A = [a_{ij}]$ and $B = [b_{ij}]$ be elements of $F^{m \times n}$, where F is a field. Then **matrix addition** is defined by setting $A + B$ equal to the matrix $C = [c_{ij}]$ defined by $c_{ij} = a_{ij} + b_{ij}$ for $i = 1, 2, \ldots, m$, and $j = 1, 2, \ldots, n$.

Note that this definition of matrix addition is an extension of that given in Section 2.2 for addition in $M_2(\mathbb{Z})$. The matrix in $F^{m \times n}$, all of whose entries are equal to the additive identity of F, is termed the **zero matrix of** $F^{m \times n}$ and is denoted by $\mathbf{0}_{m \times n}$ or simply $\mathbf{0}$.

Example 19.5

Let $F = \mathbb{R}$. Then

$$\mathbf{0}_{1 \times 1} = [0], \quad \mathbf{0}_{1 \times 2} = [0 \quad 0], \quad \text{and} \quad \mathbf{0}_{2 \times 3} = \begin{bmatrix} 0 & 0 & 0 \\ 0 & 0 & 0 \end{bmatrix}$$

We leave it to you (see Exercise 11) to show that zero matrix in $F^{m \times n}$ is an additive identity for matrix addition in $F^{m \times n}$.

Example 19.6

If

$$A = \begin{bmatrix} -1 & 2 & 0 \\ 0 & 1 & 2 \end{bmatrix} \quad \text{and} \quad B = \begin{bmatrix} 2 & -2 & 1 \\ 1 & 0 & 0 \end{bmatrix}$$

then

$$A + B = \begin{bmatrix} 1 & 0 & 1 \\ 1 & 1 & 2 \end{bmatrix}$$

If

$$S = \begin{bmatrix} 1 \\ 1 \\ 3 \end{bmatrix} \quad \text{and} \quad T = \begin{bmatrix} 2 \\ 0 \\ -1 \end{bmatrix}, \quad \text{then} \quad S + T = \begin{bmatrix} 3 \\ 1 \\ 2 \end{bmatrix}$$

Note that $A + S$ is not defined.

Matrix addition can be shown to be an associative and commutative binary operation. (See Exercise 1.) In Section 18.4, we remarked that if $\tau: F_n \to F_m$ and $\sigma: F_m \to F_k$ are linear transformations, then $\sigma \circ \tau: F_n \to F_k$ is a linear transformation. How should we combine their matrix representations if we want $M(\sigma)M(\tau) = M(\sigma \circ \tau)$; that is, how do we define matrix *multiplication* to be consistent with composition of linear transformations? Again we consider a concrete example.

Example 19.7

Let τ and σ be as in Examples 19.2 and 19.3. Then

$$M(\sigma) = \begin{bmatrix} 1 & 1 \\ 2 & 0 \end{bmatrix}, \quad M(\tau) = \begin{bmatrix} 1 & 0 \\ 1 & 1 \end{bmatrix}, \quad \text{and} \quad (\sigma \circ \tau)\left(\begin{bmatrix} x_1 \\ x_2 \end{bmatrix}\right) = \begin{bmatrix} 2x_1 + x_2 \\ 2x_1 \end{bmatrix}$$

so that

$$(\sigma \circ \tau)\left(\begin{bmatrix} 1 \\ 0 \end{bmatrix}\right) = \begin{bmatrix} 2 \\ 2 \end{bmatrix} \quad \text{and} \quad (\sigma \circ \tau)\left(\begin{bmatrix} 0 \\ 1 \end{bmatrix}\right) = \begin{bmatrix} 1 \\ 0 \end{bmatrix}$$

Therefore

$$M(\sigma \circ \tau) = \begin{bmatrix} 2 & 1 \\ 2 & 0 \end{bmatrix}$$

We hope that any definition we derive for matrix composition satisfies

$$\begin{bmatrix} 1 & 1 \\ 2 & 0 \end{bmatrix}\begin{bmatrix} 1 & 0 \\ 1 & 1 \end{bmatrix} = \begin{bmatrix} 2 & 1 \\ 2 & 0 \end{bmatrix}$$

Let us consider the general case. With $\tau: F_n \to F_m$ and $\sigma: F_m \to F_k$, let $\{e_i^n\}$, $\{e_i^m\}$, and $\{e_i^k\}$ denote the usual bases of the spaces F_n, F_m, F_k, respectively. Then

$$(\sigma \circ \tau)(\mathbf{e}_1^n) = \sigma(\tau(\mathbf{e}_1^n))$$

$$= \sigma\left(\sum_{i=1}^{m} a_{i1}\mathbf{e}_1^m\right)$$

$$= \sum_{i=1}^{m} a_{i1}\sigma(\mathbf{e}_i^m)$$

$$= \sum_{i=1}^{m}\left(a_{i1}\left(\sum_{j=1}^{k} b_{ji}\mathbf{e}_j^k\right)\right)$$

$$= \sum_{j=1}^{k}\left(\sum_{i=1}^{m} b_{ji}a_{i1}\right)\mathbf{e}_j^k$$

$$= \begin{bmatrix} \sum_{i=1}^{m} b_{1i}a_{i1} \\ \sum_{i=1}^{m} b_{2i}a_{i1} \\ \vdots \\ \sum_{i=1}^{m} b_{ki}a_{i1} \end{bmatrix}$$

Similarly,

$$(\sigma \circ \tau)(\mathbf{e}_q^n) = \begin{bmatrix} \sum_{i=1}^{m} b_{1i}a_{iq} \\ \sum_{i=1}^{m} b_{2i}a_{iq} \\ \vdots \\ \sum_{i=1}^{m} b_{ki}a_{iq} \end{bmatrix} \qquad \text{for} \quad q = 2, \ldots, n$$

Thus, we conclude

$$M(\sigma \circ \tau) = \begin{bmatrix} c_{11} & c_{12} & \cdots & c_{1n} \\ c_{21} & c_{22} & \cdots & c_{2n} \\ \vdots & \vdots & \ddots & \vdots \\ c_{k1} & c_{k2} & \cdots & c_{kn} \end{bmatrix} \qquad \text{where} \quad c_{pq} = \sum_{i=1}^{m} b_{pi}a_{iq}$$

In light of this, we make the following definition:

Definition 19.4

Let $B = [b_{pi}] \in F^{k \times m}$ and $A = [a_{iq}] \in F^{m \times n}$. We define **matrix multiplication** by setting $C = BA$, where $C = [c_{pq}] \in F^{k \times n}$ defined by

$$c_{pq} = \sum_{i=1}^{m} b_{pi} a_{iq}$$

Again our definition for matrix multiplication is consistent with the one we gave in Section 2.2 for $M_2(\mathbb{Z})$.

Example 19.8   Let

$$A = \begin{bmatrix} 1 & 0 \\ 2 & -1 \end{bmatrix} \quad \text{and} \quad B = \begin{bmatrix} 1 & 0 \\ 0 & 1 \\ 1 & 1 \end{bmatrix}$$

Then we have

$$BA = \begin{bmatrix} 1 & 0 \\ 0 & 1 \\ 1 & 1 \end{bmatrix} \begin{bmatrix} 1 & 0 \\ 2 & -1 \end{bmatrix} = \begin{bmatrix} 1 \cdot 1 + 0 \cdot 2 & 1 \cdot 0 + 0 \cdot (-1) \\ 0 \cdot 1 + 1 \cdot 2 & 0 \cdot 0 + 1 \cdot (-1) \\ 1 \cdot 1 + 1 \cdot 2 & 1 \cdot 0 + 1 \cdot (-1) \end{bmatrix}$$

$$= \begin{bmatrix} 1 & 0 \\ 2 & -1 \\ 3 & -1 \end{bmatrix}$$

Note that AB is not defined.

Example 19.9   Let

$$S = \begin{bmatrix} 1 & 1 \\ 2 & 0 \end{bmatrix} \quad \text{and} \quad T = \begin{bmatrix} 1 & 0 \\ 1 & 1 \end{bmatrix}$$

Then

$$ST = \begin{bmatrix} 1 & 1 \\ 2 & 0 \end{bmatrix} \begin{bmatrix} 1 & 0 \\ 1 & 1 \end{bmatrix} = \begin{bmatrix} 1 \cdot 1 + 1 \cdot 1 & 1 \cdot 0 + 1 \cdot 1 \\ 2 \cdot 1 + 0 \cdot 1 & 2 \cdot 0 + 0 \cdot 1 \end{bmatrix} = \begin{bmatrix} 2 & 1 \\ 2 & 0 \end{bmatrix}$$

and

$$TS = \begin{bmatrix} 1 & 0 \\ 1 & 1 \end{bmatrix} \begin{bmatrix} 1 & 1 \\ 2 & 0 \end{bmatrix} = \begin{bmatrix} 1 \cdot 1 + 0 \cdot 2 & 1 \cdot 1 + 0 \cdot 0 \\ 1 \cdot 1 + 1 \cdot 2 & 1 \cdot 1 + 1 \cdot 0 \end{bmatrix} = \begin{bmatrix} 1 & 1 \\ 3 & 1 \end{bmatrix}$$

From Example 19.9 we see that for the transformations of Example 19.7, $M(\sigma)M(\tau) = M(\sigma \circ \tau)$.

Example 19.10   Let

$$B = [1 \quad 1 \quad 3] \quad \text{and} \quad A = \begin{bmatrix} 2 \\ 0 \\ 1 \end{bmatrix}$$

Our definition then yields that

$$BA = [1 \cdot 2 + 1 \cdot 0 + 3 \cdot 1] = [5]$$

and

$$AB = \begin{bmatrix} 2 \\ 0 \\ 1 \end{bmatrix} [1 \quad 1 \quad 3] = \begin{bmatrix} 2 & 2 & 6 \\ 0 & 0 & 0 \\ 1 & 1 & 3 \end{bmatrix}$$

Let $[a_{ij}] \in F^{n \times n}$ and suppose that

$$a_{ij} = \begin{cases} 1, & i = j \\ 0, & i \neq j \end{cases}$$

where 0 and 1 are the identity and unity elements of F. Then the matrix is termed the **identity matrix of** $F^{n \times n}$ and denoted by I_n or simply I. It can be shown (see Exercise 12) that I_n is a unity for matrix multiplication in $F^{n \times n}$.

Example 19.11   If $F = \mathbb{R}$, then

$$I_1 = [1], \quad I_2 = \begin{bmatrix} 1 & 0 \\ 0 & 1 \end{bmatrix}, \quad \text{and} \quad I_3 = \begin{bmatrix} 1 & 0 & 0 \\ 0 & 1 & 0 \\ 0 & 0 & 1 \end{bmatrix}$$

By our definitions, we have shown how every linear transformation between finite-dimensional vector space can be represented by a matrix and that this representation is compatible with our definitions of transformation addition and transformation composition. What about the converse: Given a matrix with entries in a field F, does there exist a linear transformation between suitable vector spaces for which the given matrix is its matrix representation? The answer is yes.

Definition 19.5

Let $A = [a_{ij}] \in F^{m \times n}$ where F is a field. The transformation $\tau_A : F_n \to F_m$ defined by

$$\tau_A\left(\begin{bmatrix} x_1 \\ x_2 \\ \vdots \\ x_n \end{bmatrix}\right) = \begin{bmatrix} \sum_{j=1}^{n} a_{1j}x_j \\ \sum_{j=1}^{n} a_{2j}x_j \\ \vdots \\ \sum_{j=1}^{n} a_{mj}x_j \end{bmatrix}$$

that is,

$$\tau_A\left(\begin{bmatrix} x_1 \\ x_2 \\ \vdots \\ x_n \end{bmatrix}\right) = A\begin{bmatrix} x_1 \\ x_2 \\ \vdots \\ x_n \end{bmatrix}$$

is known as the **transformation associated with the matrix** A.

Example 19.12

If

$$A = \begin{bmatrix} 1 & 3 \\ -4 & 2 \end{bmatrix},$$

then

$$\tau_A\left(\begin{bmatrix} x_1 \\ x_2 \end{bmatrix}\right) = A\begin{bmatrix} x_1 \\ x_2 \end{bmatrix} = \begin{bmatrix} x_1 + 3x_2 \\ -4x_1 + 2x_2 \end{bmatrix}$$

We leave it to you to show that if one begins with a linear transformation between F_n and F_m for some field F, takes its matrix representation, and finally constructs the transformation associated with the matrix, then one obtains the original transformation. (See Exercise 2.) Similarly, if one begins with a matrix in $F^{m \times n}$, next writes down the linear transformation associated with that matrix, and finally represents that transformation as a matrix, then one obtains the original matrix. (See Exercise 3.) We can summarize these observations in the next result:

Theorem 19.1

Suppose F is a field, $\tau: F_n \rightarrow F_m$ is a linear transformation, and $A \in F^{m \times n}$. If $M(\tau)$ denotes the matrix representation of τ with respect to the usual bases, and τ_A denotes the transformation associated with the matrix A, then $\tau_{M(\tau)} = \tau$ and $M(\tau_A) = A$.

While we have investigated matrix addition and multiplication, we have not yet discussed the connection between a scalar multiple of a transformation and a scalar multiple of a matrix. Let $\tau: F_n \rightarrow F_m$ be a linear transformation, $c \in F$, and let $\sigma: F_n \rightarrow F_m$ be the mapping defined by $\sigma(\mathbf{v}) = (c\tau)(\mathbf{v}) = c(\tau(\mathbf{v}))$ for $\mathbf{v} \in F_n$. We leave it to you to show that σ is a linear transformation. (See Exercise 4.) The question of the relationship between the matrix representation of σ and the matrix representation of τ is answered by the next result:

Theorem 19.2

Let F be a field; let $\tau: F_n \rightarrow F_m$ be a linear transformation with matrix representation $M(\tau) = [a_{ij}]$; and for $c \in F$, let σ be the linear transformation defined by $\sigma(\mathbf{v}) = c(\tau(\mathbf{v}))$ for $\mathbf{v} \in F_n$. If the matrix representation of σ is $M(\sigma) = [b_{ij}]$, then $b_{ij} = ca_{ij}$ for all $i = 1, 2, \ldots, m$.

Proof

Let $\{\mathbf{e}_1^n, \mathbf{e}_2^n, \ldots, \mathbf{e}_n^n\}$ denote the usual basis of F_n. By the definition of σ, we have

$$\sigma(\mathbf{e}_i^n) = c(\tau(\mathbf{e}_i^n)) = c \begin{bmatrix} a_{1i} \\ a_{2i} \\ \vdots \\ a_{mi} \end{bmatrix} = \begin{bmatrix} ca_{1i} \\ ca_{2i} \\ \vdots \\ ca_{mi} \end{bmatrix}$$

for $i = 1, 2, \ldots, n$. By Definition 19.2, we have $M(\sigma) = [ca_{ij}]$. The result follows.   □

In view of the result in Theorem 19.2, we define the following:

Definition 19.6

Let $A = [a_{ij}] \in F^{m \times n}$ and $c \in F$. The matrix $cA = [ca_{ij}]$ is termed the **scalar multiple of** A **by** c.

Example 19.13   Let

$$A = \begin{bmatrix} 1 & 3 \\ -2 & 0 \end{bmatrix}$$

then

$$2A = 2\begin{bmatrix} 1 & 3 \\ -2 & 0 \end{bmatrix} = \begin{bmatrix} 2 & 6 \\ -4 & 0 \end{bmatrix} \quad \text{and}$$

$$-3A = (-3)\begin{bmatrix} 1 & 3 \\ -2 & 0 \end{bmatrix} = \begin{bmatrix} -3 & -9 \\ 6 & 0 \end{bmatrix}$$

Before concluding this section, we want to introduce another matrix concept that will prove useful in Section 19.4.

Definition 19.7

If $A \in F^{m \times n}$, then the **transpose of** A, denoted A^T, is the $n \times m$ matrix whose i, j-entry is the j, i-entry of A. In other words, if $B = [b_{ij}] = A^T$, then $b_{ij} = a_{ji}$.

Example 19.14

$$\begin{bmatrix} 1 & 2 & 3 \\ 4 & 5 & 6 \end{bmatrix}^T = \begin{bmatrix} 1 & 4 \\ 2 & 5 \\ 3 & 6 \end{bmatrix}, \quad [1 \quad -2 \quad 5]^T = \begin{bmatrix} 1 \\ -2 \\ 5 \end{bmatrix}, \quad \text{and}$$

$$\begin{bmatrix} 4 & -7 \\ 0 & 6 \end{bmatrix}^T = \begin{bmatrix} 4 & 0 \\ -7 & 6 \end{bmatrix}$$

In the remainder of this chapter, we apply the basic definitions of this section in a number of ways.

Exercises 19.1

1. If F is a field and $A, B, C \in F^{m \times n}$, then show
   a. $A + B = B + A$. (Matrix addition is commutative.)
   b. $(A + B) + C = A + (B + C)$. (Matrix addition is associative.)
2. Let F be a field, $\tau: F_n \to F_m$ a linear transformation, and $M(\tau)$ its matrix representation. Show $\tau_{M(\tau)} = \tau$.
3. Let F be a field, $A \in F^{m \times n}$. Show $M(\tau_A) = A$.
4. Let $\tau: F_n \to F_m$ be a linear transformation, $c \in F$, and let $\sigma: F_n \to F_m$ be defined by $\sigma(\mathbf{v}) = c(\tau(\mathbf{v}))$. Show that σ is a linear transformation.
5. Let $\tau: F_3 \to F_2$ and $\sigma: F_2 \to F_2$ be linear transformations defined by

$$\tau\left(\begin{bmatrix} x_1 \\ x_2 \\ x_3 \end{bmatrix}\right) = \begin{bmatrix} x_1 + x_2 \\ x_3 \end{bmatrix} \quad \text{and} \quad \sigma\left(\begin{bmatrix} x_1 \\ x_2 \end{bmatrix}\right) = \begin{bmatrix} x_1 \\ x_1 + x_2 \end{bmatrix}$$

Construct $M(\tau)$, $M(\sigma)$, and $M(\sigma \circ \tau)$, the matrix representations of τ, σ, $\sigma \circ \tau$. Compute $M(\sigma)M(\tau)$.

6. What is the matrix representation of the identity transformation defined by $\tau(\mathbf{v}) = \mathbf{v}$ for all $\mathbf{v} \in F_n$?

7. If σ is the linear transformation of Exercise 5 and $\rho : F_2 \to F_2$ is the linear transformation defined by

$$\rho\left(\begin{bmatrix} x_1 \\ x_2 \end{bmatrix}\right) = \begin{bmatrix} x_1 \\ x_2 - x_1 \end{bmatrix}$$

what are $M(\rho)$, $\rho\sigma$, $\sigma\rho$, $M(\rho)M(\sigma)$, and $M(\sigma)M(\rho)$?

8. Let

$$A = \begin{bmatrix} 1 & 1 \\ -1 & 0 \\ 2 & 1 \end{bmatrix} \qquad B = \begin{bmatrix} 2 & 1 \\ 0 & 1 \end{bmatrix} \qquad C = \begin{bmatrix} 0 & -1 \\ 2 & -1 \\ 1 & 1 \end{bmatrix}$$

$$D = \begin{bmatrix} 1 & 0 \\ -1 & 0 \end{bmatrix} \qquad E = \begin{bmatrix} 1 & -1 & 2 \end{bmatrix} \qquad F = \begin{bmatrix} 0 \\ 1 \\ -1 \end{bmatrix}$$

be matrices with rational entries. Compute (where possible):
a. AB b. BA c. $A + C$ d. $(A + C)B$
e. EF f. FE g. $B + 2D$

9. Let

$$A = \begin{bmatrix} 1 & 2 & 2 \\ 0 & -3 & -1 \end{bmatrix} \qquad B = \begin{bmatrix} 1 & 2 \\ -1 & 3 \end{bmatrix} \qquad C = \begin{bmatrix} 2 & 1 & 3 \end{bmatrix}$$

$$D = \begin{bmatrix} 3 \\ -1 \\ 2 \end{bmatrix} \qquad E = \begin{bmatrix} 2 & 3 \\ 0 & 1 \\ -1 & 0 \end{bmatrix} \qquad F = \begin{bmatrix} -1 & 2 & 0 \\ 3 & 4 & 1 \end{bmatrix}$$

Compute (where possible):
a. AE b. FA c. BE d. BA
e. CE f. $DC + EA$ g. $2A - 3F$ h. $E(A - 3F)$

10. Determine the matrix representation of each of the linear transformations in Exercises 8, 9, and 19 of Section 18.4.

11. Prove that $\mathbf{0}_{m \times n}$ is an additive identity in $F^{m \times n}$.

12. Verify that I_n is a multiplicative identity in $F^{n \times n}$.

13. If $A \in F_{m \times n}$, show that $I_m A = AI_n = A$.

14. Find the transpose of each of the following matrices:

$$\begin{bmatrix} 1 & -1 & 0 \\ 1 & 1 & 2 \end{bmatrix} \qquad \begin{bmatrix} 4 & -1 \\ 1 & 0 \\ 1 & 6 \end{bmatrix} \qquad \begin{bmatrix} -1 & -1 & 0 & 2 \end{bmatrix} \qquad \begin{bmatrix} 4 & 0 & 1 \\ 0 & -2 & 7 \\ 1 & 3 & 5 \end{bmatrix}$$

15. Let $A, B \in F_{m \times n}$, $C \in F^{n \times t}$, $c \in F$. Verify each of the following statements:
   a. $(A + B)^T = A^T + B^T$ b. $(cA)^T = c(A^T)$ c. $(AC)^T = C^T A^T$

*16. Write a computer program to take two matrices and compute their sum.

*17. Write a computer program to take two matrices and compute their product.

*18. Write a computer program that accepts positive integers m and n and a set of mn integers and prints the $m \times n$ matrix whose entries are the mn integers.

*19. Write a computer program that will take a matrix and multiply it by a scalar.

*20. Write a computer program that will take a linear transformation on $\mathbb{R}_3$ and return its matrix representation.

19.2

Change of Basis

In determining the matrix representation of a linear transformation on an F-vector space of dimension n, we employed the usual basis $\{e_1^n, e_2^n, \ldots, e_n^n\}$ for F^n. We chose this basis because its use made it easy to obtain the desired matrix; the image of the usual column basis vectors became the columns of our matrix. You might ask, however, what would happen if we utilized a different basis for F^n. The answer is quite involved and requires further analysis of the nature of linear transformations and bases.

     Let $\alpha = \{u_1, u_2, \ldots, u_n\}$ and $\beta = \{v_1, v_2, \ldots, v_n\}$ be two bases of an F-vector space V. Then we can express each element of β as a linear combination of elements of α:

$$v_1 = p_{11} u_1 \oplus p_{21} u_2 \oplus \cdots \oplus p_{n1} u_n$$

$$v_2 = p_{12} u_1 \oplus p_{22} u_2 \oplus \cdots \oplus p_{n2} u_n$$

$$\vdots$$

$$v_n = p_{1n} u_1 \oplus p_{2n} u_2 \oplus \cdots \oplus p_{nn} u_n$$

This representation lets us define a linear transformation $\rho: V \to V$ by $\rho(u_i) = v_i$. (See Exercise 1.) The matrix representation of this transformation $P = (p_{ij})$ is termed the **transition matrix from basis α to basis β.**

Example 19.15

Let

$$\alpha = \left\{ \begin{bmatrix} 1 \\ 0 \end{bmatrix}, \begin{bmatrix} 0 \\ 1 \end{bmatrix} \right\} \quad \text{and} \quad \beta = \left\{ \begin{bmatrix} -1 \\ 1 \end{bmatrix}, \begin{bmatrix} 0 \\ 1 \end{bmatrix} \right\}$$

Both α and β are bases of $\mathbb{R}_2$.
     Since

$$\begin{bmatrix} 1 \\ 1 \end{bmatrix} = (1) \begin{bmatrix} 1 \\ 0 \end{bmatrix} + (1) \begin{bmatrix} 0 \\ 1 \end{bmatrix} \quad \text{and} \quad \begin{bmatrix} -1 \\ 1 \end{bmatrix} = (-1) \begin{bmatrix} 1 \\ 0 \end{bmatrix} + (1) \begin{bmatrix} 0 \\ 1 \end{bmatrix}$$

we see

$$\begin{bmatrix} 1 & -1 \\ 1 & 1 \end{bmatrix}$$

is the transition matrix from α to β.

Similarly, straightforward computation shows that

$$\begin{bmatrix} 1 \\ 0 \end{bmatrix} = \frac{1}{2}\begin{bmatrix} 1 \\ 1 \end{bmatrix} + \left(-\frac{1}{2}\right)\begin{bmatrix} -1 \\ 1 \end{bmatrix} \quad \text{and} \quad \begin{bmatrix} 0 \\ 1 \end{bmatrix} = \frac{1}{2}\begin{bmatrix} 1 \\ 1 \end{bmatrix} + \frac{1}{2}\begin{bmatrix} -1 \\ 1 \end{bmatrix}$$

Therefore,

$$\begin{bmatrix} \frac{1}{2} & \frac{1}{2} \\ -\frac{1}{2} & \frac{1}{2} \end{bmatrix}$$

is the transition matrix from β to α.

You will note in our definition that the matrix $P = M(\rho)$ has for its columns the rows of scalars that appear in the preceding system of equations. The reason for this reversal of subscripts can be seen if we remember the idea of coordinates of a vector with respect to a given basis. Consider the vector v_1. Using our standard isomorphism of V with F_n, v is customarily described by coordinates given with respect to the usual basis. However, if we write v_1 in terms of the basis α, then v_1 has coordinate n-tuple

$$\begin{bmatrix} p_{11} \\ p_{21} \\ \vdots \\ p_{n1} \end{bmatrix}_\alpha$$

where the α is written beside the n-tuple to denote that we are giving the coordinates of v_1 with respect to the basis α. Similarly, we write

$$v_i = \begin{bmatrix} p_{1i} \\ p_{2i} \\ \vdots \\ p_{ni} \end{bmatrix}_\alpha$$

Therefore, writing our coordinates in terms of columns, we see that the α-coordinates of the elements of β become the columns of the transition matrix from α to β. This is consistent with our previous method of representing a linear transformation.

We have formed the transition matrix P from α to β by expressing the elements of β as linear combinations of the elements of α. Let us write

$$\begin{cases} \mathbf{u}_1 = q_{11}\mathbf{v}_1 \oplus q_{21}\mathbf{v}_2 \oplus \cdots \oplus q_{n1}\mathbf{v}_n \\ \mathbf{u}_2 = q_{12}\mathbf{v}_1 \oplus q_{22}\mathbf{v}_2 \oplus \cdots \oplus q_{n2}\mathbf{v}_n \\ \quad\vdots \\ \mathbf{u}_n = q_{1n}\mathbf{v}_1 \oplus q_{2n}\mathbf{v}_2 \oplus \cdots \oplus q_{nn}\mathbf{v}_n \end{cases}$$

Then $Q = (q_{ij})$ is the transition matrix from β to α.

How are P and Q related? Let us utilize our equations to find out:

$$\mathbf{v}_j = \sum_{k=1}^{n} p_{kj}\mathbf{u}_k = \sum_{k=1}^{n} p_{kj}\left(\sum_{i=1}^{n} q_{ik}\mathbf{v}_i\right) = \sum_{i=1}^{n}\left(\sum_{k=1}^{n} q_{ik}p_{kj}\right)\mathbf{v}_i$$

using the commutative and distributive properties of F and V. Since β is a basis, we know that each $\mathbf{v}_i$ can be written in only one way. Therefore

$$\sum_{k=1}^{n} q_{ik}p_{kj} = \begin{cases} 0, & j \neq i \\ 1, & j = i \end{cases}$$

But

$$\sum_{k=1}^{n} q_{ik}p_{kj}$$

is the i, j-entry in the matrix QP. Therefore $QP = I$. A similar computation shows $PQ = I$. (See Exercise 3.)

We can summarize our discussion in the following result:

Theorem 19.3

If P is the transition matrix of an F-vector space V from basis α to basis β, then P is nonsingular, and P^{-1} is the transition matrix of V from basis β to basis α.

Example 19.16

In Example 19.15,

$$P = \begin{bmatrix} 1 & -1 \\ 1 & 1 \end{bmatrix} \quad \text{and} \quad Q = \begin{bmatrix} \frac{1}{2} & \frac{1}{2} \\ -\frac{1}{2} & \frac{1}{2} \end{bmatrix}$$

Then

$$PQ = \begin{bmatrix} 1 & -1 \\ 1 & 1 \end{bmatrix}\begin{bmatrix} \frac{1}{2} & \frac{1}{2} \\ -\frac{1}{2} & \frac{1}{2} \end{bmatrix} = \begin{bmatrix} 1 & 0 \\ 0 & 1 \end{bmatrix}$$

Example 19.17

One can utilize Theorem 19.3 to simplify the process of computing a transition matrix. Let

$$\alpha = \left\{ \begin{bmatrix} 2 \\ 1 \end{bmatrix}, \begin{bmatrix} 0 \\ 2 \end{bmatrix} \right\} \quad \text{and} \quad \beta = \left\{ \begin{bmatrix} 1 \\ 1 \end{bmatrix}, \begin{bmatrix} -1 \\ 1 \end{bmatrix} \right\}$$

Therefore, α and β are both bases of $\mathbb{R}_2$. By Example 19.15 and Theorem 19.3, we know that

$$\begin{bmatrix} \frac{1}{2} & \frac{1}{2} \\ -\frac{1}{2} & \frac{1}{2} \end{bmatrix}$$

is the transition matrix from the usual basis to α and

$$\begin{bmatrix} 2 & 0 \\ 1 & 2 \end{bmatrix}$$

is the transition matrix from β to the usual basis. Hence, since transition matrices change coordinates and a composition of coordinate changes is a coordinate change,

$$\begin{bmatrix} \frac{1}{2} & \frac{1}{2} \\ -\frac{1}{2} & \frac{1}{2} \end{bmatrix} \begin{bmatrix} 2 & 0 \\ 1 & 2 \end{bmatrix} = \begin{bmatrix} \frac{3}{2} & 1 \\ -\frac{1}{2} & 1 \end{bmatrix}$$

is the transition matrix from β to α. We leave it to you to verify this statement directly. (See Exercise 4.)

While we have investigated the nature of transition matrices, we have as yet not shown any practical application. One important application is the simplification of the computation of change of coordinates. Given arbitrary vectors written in the usual coordinates, one is often faced with laborious computations to determine the coordinates of those vectors with respect to another basis. The following theorem shows that the correct transition matrix eliminates much of the work.

Theorem 19.4

Let $\alpha = \{\mathbf{u}_1, \mathbf{u}_2, \ldots, \mathbf{u}_n\}$ and $\beta = \{\mathbf{v}_1, \mathbf{v}_2, \ldots, \mathbf{v}_n\}$ be bases of an F-vector space V and $\mathbf{v} \in V$. If

$$\mathbf{v} = \begin{bmatrix} a_1 \\ a_2 \\ \vdots \\ a_n \end{bmatrix}_\alpha \quad \text{and} \quad \mathbf{v} = \begin{bmatrix} b_1 \\ b_2 \\ \vdots \\ b_n \end{bmatrix}_\beta$$

then

$$\begin{bmatrix} a_1 \\ a_2 \\ \vdots \\ a_n \end{bmatrix} = P \begin{bmatrix} b_1 \\ b_2 \\ \vdots \\ b_n \end{bmatrix}$$

where P is the transition matrix from α to β.

Proof

By our hypothesis,

$$\mathbf{v} = \sum_{i=1}^{n} b_i \mathbf{v}_i = \sum_{i=1}^{n} b_i \left(\sum_{j=1}^{n} p_{ji} \mathbf{u}_j \right) = \sum_{j=1}^{n} \left(\sum_{i=1}^{n} p_{ji} b_i \right) \mathbf{u}_j$$

But $\mathbf{v} = \sum_{j=1}^{n} a_j \mathbf{u}_j$. By uniqueness of expression,

$$a_j = \sum_{i=1}^{n} p_{ji} b_i$$

Note that the expression on the right side of this equality is the jth row of the matrix

$$P \begin{bmatrix} b_1 \\ b_2 \\ \vdots \\ b_n \end{bmatrix}$$

Since each entry of

$$\begin{bmatrix} a_1 \\ a_2 \\ \vdots \\ a_n \end{bmatrix}$$

is equal to the corresponding entry of

$$P \begin{bmatrix} b_1 \\ b_2 \\ \vdots \\ b_n \end{bmatrix}$$

the result follows.   □

Example 19.18   Let

$$\alpha = \left\{ \begin{bmatrix} 1 \\ 0 \end{bmatrix}, \begin{bmatrix} 0 \\ 1 \end{bmatrix} \right\} \quad \text{and} \quad \beta = \left\{ \begin{bmatrix} 1 \\ 1 \end{bmatrix}, \begin{bmatrix} -1 \\ 1 \end{bmatrix} \right\}$$

Then α and β are bases of $\mathbb{R}_2$. If

$$\begin{bmatrix} 2 \\ 3 \end{bmatrix}_\alpha \in \mathbb{R}_2$$

what are the coordinates of

$$\begin{bmatrix} 2 \\ 3 \end{bmatrix}_\alpha$$

in the β basis? To answer this question, we must find a and b such that

$$\begin{bmatrix} 2 \\ 3 \end{bmatrix} = a\begin{bmatrix} 1 \\ 1 \end{bmatrix} + b\begin{bmatrix} -1 \\ 1 \end{bmatrix} = \begin{bmatrix} a-b \\ a+b \end{bmatrix}$$

We see that $a - b = 2$ and $a + b = 3$, so that $a = \frac{5}{2}$ and $b = \frac{1}{2}$. Thus

$$\begin{bmatrix} 2 \\ 3 \end{bmatrix}_\alpha = \begin{bmatrix} \frac{5}{2} \\ \frac{1}{2} \end{bmatrix}_\beta$$

In Example 19.15, we found that

$$P = \begin{bmatrix} \frac{1}{2} & \frac{1}{2} \\ -\frac{1}{2} & \frac{1}{2} \end{bmatrix}$$

was the transition matrix from α to β. Note that

$$\begin{bmatrix} \frac{1}{2} & \frac{1}{2} \\ -\frac{1}{2} & \frac{1}{2} \end{bmatrix}\begin{bmatrix} 2 \\ 3 \end{bmatrix} = \begin{bmatrix} \frac{5}{2} \\ \frac{1}{2} \end{bmatrix}$$

Another application of the transition matrix concept is a method to find inverses of certain matrices.

Example 19.19

To find the inverse of

$$P = \begin{bmatrix} 2 & 1 \\ -1 & 3 \end{bmatrix}$$

we may think of P as a transition matrix that maps

$$\begin{bmatrix} 1 \\ 0 \end{bmatrix} \text{ to } \begin{bmatrix} 2 \\ -1 \end{bmatrix} \quad \text{and} \quad \begin{bmatrix} 0 \\ 1 \end{bmatrix} \text{ to } \begin{bmatrix} 1 \\ 3 \end{bmatrix}$$

To find this inverse, we must work with computations to express

$$\begin{bmatrix} 1 \\ 0 \end{bmatrix} \quad \text{and} \quad \begin{bmatrix} 0 \\ 1 \end{bmatrix}$$

in terms of

$$\begin{bmatrix} 2 \\ -1 \end{bmatrix} \quad \text{and} \quad \begin{bmatrix} 1 \\ 3 \end{bmatrix}$$

We find

$$\begin{bmatrix} 1 \\ 0 \end{bmatrix} = \frac{3}{7}\begin{bmatrix} 2 \\ -1 \end{bmatrix} + \frac{1}{7}\begin{bmatrix} 1 \\ 3 \end{bmatrix} \quad \text{and} \quad \begin{bmatrix} 0 \\ 1 \end{bmatrix} = -\frac{1}{7}\begin{bmatrix} 2 \\ -1 \end{bmatrix} + \frac{2}{7}\begin{bmatrix} 1 \\ 3 \end{bmatrix}$$

Thus

$$P^{-1} = \begin{bmatrix} \frac{3}{7} & -\frac{1}{7} \\ \frac{1}{7} & \frac{2}{7} \end{bmatrix}$$

Of course, this method only works if the matrix at hand has an inverse. We saw in Theorem 19.3 that a transition matrix must have a multiplicative inverse.

We now turn to our major application of the transition matrix and utilize the ideas we have developed to determine the relationship between representations of a linear transformation with respect to different bases of the underlying spaces.

Let $\tau: F_n \to F_m$ be a linear transformation, $\alpha = \{\mathbf{u}_1, \mathbf{u}_2, \ldots, \mathbf{u}_n\}$ be a basis of F_n, and $\alpha' = \{\mathbf{u}'_1, \mathbf{u}'_2, \ldots, \mathbf{u}'_m\}$ be a basis of F_m. Suppose

$$\begin{cases} \tau(\mathbf{u}_1) = a_{11}\mathbf{u}'_1 \oplus a_{21}\mathbf{u}'_2 \oplus \cdots \oplus a_{m1}\mathbf{u}'_m \\ \tau(\mathbf{u}_2) = a_{12}\mathbf{u}'_1 \oplus a_{22}\mathbf{u}'_2 \oplus \cdots \oplus a_{m2}\mathbf{u}'_m \\ \qquad\qquad\qquad \vdots \\ \tau(\mathbf{u}_n) = a_{1n}\mathbf{u}'_1 \oplus a_{2n}\mathbf{u}'_2 \oplus \cdots \oplus a_{mn}\mathbf{u}'_m \end{cases}$$

We define the **matrix of τ with respect to α and α'** to be

$$\begin{bmatrix} a_{11} & a_{12} & \cdots & a_{1n} \\ a_{21} & a_{22} & \cdots & a_{2n} \\ \vdots & \vdots & \ddots & \vdots \\ a_{m1} & a_{m2} & \cdots & a_{mn} \end{bmatrix}$$

You should check that this is consistent with our definition of the matrix representation of τ with respect to the usual bases. To properly utilize this matrix representation, remember that the vector from the domain is in α-coordinates and the resulting vector is in α'-coordinates. This matrix is not, in general, the transition matrix from α to α'.

| Example 19.20 |

Consider the basis

$$\alpha = \left\{ \begin{bmatrix} 1 \\ 1 \end{bmatrix}, \begin{bmatrix} -1 \\ 1 \end{bmatrix} \right\}$$

for $\mathbb{R}_2$ and the basis

$$\alpha' = \left\{ \begin{bmatrix} 3 \\ 1 \\ 0 \end{bmatrix}, \begin{bmatrix} 2 \\ 0 \\ 1 \end{bmatrix}, \begin{bmatrix} 1 \\ 1 \\ 1 \end{bmatrix} \right\}$$

for $\mathbb{R}_3$. If $\tau: \mathbb{R}_2 \to \mathbb{R}_3$ is defined by

$$\tau\left(\begin{bmatrix} 1 \\ 1 \end{bmatrix}\right) = 2 \begin{bmatrix} 3 \\ 1 \\ 0 \end{bmatrix} + (-1) \begin{bmatrix} 2 \\ 0 \\ 1 \end{bmatrix} + (1) \begin{bmatrix} 1 \\ 1 \\ 1 \end{bmatrix}$$

and

$$\tau\left(\begin{bmatrix} -1 \\ 1 \end{bmatrix}\right) = 1 \begin{bmatrix} 3 \\ 1 \\ 0 \end{bmatrix} + 0 \begin{bmatrix} 2 \\ 0 \\ 1 \end{bmatrix} + 2 \begin{bmatrix} 1 \\ 1 \\ 1 \end{bmatrix}$$

then the matrix of τ with respect to α and α' is

$$\begin{bmatrix} 2 & 1 \\ -1 & 0 \\ 1 & 2 \end{bmatrix}$$

Thus for $\begin{bmatrix} 2 \\ -1 \end{bmatrix}_\alpha$,

$$\tau\left(\begin{bmatrix} 2 \\ -1 \end{bmatrix}_\alpha\right) = \begin{bmatrix} 2 & 1 \\ -1 & 0 \\ 1 & 2 \end{bmatrix} \begin{bmatrix} 2 \\ -1 \end{bmatrix}_\alpha = \begin{bmatrix} 3 \\ -2 \\ 0 \end{bmatrix}_{\alpha'}$$

How are representations of the same transformation with respect to different bases related? Utilizing our work with transition matrices, we answer the question with the following result:

Theorem 19.5

Let $\tau: V \to V'$ be a linear transformation, P be the transition matrix from α to β, Q be the transition matrix from β' to α', A the matrix representation of τ with respect to α and α', and B the matrix representation of τ with respect of β and β'. Then $B = QAP$.

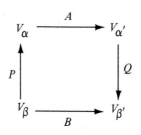

Figure 19.2

Proof

To facilitate the proof, we refer you to Figure 19.2. Let

$$\alpha = \{\mathbf{u}_1, \mathbf{u}_2, \ldots, \mathbf{u}_n\} \qquad \beta = \{\mathbf{v}_1, \mathbf{v}_2, \ldots, \mathbf{v}_n\}$$
$$\alpha' = \{\mathbf{u}'_1, \mathbf{u}'_2, \ldots, \mathbf{u}'_m\} \qquad \beta' = \{\mathbf{v}'_1, \mathbf{v}'_2, \ldots, \mathbf{v}'_m\}$$

Then

$$\tau(\mathbf{v}_j) = \tau\left(\sum_{k=1}^{n} p_{kj}\mathbf{u}_k\right)$$

$$= \sum_{k=1}^{n} p_{kj}\tau(\mathbf{u}_k)$$

$$= \sum_{k=1}^{n} p_{kj}\left(\sum_{i=1}^{m} a_{ik}\mathbf{u}'_i\right)$$

$$= \sum_{i=1}^{m} \left(\sum_{k=1}^{n} a_{ik}p_{kj}\right)\mathbf{u}'_i$$

$$= \sum_{i=1}^{m} \left(\sum_{k=1}^{n} a_{ik}p_{kj}\right)\left(\sum_{w=1}^{m} q_{wi}\mathbf{v}'_w\right)$$

$$= \sum_{w=1}^{m} \sum_{i=1}^{m} \sum_{k=1}^{n} (q_{wi}a_{ik}p_{kj})\mathbf{v}'_w$$

But $\tau(\mathbf{v}_j) = \sum_{w=1}^{m} b_{wj}\mathbf{v}'_w$. By uniqueness of expression,

$$b_{wj} = \sum_{i=1}^{m} \sum_{k=1}^{n} q_{wi}a_{ik}p_{kj}$$

Since the term on the right side of this equation is the wj-term of the matrix QAP, the result follows.   ◻

Corollary 19.1

Let τ be a linear transformation on F_n and $M(\tau)$ its matrix representation with respect to the usual basis of F_n. If α is another basis of F_n, P is the transition matrix from the usual basis to α, and B is the matrix of τ with respect to the basis α, then $B = P^{-1}M(\tau)P$.

Proof

The result follows immediately from the theorem if we simply note that if P is the transition matrix from the usual basis to α, then $Q = P^{-1}$ is the transition matrix from α to the usual basis.   ◻

Example 19.21

Consider the mapping $\tau\colon \mathbb{R}_2 \to \mathbb{R}_3$ defined by

$$\tau\left(\begin{bmatrix} x \\ y \end{bmatrix}\right) = \begin{bmatrix} 5x \\ 3x \\ x-y \end{bmatrix}, \qquad M(\tau) = \begin{bmatrix} 5 & 0 \\ 3 & 0 \\ 1 & -1 \end{bmatrix}$$

If we want to find the representation of τ with respect to the bases

$$\alpha = \left\{ \begin{bmatrix} 1 \\ 1 \end{bmatrix}, \begin{bmatrix} 1 \\ -1 \end{bmatrix} \right\} \quad \text{and} \quad \alpha' = \left\{ \begin{bmatrix} 3 \\ 1 \\ 0 \end{bmatrix}, \begin{bmatrix} 2 \\ 0 \\ 1 \end{bmatrix}, \begin{bmatrix} 1 \\ 1 \\ 1 \end{bmatrix} \right\}$$

we first find the transition matrices. We leave it to you in Exercise 5 to show

$$P = \begin{bmatrix} 1 & 1 \\ 1 & -1 \end{bmatrix}$$

is the transition matrix on $\mathbb{R}_2$ from the usual basis to α and

$$Q = \begin{bmatrix} \frac{1}{4} & \frac{1}{4} & -\frac{1}{2} \\ \frac{1}{4} & -\frac{3}{4} & \frac{1}{2} \\ -\frac{1}{4} & \frac{3}{4} & \frac{1}{2} \end{bmatrix}$$

is the transition matrix from α' to the usual basis. Then the matrix representation of τ with respect to α and α' is

$$QM(\tau)P = \begin{bmatrix} \frac{1}{4} & \frac{1}{4} & -\frac{1}{2} \\ \frac{1}{4} & -\frac{3}{4} & \frac{1}{2} \\ -\frac{1}{4} & \frac{3}{4} & \frac{1}{2} \end{bmatrix} \begin{bmatrix} 5 & 0 \\ 3 & 0 \\ 1 & -1 \end{bmatrix} \begin{bmatrix} 1 & 1 \\ 1 & -1 \end{bmatrix} = \begin{bmatrix} 2 & 1 \\ -1 & 0 \\ 1 & 2 \end{bmatrix}$$

(Compare with Example 19.20.)

While Theorem 19.5 answers the question on matrix representations that motivated this section, the theorem itself is difficult to state. We can, however, make a definition that will enable us to state the connection between various matrix representations of a given linear transformation in a more convenient manner.

Definition 19.8

Let F be a field and A, B be elements of $F^{m \times n}$. Then A is said to be **equivalent** to B if there exist invertible matrices $Q \in F^{m \times m}$ and $P \in F^{n \times n}$ such that $B = QAP$.

We leave the proof of Theorem 19.6 as Exercise 10.

Theorem 19.6

Matrix equivalence is an equivalence relation on $F^{m \times n}$.

With the aid of Definition 19.8, we now are able to describe matrices representing a linear transformation with respect to different bases: If A is the matrix representing a linear transformation with respect to one pair of bases and B is the matrix representing that transformation with respect to another pair of bases, then A and B are equivalent.

Exercises 19.2

1. Suppose $\{\mathbf{u}_1, \mathbf{u}_2, \ldots, \mathbf{u}_n\}$ is a basis of F_n, F a field, and $\{\mathbf{v}_1, \mathbf{v}_2, \ldots, \mathbf{v}_n\} \subseteq F_n$. Define $\tau: F_n \to F_n$ by

$$\tau\left(\sum_{i=1}^{n} a_i \mathbf{u}_i\right) = \sum_{i=1}^{n} a_i \mathbf{v}_i$$

Show that τ is a linear transformation.

2. Let $\{\mathbf{u}_1, \mathbf{u}_2, \ldots, \mathbf{u}_n\}$ and $\{\mathbf{v}_1, \mathbf{v}_2, \ldots, \mathbf{v}_n\}$ be bases of F_n. Show that there is but one linear transformation $\tau: F_n \to F_n$ with the property that $\tau(\mathbf{u}_i) = \mathbf{v}_i$ for $i = 1, 2, \ldots, n$.

3. In the discussion preceding Theorem 19.3, verify $PQ = I$.

4. Use the definition of transition matrix to prove directly that

$$\begin{bmatrix} \frac{3}{2} & 1 \\ -\frac{1}{2} & 1 \end{bmatrix}$$

is the transition matrix from

$$\beta = \left\{ \begin{bmatrix} 1 \\ 1 \end{bmatrix}, \begin{bmatrix} -1 \\ 1 \end{bmatrix} \right\} \quad \text{to} \quad \alpha = \left\{ \begin{bmatrix} 2 \\ 1 \end{bmatrix}, \begin{bmatrix} 0 \\ 2 \end{bmatrix} \right\}$$

5. Verify that

$$\begin{bmatrix} 1 & 1 \\ 1 & -1 \end{bmatrix}$$

is the transition matrix from the usual basis of $\mathbb{R}_2$ to

$$\alpha' = \left\{ \begin{bmatrix} 1 \\ 1 \end{bmatrix}, \begin{bmatrix} 1 \\ -1 \end{bmatrix} \right\}$$

and that

$$Q = \begin{bmatrix} \frac{1}{4} & \frac{1}{4} & -\frac{1}{2} \\ \frac{1}{4} & -\frac{3}{4} & \frac{1}{2} \\ -\frac{1}{4} & \frac{3}{4} & \frac{1}{2} \end{bmatrix}$$

is the transition matrix from

$$\alpha' = \left\{ \begin{bmatrix} 3 \\ 1 \\ 0 \end{bmatrix}, \begin{bmatrix} 2 \\ 0 \\ 1 \end{bmatrix}, \begin{bmatrix} 1 \\ 1 \\ 1 \end{bmatrix} \right\}$$

to the usual basis of $\mathbb{R}_3$.

6. Let α denote the usual basis in $\mathbb{R}^2$ and β denote the basis $\{[1, 2], [2, -1]\}$. Find the transition matrix Q from the new β to the old α. What is PQ? What is $[2, -1]$ in the new basis β?

7. Find the transition matrix P from the usual base of $\mathbb{R}_3$ to

$$\beta = \left\{ \begin{bmatrix} 1 \\ 0 \\ 0 \end{bmatrix}, \begin{bmatrix} 0 \\ 1 \\ 1 \end{bmatrix}, \begin{bmatrix} -1 \\ 0 \\ 1 \end{bmatrix} \right\}$$

If

$$\mathbf{v} \in \mathbb{R}_3 \quad \text{and} \quad \mathbf{v}_\beta = \begin{bmatrix} 6 \\ 7 \\ 8 \end{bmatrix}$$

then what is $\mathbf{v}_\alpha$?

8. Suppose

$$P(\theta) = \begin{bmatrix} \cos \theta & -\sin \theta \\ \sin \theta & \cos \theta \end{bmatrix}$$

is a transition matrix from the usual basis of $\mathbb{R}_2$ to a new basis. Determine what form an arbitrary vector

$$\begin{bmatrix} x \\ y \end{bmatrix}$$

has in the new coordinate system. If $\theta = 45°$, how has the coordinate system changed? Compute $P(\gamma)P(\psi)$.

9. Compute the transition matrix from

$$\alpha = \{[1, 2], [3, -1]\} \quad \text{to} \quad \beta = \{[1, 4], [2, 2]\}$$

10. Prove that matrix equivalence is an equivalence relation.

11. Prove that if $\tau: V \to V$ is a vector space isomorphism, then its inverse τ^{-1} is a vector space isomorphism.

12. a. Find a matrix

$$A = \begin{bmatrix} a & b \\ c & d \end{bmatrix} \in \mathbb{Q}^{2 \times 2}$$

such that

$$A \begin{bmatrix} 2 & 3 \\ 1 & -2 \end{bmatrix} = \begin{bmatrix} 1 & 0 \\ 0 & 1 \end{bmatrix}$$

b. Let

$$\alpha = \left\{ \begin{bmatrix} 2 \\ 1 \end{bmatrix}, \begin{bmatrix} 3 \\ -2 \end{bmatrix} \right\}$$

Show that α is a basis of $\mathbb{Q}_2$.

c. Find the transition matrix from the usual basis of $\mathbb{Q}_2$ to α.

d. Let $\tau: \mathbb{Q}_2 \to \mathbb{Q}_2$ defined by

$$\tau\left(\begin{bmatrix} x \\ y \end{bmatrix} \right) = \begin{bmatrix} 2x + y \\ x - y \end{bmatrix}$$

Find the matrix of τ with respect to the usual basis of $\mathbb{Q}_2$.

e. Find the matrix of τ with respect to the basis α.

f. Compute $AM(\tau)A^{-1}$.

13. Let $\alpha = \{\mathbf{v}_1, \mathbf{v}_2, \ldots, \mathbf{v}_n\}$ be a basis of an F-vector space V.

a. Prove that the matrix representation with respect to α of 0, the zero transformation on V defined by $0(\mathbf{v}) = \mathbf{0}$, is $0_{n \times n}$.

b. Prove that the matrix representation with respect to α of 1_V, the identity transformation of V, is I_n.

19.3

Systems of Linear Equations

In this section we will consider the use of matrix methods to solve systems of equations. We shall restrict ourselves to systems of linear equations whose coefficients lie in a field F.

Consider the system

$$\begin{cases} a_{11}x_1 + a_{12}x_2 + \cdots + a_{1n}x_n = b_1 \\ a_{21}x_1 + a_{22}x_2 + \cdots + a_{2n}x_n = b_2 \\ \vdots \qquad \vdots \qquad \ddots \qquad \vdots \qquad \vdots \\ a_{m1}x_1 + a_{m2}x_2 + \cdots + a_{mn}x_n = b_m \end{cases}$$

where $a_{ij}, b_i \in F$, $1 \leqslant i \leqslant m$, $1 \leqslant j \leqslant n$. To enable us to construct a matrix system to use in solving the preceding simultaneous linear equations, we will first consider the preceding system as representing equality of two vectors in F_m:

$$\begin{bmatrix} a_{11}x_1 + a_{12}x_2 + \cdots + a_{1n}x_n \\ a_{21}x_1 + a_{22}x_2 + \cdots + a_{2n}x_n \\ \vdots \quad\quad \vdots \quad\quad \ddots \quad\quad \vdots \\ a_{m1}x_1 + a_{m2}x_2 + \cdots + a_{mn}x_n \end{bmatrix} = \begin{bmatrix} b_1 \\ b_2 \\ \vdots \\ b_m \end{bmatrix}$$

Using the properties of matrix multiplication, we can rewrite this as

$$\begin{bmatrix} a_{11} & a_{12} & \cdots & a_{1n} \\ a_{21} & a_{22} & \cdots & a_{2n} \\ \vdots & \vdots & \ddots & \vdots \\ a_{m1} & a_{m2} & \cdots & a_{mn} \end{bmatrix} \begin{bmatrix} x_1 \\ x_2 \\ \vdots \\ x_n \end{bmatrix} = \begin{bmatrix} b_1 \\ b_2 \\ \vdots \\ b_m \end{bmatrix}$$

or, to simplify notation, $AX = B$, where $A = [a_{ij}] \in F^{m \times n}$ and $B \in F^{m \times 1}$. Let us rewrite our original system in terms of column vectors:

$$\begin{bmatrix} a_{11} \\ a_{21} \\ \vdots \\ a_{m1} \end{bmatrix} x_1 + \begin{bmatrix} a_{12} \\ a_{22} \\ \vdots \\ a_{m2} \end{bmatrix} x_2 + \cdots + \begin{bmatrix} a_{1n} \\ a_{2n} \\ \vdots \\ a_{mn} \end{bmatrix} x_n = \begin{bmatrix} b_1 \\ b_2 \\ \vdots \\ b_n \end{bmatrix}$$

In this form, we see that the solution to a system of linear equations is related to the question of whether the vector on the right can be written as a linear combination of the vectors on the left. If the columns of the matrix A are written as A^1, A^2, ..., A^n, then our observation has an immediate consequence:

| Theorem 19.7 | A system of linear equations $AX = B$ has a solution if and only if the dimension of $\langle A^1, A^2, \dots, A^n, B \rangle$ equals the dimension of $\langle A^1, A^2, \dots, A^n \rangle$. |

This result relates systems of linear equations to the theory of subspaces and dimension. The solution to systems of equations is also intimately connected to the theory of linear transformations. Suppose that X_1 and X_2 are two solutions to $AX = B$. Then

$$A(X_1 - X_2) = AX_1 - AX_2 = B - B = 0$$

the zero vector in $F^{m \times 1}$. Thus $X_1 - X_2$ is an element of the kernel of the transformation τ_A defined by $\tau_A(X) = AX$.

By this observation, we conclude that if X_1 is any vector in F_n such that $AX_1 = B$, then all solutions have the form $X_1 + K$ where $K \in \ker \tau_A$. Thus the possible solutions for the vector $X \in F_n$ are the elements of a coset of F_n (as an abelian group). We can summarize the theory of systems of linear equations of the form $AX = B$ as follows:

1. There may be no solutions: B is not contained in the rank space of τ_A.
2. There may be a unique solution: τ_A is an injection; that is, $\ker \tau_A = \{\mathbf{0}\}$.
3. There are as many solutions as there are elements in $\ker \tau_A$. (We note in the case where F is infinite that there may be an infinite number of solutions.)

By the preceding observations, a logical place to start our investigation of the systems $AX = B$ is to consider $B = \mathbf{0}$ and determine the kernel of τ_A. We are interested in specific vectors in the domain space. Whatever method we use to determine these vectors must not alter the rank space of τ_A. To facilitate this process, we will use the concept of elementary operations.

Definition 19.9

Let $S = \{\mathbf{v}_1, \mathbf{v}_2, \ldots, \mathbf{v}_n\}$ be an ordered subset of V; that is, the elements of S are in a given order. An **elementary operation** with domain S is any one of the following:

$\mathscr{E}$: Interchange the position of any two vectors in the list.
$\mathscr{M}$: Multiply one vector in S by the nonzero element of F.
$\mathscr{A}$: Replace one vector in S by the sum of that vector and a scalar multiple of another vector in S.

The usefulness of the elementary operations hinges upon the following result.

Theorem 19.8

Let $S = \{\mathbf{v}_1, \mathbf{v}_2, \ldots, \mathbf{v}_n\}$ be a subset of V, and let $S' = \{\mathbf{u}_1, \mathbf{u}_2, \ldots, \mathbf{u}_n\}$ be a subset of V that results from the application of an elementary operation to S. Then the span of S is equal to the span of S'.

Proof

If the elementary operation is one of type $\mathscr{E}$, then the result holds, since

$$\langle \mathbf{v}_1, \ldots, \mathbf{v}_i, \ldots, \mathbf{v}_j, \ldots, \mathbf{v}_n \rangle = \langle \mathbf{v}_1, \ldots, \mathbf{v}_j, \ldots, \mathbf{v}_i, \ldots, \mathbf{v}_n \rangle$$

Similarly, if the operation is of type $\mathscr{M}$, then we leave it to you in Exercise 1 to show that

$$\langle \mathbf{v}_1, \ldots, \mathbf{v}_i, \ldots, \mathbf{v}_n \rangle = \langle \mathbf{v}_1, \ldots, a\mathbf{v}_i, \ldots, \mathbf{v}_n \rangle \qquad \text{for} \quad a \in F - \{0\}$$

Finally, if the operation is of type $\mathscr{A}$, say $\mathbf{u}_i = \mathbf{v}_i \oplus a\mathbf{v}_j$, then

$$\{\mathbf{u}_1, \ldots, \mathbf{u}_i, \ldots, \mathbf{u}_n\} = \{\mathbf{v}_1, \mathbf{v}_{i-1}, \mathbf{v}_i \oplus a\mathbf{v}_j, \mathbf{v}_{i+1}, \ldots, \mathbf{v}_n\}$$

Clearly,

$$\langle \mathbf{v}_1, \ldots, \mathbf{v}_{i-1}, \mathbf{v}_i \oplus a\mathbf{v}_j, \mathbf{v}_{i+1}, \ldots, \mathbf{v}_n \rangle \subseteq \langle \mathbf{v}_1, \ldots, \mathbf{v}_i, \ldots, \mathbf{v}_n \rangle$$

Conversely,

$$\mathbf{v}_i = (\mathbf{v}_i \oplus a\mathbf{v}_j) \ominus a\mathbf{v}_j$$

and Theorem 18.7 implies that

$$\mathbf{v}_i \in \langle \mathbf{v}_1, \ldots, \mathbf{v}_{i-1}, \mathbf{v}_i \oplus a\mathbf{v}_j, \mathbf{v}_{i+1}, \ldots, \mathbf{v}_n \rangle$$

so that

$$\langle \mathbf{v}_1, \ldots, \mathbf{v}_i, \ldots, \mathbf{v}_n \rangle \subseteq \langle \mathbf{v}_1, \ldots, \mathbf{v}_{i-1}, \mathbf{v}_i \oplus a\mathbf{v}_j, \mathbf{v}_{i+1}, \ldots, \mathbf{v}_n \rangle$$

The result follows. $\qquad\qquad\qquad\qquad\qquad\qquad\qquad\qquad\qquad\quad$ □

This theorem assures us that elementary operations with domain S are actually operations on the span of S. In the event $S = \{\mathbf{v}_1, \mathbf{v}_2, \ldots, \mathbf{v}_n\}$ is a basis of V, we have that the set S' obtained from S by an elementary operation is also a basis of V since S' is also a spanning set containing n nonzero vectors.

Corollary 19.2

Let $S = \{\mathbf{v}_1, \mathbf{v}_2, \ldots, \mathbf{v}_n\}$ be a basis of V, and let S' be the subset of V obtained by applying an elementary operation to the elements of S. Then S' is a basis of V.

In light of this result, it is appropriate to develop a notation for elementary operations as applied to a basis $\{\mathbf{v}_1, \mathbf{v}_2, \ldots, \mathbf{v}_n\}$ of V. We will begin by defining

$$\mathscr{E}_{ij}(\mathbf{v}_k) = \begin{cases} \mathbf{v}_k, & k \neq i, j \\ \mathbf{v}_j, & k = i \\ \mathbf{v}_i, & k = j \end{cases}$$

Then $\mathscr{E}_{ij}$ is an elementary operation of type $\mathscr{E}$, which has the effect of interchanging the positions of $\mathbf{v}_i$ and $\mathbf{v}_j$ in our ordered basis.

Example 19.22

If $S = \{[1, 0, 2], [2, 1, -1], [3, -1, 0]\}$, then

$$\mathscr{E}_{13}(S) = \{[3, -1, 0], [2, 1, -1], [1, 0, 2]\}$$

and

$$\mathscr{E}_{12}(S) = \{[2, 1, -1], [1, 0, 2], [3, -1, 0]\}$$

Next consider the operation $\mathscr{M}$ applied to our fixed ordered basis $\{v_1, v_2, \ldots, v_n\}$. For $a \in F - \{0\}$, define

$$\mathscr{M}_{(a)i}(v_k) = \begin{cases} v_k & \text{if } k \neq j \\ av_i & \text{if } k = i \end{cases}$$

Then $\mathscr{M}_{(a)i}$ is an elementary operation of type $\mathscr{M}$, which has the effect of replacing the vector v_i by the vector av_i.

Example 19.23

If $S = \{[1, 0, 2], [2, 1, -1], [3, -1, 0]\}$, then

$$\mathscr{M}_{(-2)2}(S) = \{[1, 0, 2], [-4, -2, 2], [3, -1, 0]\}$$

and

$$\mathscr{M}_{(3)3}(S) = \{[1, 0, 2], [2, 1, -1], [9, -3, 0]\}$$

Finally, if $a \in F$, $i \neq j$, define

$$\mathscr{A}_{(a)i+j}(v_k) = \begin{cases} v_i & \text{if } k \neq j \\ av_i \oplus v_j & \text{if } k = j \end{cases}$$

$\mathscr{A}_{(a)i+j}$ is an operation of type α, which has the effect of replacing the vector v_j by the vector $av_i \oplus v_j$.

Example 19.24

If $S = \{[1, 0, 2], [2, 1, -1], [3, -1, 0]\}$, then

$$\mathscr{A}_{(-2)2+3}(S) = \{[1, 0, 2], [2, 1, -1], [-1, -3, 2]\}$$

and

$$\mathscr{A}_{(1)2+1}(S) = \{[3, 1, 1], [2, 1, -1], [3, -1, 0]\}$$

We can now proceed to investigate the effect of an elementary operation on an ordered basis of a vector space V. It is desirable to utilize our knowledge of linear transformations; however, elementary operations, as defined, are not linear transformations. To extend our definition of elementary operations so as to obtain a linear transformation, we need a preliminary result.

| Lemma 19.1 |

Let $\{v_1, v_2, \ldots, v_n\}$ be a basis of V and $\{w_1, w_2, \ldots, w_i\}$ be a subset of V. If τ is a mapping on V defined by

$$\tau\left(\sum_{i=1}^{n} a_i v_i\right) = \sum_{i=1}^{n} a_i w_i$$

for $a_i \in F$, $i = 1, 2, \ldots, n$, then τ is a linear transformation such that $\tau(v_i) = w_i$ for $i = 1, 2, \ldots, n$.

Proof

Since $\{v_1, v_2, \ldots, v_n\}$ spans V, every element of V can be expressed in the form

$$\sum_{i=1}^{n} a_i v_i$$

for some $a_i \in F$, $i = 1, 2, \ldots, n$. Therefore τ is defined for all vectors in V. The linearity of τ follows from the definition. (See Exercise 2.) Finally,

$$\tau(v_i) = \tau(0v_1 \oplus \cdots \oplus 0v_{i-1} \oplus 1v_i \oplus 0v_{i+1} \oplus \cdots \oplus 0v_n)$$
$$= 0w_i \oplus \cdots \oplus 0w_{i-1} \oplus 1w_i \oplus 0w_{i+1} \oplus \cdots \oplus 0w_n$$
$$= w_i \qquad\qquad \square$$

Thus, an elementary operation can be considered to be a linear transformation if we require that it preserves vector addition and scalar multiplication.

| Definition 19.10 |

Let $S = \{v_1, v_2, \ldots, v_n\}$ be basis of V and $\tau: V \to V$ be an elementary operation with domain S with the property that

$$\tau\left(\sum_{i=1}^{n} a_i v_i\right) = \sum_{i=1}^{n} a_i \tau(v_i)$$

for all $a_i \in F$, $i = 1, 2, \ldots, n$. Then τ is termed an **elementary transformation on** V.

Henceforth we will use the symbols $\mathscr{E}_{ij}$, $\mathscr{M}_{(a)i}$, and $\mathscr{A}_{(a)i+j}$ to denote elementary transformations of V. One can easily see that $\mathscr{E}_{ii}$ is the identity transformation on V and $\mathscr{E}_{ij} = \mathscr{E}_{ji}$. Moreover, $\mathscr{M}_{(a)i}$ is the identity if and only if $a = 1$ (see Exercise 3), and $\mathscr{A}_{(a)i+j}$ is the identity transformation on V if and only if $a = 0$. (See Exercise 4.)

If we apply Corollary 19.2 to elementary transformations, we conclude that each elementary transformation on V is an isomorphism, and hence, invertible. It is interesting to note that the inverse of an elementary transformation is another elementary transformation. (See Exercises 5, 6, 7.)

To see how we may use elementary operations to solve a system of equations, let us assume we have a system of equations $AX = B$ that has the following special form:

$$\begin{bmatrix} 1 & 0 & 0 & \cdots & 0 & c_{1,k+1} & \cdots & c_{1,n} \\ 0 & 1 & 0 & \cdots & 0 & c_{2,k+1} & \cdots & c_{2,n} \\ \vdots & \vdots & \vdots & \vdots & \vdots & \vdots & \vdots & \vdots \\ 0 & 0 & 0 & \cdots & 1 & c_{k,k+1} & \cdots & c_{k,n} \\ 0 & 0 & 0 & \cdots & 0 & 0 & \cdots & 0 \\ \vdots & \vdots & \vdots & \vdots & \vdots & \vdots & \vdots & \vdots \\ 0 & 0 & 0 & \cdots & 0 & 0 & \cdots & 0 \end{bmatrix} \begin{bmatrix} x_1 \\ x_2 \\ \vdots \\ x_n \end{bmatrix} = \begin{bmatrix} 0 \\ 0 \\ \vdots \\ 0 \end{bmatrix}$$

If we apply the underlying transformation τ_A to the usual basis of F_n, we see that k is the dimension of the rank space of τ_A. By Theorem 18.19, $n - k$ must be the dimension of the kernel of τ_A.

We can obtain these results another way by changing this system back to equation form:

$$\begin{cases} x_1 & +c_{1,k+1}x_{k+1} + \cdots + c_{1,n}x_n = 0 \\ & x_2 & +c_{2,k+1}x_{k+1} + \cdots + c_{2,n}x_n = 0 \\ \vdots & \vdots & \vdots \\ & & x_k & +c_{k,k+1}x_{k+1} + \cdots + c_{k,n}x_n = 0 \end{cases}$$

We can rewrite these equations as follows

$$\begin{cases} x_1 = -c_{1,k+1}x_{k+1} - \cdots - c_{1,n}x_n \\ x_2 = -c_{2,k+1}x_{k+1} - \cdots - c_{2,n}x_n \\ \vdots \qquad \vdots \qquad \vdots \\ x_k = -c_{k,k+1}x_{k+1} - \cdots - c_{k,n}x_n \end{cases}$$

In this form, we see that a vector in the kernel of τ_A can be written as

$$\begin{bmatrix} -c_{1,k+1}x_{k+1} - \cdots - c_{1,n}x_n \\ -c_{2,k+1}x_{k+1} - \cdots - c_{2,n}x_n \\ \vdots \\ -c_{k,k+1}x_{k+1} - \cdots - c_{k,n}x_n \\ x_{k+1} \\ \vdots \\ x_n \end{bmatrix}$$

We can rewrite this vector in the following form:

$$(x_{k+1})\begin{bmatrix} -c_{1,k+1} \\ -c_{2,k+1} \\ \vdots \\ -c_{k,k+1} \\ 1 \\ 0 \\ \vdots \\ 0 \end{bmatrix} + \cdots + (x_n)\begin{bmatrix} -c_{1,n} \\ -c_{2,n} \\ \vdots \\ -c_{k,n} \\ 0 \\ 0 \\ \vdots \\ 1 \end{bmatrix}$$

These vectors that span the kernel of τ_A are clearly linearly independent; therefore, they must be a basis for the kernel (null space).

A system of equations having the special matrix representation as just shown is easily solved. We use this as motivation for the following definition and discussion.

Definition 19.11

Let $A = [a_{ij}] \in F^{m \times n}$. A is said to be in **reduced row echelon form** if it satisfies each of the following:

i. The first nonzero element in any row is 1.
ii. If the first nonzero element in row i is a_{ij}, then $a_{kj} = 0$ for $k \neq i$.
iii. If the first nonzero element in row i is a_{ij} and the first nonzero entry of row $i + 1$ is $a_{i+1,k}$, then $k > j$.
iv. If row k is all zeros, then each row below it consists of all zeros.

Example 19.25 The following matrices are in reduced row echelon form:

$$
\begin{bmatrix} 1 & 0 & 0 \\ 0 & 1 & 0 \\ 0 & 0 & 0 \end{bmatrix}
\quad
\begin{bmatrix} 1 & 2 & 0 \\ 0 & 0 & 1 \\ 0 & 0 & 0 \\ 0 & 0 & 0 \end{bmatrix}
\quad
\begin{bmatrix} 1 & 0 & 4 & 0 \\ 0 & 1 & 3 & 0 \\ 0 & 0 & 0 & 1 \end{bmatrix}
$$

$$
\begin{bmatrix} 1 & 2 & 4 & 0 \\ 0 & 0 & 0 & 0 \\ 0 & 0 & 0 & 0 \end{bmatrix}
\quad
\begin{bmatrix} 1 & 3 & 0 & 5 & 0 & 7 \\ 0 & 0 & 1 & 4 & 0 & 8 \\ 0 & 0 & 0 & 0 & 1 & 6 \end{bmatrix}
$$

Up to this point, we have applied elementary operations to spanning sets for vector subspaces. Henceforth, we shall utilize them on the rows of the matrices representing systems of linear equations in order to obtain a reduced row echelon form.

Definition 19.12

Two matrices A, $B \in F^{m \times n}$ are **row equivalent** if there exists a sequence of elementary operations taking A to B.

If A is taken to B by an elementary operation, we write $A \sim B$. The symbol $\sim$ means **is equivalent to**.

Example 19.26 Since

$$
\begin{bmatrix} 1 & 2 & -1 & 0 \\ 2 & 3 & 1 & 1 \\ 1 & 1 & 2 & 1 \end{bmatrix}
\sim
\begin{bmatrix} 1 & 2 & -1 & 0 \\ 2 & 3 & 1 & 1 \\ 0 & -1 & 3 & 1 \end{bmatrix}
\sim
\begin{bmatrix} 1 & 2 & -1 & 0 \\ 0 & -1 & 3 & 1 \\ 0 & -1 & 3 & 1 \end{bmatrix}
$$

$$
\sim
\begin{bmatrix} 1 & 2 & -1 & 0 \\ 0 & -1 & 3 & 1 \\ 0 & 0 & 0 & 0 \end{bmatrix}
\sim
\begin{bmatrix} 1 & 0 & 5 & 2 \\ 0 & -1 & 3 & 1 \\ 0 & 0 & 0 & 0 \end{bmatrix}
$$

$$
\sim
\begin{bmatrix} 1 & 0 & 5 & 2 \\ 0 & 1 & -3 & -1 \\ 0 & 0 & 0 & 0 \end{bmatrix}
$$

we see that

$$\begin{bmatrix} 1 & 2 & -1 & 0 \\ 2 & 3 & 1 & 1 \\ 1 & 1 & 2 & 1 \end{bmatrix} \quad \text{and} \quad \begin{bmatrix} 1 & 0 & 5 & 2 \\ 0 & 1 & -3 & -1 \\ 0 & 0 & 0 & 0 \end{bmatrix}$$

are row equivalent.

We know by Theorem 19.8 that elementary operations do not change the span and, hence, the dimension of the span of a set of vectors. This assures us that if two matrices are row equivalent, then they must have the same number of independent rows.

In Example 19.26, the final matrix was in reduced row echelon form. Given a matrix A, it is always possible to find a reduced row echelon matrix that is row equivalent to A:

Theorem 19.9

Every matrix is row equivalent to a matrix in reduced row echelon form.

Proof

Let $A = [a_{ij}] \in F^{m \times n}$. If A is the zero matrix, then A is already in reduced row echelon form. We may assume that A is a nonzero matrix. Consider the elements in the first column, a_{i1}, $i = 1, 2, \ldots, n$. If any of these elements is nonzero, use elementary row interchanges (if necessary) to get that row with a nonzero first element to be the first row. Then multiply the first row of the new matrix (which is row equivalent to the first matrix) by the multiplicative inverse of its first entry. We now have a matrix that is row equivalent to the first and has the form:

$$\begin{bmatrix} 1 & * & * & * & \cdots \\ * & * & * & * & \cdots \\ \vdots & \vdots & \vdots & \vdots & \vdots \end{bmatrix}$$

Having obtained this form, we now add appropriate multiples of the first row to the succeeding rows and obtain a matrix that has the form

$$\begin{bmatrix} 1 & * & * & * & \cdots \\ 0 & * & * & * & \cdots \\ 0 & * & * & * & \cdots \\ \vdots & \vdots & \vdots & \vdots & \vdots \end{bmatrix}$$

If the elements in the first column are all zeros, then consider the second column and utilize the same procedure. If the second column is all zeros, then consider the third, and so on until a 1 is obtained in the first row. By this procedure, we obtain a matrix that, in general, has the form

$$
\begin{bmatrix}
0 & 0 & \cdots & 1 & * & * & * \\
0 & 0 & \cdots & 0 & * & * & * \\
0 & 0 & \cdots & 0 & * & * & * \\
\vdots & \vdots & \vdots & \vdots & \vdots & \vdots & \vdots \\
0 & 0 & \cdots & 0 & * & * & *
\end{bmatrix}
$$

Now, by using the principles of mathematical induction on the number of rows of A, we assume we have obtained a matrix in which: the first nonzero element in each of the first k rows is 1; there are zeros above and below each of these first nonzero entries; and the 1's move progressively to the right with each row.

$$
\begin{bmatrix}
0 & \cdots & 1 & \cdots & 0 & \cdots & 0 & * & * \\
0 & \cdots & 0 & \cdots & 1 & \cdots & 0 & * & * \\
\vdots & \vdots & \vdots & \vdots & \vdots & \vdots & \vdots & \vdots & \vdots \\
0 & \cdots & 0 & \cdots & 0 & \cdots & 0 & * & * \\
0 & \cdots & 0 & \cdots & 0 & \cdots & 0 & * & * \\
\vdots & \vdots & \vdots & \vdots & \vdots & \vdots & \vdots & \vdots & \vdots \\
0 & \cdots & 0 & \cdots & 0 & \cdots & 0 & * & *
\end{bmatrix}
$$

Assume the first nonzero element in the kth row is in the jth column. Then consider the elements $a_{i,\,j+1}$ for $i > k$. If one of these elements is nonzero, use row interchanges to place it in position $a_{k+1,\,j+1}$, then multiply by the inverse of the element, and use multiples of the row added to other rows to obtain zeros for all the other entries in that column. If the elements $a_{i,\,j+1}$ for $i > k$ are all zeros, then move to the first column on the right that contains a nonzero entry among the last $m - k$ terms. Repeat the process just given to obtain a 1 as the first nonzero entry in row $k + 1$ and, as before, obtain zeros for all other entries in that column. If no such nonzero entry exists, we are done.

By mathematical induction, we conclude that the original matrix is row equivalent to a matrix in reduced row echelon form.   □

The procedure used in the proof of the preceding result is known as **Gauss-Jordan Elimination.**[1]

Example 19.27

The following example illustrates the Gauss-Jordan elimination procedure as applied to a matrix to determine its reduced row echelon form:

$$\begin{bmatrix} 0 & 2 & 1 & 3 \\ 2 & 2 & 0 & 4 \\ 0 & 4 & 2 & 6 \end{bmatrix} \sim \begin{bmatrix} 2 & 2 & 0 & 4 \\ 0 & 2 & 1 & 3 \\ 0 & 4 & 2 & 6 \end{bmatrix} \sim \begin{bmatrix} 1 & 1 & 0 & 2 \\ 0 & 2 & 1 & 3 \\ 0 & 4 & 2 & 6 \end{bmatrix}$$

$$\sim \begin{bmatrix} 1 & 1 & 0 & 2 \\ 0 & 1 & \frac{1}{2} & \frac{3}{2} \\ 0 & 4 & 2 & 6 \end{bmatrix} \sim \begin{bmatrix} 1 & 0 & -\frac{1}{2} & \frac{1}{2} \\ 0 & 1 & \frac{1}{2} & \frac{3}{2} \\ 0 & 0 & 0 & 0 \end{bmatrix}$$

The Gauss-Jordan elimination procedure can be used to solve systems of linear equations. The definition of elementary operations was motivated by procedures involved in solving a system of linear equations. Given a system of linear equations, the solution to the system is unchanged if we interchange the position of two of the equations, multiply any equation by a nonzero constant, or replace an equation by the sum of that equation and a multiple of another equation. Elementary operations applied to a system of equations or, more easily, to the rows of the matrix representing the system, yield another system with the same solution as the original. The Gauss-Jordan procedure produces a system whose solution is easily obtained. Consider the following examples:

Example 19.28

Solve

$$\begin{cases} x_1 - x_2 + 2x_3 = 0 \\ 2x_1 + x_2 - x_3 = 0 \\ x_1 + 2x_2 - x_3 = 0 \end{cases}$$

[1] C. Jordan (1838–1922) was a French mathematician whose *Traite de Substitutions*, published in 1870, led to full understanding of the importance of Galois' work. One of the most famous results in mathematics bears his name; the Jordan Curve Theorem, simply stated, asserts that every simple closed curve in the plane has an inside and an outside.

We can represent this system by the matrix

$$\begin{bmatrix} 1 & -1 & 2 \\ 2 & 1 & -1 \\ 1 & 2 & -1 \end{bmatrix}$$

Then

$$\begin{bmatrix} 1 & -1 & 2 \\ 2 & 1 & -1 \\ 1 & 2 & -1 \end{bmatrix} \sim \begin{bmatrix} 1 & -1 & 2 \\ 0 & 3 & -5 \\ 0 & 3 & -3 \end{bmatrix} \sim \begin{bmatrix} 1 & -1 & 2 \\ 0 & 3 & -5 \\ 0 & 0 & 2 \end{bmatrix} \sim \begin{bmatrix} 1 & 0 & 0 \\ 0 & 1 & 0 \\ 0 & 0 & 1 \end{bmatrix}$$

Since this matrix has 3 independent columns, the dimension of the rank space is 3. Because the dimension of the domain space is 3, we know by Theorem 18.19 that the kernel has dimension 0. Therefore the solution to this system of equations must be $x_1 = 0$, $x_2 = 0$, $x_3 = 0$.

Example 19.29   Solve

$$\begin{cases} x_1 - x_2 \qquad + x_4 = 0 \\ 2x_1 + x_2 + x_3 - x_4 = 0 \\ x_1 + 2x_2 + x_3 - 2x_4 = 0 \end{cases}$$

Since

$$\begin{bmatrix} 1 & -1 & 0 & 1 \\ 2 & 1 & 1 & -1 \\ 1 & 2 & 1 & -2 \end{bmatrix} \sim \begin{bmatrix} 1 & -1 & 0 & 1 \\ 0 & 3 & 1 & -3 \\ 0 & 3 & 1 & -3 \end{bmatrix}$$

$$\sim \begin{bmatrix} 1 & -1 & 0 & 1 \\ 0 & 1 & \frac{1}{3} & -1 \\ 0 & 0 & 0 & 0 \end{bmatrix} \sim \begin{bmatrix} 1 & 0 & \frac{1}{3} & 0 \\ 0 & 1 & \frac{1}{3} & -1 \\ 0 & 0 & 0 & 0 \end{bmatrix}$$

we have

$$\begin{cases} x_1 \qquad + \frac{1}{3}x_3 \qquad = 0 \\ x_2 + \frac{1}{3}x_3 - x_4 = 0 \end{cases}$$

Looking at the echelon form of the matrix, we realize first that the dimension of the rank space is 2. Since the dimension of the domain space is 4, we know by Theorem 18.19 that the dimension of the null space is 2. Let us determine the null space directly. We have

$$\begin{cases} x_1 = -\tfrac{1}{3}x_3 \\ x_2 = -\tfrac{1}{3}x_3 + x_4 \\ x_3 = x_3 \\ x_4 = x_4 \end{cases}$$

Thus x_3 and x_4 can range over all numbers independently, while x_1 and x_2 are restricted by their relationships to x_3 and x_4.

A vector in the kernel has the form

$$\begin{bmatrix} -\tfrac{1}{3}x_3 \\ -\tfrac{1}{3}x_3 + x_4 \\ x_3 \\ x_4 \end{bmatrix} = \begin{bmatrix} -\tfrac{1}{3}x_3 + 0x_4 \\ -\tfrac{1}{3}x_3 + x_4 \\ x_3 + 0x_4 \\ 0x_3 + x_4 \end{bmatrix} = x_3 \begin{bmatrix} -\tfrac{1}{3} \\ -\tfrac{1}{3} \\ 1 \\ 0 \end{bmatrix} + x_4 \begin{bmatrix} 0 \\ 1 \\ 0 \\ 1 \end{bmatrix}$$

and basis vectors for the kernel are

$$\begin{bmatrix} -\tfrac{1}{3} \\ -\tfrac{1}{3} \\ 1 \\ 0 \end{bmatrix} \quad \text{and} \quad \begin{bmatrix} 0 \\ 1 \\ 0 \\ 1 \end{bmatrix}$$

These last two examples involved systems of linear equations of the form $AX = B$, where the column of scalars B is the zero vector. Such systems are termed **homogeneous** and involve the determination of the kernel of the associated transformation τ_A. Now let us turn to systems that are not homogeneous. If B is not the zero vector, then the system is called **nonhomogeneous**. Such a system can be represented by a single matrix, called a **partitioned matrix**. For example, the system

$$\begin{cases} x_1 - x_2 + x_3 = 2 \\ 2x_1 - x_2 - x_3 = -1 \\ x_1 \quad\quad - 2x_3 = -3 \end{cases}$$

can be written in partitioned matrix form as

$$\begin{bmatrix} 1 & -1 & 1 & \vdots & 2 \\ 2 & -1 & -1 & \vdots & -1 \\ 1 & 0 & -2 & \vdots & -3 \end{bmatrix}$$

The dotted line represents the equality sign in the system. This matrix is termed the **augmented matrix** of the system and is obtained by augmenting or adding the column of scalars

$$\begin{bmatrix} 2 \\ -1 \\ -3 \end{bmatrix}$$

to the **coefficient matrix**

$$\begin{bmatrix} 1 & -1 & 1 \\ 2 & -1 & -1 \\ 1 & 0 & 2 \end{bmatrix}$$

Nonhomogeneous systems can be solved by applying the Gauss-Jordan elimination procedure to the augmented matrix of the system.

Example 19.30   Solve

$$\begin{cases} x_1 - x_2 + x_3 = 2 \\ 2x_1 - x_2 - x_3 = -1 \\ x_1 \quad\quad - 2x_3 = -3 \end{cases}$$

By the Gauss-Jordan elimination procedure, we have

$$\begin{bmatrix} 1 & -1 & 1 & \vdots & 2 \\ 2 & -1 & -1 & \vdots & -1 \\ 1 & 0 & -2 & \vdots & -3 \end{bmatrix} \sim \begin{bmatrix} 1 & -1 & 1 & \vdots & 2 \\ 0 & 1 & -3 & \vdots & -5 \\ 0 & 1 & -3 & \vdots & -5 \end{bmatrix} \sim \begin{bmatrix} 1 & 0 & -2 & \vdots & -3 \\ 0 & 1 & -3 & \vdots & -5 \\ 0 & 0 & 0 & \vdots & 0 \end{bmatrix}$$

Therefore, our original system is equivalent to the system

$$\begin{cases} x_1 - 2x_3 = -3 \\ x_2 - 3x_3 = -5 \end{cases}$$

Any solution must be of the form

$$\begin{bmatrix} x_1 \\ x_2 \\ x_3 \end{bmatrix} = \begin{bmatrix} 2x_3 - 3 \\ 3x_3 - 5 \\ x_3 \end{bmatrix} = x_3 \begin{bmatrix} 2 \\ 3 \\ 1 \end{bmatrix} + \begin{bmatrix} -3 \\ -5 \\ 0 \end{bmatrix}$$

Note that the kernel of the associated homogeneous transformation is

$$\left\langle \begin{bmatrix} 2 \\ 3 \\ 1 \end{bmatrix} \right\rangle$$

so that the solution to this nonhomogeneous system is a coset of the subgroup of $\mathbb{R}_3$ determined by the kernel. Setting $x_3 = 0$, we find that

$$\begin{bmatrix} -3 \\ 5 \\ 0 \end{bmatrix}$$

is a **particular solution** to the system. The general solution to a nonhomogeneous system of linear equations can always be expressed as a particular solution plus the solution set of the associated homogeneous set of equations.

Example 19.31

Solve

$$\begin{cases} x_1 - x_2 \quad\;\; = 3 \\ 2x_1 - x_2 + x_3 = 2 \\ x_1 + x_2 - x_3 = 0 \end{cases}$$

Since

$$\begin{bmatrix} 1 & -1 & 0 & \vdots & 3 \\ 2 & -1 & 1 & \vdots & 2 \\ 1 & 1 & -1 & \vdots & 0 \end{bmatrix} \sim \begin{bmatrix} 1 & -1 & 0 & \vdots & 3 \\ 0 & 1 & 1 & \vdots & -4 \\ 0 & 2 & -1 & \vdots & -3 \end{bmatrix} \sim \begin{bmatrix} 1 & -1 & 0 & \vdots & 3 \\ 0 & 1 & 1 & \vdots & -4 \\ 0 & 0 & -3 & \vdots & 5 \end{bmatrix}$$

$$\sim \begin{bmatrix} 1 & -1 & 0 & \vdots & 3 \\ 0 & 1 & 0 & \vdots & -\frac{7}{3} \\ 0 & 0 & 1 & \vdots & -\frac{5}{3} \end{bmatrix} \sim \begin{bmatrix} 1 & 0 & 0 & \vdots & \frac{2}{3} \\ 0 & 1 & 0 & \vdots & -\frac{7}{3} \\ 0 & 0 & 1 & \vdots & -\frac{5}{3} \end{bmatrix}$$

we have that

$$\begin{bmatrix} x_1 \\ x_2 \\ x_3 \end{bmatrix} = \begin{bmatrix} \frac{2}{3} \\ -\frac{7}{3} \\ -\frac{5}{3} \end{bmatrix} = \begin{bmatrix} \frac{2}{3} \\ -\frac{7}{3} \\ -\frac{5}{3} \end{bmatrix} + x \begin{bmatrix} 0 \\ 0 \\ 0 \end{bmatrix}$$

is the only solution to the original system.

Example 19.32   Solve

$$\begin{cases} x_1 - 2x_2 + x_3 = 1 \\ x_1 - x_2 - 2x_3 = 0 \\ x_1 \qquad\;\; - 5x_3 = 2 \end{cases}$$

Now

$$\begin{bmatrix} 1 & -2 & 1 & \vdots & 1 \\ 1 & -1 & -2 & \vdots & 0 \\ 1 & 0 & -5 & \vdots & 2 \end{bmatrix} \sim \begin{bmatrix} 1 & -2 & 1 & \vdots & 1 \\ 0 & 1 & -3 & \vdots & -1 \\ 0 & 2 & -6 & \vdots & 1 \end{bmatrix} \sim \begin{bmatrix} 1 & -2 & 1 & \vdots & 1 \\ 0 & 1 & -3 & \vdots & -1 \\ 0 & 0 & 0 & \vdots & 3 \end{bmatrix}$$

At this point, we can see that the vector

$$\begin{bmatrix} 1 \\ -1 \\ 3 \end{bmatrix}$$

is not in the subspace spanned by the vectors

$$\begin{bmatrix} 1 \\ 0 \\ 0 \end{bmatrix}, \quad \begin{bmatrix} -2 \\ 1 \\ 0 \end{bmatrix}, \quad \begin{bmatrix} 1 \\ -3 \\ 0 \end{bmatrix}$$

and, therefore, not in the rank space of the original transformation. By Theorem 19.7, this system has no solution. But we can show this directly since the final matrix corresponds to the system

$$\begin{cases} x_1 - 2x_2 + x_3 = 1 \\ \qquad\;\; x_2 - 3x_3 = 1 \\ \qquad\qquad\quad\; 0 = 3 \end{cases}$$

We argue that if our original system had a solution, then 0 must equal 3. Therefore no solution exits.

In this section, we have explored just a part of the rich theory involving systems of linear equations, which can be found in any text on linear algebra. We hope that we have shown you how this material is connected to some of the basic ideas in abstract algebra, such as cosets, linear transformations or vector space homomorphisms, and linear dependence and independence. In the next section, we turn to a much different application, namely, matrix group coding.

1. Prove Theorem 19.8 for an elementary operation of type $\mathcal{M}$.

2. Verify that the mapping τ of Lemma 19.1 is a linear transformation.

3. Prove $\mathcal{M}_{(a)i}$ is the identity transformation if and only if $a = 1$.

4. Prove that $\mathcal{A}_{(a)i+j}$ is the identity transformation if and only if $a = 0$.

5. Prove that the inverse of $\mathcal{E}_{ij}$ is $\mathcal{E}_{ji}$.

6. Verify that the inverse of $\mathcal{M}_{(a)i}$ is $\mathcal{M}_{(a^{-1})i}$.

7. Show that the inverse of $\mathcal{A}_{(a)i+j}$ is $\mathcal{A}_{(-a)i+j}$.

8. For each of the following matrices, determine the row reduced echelon form:

a. $\begin{bmatrix} 1 & 3 & 2 \\ -1 & 2 & -1 \end{bmatrix}$

b. $\begin{bmatrix} 1 & 2 & 0 \\ 3 & -1 & 7 \\ -1 & 3 & -5 \end{bmatrix}$

c. $\begin{bmatrix} 1 & 0 & -1 & 1 \\ 3 & 1 & 2 & 0 \\ 2 & 3 & 4 & 3 \end{bmatrix}$

d. $\begin{bmatrix} 1 & -1 & 1 \\ 4 & 1 & 2 \\ -1 & 0 & 3 \\ 2 & 3 & -1 \end{bmatrix}$

e. $\begin{bmatrix} 4 & 3 & 1 \\ 1 & 1 & 3 \\ 2 & 2 & 2 \end{bmatrix}$

f. $\begin{bmatrix} 1 & 2 \\ 3 & 1 \\ -1 & -3 \end{bmatrix}$

9. Let $\tau: F_3 \to F_3$ be an elementary operation on the usual basis. Determine the matrix representation of τ with respect to the usual basis of F_3 for τ equal to
   a. $\mathcal{E}_{13}$ b. $\mathcal{M}_{(2)3}$ c. $\mathcal{A}_{2(1)+2}$

10. Verify that the system below has an infinite number of solutions in $\mathbb{R}^3$ but only a finite number of solutions in $(\mathbb{Z}_3)^3$.

$$\begin{cases} 2x_1 - x_2 + x_3 = 1 \\ x_1 + x_2 + x_3 = -1 \end{cases}$$

11. Use elementary operations to determine the linear dependence or independence of each of the following sets of vectors.

a. $\begin{bmatrix} 1 \\ 2 \\ -2 \end{bmatrix}$, $\begin{bmatrix} 2 \\ 0 \\ 2 \end{bmatrix}$, $\begin{bmatrix} 2 \\ 3 \\ -1 \end{bmatrix}$

b. $\begin{bmatrix} 1 \\ 3 \\ -2 \end{bmatrix}$, $\begin{bmatrix} 1 \\ 0 \\ 2 \end{bmatrix}$, $\begin{bmatrix} 2 \\ 3 \\ -1 \end{bmatrix}$

c.
$$\begin{bmatrix} 1 \\ 3 \\ -1 \\ 2 \end{bmatrix}, \begin{bmatrix} -1 \\ 1 \\ 2 \\ -1 \end{bmatrix}, \begin{bmatrix} 0 \\ 1 \\ 2 \\ -1 \end{bmatrix}, \begin{bmatrix} -1 \\ 2 \\ 1 \\ 2 \end{bmatrix}$$
d. $[3, 1, 2], [2, -1, 1], [1, -1, 2]$

12. Use the Gauss-Jordan elimination procedure to solve the following systems of equations:

a. $\begin{cases} 5x + 7y + 8z = 0 \\ 10x + 9y + 6z = 0 \\ 10y + 5z = 0 \end{cases}$

b. $\begin{cases} x + 2y \quad\quad = 3 \\ 3x + 5y \quad\quad = 1 \\ 7x + 9y + z = 2 \end{cases}$

c. $\begin{cases} x + 5y + 9z = 0 \\ 2x + 6y + 10z = 1 \\ 3x + 7y + 11z = 2 \\ 4x + 8y + 12z = 0 \end{cases}$

d. $\begin{cases} x_1 + x_2 \quad\quad = 5 \\ x_2 + x_3 = 6 \\ x_1 \quad\quad - x_3 = 4 \end{cases}$

e. $\begin{cases} x_1 + 2x_2 - 3x_3 = 4 \\ 2x_1 - 3x_2 + x_3 = 7 \end{cases}$

f. $\begin{cases} x^2 + 2y^2 = 3 \\ x^2 - y^2 = -4 \end{cases}$

13. Due to the rising cost of food, you determine that there are but three types of food that you can afford: applesauce (A), beans (B), and carp (C). In one ounce portions: applesauce has 1 unit of protein, 4 units of carbohydrates, and 3 units of fat; beans have 2, 5, and 3, respectively; and carp has 2, 2, and 0. You need 11 units of protein, 20 units of carbohydrates, and 9 units of fat per day.
   a. Find all possible amounts (in ounces) of the three foods providing precisely these amounts of needed nutrients.
   b. If A costs 40¢/lb, and B and C each cost 20¢/lb, is there a solution costing exactly $1.40?

14. If $A \in F^{m \times n}$, the **rank** of A is defined to be the maximum number of independent rows of A. Prove:
   a. If A is row equivalent to B, then the rank of A is equal to the rank of B.
   b. If A is in row-reduced echelon form, then the rank of A is the number of nonzero rows of A.

15. Prove that row equivalence is an equivalence relation on $F^{m \times n}$.

*16. Write a program to implement the Gauss-Jordan procedure on a given $m \times n$ matrix for $1 \leqslant m, n \leqslant 5$.

19.4

Block Codes

In Chapter 12, we investigated coding theory and saw how group theory can aid in the detection and correction of errors. Now that we have discussed vector spaces and matrices, we can extend our earlier work and investigate coding schemes in which the encoding function $f: B(m) \rightarrow B(n)$ is defined by an $m \times n$ matrix whose entries lie in $\mathbb{Z}_2$. These (m, n) **block** or **matrix codes** were first investigated by R. W. Hamming in 1950.

Recall the parity-check code that added a parity-check digit to each block of code words. For example, the parity check-code $f: B(4) \rightarrow B(5)$ has the property that $f(1000) = 10001$ and $f(1111) = 11110$. We can represent the action of this code by the formula

$$f(x_1, x_2, x_3, x_4) = [x_1, x_2, x_3, x_4, x_1 + x_2 + x_3 + x_4]$$

Note that this formula has an alternate representation in terms of matrix multiplication; namely, if

$$G = \begin{bmatrix} 1 & 0 & 0 & 0 & 1 \\ 0 & 1 & 0 & 0 & 1 \\ 0 & 0 & 1 & 0 & 1 \\ 0 & 0 & 0 & 1 & 1 \end{bmatrix}$$

then

$$f(x_1, x_2, x_3, x_4) = [x_1, x_2, x_3, x_4]G$$

$$= [x_1, x_2, x_3, x_4] \begin{bmatrix} 1 & 0 & 0 & 0 & 1 \\ 0 & 1 & 0 & 0 & 1 \\ 0 & 0 & 1 & 0 & 1 \\ 0 & 0 & 0 & 1 & 1 \end{bmatrix}$$

$$= [x_1, x_2, x_3, x_4, x_1 + x_2 + x_3 + x_4]$$

Actually, if I_4 denotes the 4×4 identity matrix, then G can be written in partitioned form as

$$G = \left[\begin{array}{c:c} I_4 & \begin{matrix} 1 \\ 1 \\ 1 \\ 1 \end{matrix} \end{array} \right] = \left[\begin{array}{cccc:c} 1 & 0 & 0 & 0 & 1 \\ 0 & 1 & 0 & 0 & 1 \\ 0 & 0 & 1 & 0 & 1 \\ 0 & 0 & 0 & 1 & 1 \end{array} \right]$$

It is a straightforward calculation to show if I_m is the $m \times m$ identity matrix and $f: B(m) \rightarrow B(m+1)$ is the parity-check code, then

$$f(x_1, x_2, \ldots, x_m) = [x_1, x_2, \ldots, x_m]G$$

$$= [x_1, x_2, \ldots, x_m] \left[\begin{array}{c|c} & 1 \\ & 1 \\ I_m & \vdots \\ & 1 \end{array} \right]$$

$$= [x_1, x_2, \ldots, x_m] \left[\begin{array}{ccccc|c} 1 & 0 & 0 & \cdots & 0 & 1 \\ 0 & 1 & 0 & \cdots & 0 & 1 \\ \vdots & \vdots & \vdots & \ddots & \vdots & \vdots \\ 0 & 0 & 0 & \cdots & 1 & 1 \end{array} \right]$$

$$= [x_1, x_2, \ldots, x_m, x_1 + x_2 + \cdots + x_m]$$

Thus any parity-check code is a **matrix code**, that is, one that can be realized by matrix multiplication.

How do group codes fit into this scheme? Consider the following group code:

$000 \rightarrow 0000000$ $\quad$ $100 \rightarrow 1000111$

$001 \rightarrow 0011100$ $\quad$ $101 \rightarrow 1011011$

$010 \rightarrow 0100011$ $\quad$ $110 \rightarrow 1100100$

$011 \rightarrow 0111111$ $\quad$ $111 \rightarrow 1111000$

This code $f: B(3) \rightarrow B(7)$ is also a matrix code. The matrix

$$G = \left[\begin{array}{ccccccc} 1 & 0 & 0 & 0 & 1 & 1 & 1 \\ 0 & 1 & 0 & 0 & 0 & 1 & 1 \\ 0 & 0 & 1 & 1 & 1 & 0 & 0 \end{array} \right] = \left[\begin{array}{c|cccc} & 0 & 1 & 1 & 1 \\ I_3 & 0 & 0 & 1 & 1 \\ & 1 & 1 & 0 & 0 \end{array} \right]$$

yields this code by the formula

$$f(x_1, x_2, x_3) = [x_1, x_2, x_3]G$$

$$= [x_1, x_2, x_3, x_3, x_1 + x_3, x_1 + x_2, x_1 + x_2]$$

The important idea here is that any $m \times n$ matrix G with entries in $\mathbb{Z}_2$ gives rise to a group code from $B(m)$ to $B(n)$ by the formula

$$f(x_1, x_2, \ldots, x_m) = [x_1, x_2, \ldots, x_m]G$$

To prove this, we must show that f defined in our last example is actually a group code; that is, if

$$X = [x_1, x_2, \ldots, x_m] \quad \text{and} \quad Y = [y_1, y_2, \ldots, y_m]$$

then $f(X + Y) = f(X) + f(Y)$. But this follows from the basic properties of matrix multiplication; namely,

$$f(X + Y) = (X + Y)G = XG + YG = f(X) + f(Y)$$

Conversely if $f: B(m) \rightarrow B(n)$ is a group code, then there exists an $m \times n$ matrix G with entries in $\mathbb{Z}_2$ such that $f(X) = XG$. To find this **generating matrix** G, simply take the basis unit vectors $100 \cdots 0, 010 \cdots 0, \ldots, 00 \cdots 01$ in $B(m)$ and check their images. These images make up the rows of G.

You might wonder why one would want to use matrices to implement a code. The answer lies in the decoding of transmitted blocks. We showed you how to correct errors and decode blocks using coset leaders. There is an alternate method that utilizes matrices. Consider the following example.

Suppose $f: B(3) \rightarrow B(6)$ is a group code given by the generating matrix

$$G = \begin{bmatrix} 1 & 0 & 0 & 1 & 1 & 1 \\ 0 & 1 & 0 & 0 & 1 & 1 \\ 0 & 0 & 1 & 1 & 0 & 1 \end{bmatrix}$$

Then

$$f(x_1, x_2, x_3) = [x_1, x_2, x_3] \begin{bmatrix} 1 & 0 & 0 & 1 & 1 & 1 \\ 0 & 1 & 0 & 0 & 1 & 1 \\ 0 & 0 & 1 & 1 & 0 & 1 \end{bmatrix}$$

$$= [x_1, x_2, x_3, x_1 + x_3, x_1 + x_2, x_1 + x_2 + x_3]$$

Suppose $[x_1, x_2, x_3]$ is encoded, transmitted, and $[y_1, y_2, y_3, y_4, y_5, y_6]$ is received. If no error is made in transmission, then

$$y_4 = y_1 + y_3, \qquad y_5 = y_1 + y_2, \qquad y_6 = y_1 + y_2 + y_3$$

These equations form a system of three equations, which can be written as

$$\begin{cases} y_1 \quad\ \ + y_3 + y_4 \qquad\qquad = 0 \\ y_1 + y_2 + \qquad\qquad y_5 \quad\ = 0 \\ y_1 + y_2 + y_3 + \qquad\quad + y_6 = 0 \end{cases}$$

since the entries are in $\mathbb{Z}_2$ and $-1 \equiv 1 \pmod 2$. This system has matrix representation

$$
\begin{bmatrix} 1 & 0 & 1 & 1 & 0 & 0 \\ 1 & 1 & 0 & 0 & 1 & 0 \\ 1 & 1 & 1 & 0 & 0 & 1 \end{bmatrix}
\begin{bmatrix} y_1 \\ y_2 \\ y_3 \\ y_4 \\ y_5 \\ y_6 \end{bmatrix}
=
\begin{bmatrix} 0 \\ 0 \\ 0 \end{bmatrix}
$$

The matrix

$$
H = \begin{bmatrix} 1 & 0 & 1 & 1 & 0 & 0 \\ 1 & 1 & 0 & 0 & 1 & 0 \\ 1 & 1 & 1 & 0 & 0 & 1 \end{bmatrix}
$$

is called **parity-check matrix**.

The key to error detection and correction using matrices is the fact that if $[x_1, x_2, x_3]$ is encoded and transmitted and $Y = [y_1, y_2, y_3, y_4, y_5, y_6]$ is received with *no* error, then

$$
HY^T = \begin{bmatrix} 0 \\ 0 \\ 0 \end{bmatrix}
$$

Suppose 111 is the word to be sent and $f(111) = [1, 1, 1]G = [1, 1, 1, 0, 0, 1]$ or 111001 is the word transmitted. If the word 101001 is received, then the error message is 010000. Note that

$$
H \begin{bmatrix} 1 \\ 0 \\ 1 \\ 0 \\ 0 \\ 1 \end{bmatrix}
=
\begin{bmatrix} 1 & 0 & 1 & 1 & 0 & 0 \\ 1 & 1 & 0 & 0 & 1 & 0 \\ 1 & 1 & 1 & 0 & 0 & 1 \end{bmatrix}
\begin{bmatrix} 1 \\ 0 \\ 1 \\ 0 \\ 0 \\ 1 \end{bmatrix}
=
\begin{bmatrix} 0 \\ 1 \\ 1 \end{bmatrix}
$$

The answer or **syndrome**

$$\begin{bmatrix} 0 \\ 1 \\ 1 \end{bmatrix}$$

is not

$$\begin{bmatrix} 0 \\ 0 \\ 0 \end{bmatrix}$$

This tells us that we have an error in transmission. But

$$\begin{bmatrix} 0 \\ 1 \\ 1 \end{bmatrix} = H^2$$

the second column of H. Since

$$H \begin{bmatrix} 0 \\ 1 \\ 0 \\ 0 \\ 0 \\ 0 \end{bmatrix}$$

is also equal to

$$\begin{bmatrix} 0 \\ 1 \\ 1 \end{bmatrix}$$

the error pattern 010000 produces the same syndrome.

Single errors are more likely to occur than double or triple errors. It seems reasonable, therefore, to assume that a single error has occurred in the second entry of the transmitted word. Therefore we correct the received word to $101001 + 010000 = 111001$ and decode as 111. This is called the **maximum-likelihood decoding principle**.

You have to admit that this procedure is easier to use than our previous method developed in Chapter 12 involving the construction of various coset leaders, especially if we are utilizing a computer language that incorporates matrix multiplication. But is this method foolproof? The answer is no. Consider the following situation.

Suppose 111 is encoded as 111001, and 010001 is received. Then

$$H(010001)^T = \begin{bmatrix} 0 \\ 1 \\ 0 \end{bmatrix} = H^5$$

is the syndrome and the maximum-likelihood decoding method would have us correct 010001 as 010011. The message would then be **incorrectly** decoded as 010.

Well, then, when does this matrix method work? The answer is contained in the following result:

Theorem 19.10

Suppose $f: B(m) \to B(n)$ is a code with generator matrix G and corresponding parity-check matrix H. Then all single errors in transmission can be detected and corrected if and only if the columns of H are distinct and nonzero.

Proof

Suppose first that the columns of H are distinct and nonzero. If a word is transmitted as X and received as Y, then $Y = X + E$, where E is the error pattern. By Exercise 15 of Section 19.1, the syndrome will be

$$HY^T = H(X + E)^T = HX^T + HE^T = HE^T$$

since HX^T is the zero column vector. Suppose the error pattern E corresponds to a single error, say in the ith position. Then

$$i\text{th place} \\ \downarrow \\ E = E_i = [0, 0, \ldots, 0, 1, 0, 0, 0, 0, 0]$$

and $HE^T = HE_i^T = H^i$, the ith column of H. (See Exercise 13.) Since the columns are nonzero, this error will be detected. Since the columns are distinct, we know what single location in Y to change. Therefore, we can correct all single errors.

Conversely, suppose some columns of H are identical or equal to the zero column. If $H^i = H^j$ for some $i \neq j$, then a single error in the ith location produces the syndrome $H^i = H^j$. Since we do not know whether to correct the value in the ith location or the jth location, the word cannot be accurately decoded. On the other hand, if H^i is the zero vector, then the syndrome corresponding to a single error in the ith location is H^i, the zero vector. But an error-free transmission also produces the zero vector. Therefore we could not detect such an error.   □

We summarize our matrix decoding scheme as follows: Suppose H is the parity-check matrix corresponding to a matrix code given by the generator matrix G. If the columns of H are nonzero, let $S = HY^T$ be the syndrome corresponding to a transmitted word Y.

1. If S is the zero vector, assume no error and decode, accordingly, by removing the parity-check digits.
2. If $S = H^i$, the ith column of H, assume that there is a single error in the ith entry of Y. Correct and decode.
3. If S is a nonzero vector not equal to any column of H, then there are at least two errors in transmission. Do not decode.

If the probability p of an error in transmission is small, then the scheme just described is quite reliable, since double or triple errors seldom occur.

Recall that the ability of a group code to detect and correct errors depends upon the Hamming distance between code words. In defining a code $f: B(m) \to B(n)$ by a generator matrix G, the usual basis vectors are taken by f to the rows of the matrix G. Since every word in $B(m)$ is a linear combination of the unit vectors, the images of these words in $B(m)$ will be linear combinations of the rows of G. It may happen that the distance between these code words is large enough so that we can detect or correct more than just single errors. In this case, we could resort to the coset decoding techniques outlined in Chapter 12.

A few other comments should be made about block codes. The first is that our discussion has so far centered on parity-check codes of the form $f: B(m) \to B(n)$ given by a generator matrix G whose leftmost part consists of an $m \times m$ identity matrix. These codes have the advantage that each encoded word has the original m-digit code word as its first m coordinates. Decoding in this case simply amounts to "lopping off" the parity-check digits from the corrected word. It is not necessary, however, that the generator matrix have an identity submatrix on the left. Actually any $m \times n$ matrix G with entries in $\mathbb{Z}_2$ can serve as a generating matrix for a group code. This matrix G will encode the elements of $B(m)$ as n-tuples, and the resulting equations will give rise to a homogeneous system of equations with corresponding matrix H. Theorem 12.3 and our decoding scheme still apply.

A second observation on block codes concerns their origin. In our discussion we started with a generator matrix G and derived a parity-check matrix H. We can go the other way. For example, suppose our matrix H is

$$\begin{bmatrix} 1 & 1 & 1 & 0 & 0 \\ 0 & 1 & 0 & 1 & 0 \\ 1 & 1 & 0 & 0 & 1 \end{bmatrix}$$

Then

$$HY^T = \begin{bmatrix} 0 \\ 0 \\ 0 \end{bmatrix}$$

yields the system

$$\begin{cases} y_1 + y_2 + y_3 && = 0 \\ y_2 && + y_4 && = 0 \\ y_1 + y_2 && + y_5 = 0 \end{cases}$$

Therefore, in $\mathbb{Z}_2$, $y_3 = y_1 + y_2$, $y_4 = y_2$ and $y_5 = y_1 + y_2$, and our code $f: B(2) \to B(3)$ is given by

$$f(x_1, x_2) = (x_1, x_2, x_1 + x_2, x_2, x_1 + x_2)$$

with generator matrix

$$G = \begin{bmatrix} 1 & 0 & 1 & 0 & 1 \\ 0 & 1 & 1 & 1 & 1 \end{bmatrix}$$

The connection between a parity-check matrix H and its corresponding generator matrix G is given in Exercise 10.

In Chapter 21, we will investigate block codes further and see how polynomials relate to the detection and correction of transmitted messages.

Exercises 19.4

1. A code $f: B(2) \to B(5)$ is given by the generator matrix

$$G = \begin{bmatrix} 1 & 0 & 1 & 0 & 1 \\ 0 & 1 & 1 & 1 & 1 \end{bmatrix}$$

Determine all the code words in $B(5)$.

2. A code $f: B(4) \to B(7)$ is given by the generator matrix

$$G = \begin{bmatrix} 1 & 0 & 0 & 0 & 1 & 1 & 1 \\ 0 & 1 & 0 & 0 & 0 & 1 & 1 \\ 0 & 0 & 1 & 0 & 1 & 0 & 1 \\ 0 & 0 & 0 & 1 & 1 & 1 & 0 \end{bmatrix}$$

Determine all the code words in $B(7)$.

3. Find a generator matrix for the group code $f: B(3) \to B(5)$ given by the formula
$f(x_1, x_2, x_3) = [x_1, x_2, x_3, x_1 + x_2, x_2 + x_3]$.

4. For the group codes in Exercises 1 and 2, find the corresponding parity-check matrices. Which code(s) correct(s) all single errors?

5. Find the generator and parity-check matrices for the triple-repetition code discussed in Section 12.2.

6. A code $f: B(3) \rightarrow B(7)$ is given by the generator matrix

$$G = \begin{bmatrix} 1 & 0 & 0 & 1 & 1 & 1 \\ 0 & 1 & 0 & 1 & 1 & 1 \\ 0 & 0 & 1 & 0 & 1 & 1 \end{bmatrix}$$

A word is encoded, transmitted, and received as 001100. What is the syndrome? Correct and decode the original word.

7. A group code $f: B(4) \rightarrow B(7)$ has parity-check matrix

$$H = \begin{bmatrix} 1 & 0 & 1 & 1 & 1 & 0 & 0 \\ 1 & 1 & 1 & 0 & 0 & 1 & 0 \\ 0 & 1 & 1 & 1 & 0 & 0 & 1 \end{bmatrix}$$

Use the decoding scheme discussed in the text to decode the message

0010011   1011010   111111   100001

8. A group code $f: B(3) \rightarrow B(6)$ is defined by the generator matrix

$$G = \begin{bmatrix} 1 & 0 & 0 & 0 & 1 & 1 \\ 0 & 1 & 0 & 1 & 0 & 1 \\ 0 & 0 & 1 & 1 & 1 & 0 \end{bmatrix}$$

Use the decoding scheme discussed in the text to decode (if possible) the message

110001   111110   001100   101001

9. Prove that a parity-check matrix H has distinct nonzero columns and if and only if the minimum distance between code words is at least three.

10. Suppose $f: B(m) \rightarrow B(n)$ is a group code with generator matrix G, $r = n - m$, and $f(x_1, x_2, \ldots, x_m) = [x_1, x_2, \ldots, x_m, z_1, z_2, \ldots, z_r]$.
    a. Show that G has the form $G = [I_m \vdots A]$, where A is an $m \times r$ matrix.
    b. Suppose H is the corresponding parity-check matrix. Verify that H has the form $H = [B \vdots I_r]$, where $B = [b_{ij}]$ is the $r \times m$ matrix given by $b_{ij} = a_{ji}$. (In the terminology of matrix algebra, this says $H = [A^T \vdots I_r]$.)

11. Suppose H is an $(n - m) \times m$ parity-check matrix for a code $f: B(m) \rightarrow B(n)$. Set $r = n - m$. Prove that the maximum number of distinct nonzero columns of H is $2^r - 1$. Conclude that such a code whose parity-check matrix has all possible distinct nonzero columns must be one in which $m = 2^r - r - 1$ and $n = 2^r - 1$. (Such codes are termed **Hamming codes**, and they were first investigated by R. W. Hamming in 1950.)

12. If $f: B(m) \to B(n)$ is a block code with parity-check matrix H, prove that the set of all n-digit words Y, such that HY^T is the $n - m$ column matrix of zeros, forms a subspace of $B(n)$.

13. Verify that if H is an $r \times n$ matrix and E_i is a $1 \times n$ matrix with a 1 in the ith location as the only nonzero entry, then

$$H(E_i)^T = H \begin{bmatrix} 0 \\ \vdots \\ 0 \\ 1 \\ 0 \\ \vdots \\ 0 \end{bmatrix} \leftarrow i\text{th coordinate}$$

is equal to the ith column of H.

20

The Theory

of Field Extensions

Dedekind was probably the first mathematician to give an explicit definition of a field: a commutative ring whose nonidentity elements form an abelian group under the second ring operation. However, the concept of a field was implicitly contained in the work of Galois and Abel. Their ideas were responsible for much of the contents of this chapter.

Recall that Abel and Galois were both concerned with the problem of developing a formula for solving an arbitrary polynomial equation of degree greater than four. In this chapter, we will investigate this same question, calling upon many results from previous chapters. In our discussion, we will look more deeply at the structure of fields, both infinite and finite. First we will investigate field extensions, specifically, finite algebraic extensions. Next we define the concept of splitting fields. Finally, we consider a most interesting application of the theory of field extensions to euclidean constructions. This chapter will set the stage for the material on finite fields and Galois Theory in the chapters ahead.

Throughout this discussion, F will denote a field with identity 0 and unity 1.

20.1

Finite Algebraic Extensions

In Chapter 14, we investigated rings and their subrings. In this chapter, we wish to focus our attention on those subrings of F which are themselves fields. It is clear from the example of the rational field $\mathbb{Q}$, and the integers $\mathbb{Z}$, that not every subring of a field need be a subfield. For a subring F of a field E to be a subfield of E, the subring F must contain the unity of E and the *multiplicative* inverse of each of its elements. (See Exercise 1.) The next definition shows how we can consider the relationship between F and E in another way.

Definition 20.1

> An **extension field** of a field F is any field E containing a subfield isomorphic of F.

A field is an extension of any one of its subfields, including itself. By our identification of $\mathbb{R}$ with its corresponding subfield in $\mathbb{C}$, we can consider $\mathbb{C}$ to be an extension of $\mathbb{R}$. If F_1 and F_2 are subfields of F, then not only is $F_1 \cap F_2$ a *subring* of F by Exercise 10 of Section 14.1, but it can be shown quite easily that $F_1 \cap F_2$ is another *subfield* of F. It is a straightforward exercise to extend this result to prove that the intersection of all the subfields of a field F is again a subfield of F. This field is the smallest subfield of F and is termed the **prime subfield** of F. For example, the prime subfields of both the real number field $\mathbb{R}$ and the complex number field $\mathbb{C}$ are isomorphic to the field of rational numbers $\mathbb{Q}$. The details of the proofs of these assertions are left to you as exercises at the end of this section.

Let E be an extension of F and $\alpha \in E$. Define $F(\alpha)$ to be the intersection of all subfields of E containing both F and α. Because of our comments in the preceding paragraph, $F(\alpha)$ is the smallest field containing both F and α. $F(\alpha)$ is termed a **simple extension of F by** α and is said to be obtained by **adjoining** α **to** F.

There is another way to describe $F(\alpha)$. If $f(x) \in F[x]$, then $f(\alpha)$ is contained in $F(\alpha)$. If $f(\alpha) \neq 0$, then $1/f(\alpha)$ is also contained in $F(\alpha)$. The set

$$\left\{ \frac{f(\alpha)}{g(\alpha)} \mid f(x), g(x) \in F[x], g(\alpha) \neq 0 \right\}$$

can be shown to be a field contained in $F(\alpha)$. (See Exercise 6.) Since $F(\alpha)$ is the smallest field containing F and α, $F(\alpha)$ is precisely this set.

Example 20.1

Identify $\mathbb{R}$ with the corresponding subfield of $\mathbb{C}$, and consider $\mathbb{R}(i)$. Every element of $\mathbb{C}$ can be written as $a + bi$ for a, $b \in \mathbb{R}$. Therefore, the smallest subfield of $\mathbb{C}$ containing both $\mathbb{R}$ and i is $\mathbb{C}$ itself; that is, $\mathbb{R}(i) = \mathbb{C}$. This gives us one characterization of $\mathbb{C}$. Secondly, recall that in Chapter 15 we saw how $\mathbb{C}$

is isomorphic to the factor ring $\mathbb{R}[x]/(x^2 - 1)$. Finally, if $f(x) \in \mathbb{R}[x]$, then $f(i)$ can be written in the form $a + bi$, so that

$$\left\{ \frac{f(i)}{g(i)} \,\Big|\, f, g \in \mathbb{R}[x] \right\} = \left\{ \frac{a + bi}{c + di} \,\Big|\, a, b, c, d \in \mathbb{R} \right\}$$

But the multiplicative inverse of $c + di$ is another element in $\mathbb{C}$ of the form $r + si$. Therefore,

$$\left\{ \frac{a + bi}{c + di} \,\Big|\, a, b, c, d \in \mathbb{R} \right\} = \{(a + bi)(r + si) \,|\, a, b, r, s \in \mathbb{R}\}$$

$$= \{u + vi \,|\, u, v \in \mathbb{R}\}$$

$$= \mathbb{C}$$

Example 20.2

Let $F = \mathbb{Q}$, $\alpha = \sqrt{2} \in \mathbb{R}$. Then $K = \{a + b\sqrt{2} \,|\, a, b \in \mathbb{Q}\}$ is a field with identity $0 = 0 + 0\sqrt{2}$ and unity $1 = 1 + 0\sqrt{2}$. (See Exercise 9.) Since $\mathbb{Q}(\sqrt{2})$ contains all elements of the form $a + b\sqrt{2}$, $K \subseteq \mathbb{Q}(\sqrt{2})$. Because $\mathbb{Q}(\sqrt{2})$ is the smallest field containing $\mathbb{Q}$ and $\sqrt{2}$, we have $\mathbb{Q}(\sqrt{2}) \subseteq K$. Thus, $\mathbb{Q}(\sqrt{2}) = K$.

We can extend our definition inductively. If $\alpha_1, \alpha_2, \ldots, \alpha_n \in E$, we define $F(\alpha_1, \alpha_2, \ldots, \alpha_n)$ to be the intersection of all subfields of E containing $F(\alpha_1, \alpha_2, \ldots, \alpha_{n-1})$ and α_n. In this way, $F(\alpha_1, \alpha_2) = (F(\alpha_1))(\alpha_2)$.

Example 20.3

What is the field $\mathbb{Q}(\sqrt{2}, i)$? The set K of all complex numbers of the form $a + b\sqrt{2} + ci + di\sqrt{2}$ for $a, b, c, d \in \mathbb{Q}$ is certainly contained in $\mathbb{Q}(\sqrt{2}, i)$. It is not difficult to see that K is a subring of $\mathbb{C}$. What is $(a + b\sqrt{2} + ci + di\sqrt{2})^{-1}$? We leave it to you to verify in Exercise 12 that

$$(a + b\sqrt{2} + ci + di\sqrt{2})^{-1}$$

$$= \frac{a + b\sqrt{2} - ci - di\sqrt{2}}{(a + b\sqrt{2})^2 + (c + d\sqrt{2})^2}$$

$$= \frac{a + b\sqrt{2} - ci - di\sqrt{2}}{(a^2 + c^2 + 2b^2 + 2d^2) + 2(ab + cd)\sqrt{2}}$$

$$= \frac{(a + b\sqrt{2} - ci - di\sqrt{2})[(a^2 + c^2 + 2b^2 + 2d^2) - 2(ab + cd)\sqrt{2}]}{(a^2 + c^2 + 2b^2 + 2d^2)^2 - 8(ab + cd)^2}$$

But this is another element of K. Therefore K is a field and $\mathbb{Q}(\sqrt{2}, i) = K$.

We have previously remarked that if F is a subfield of E, then E can be thought of as an F-vector space.

Definition 20.2

If E is an extension field of F, then the **degree of E over F** is the dimension of E over F; that is, $\dim_F(E)$. This degree is denoted by $[E:F]$. If $[E:F]$ is finite, then E is termed a **finite extension of F**.

Note $[\mathbb{C}:\mathbb{R}] = 2$ since $\{1, i\}$ is a basis of $\mathbb{C}$ over $\mathbb{R}$. Similarly, $[\mathbb{Q}(\sqrt{2}):\mathbb{Q}] = 2$. Not all extensions are finite, however. In Chapter 17 we proved that if F is a field, then $F[x]$ is an integral domain, a PID, in fact. Consider the field of quotients of $F[x]$. This field of quotients consists of all quotients of polynomials $f(x)/g(x)$, where $g(x) \neq 0$. This field is an infinite extension of F. (See Exercise 10.) Inasmuch as this field is the smallest extension of F containing x, we denote this extension by $F(x)$. $F(x)$ is known as the **field of rational functions of F**.

We will center our study of field extensions on finite extensions. In general, they are easier to analyze and the concept of degree lends itself to proofs by induction.

Theorem 20.1

Let E be a finite extension of K, and K be a finite extension of F. Then E is a finite extension of F and $[E:F] = [E:K][K:F]$.

Proof

Let $\alpha_1, \alpha_2, \ldots, \alpha_n$ be a basis of E over K and $\beta_1, \beta_2, \ldots, \beta_m$ be a basis of K over F. Our proof consists of showing that the mn elements $\alpha_i\beta_j$ form a basis of E over F.

Let $\gamma \in E$. Then there exist $c_1, c_2, \ldots, c_n \in K$ such that

$$\gamma = \sum_{i=1}^{n} c_i\alpha_i$$

Since each $c_i \in K$, there exist $d_{i1}, d_{i2}, \ldots, d_{im} \in F$ such that

$$c_i = \sum_{j=1}^{m} d_{ij}\beta_j$$

But then

$$\gamma = \sum_{i=1}^{n} \left(\sum_{j=1}^{m} d_{ij}\beta_j \right)\alpha_i = \sum_{i=1,j=1}^{n,m} d_{ij}(\alpha_i\beta_j)$$

so that the set $\{\alpha_i\beta_j\}$ spans E.

To show that $\{\alpha_i\beta_j\}$ is an independent set, suppose that

$$0 = \sum_{i=1,j=1}^{n,m} d_{ij}(\alpha_i\beta_j)$$

Then

$$0 = \sum_{i=1}^{n} \left(\sum_{j=1}^{m} d_{ij}\beta_j \right)\alpha_i$$

The independence of $\{\alpha_1, \alpha_2, \ldots, \alpha_n\}$ over K implies that

$$0 = \sum_{j=1}^{m} d_{ij}\beta_j$$

for all i. But now, for each i, the independence of $\{\beta_1, \beta_2, \ldots, \beta_m\}$ over F forces $d_{ij} = 0$ for all j. Thus $\{\alpha_i\beta_j\}$ is an independent set and, hence, a basis.   □

Note in Example 20.3 that $\{1, \sqrt{2}\}$ is a basis of $\mathbb{Q}(\sqrt{2})$ over $\mathbb{Q}$, $\{1, i\}$ is a basis of $\mathbb{Q}(\sqrt{2}, i)$ over $\mathbb{Q}(\sqrt{2})$, and $\{1, \sqrt{2}, i, i\sqrt{2}\}$ is a basis of $\mathbb{Q}(\sqrt{2}, i)$ over $\mathbb{Q}$.

The next result follows immediately from Theorem 20.1 by noting that a finite extension of F is also a finite extension on any subfield containing F.

Corollary 20.1

If $E = F(\alpha_1, \alpha_2, \ldots, \alpha_n\}$ is a finite extension of F, then

$$[E:F] = [E:F(\alpha_1, \alpha_2, \ldots, \alpha_{n-1})] \cdots [F(\alpha_1, \alpha_2):F(\alpha_1)][F(\alpha_1):F]$$

Example 20.4

Consider $E = \mathbb{Q}(\sqrt{2}, \sqrt{5}, \sqrt{10})$. We have already seen that $[\mathbb{Q}(\sqrt{2}):\mathbb{Q}] = 2$. Now $\sqrt{5} \notin \mathbb{Q}(\sqrt{2})$, so that $[\mathbb{Q}(\sqrt{2}, \sqrt{5}):\mathbb{Q}] \geqslant 2$. Since 1 and $\sqrt{5}$ form a basis of $\mathbb{Q}(\sqrt{2}, \sqrt{5})$ over $\mathbb{Q}(\sqrt{2})$, the proof of Theorem 20.1 shows that

$$\{1, \sqrt{2}, \sqrt{5}, \sqrt{2}\sqrt{5} = \sqrt{10}\}$$

is a basis of $\mathbb{Q}(\sqrt{2}, \sqrt{5})$ over $\mathbb{Q}$. Now $\sqrt{10} \in \mathbb{Q}(\sqrt{2}, \sqrt{5})$, so that $[E:\mathbb{Q}] = 4$.

So far we have concerned ourselves with finite and simple extensions. There are, however, many other types of extensions. One is especially important since it is useful in describing extensions that are both finite and simple.

Definition 20.3

> Let E be an extension of F and $\alpha \in E$. α is said to be **algebraic over** F if there exists a nonconstant polynomial $f(x) \in F[x]$ such that $f(\alpha) = 0$. If no such polynomial exists, then α is said to be **transcendental over** F. If every element of E is algebraic over F, then E is termed an **algebraic extension of** F; otherwise, E is called a **transcendental extension of** F.

Every element $a \in F$ is the root of the polynomial $f(x) = x - a$. Thus, F is an algebraic extension of itself. The real numbers π and e are both transcendental over $\mathbb{Q}$, so that $\mathbb{R}$ is a transcendental extension of $\mathbb{Q}$.[1] $\mathbb{C}$ is an algebraic extension of $\mathbb{R}$. To see this, let $a + bi \in \mathbb{C}$. Then $a + bi$ is a root of

$$f(x) = (x - (a + bi))(x - (a - bi)) = x^2 - 2ax + (a^2 + b^2)$$

a polynomial in $\mathbb{R}[x]$.

The next result explores the connection between algebraic and finite extensions.

Theorem 20.2

Let E be a finite extension of F. Then E is an algebraic extension of F.

Proof

Let $n = [E:F]$ and $\alpha \in E$. We must find a nonzero polynomial $f(x) \in F[x]$ with the property that $f(\alpha) = 0$. Since $n = [E:F]$, the elements $1, \alpha, \alpha^2, \ldots, \alpha^n$ must be linearly dependent over F; that is, there exist $a_0, a_1, a_2, \ldots, a_n \in F$ such that $a_0 + a_1\alpha + \cdots + a_n\alpha^n = 0$. If we set

$$f(x) = a_n x^n + a_{n-1}x^{n-1} + \cdots + a_1 x + a_0$$

we have found a polynomial $f(x)$ in $F[x]$ such that $f(\alpha) = 0$. Thus α is algebraic over F. $\qquad\square$

Let α be an element of E algebraic over F. The definition of *algebraic* implies that the set $W = \{f(x) \in F[x] \mid f(\alpha) = 0\}$ is nonempty. Moreover, W is a nontrivial ideal of the principal ideal domain $F[x]$. Therefore, there exists at least one polynomial that generates W. Since any multiple of a generator by a nonzero constant in F is another generator, we may choose a monic polynomial as a generator of W.

Definition 20.4

Let E be an extension of F and $\alpha \in E$ be algebraic over F. The **minimal polynomial of α over** F, denoted $m_{\alpha, F}(x)$, is the monic polynomial of least degree such that $m_{\alpha, F}(x) = 0$. The degree of $m_{\alpha, F}(x)$ is the **degree of α over** F.

We saw in Chapter 16 that a generator of an ideal W in $F[x]$ must have minimal degree among the nonzero elements of W. If there were more than one monic generator of W, then the difference of these polynomials would be an element of W of less degree, contrary to our choice of a generator. Therefore, a monic generator must be unique and justifies our description of $m_{\alpha, F}(x)$ as **the** monic polynomial. Since F is a field, $m_{\alpha, F}(x)$ must always be

---

[1]Charles Hermite proved that e is a transcendental number in 1873. A German mathematician, Ferdinand Lindemann, proved that π is transcendental in 1882.

irreducible over F. (See Exercise 16.) The minimal polynomial of $\sqrt{2}$ over $\mathbb{Q}$ is $x^2 - 2$. Hence $\sqrt{2}$ has degree 2 over $\mathbb{Q}$. But the minimal polynomial of $\sqrt{2}$ over $\mathbb{R}$ is $x - \sqrt{2}$. Thus $\sqrt{2}$ has degree 1 over $\mathbb{R}$. These examples point out the necessity of keeping track of the base field F.

Theorem 20.3

If E is an extension of F and $\alpha \in E$ is algebraic over F of degree n, then $[F(\alpha):F] = n$.

Proof

We claim that $\Gamma = \{1, \alpha, \alpha^2, \ldots, \alpha^{n-1}\}$ is a basis of $F(\alpha)$ over F. We first show that Γ is independent over F. If

$$\sum_{i=1}^{n-1} a_i \alpha^i = 0$$

then $f(x) = a_{n-1}x^{n-1} + \cdots + a_1 x + a_0$ is a polynomial of degree at most $n - 1$, which vanishes at α. But the fact that the degree of α over F is n implies that $\deg(m_{\alpha,F}(x)) = n$. This forces $f(x)$ to be the zero polynomial. Therefore $a_0 = a_1 = \cdots = a_{n-1} = 0$. Thus Γ is independent over F.

Clearly, $\Gamma \subseteq F(\alpha)$. To show that Γ spans $F(\alpha)$, we recall that $F(\alpha)$ is the smallest field containing both F and α. If

$$m(x) = m_{\alpha,F}(x) = b_0 + b_1 x + \cdots + b_n x^n$$

let F' be the set of all elements of the form $g(\alpha)$, where $g(x) \in F[x]$ and $\deg(g(x)) < n$. Obviously, F' is closed under polynomial addition and subtraction. If $g(\alpha)$, $h(\alpha) \in F'$, then $g(x)h(x) \in F[x]$, and the Polynomial Division Algorithm (Theorem 17.3) implies that

$$g(x)h(x) = q(x)m(x) + r(x)$$

where $\deg(r(x)) < n$. But then

$$g(\alpha)h(\alpha) = q(\alpha)m(\alpha) + r(\alpha) = q(x)0 + r(\alpha) = r(\alpha)$$

so that F' is closed under multiplication. Let $g(\alpha) \in F'$. Since $m(x)$ is irreducible over F and $\deg(g(x)) < n$, we have $(m(x), g(x)) = 1$. By Theorems 16.3 and 17.4, there exist $s(x), t(x) \in F[x]$ such that $1 = s(x)m(x) + t(x)g(x)$. But then

$$1 = s(\alpha)0 + t(\alpha)g(\alpha) = t(\alpha)g(\alpha)$$

If $t(x) = u(x)m(x) + v(x)$ where $\deg(v(x)) < n$, then $t(\alpha) = u(\alpha)0 + v(\alpha) = v(\alpha)$, so that $v(\alpha)$ is the inverse of $g(\alpha)$ and $v(\alpha) \in F'$. Therefore we have shown that every nonzero element of F' has an inverse in F'. Thus F' is a field, $F' \subseteq F(\alpha)$, and F' contains F and α. This forces $F(\alpha) = F'$, so $F(\alpha)$ is spanned by Γ. The result follows.   □

| Corollary 20.2 |

If α is algebraic over F and the degree of α over F is n, then

$$F(\alpha) = \{a_0 + a_1\alpha + \cdots + a_{n-1}\alpha^{n-1} \mid a_i \in F\}$$

This last corollary generalizes our observation that

$$\mathbb{R}(i) = \{a + bi \mid a, b \in \mathbb{R}\}$$

and $\mathbb{Q}(\sqrt{2}) = \{a + b\sqrt{2} \mid a, b \in \mathbb{Q}\}$. We now generalize our theorem to extensions of form $F(\alpha, \beta)$.

| Corollary 20.3 |

If $\alpha_1, \alpha_2, \ldots, \alpha_n$ are algebraic over F, then $F(\alpha_1, \alpha_2, \ldots, \alpha_n)$ is a finite extension of F.

Proof

We use induction on n. The case $n = 1$ is covered by the theorem. Suppose the result is true for $n = k$. Since α_{k+1} is algebraic over F, α_{k+1} is algebraic over $F(\alpha_1, \alpha_2, \ldots, \alpha_k)$. (Why?) Thus, $F(\alpha_1, \alpha_2, \ldots, \alpha_{k+1})$ is a finite extension of $F(\alpha_1, \alpha_2, \ldots, \alpha_k)$. By induction, $F(\alpha_1, \alpha_2, \ldots, \alpha_k)$ is a finite extension of F. We complete the proof by applying Theorem 20.1. □

Theorem 20.1 states that a finite extension of a finite extension is a finite extension. A similar result holds for algebraic extensions.

| Corollary 20.4 |

If E is an algebraic extension of K and if K is an algebraic extension of F, then E is an algebraic extension of F.

Proof

Let $\alpha \in E$. We must show that $F(\alpha)$ is an algebraic extension of F. Suppose α has degree n over K. Therefore we can find elements $a_0, a_1, \ldots, a_n \in K$ such that $0 = a_0 + a_1\alpha + \cdots + a_n\alpha^n$. By Corollary 20.3, $F(a_0, a_1, \ldots, a_n)$ is a finite extension of F. Moreover, $F(a_0, a_1, \ldots, a_n, \alpha)$ is a finite extension of $F(a_0, a_1, \ldots, a_n)$, since whenever α is algebraic over F, α is also algebraic over $F(a_0, a_1, \ldots, a_n)$. Therefore, $F(\alpha)$ is a finite extension of F by Theorem 20.1. The result follows from Theorem 20.2. □

Suppose E is a finite extension of F. If $\alpha \in E$ is algebraic over F, then let $m(x) = m_{\alpha, F}(x)$. Suppose $\beta \in E$, $\beta \neq \alpha$, and $m(\beta) = 0$. How is $F(\beta)$ related to $F(\alpha)$?

Let us look at an example. The minimal polynomial of i over $\mathbb{Q}$ is $x^2 + 1$. The complex number $-i$ is another root of this polynomial. By Corollary 20.2, $\mathbb{Q}(i) = \{a + bi \mid a, b \in \mathbb{Q}\}$, while

$$\mathbb{Q}(-i) = \{a + b(-i) \mid a, b \in \mathbb{Q}\} = \{a - bi \mid a, b \in \mathbb{Q}\}$$

Clearly, $\mathbb{Q}(i)$ and $\mathbb{Q}(-i)$ are isomorphic as fields; in fact, they contain the same elements. The next result generalizes our observation:

Theorem 20.4

Let $p(x)$ be an irreducible polynomial in $F[x]$ of degree n. Suppose further that $p(\alpha) = p(\beta) = 0$ where α and β lie in some extension of F. Then $F(\alpha) \cong F(\beta)$.

Proof

Let us assume $p(x)$ is monic; otherwise, replace $p(x)$ by $p(x)$ divided by its leading coefficient. Then $p(x) = m_{\alpha, F}(x)$ since $p(\alpha) = 0$ implies $m_{\alpha, F}(x)$ divides $p(x)$, an irreducible polynomial. By Corollary 20.2,

$$F(\alpha) = \{a_0 + a_1\alpha + \cdots + a_{n-1}\alpha^{n-1} \mid a_i \in F\}$$

Similarly, $p(x) = m_{\beta, F}(x)$ and $F(\beta) = \{a_0 + a_1\beta + \cdots + a_{n-1}\beta^{n-1} \mid a_i \in F\}$. We will let you prove that the mapping $\phi\colon F(\alpha) \to F(\beta)$ defined by

$$\phi(a_0 + a_1\alpha + \cdots + a_{n-1}\alpha^{n-1}) = a_0 + a_1\beta + \cdots + a_{n-1}\beta^{n-1}$$

is an isomorphism of fields. (See Exercise 19.) □

In the next section, we investigate the following problem: Given $f(x) \in F[x]$, find the smallest field E containing F and not just one, but all roots of $f(x)$.

Exercises 20.1

1. Prove: A nonempty subset F of a field E with $|F| \geqslant 2$ is a subfield of E if and only if $a, b \in F - \{0\}$ implies $a - b, ab^{-1} \in F$.

2. Verify that if F_1 and F_2 are subfields of a field F, then $F_1 \cap F_2$ is a subfield of F.

3. Prove that the prime subfield of a field F is the smallest subfield in F.

4. Show that $\mathbb{Q}$ has no proper subfields.

5. We said in Theorem 14.9 that every ring with unity contains a subring which is isomorphic either to $\mathbb{Z}_n$ for some n or isomorphic to $\mathbb{Z}$. Prove that if F is a field containing $\mathbb{Z}$ isomorphically, then the prime subfield of F is isomorphic to $\mathbb{Q}$.

6. Let E be an extension of F and $\alpha \in E$. Verify that the set of all quotients of the form $p(\alpha)/q(\alpha)$, where $p(x), q(x) \in F[x]$ and $g(\alpha) \neq 0$, is a field containing F and α.

7. Let F be a subfield of E and $S \subseteq E$. Prove that the intersection of all subfields of E containing both F and S is the smallest subfield of E containing F and S.

8. Find all subfields of $\mathbb{Q}(\sqrt{2}, \sqrt{5})$.

9. Prove that $K = \{a + b\sqrt{2} \mid a, b \in \mathbb{Q}\}$ is a field.

10. Prove that $F(x)$, the field of all rational functions of F, is an infinite extension of F.

11. Prove that if E is an extension of F such that $[E:F] = 1$, then $E = F$.

12. Verify the assertion in Example 20.3 that

$$(a + b\sqrt{2} + ci + di\sqrt{2})^{-1}$$

$$= \frac{(a + b\sqrt{2} - ci - di\sqrt{2})[(a^2 + c^2 + 2b^2 + 2d^2) - 2(ab + cd)\sqrt{2}]}{(a^2 + c^2 + 2b^2 + 2d^2)^2 - 8(ab + cd)^2}$$

13. Compute $[\mathbb{Q}\sqrt{3}, \sqrt{7}, \sqrt{336}):\mathbb{Q}]$.

14. Find a basis of $\mathbb{Q}(\sqrt[3]{2})$ over $\mathbb{Q}$.

15. Prove that $m_{\alpha, F}(x)$ is unique.

16. If α is algebraic over F, prove that $m_{\alpha, F}(x)$ is irreducible over F.

17. Let K be an algebraic extension of F, and let E be an algebraic extension of K. If $\alpha \in E$, prove that $m_{\alpha, E}(x)$ divides $m_{\alpha, F}(x)$.

18. Find the minimal polynomial of $\sqrt[3]{2}$ over $\mathbb{Q}$, over $\mathbb{R}$, and over $\mathbb{C}$.

19. If α and β are roots of the irreducible monic polynomial $p(x) \in F[x]$ of degree n, prove that the mapping defined by

$$\phi(a_0 + a_1\alpha + \cdots + a_{n-1}\alpha^{n-1}) = a_0 + a_1\beta + \cdots + a_{n-1}\beta^{n-1}$$

is an isomorphism between $F(\alpha)$ and $F(\beta)$.

20. Suppose F, K, and E are fields such that $F \subseteq K \subseteq E$. If $\alpha \in E$ is algebraic over F, prove that α is algebraic over K.

21. Show that if α and β lie in some extension of F, then $F(\alpha, \beta) = F(\beta, \alpha)$.

22. Suppose α is algebraic of prime degree over F. Prove $F(\alpha)$ does not contain a subfield K such that $F \subset K \subset F(\alpha)$.

23. Let E be an extension of F.
    a. If $\alpha, \beta \in E$ are algebraic over F, show that $\alpha + \beta$ and $\alpha\beta$ are also algebraic over F.
    b. Suppose α is algebraic over F and $m_{\alpha, F}(x)$ has degree n. Prove that α^{-1} is algebraic over F and that $m_{\alpha^{-1}, F}(x)$ has degree n.
    c. Verify that the set K of all elements in E algebraic over F is a field. (The set of all numbers algebraic over F is termed the **algebraic closure** of F. The Fundamental Theorem of Algebra asserts that $\mathbb{C}$ is its own algebraic closure.)

20.2

Splitting Fields

In this section, we show how, given any polynomial in $F[x]$, it is always possible to find an extension field E of F such that all the roots of that polynomial lie in E, but in no proper subfield of E. In the course of this analysis, we will examine some of the properties of such fields.

Let us begin with an irreducible polynomial $p(x)$ in $F[x]$. Unless $\deg(p(x)) = 1$, no root of $p(x)$ lies in F. First we will determine whether there exists some extension E of F containing one root of $p(x)$, and, if so, how large is this extension? In case $p(x) \in \mathbb{Q}[x]$, the Fundamental Theorem of Algebra assures us that $p(x)$ has a root in $\mathbb{C}$. But do we need to take $E = \mathbb{C}$? The answer is no.

Theorem 20.5   Kronecker

Let $p(x)$ be a nonconstant irreducible polynomial in $F[x]$. There exists an extension E of F in which $p(x)$ has a root. Moreover, E can be found so that $[E:F] = \deg(p(x))$.

Proof

Let $p(x) = c_n x^n + c_{n-1} x^{n-1} + \cdots + c_1 x + c_0$, $c_n \neq 0$, and set $I = (p(x))$, the set of all multiples of $p(x)$ in $F[x]$. I is an ideal in $F[x]$, which is a PID by Theorem 17.4, and the quotient ring $F[x]/I$ is a commutative ring with unity. Recall that this ring $F[x]/I$ contains F (isomorphically) as a subring. We will identify F with this subring by equating each $c \in F$ with the coset $c + I$. Then the polynomial $f(x) \in F[x]$ can be identified with a polynomial in $F[x]/I$ by replacing each coefficient of $f(x)$ with its corresponding coset in $F[x]/I$. Consider the coset $x + I$. Since

$$p(x + I) = c_n(x + I)^n + c_{n-1}(x + I)^{n-1} + \cdots + c_1(x + I) + c_0$$
$$= c_n(x^n + I) + c_{n-1}(x^{n-1} + I) + \cdots + c_1(x + I) + c_0$$
$$= (c_n x^n + c_{n-1} x^{n-1} + \cdots + c_1 x + c_0) + I$$
$$= p(x) + I = I = 0 + I$$

the coset $x + I$ is a root of $p(x)$ in $F[x]/I$.

We claim $F[x]/I$ is a field. One way to verify this assertion is to prove I is a maximal ideal (see Exercise 2) and apply Theorem 15.3. Instead we will show directly that each nonidentity element of $F[x]/I$ has a multiplicative inverse. Let $f(x) \in F[x]$; the Polynomial Division Algorithm states that there exist $q(x), r(x) \in F[x]$ such that $f(x) = p(x)q(x) + r(x)$, where $r(x) = 0$ or $\deg(r(x)) < n$. Since

$$f(x) + I = (r(x) + p(x)q(x)) + I = (r(x) + p(x)q(x)) + (p(x)) = r(x) + I$$

we can assume that every element of $F[x]/I$ is of the form $r(x) + I$, where $r(x) = 0$ or $\deg(r(x)) < n$.

Now let $r(x) + I$ be any such nonidentity element of $F[x]/I$. Since $p(x)$ is irreducible and $\deg(r(x)) < n$, we have that $r(x)$ and $p(x)$ are relatively prime. By Theorem 16.3, there exist $s(x), t(x) \in F[x]$ such that $1 = s(x)r(x) + t(x)p(x)$. But then,

$$1 + I = (s(x)r(x) + t(x)p(x)) + (p(x))$$
$$= s(x)r(x) + (t(x)p(x)) + (p(x)))$$
$$= s(x)r(x) + (p(x))$$
$$= s(x)r(x) + I$$
$$= (s(x) + I)(r(x) + I)$$

so that $s(x) + I$ is the inverse of $r(x) + I$ in $F[x]/I$. Hence $r(x) + I$ is invertible and every nonidentity element of $F[x]/I$ has an inverse.

We allow you to verify that the cosets $\{1 + I, x + I, x^2 + I, \ldots, x^{n-1} + I\}$ form a basis of $F[x]/I$ over F. (See Exercise 1.) Thus, $[F[x]/I : F] = n$.

To complete the proof, we need to prove that F is a subfield of $F[x]/I$. Actually, this is not true; F is not a subfield of $F[x]/I$. Instead, we prove that F is not only isomorphic to a **subring** of $F[x]/I$ but that F is isomorphic to a

subfield of $F[x]/I$. As we noted earlier, the mapping $\phi\colon F \to F[x]/I$, defined by $\phi(a) = a + I$, is an injective homomorphism from F onto the set F' of all cosets of $F[x]/I$ of the form $a + I$ for $a \in F$. Therefore F' is a subfield of $F[x]/I$ isomorphic to F. The result follows. $\qquad\qquad\square$

Corollary 20.5

If F is a field and $p(x)$ is irreducible in $F[x]$, then the quotient ring $F[x]/(p(x))$ is a field.

Suppose $f(x)$ is a nonconstant polynomial in $F[x]$. If f is not irreducible, then $f(x)$ has an irreducible factor $p(x) \in F[x]$ and $\deg(p(x)) < \deg(f(x))$. If we apply our result to the factor $p(x)$, we obtain the following corollary:

Corollary 20.6

Let $f(x) \in F[x]$. Then there exists an extension E of F in which $f(x)$ has a root; moreover, $[E\colon F] \leqslant \deg(f(x))$.

Very often one is interested in a field containing not just *one* root of a polynomial but *all* roots of the polynomial. This concept utilizes the next definition. It is one of the most important in this chapter.

Definition 20.5

> Let $f(x) \in F[x]$. If E is an extension of F, then $f(x)$ **splits** in E if $f(x)$ can be factored in $E[x]$ as a product of linear factors. E is said to be a **splitting field for $f(x)$ over** F if $f(x)$ splits in E but not in any proper subfield of E.

If $f(x) \in F[x]$ and all the roots of $f(x)$ are contained in F, then clearly F is a splitting field for $f(x)$ over F. In this way, $\mathbb{R}$ is a splitting field for $x^2 - 2$ over $\mathbb{R}$. On the other hand, $\mathbb{Q}(\sqrt{2})$ is a splitting field for $x^2 - 2$ over $\mathbb{Q}$, since $\mathbb{Q}(\sqrt{2})$ contains both $\sqrt{2}$ and $-\sqrt{2}$ and any field that contains $\sqrt{2}$ and $\mathbb{Q}$ must contain $\mathbb{Q}(\sqrt{2})$.

In order to find a splitting field for a polynomial $f(x) \in F[x]$, we can use Corollary 20.6. By this result, there exists a field E_1 that contains F and also contains a root α_1 of $f(x)$. Moreover, $[E_1\colon F] \leqslant n = \deg(f(x))$. If E_1 does not contain all the roots of $f(x)$, then $f(x) = (x - \alpha_1)f_1(x)$, where $f_1 \in E_1[x]$ has degree $n - 1$, and any root of f_1 is a root of f. Again, applying Corollary 20.6 we know there exists a field extension E_2 of E_1 (and, hence, of F) that contains another root α_2 of $f_1(x)$ with $[E_2\colon E_1] \leqslant n - 1$. By Theorem 20.1,

$$[E_2\colon F] = [E_2\colon E_1][E_1\colon F] \leqslant (n - 1)(n)$$

If $f_1(x)$ splits in E_2, we are done; otherwise, $f_1(x) = (x - \alpha_2)f_2(x)$ where $\deg(f_2(x)) = n - 2$. Proceeding in this way, we eventually obtain an extension field E of F containing all the roots of f with the property

$$[E:F] \leqslant n(n-1)(n-2) \cdots 2 \cdot 1 = n!$$

We list our result in Theorem 20.6; the details of a proof utilizing induction on $\deg(f(x))$ are left to you in Exercise 7.

Theorem 20.6   Let $f(x) \in F[x]$ and $\deg(f(x)) = n \geqslant 1$. Then there exists an extension field of F of degree at most $n!$ in which $f(x)$ splits.

By this result, we know that if $f(x) \in F[x]$ has degree n, then there exists a splitting field E of f over F with $[E:F] \leqslant n!$

Example 20.5   Let $f(x) = x^2 - x + 1 \in Q[x]$. The roots of $f(x)$ are

$$\frac{1+\sqrt{-3}}{2} \quad \text{and} \quad \frac{1-\sqrt{-3}}{2}$$

Then

$$E = Q\left(\frac{1}{2} + \frac{i\sqrt{3}}{2}, \frac{1}{2} - \frac{i\sqrt{3}}{2}\right)$$

is a splitting field of $f(x)$ over Q. What is this field?

Consider the field $Q(i\sqrt{3}) = \{a + bi\sqrt{3} \mid a, b \in Q\}$. Clearly $E \subseteq Q(i\sqrt{3})$. Moreover,

$$i\sqrt{3} = \left(\frac{1}{2} + \frac{i\sqrt{3}}{2}\right) - \left(\frac{1}{2} - \frac{i\sqrt{3}}{2}\right)$$

so that $Q(i\sqrt{3}) \subseteq E$. Thus, $E = Q(i\sqrt{3})$ is a splitting field for $f(x)$. Note that $[Q(i\sqrt{3}):Q] = 2$.

Example 20.6   Let $f(x) = x^3 - 5 \in Q[x]$. By De Moivre's Theorem, the roots of $f(x)$ are

$$\sqrt[3]{5}, \quad \omega\sqrt[3]{5}, \quad \text{and} \quad \omega^2\sqrt[3]{5} \quad \text{where} \quad \omega = -\frac{1}{2} + \frac{i\sqrt{3}}{2}$$

Let E be a splitting field for $f(x)$ over Q. Then $\sqrt[3]{5} \in E$ and $Q(\sqrt[3]{5})$ is a subfield of E containing one but not all the roots of $f(x)$, since $Q(\sqrt[3]{5})$ contains no imaginary numbers. Now $[E:Q] \leqslant 3! = 6$, and $[Q(\sqrt[3]{5}):Q] = 3$. (See Exercise 3.) Thus, $[E:Q]$ is a number greater than 3, less than or equal to 6, and divisible by 3. This forces $[E:Q] = 6$. In Exercise 4, you are asked to show

$$E = Q(\sqrt[3]{5}, i\sqrt{3})$$
$$= \{a + b\sqrt[3]{5} + c\sqrt[3]{25} + di\sqrt{3} + ei\sqrt[3]{5}\sqrt{3} + fi\sqrt[3]{25}\sqrt{3} \mid a, b, c, d, e, f \in Q\}$$

Suppose $f(x) \in F[x]$ and $f(x)$ factors as $(x - \alpha)^m (x - \beta)^n$ in $E[x]$ where E is some extension field of F. Then $F(\alpha, \beta) = (F(\alpha))(\beta)$ is a splitting field for $f(x)$ over F, since $F(\alpha, \beta)$ is the smallest field containing F, α, and β. But $F(\beta, \alpha) = (F(\beta))(\alpha)$ is also a splitting field for $f(x)$ over F. How are these two splitting fields for $f(x)$ related? The next result shows that the splitting field of a polynomial over a given field is unique up to isomorphism.

Theorem 20.7

Let $f(x)$ be a nonconstant polynomial in $F[x]$. If E and E' are splitting fields for $f(x)$ over F, then E and E' are isomorphic by a mapping that fixes every element of F.

Proof

Let us induct on the degree of $f(x)$. If $\deg(f(x)) = 1$, then $f(x)$ is linear and F is the unique splitting field for $f(x)$ over F. Assume the result true for polynomials of degree k, and suppose $\deg(f(x)) = k + 1$. Let α be a root of $f(x)$. Then $f(x) = (x - \alpha)f_1(x)$, where $f_1(x)$ is a polynomial of degree k in $F(\alpha)[x]$. Now E and E' are splitting fields for $f_1(x)$ over $F(\alpha)$. This is true since E and E' clearly contain all the roots of $f_1(x)$; moreover, if $f_1(x)$ splits in any proper subfield of E or E' containing $F(\alpha)$, then $f(x)$ splits in that subfield. This is contrary to the assumption that E and E' are splitting fields for $f(x)$ over F. Applying our induction hypothesis to $f_1(x)$, we conclude that E and E' are isomorphic by a mapping fixing $F(\alpha)$ and, hence, F. $\square$

The proof of the preceding can be modified to yield the following result. We leave the details as Exercise 8.

Corollary 20.7

Let $\phi: F \to F_1$ be an isomorphism of fields. Suppose $f(x) \in F[x]$ and $f_1(x)$ is the corresponding polynomial in $F_1[x]$. If E is a splitting field for $f(x)$ over F and E_1 is a splitting field for $f_1(x)$ over F_1, then E and E_1 are isomorphic by a mapping equivalent to ϕ on F.

There is a question that sometimes arises in a discussion of splitting fields: "Can an irreducible polynomial have a multiple root?" If the answer is yes, how does this affect the splitting field?

Example 20.7

Let $F = \mathbb{Z}_2(x)$, the field of rational functions over $\mathbb{Z}_2$, that is, the set of all ratios of polynomials in $\mathbb{Z}_2[x]$. Consider $f(t) = t^2 - x \in F[t]$. We claim $f(t)$ is irreducible over F. Let $p(x)/q(x) \in F$ where $(p(x), q(x)) = 1$. If $p(x)/q(x)$ is a root of $f(t)$ in F, then

$$0 = \left(\frac{p(x)}{q(x)}\right)^2 - x = \frac{p^2(x)}{q^2(x)} - x$$

Thus, $p(x)^2 = xq(x)^2$. Since x is a factor of $p(x)^2$, it follows that x divides $p(x)$.

But then $p(x) = xp_1(x)$ implies $xp_1(x)^2 = q(x)^2$, so that x is a factor of $q(x)^2$. This forces x to be also a factor of $q(x)$, contradicting the assumption $(p(x), q(x)) = 1$. We conclude that $f(t)$ is irreducible over F.

Suppose ξ is a root of $f(t)$ in some extension field of F. Then $0 = f(\xi) = \xi^2 - x$, so that $\xi^2 = x$ and

$$(t - \xi)^2 = t^2 - 2t\xi + \xi^2$$

$$= t^2 + \xi^2$$

$$= t^2 + x$$

$$= t^2 - x = f(t) \qquad \text{since } 2 \equiv 0 \text{ (modulo 2)}$$

Therefore, $f(t)$ is an irreducible polynomial possessing a multiple root.

Admittedly, this example is strange, but it does point out that an irreducible polynomial can have a multiple root. In this case, adjoining a single root to $\mathbb{Z}_2(x)$ produces a splitting field. Note that this example involves a field of finite characteristic, an idea we first encountered in Exercise 21 of Section 13.2. For completeness, we recall that definition.

Definition 20.6

Let R be a ring with identity 0 and unity 1. If there exists no positive integer n such that $n1 = 0$, then R is said to have **characteristic** 0. Otherwise, the smallest positive integer n with the property that $n1 = 0$ is termed the **characteristic of** R, denoted char(R).

We will investigate fields that do not have characteristic 0 in the next chapter. For now, simply note the field $\mathbb{Z}_2(x)$ of our example is a field of characteristic 2, which contains an irreducible polynomial possessing a multiple root. This cannot happen in a field of characteristic 0.

Theorem 20.8

Let $p(x)$ be an irreducible polynomial in $F[x]$ where char(F) = 0. Then $p(x)$ has no multiple roots.

Proof

Suppose ξ is a multiple root of $p(x)$. Then $p(x) = (x - \xi)^n q(x)$, where $n \geqslant 2$ and $q(x) \in F(\xi)[x]$. From introductory calculus, we calculate the formal derivative

$$p'(x) = n(x - \xi)^{n-1}q(x) + (x - \xi)^n q'(x) = (x - \xi)p_0(x)$$

for $p_0(x) \in F(\xi)[x]$. Thus ξ is also a root of $p'(x)$.

Since $p(x)$ is irreducible and $n \geqslant 2$, then $p(x)$ and $p'(x)$ are relatively prime in $F[x]$. By Theorem 16.3, there exist $s(x)$, $t(x) \in F[x]$ such that $1 = s(x)p(x) + t(x)p'(x)$. But then $1 = s(\xi)p(\xi) + t(\xi)p'(\xi) = 0$, an obvious contradiction. We conclude that $p(x)$ can have no multiple root. $\qquad \square$

Motivated by this result, we make the following definition:

Definition 20.7

A polynomial $f(x) \in F[x]$ is **separable** if it has no multiple roots. An element α in an extension E of F is **separable over** F if it is the root of a separable polynomial in $F[x]$. E is a **separable extension** of F if every element of E is separable over F.

If α is algebraic over F where $\text{char}(F) = 0$, then the minimal polynomial of α over F is irreducible. By Theorem 20.8, this minimal polynomial has no multiple roots; thus α is separable over F.

Corollary 20.8

If E is a finite extension of F where $\text{char}(F) = 0$, then E is a separable extension of F.

We conclude this section with a result that gives a condition under which a finite extension is a simple extension.

Theorem 20.9

Let F be a field of characteristic 0. If $E = F(\alpha, \beta)$ is a finite extension of F, then there exists $\gamma \in E$ such that $E = F(\gamma)$.

Proof

Since E is a finite field extension of F, Theorem 20.2 implies α and β are algebraic over F. Let $p(x)$ be the minimal polynomial of α over F and $q(x)$ be the minimal polynomial of β over F. Let $\alpha = \alpha_1, \alpha_2, \ldots, \alpha_n$ be the roots of $p(x)$ (the α_i's are distinct by Theorem 20.8) and $\beta = \beta_1, \beta_2, \ldots, \beta_m$ be the roots of $q(x)$. We can choose the α_i's and β_j's to lie in some splitting field of $p(x)q(x) \in F[x]$.

Since $\text{char}(F) = 0$, F must have an infinite number of distinct elements. (See Exercise 9.) Choose $c \in F$ such that

$$c \neq \frac{\alpha_i - \alpha_j}{\beta_k - \beta_\ell} \qquad \text{for any} \quad i \neq j,\ k \neq \ell$$

If $\gamma = \alpha + c\beta$, we claim $F(\gamma) = F(\alpha, \beta)$.

Clearly, $F(\gamma) \subseteq F(\alpha, \beta)$. Since $q(x) \in F[x]$, $q(x)$ is also a polynomial in $F(\gamma)[x]$. Let $h(x) = p(\gamma - cx)$. Then $h(\beta) = p(\gamma - c\beta) = p(\alpha) = 0$. Now $q(\beta) = 0$, so that $h(x)$ and $q(x)$ have a nontrivial greatest common divisor $d(x)$ in $F(\gamma)[x]$. Choose $d(x)$ monic. Now $d(\beta) = 0$. If ξ is another root of $d(x)$, then $d(x) \mid q(x)$ implies $\xi = \beta_i$, $i \neq 1$. Then $d(\beta_i) = 0$ implies $h(\beta_i) = 0$ and $\gamma = c\beta_i$ is a root of $p(x)$. But $\{\alpha_1, \alpha_2, \ldots, \alpha_n\}$ is a complete set of roots of $p(x)$. Thus

$$\gamma - c\beta_i = (\alpha + c\beta) - c\beta_i = \alpha_j \qquad j > 1$$

Then

$$c = \frac{\alpha_i - \alpha}{\beta - \beta_i}$$

contrary to the choice of c. Therefore, β must be the only root of $d(x)$. Since every root of $d(x)$ is a root of $q(x)$, and $q(x)$ has no multiple roots, $d(x) = x - \beta$. But $d(x) \in F(\gamma)[x]$, so that $\beta \in F(\gamma)$ and $\alpha = \gamma - c\beta \in F(\gamma)$, which implies that $F(\alpha, \beta) \subseteq F(\gamma)$. The result follows. □

The element γ of Theorem 20.9 is termed a **primitive element of E over F.**

Example 20.8

Let $E = \mathbb{Q}(\sqrt{2}, \sqrt{5})$. We will take $c = -1$, $\gamma = \alpha + c\beta = \sqrt{2} - \sqrt{5}$, and verify that γ is a primitive of E over $\mathbb{Q}$. Since $\gamma = \sqrt{2} - \sqrt{5}$, we have

$$\gamma^2 = 7 - 2\sqrt{10}$$

$$\gamma^3 = 17\sqrt{2} - 11\sqrt{5}$$

$$\gamma^4 = 89 - 28\sqrt{10}$$

Now $\gamma^4 - 14\gamma^2 + 9 = 0$, so that γ is a root of $p(x) = x^4 - 14x^2 + 9$. We will let you prove that $p(x)$ is the minimal polynomial of γ over $\mathbb{Q}$ (See Exercise 11.) and $[\mathbb{Q}(\gamma):\mathbb{Q}] = 4$. Since $\mathbb{Q}(\gamma) \subseteq \mathbb{Q}(\sqrt{2}, \sqrt{5})$ and $[\mathbb{Q}(\sqrt{2}, \sqrt{5}):\mathbb{Q}] = 4$, we conclude $\mathbb{Q}(\gamma) = \mathbb{Q}(\sqrt{2}, \sqrt{5})$. In particular,

$$\sqrt{2} = \tfrac{1}{6}(\gamma^3 - 11\gamma) \qquad \text{and} \qquad \sqrt{5} = \tfrac{1}{6}(\gamma^3 - 17\gamma)$$

Induction can be used to obtain the following corollary from Theorem 20.9. (See Exercise 10.)

Corollary 20.9

Any finite extension of a field of characteristic 0 is simple.

Both Theorem 20.9 and Corollary 20.7 refer to fields of characteristic 0. What about an extension E of a field F where $\mathrm{char}(F) = p > 0$? Under what conditions does an extension of F have a primitive element? If E is a finite field of characteristic p, we will show in the next chapter that $(E - \{0\}, \cdot)$ is a cyclic group, and therefore, $E - \{0\} = \langle \gamma \rangle$ for some $\gamma \in E - \{0\}$. In this case, E is isomorphic to $\mathbb{Z}_p(\gamma)$.

There remains the question of what happens when E is an infinite field of finite characteristic p. For example, we may have $E = \mathbb{Z}_2(x)$. In the proof of Theorem 20.9, the key to determining a primitive element γ was in choosing an element c such that

$$c \neq \frac{\alpha_i - \alpha_j}{\beta_k - \beta_\ell}$$

It can be shown that if there exists $f(x) \in F[x]$ such that

1. F is a subfield of E,
2. $[E:F]$ is finite,
3. $f(x)$ splits over E, and
4. there are but a finite number of fields intermediate to F and E.

then such an element can be found.

From a different point of view, we have seen that any finite extension of a field of characteristic 0 is separable. Is Theorem 20.9 still true if we replace the requirement that F be a field of characteristic 0 with the hypothesis that F be separable? The answer is yes, but we will leave the proof to more advanced texts.[2]

In the next section, we want to show you how the theory of field extensions developed here can solve some problems in geometry that resisted solutions for thousands of years.

Exercises 20.2

1. If $p(x) \in F[x]$ is an irreducible monic polynomial of degree n and $I = (p(x))$, then show that
    $$\{1 + I, x + I, \ldots, x^{n-1} + I\}$$
    is a basis of $F[x]/I$ over F.

2. Prove that if F is a field and $p(x)$ is irreducible in $F[x]$, then $(p(x))$ is a maximal ideal in $F[x]$.

3. Prove that $[\mathbb{Q}(\sqrt[3]{5}):\mathbb{Q}] = 3$.

4. Prove that $\{1, \sqrt[3]{5}, \sqrt[3]{25}, i\sqrt{3}, i\sqrt{3}\sqrt[3]{5}, i\sqrt{3}\sqrt[3]{25}\}$ is a basis for $\mathbb{Q}(\sqrt[3]{5}, i\sqrt{3})$ over $\mathbb{Q}$.

5. Find a splitting field for
    a. $x^2 - x + 1$ over $\mathbb{Q}(i)$ b. $x^2 - 3$ over $\mathbb{Q}$
    c. $x^3 + 3x + 1$ over $\mathbb{Q}$ d. $x^3 + x^2 + x + 1$ over $\mathbb{Q}$
    e. $x^3 - 4$ over $\mathbb{Q}$ f. $x^4 - 9$ over $\mathbb{Q}$
    g. $x^4 + 9$ over $\mathbb{Q}$ h. $x^4 - x^2 - 2$ over $\mathbb{Q}$
    i. $x^2 + i$ over $\mathbb{C}$ j. $x^3 - 8$ over $\mathbb{Q}$

6. If E is splitting field for $f(x) \in \mathbb{R}[x]$, prove that $[E:\mathbb{R}] \leqslant 2$.

7. Use induction on the degree of $f(x)$ to prove Theorem 20.6.

8. If $\phi: F \to F_1$ is an isomorphism of fields, $f(x) \in F[x]$, and $f_1(x)$ is the corresponding polynomial in $F_1[x]$, then show that any splitting field for $f(x)$ over F is isomorphic to any splitting field of $f_1(x)$ over F_1 by a mapping equivalent to ϕ on F.

9. Verify that if $\mathrm{char}(F) = 0$, then F is an infinite field.

---

[2]See Nathan Jacobson's *Basic Algebra I*, San Francisco: W.F. Freeman and Company, 1974, p. 279.

10. Prove that any finite extension of a field of characteristic 0 is a simple extension.

11. Verify that $x^4 - 14x^2 + 9$ is the minimal polynomial of $\sqrt{2} - \sqrt{5}$ over $\mathbb{Q}$.

12. Prove that $\sqrt{2} + \sqrt{5}$ is also a primitive element of $\mathbb{Q}(\sqrt{2}, \sqrt{5})$ over $\mathbb{Q}$. How does the minimal polynomial of $\sqrt{2} + \sqrt{5}$ compare to that of $\sqrt{2} - \sqrt{5}$?

13. Find a primitive of the following extensions:
    a. $\mathbb{Q}(i, \sqrt{2})$
    b. $\mathbb{R}(i, \sqrt{2})$
    c. $\mathbb{Q}(\sqrt{2}, \sqrt{3})$
    d. $\mathbb{R}(\sqrt{2}, \sqrt{3})$
    e. $\mathbb{Q}(\sqrt{-2}, \sqrt{-3})$
    f. $\mathbb{Z}_2(i, \sqrt{3})$
    g. $\mathbb{Q}(\sqrt{2}, \sqrt{3}, i)$
    h. $\mathbb{Q}(\sqrt{2}, \sqrt{-2})$

14. Find the minimal polynomial over $\mathbb{Q}$ of the following elements:
    a. $\sqrt{2} - i$
    b. $\sqrt{2} - \sqrt{-2}$
    c. $\sqrt{7} - i$
    d. $\sqrt{2} - \sqrt{3}$
    e. $\sqrt{2} - \sqrt{3} - i$
    f. $\sqrt{7} - \sqrt{14}$

15. If E is an extension of F where $\operatorname{char}(F) = 0$, then prove $\operatorname{char}(E) = 0$.

16. Let F be a field of characteristic 0, $f(x) \in F[x]$, and $f'(x)$ denote the formal derivative of the polynomial $f(x) \in F[x]$. Prove that $f(x)$ has no multiple roots if and only if $(f(x), f'(x)) = 1$.

17. Let T be an integral domain containing the domain D as a subring. An element $\alpha \in T$ is **integral over** D if there exists a monic polynomial $f(x) \in D[x]$ such that $f(\alpha) = 0$.
    a. Why is $2/3 \in \mathbb{Q}$ not integral over $\mathbb{Z}$?
    b. Prove that the set of elements of T integral over D form a subring of T.
    c. Suppose T is not only an integral domain but also a field. Verify that if $\beta \in T$ is algebraic over the field of quotients of D, then there exists $d \in D$ such that $d\beta$ is integral over D.

20.3

Geometric Constructions

Almost all students have encountered problems involving geometric constructions using only a compass and a straightedge, that is, a ruler without markings. Common constructions include finding the perpendicular bisector of a given line segment, drawing a line parallel to a certain line, and the bisection of a given angle. If you are supplied with a unit length, then it is possible to produce line segments of any positive integer length. For integer lengths a and b, there exist methods whereby one can construct, using straightedge and compass alone, line segments of length $a + b$, a/b, and $\sqrt{a}$. Some examples of such methods employing the theory of similar triangles are pictured in Figure 20.1.

Let us say that a real number α is **constructible** if we can draw a line segment of length α or $-\alpha$ by a finite sequence of straightedge and compass constructions. In a similar way, an angle θ or $-\theta$ is constructible if we can draw two line segments that form θ. In other words, an angle $\theta > 0$ is

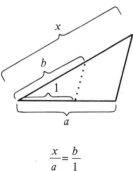

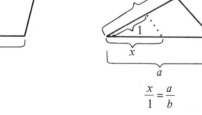

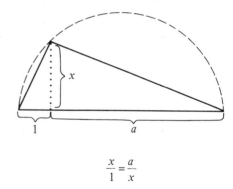

$$\frac{x}{a} = \frac{b}{1}$$

$$\frac{x}{1} = \frac{a}{b}$$

$$\frac{x}{1} = \frac{a}{x}$$

Figure 20.1

constructible if we can construct line segments to form a triangle ABC such that $\angle ABC = \theta$.

Let K be the set of all constructible numbers. From the preceding discussion, we see that K must contain $\mathbb{Q}$ and the square root of each positive element of $\mathbb{Q}$; in fact K contains the square root of each positive element of K. Since K is a subset of $\mathbb{R}$ containing not only 0 and 1 but also the sum, product, and the additive and multiplicative inverses of each of its elements, K is a subfield of $\mathbb{R}$. In this section, we want to investigate K and relate this field to questions involving the "three classical problems of antiquity."

Plutarch reported that the philosopher Anaxagoras, while in prison, occupied himself with the problem of "squaring the circle." This has come to be described as finding a square whose area is exactly equal to the area of a given circle. The square or, equivalently, the length of its side, must be constructed by the use of straightedge and compass alone.

Around the time of the death of Anaxagoras in 428 B.C., a plague was reported to have devastated the population near Athens. A delegation of Athenians was sent to the oracle of Apollo at Delos to inquire how to avert the plague. The reply they received instructed them to double the size of the altar to Apollo. The altar was cubical in shape. The Athenians dutifully doubled the dimensions of the altar, which resulted in an eight-fold increase in volume. When the plague struck Athens, the citizens realized that they must have misinterpreted the reply. The "Delian problem" became the following: Given the edge of a cube, use straightedge and compass alone to find the edge of a second cube having twice the volume of the first.

At approximately the same time, there arose a third problem that has fascinated would-be mathematicians to the present day. Given an angle, construct an angle one-third its original size using only a straightedge and a compass. Innumerable solutions have been presented throughout the years purporting to show how such a construction can be done. Most of these "solutions" erroneously assume that $\angle ABC$ can be trisected if and only if the chord AC can be trisected. (See Figure 20.2.)

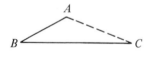

Figure 20.2

We will prove that each of these three problems cannot be solved, that each is an impossible construction according to the rules involving the use of straightedge and compass. To this end, suppose F is a field of constructible numbers, so that F is a subfield of K, the field of *all* constructible numbers. Consider the set of points (α, β), where $\alpha, \beta \in F$. If (α_1, β_1) and (α_2, β_2) are elements of F^2, then we know from elementary geometry that the line through these two points satisfies either the equation

$$x = \alpha \qquad \text{for} \quad \alpha \in F$$

or the equation

$$y = \mu x + \delta \qquad \text{for} \quad \mu = \frac{\beta_2 - \beta_1}{\alpha_2 - \alpha_1} \quad \text{and} \quad \delta = \beta_1 - \mu \alpha_1$$

In either case, the line can be put in the form $\alpha x + \beta y = \gamma$ where $\alpha, \beta, \gamma \in F$. Such a line is said to be a **line in** F. Any circle having center $(\alpha, \beta) \in F^2$ and radius ρ in F satisfies the equation

$$(x - \alpha)^2 + (y - \beta)^2 = \rho^2$$

This equation can be rewritten as

$$x^2 + y^2 + \delta x + \lambda y + \eta = 0$$

where $\delta, \lambda, \eta \in F$. Any equation of this form is termed a **circle in** F.

What happens in a construction involving a straightedge and a compass? The straightedge allows you to produce, from two points in F^2, the line in F passing through these two points. The compass permits you to construct from two points in F^2 a circle in F centered at one of the points with radius equal to the distance between the original pair of points. The use of both straightedge and compass permits you to obtain points in $\mathbb{R}^2$ corresponding to the intersection of two lines in F, a line in F and a circle in F, or two circles in F. The important thing here is that the points of intersection of these geometric objects are not just points in $\mathbb{R}^2$, but are related to F in a special way.

| Theorem 20.10 |

Let F be a field consisting of constructible numbers. The following results hold:

    i. The intersection of two distinct nonparallel lines in F is a point in F^2.

    ii. If a line in F and a circle in F intersect nontrivially, then the intersection is a point either in F^2 or in F_1^2, where F_1 is the field $F(\sqrt{\tau})$ for some $\tau \in F$.

    iii. If two circles in F intersect nontrivially, then the intersection is a point either in F^2 or in F_1^2, where F_1 is the field $F(\sqrt{\tau})$ for some $\tau \in F$.

Proof

We will treat the first two assertions and leave the proof of the third to you as Exercise 3. Suppose

$$\alpha_1 x + \beta_1 y = \gamma_1 \qquad \text{and} \qquad \alpha_2 x + \beta_2 y = \gamma_2$$

are two lines in F. Since the two lines are nonparallel, the vector $[\alpha_2, \beta_2]$ is not a multiple of the vector $[\alpha_1, \beta_1]$ and $\delta = \alpha_1 \beta_2 - \alpha_2 \beta_1 \neq 0$. (See Exercise 1.) The solution to this system of two equations in two unknowns must satisfy

$$x = \frac{\beta_2 \alpha_1 - \beta_1 \alpha_2}{\delta} \qquad \text{and} \qquad y = \frac{\alpha_1 \beta_2 - \alpha_2 \beta_1}{\delta}$$

Since F is a field, both these values lie in F. This verifies the first assertion.

To establish the second statement, let the line and the circle have equations

$$\alpha x + \beta y = \gamma \qquad \text{and} \qquad x^2 + y^2 + \delta x + \lambda y + \eta = 0$$

where $\alpha, \beta, \gamma, \delta, \lambda, \eta \in F$. If $\alpha \neq 0$, then

$$x = \frac{\gamma - \beta y}{\alpha} \qquad \text{and} \qquad \left(\frac{\gamma - \beta y}{\alpha}\right)^2 + y^2 + \delta\left(\frac{\gamma - \beta y}{\alpha}\right) + \lambda y + \eta = 0$$

becomes

$$\left(\frac{\beta^2}{\alpha^2} + 1\right) y^2 + \left(\lambda - \frac{2\beta\gamma}{\alpha^2} - \frac{\delta\beta}{\alpha}\right) y + \left(\frac{\gamma^2}{\alpha^2} + \frac{\delta\gamma}{\alpha} + \eta\right) = 0$$

This is a quadratic equation with coefficients in the field F. If we relabel these coefficients and write

$$\alpha_1 y^2 + \beta_1 y + \gamma_1 = 0$$

then we can solve for y by the quadratic formula. If the line and circle intersect nontrivially, then the solutions to this equation lie in the field $F(\sqrt{\tau})$, where $\tau = \beta_1^2 - 4\alpha_1\gamma_1$ and $0 \leqslant \tau$. If τ is a perfect square in F, then y and $x = (\gamma - \beta y)/\alpha$ are also values in F. If τ is not a perfect square, then $F_1 = F(\sqrt{\tau})$ is a quadratic extension of F, that is, an extension of degree two, containing the point of intersection. This proves the second assertion in the case $\alpha \neq 0$. We leave the case where $\alpha = 0$ to you in Exercise 2. □

Now suppose we start with a unit length and construct with compass and straightedge all the values in $\mathbb{Q}$. If we then perform a sequence of constructions resulting in an element α, how is this element related to $\mathbb{Q}$? At each stage of our procedure, we produce either a point whose components have already been constructed, or a point whose components are square roots of nonnegative values previously constructed. Let $\sqrt{\tau_1}, \sqrt{\tau_2}, \ldots, \sqrt{\tau_n}$

be the new values obtained in a construction procedure. Then

$$\tau_1 \in \mathbb{Q}, \quad \tau_2 \in \mathbb{Q}(\sqrt{\tau_1}), \quad \tau_3 \in \mathbb{Q}(\sqrt{\tau_1})(\sqrt{\tau_2}) = \mathbb{Q}(\sqrt{\tau_1}, \sqrt{\tau_2})$$

and so on. Since the value of α lies in our final field $\mathbb{Q}(\sqrt{\tau_1}, \sqrt{\tau_2}, \ldots, \sqrt{\tau_n})$, $\mathbb{Q}(\alpha)$ must be a subfield of this field. The degree of α over $\mathbb{Q}$ is equal to $[\mathbb{Q}(\alpha):\mathbb{Q}]$ by Theorem 20.3. According to Theorem 20.1 and Corollary 20.1, $[\mathbb{Q}(\alpha):\mathbb{Q}]$ divides $[\mathbb{Q}(\sqrt{\tau_1}, \sqrt{\tau_2}, \ldots, \sqrt{\tau_n}):\mathbb{Q}]$, which must be a power of 2. This is an important result.

Theorem 20.11

If a number α is constructible using straightedge and compass alone, then $[\mathbb{Q}(\alpha):\mathbb{Q}] = 2^k$ where k is a nonnegative integer.

With this result, we can tackle our three classical problems. Let us begin with the Delian problem.

Corollary 20.10

Suppose there is a cube whose side is a constructible length. It is impossible to construct the side of a second cube whose volume is twice that of the first.

Proof
Suppose that such a process does exist. Then it would be possible to double the volume of a cube whose side has unit length. This means that we would be able to construct a side of length α such that $\alpha^3 = 2$. But $\sqrt[3]{2}$ is not constructible, since $[\mathbb{Q}(\sqrt[3]{2}):\mathbb{Q}] = 3$, and 3 is not a power of 2. $\square$

We will show shortly that not every angle can be trisected using compass and straightedge. We are not saying, though, that **no** angle can be trisected. For example, you should be able to trisect 90° and 135°. (See Exercise 4.) To show that no general trisection method exists, we need only produce one angle that cannot be trisected.

Figure 20.3 shows that if we could trisect 60° to obtain an angle of measure 20°, then we would also be able to construct a length equal in value to cos 20°. But $x = \cos 20°$ is a root of $p(x) = 8x^3 - 6x - 1$, which is irreducible over $\mathbb{Q}$. (See Exercise 5). Then Theorem 20.3 implies that $[\mathbb{Q}(\cos 20°):\mathbb{Q}] = 3$. By Theorem 20.11, we obtain the following corollary.

Corollary 20.11

No method exists that allows the use of straightedge and compass alone to construct a second angle of measure $\theta/3$ from a given angle of measure θ.

Finally, we turn to the oldest of our problems, concerning the constructibility of a square whose area is equal to that of a given circle. Let us begin with a circle of radius 1. If we could "square the circle," we could find a

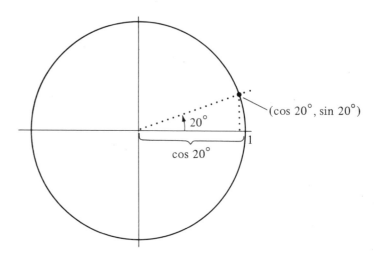

Figure 20.3

square of area π and side $\sqrt{\pi}$. But then $\pi = (\sqrt{\pi})^2$ would be constructible and $[\mathbb{Q}(\pi):\mathbb{Q}]$ would be a power of 2 by Theorem 20.11. But F. Lindemann in 1882 proved that π is not algebraic over $\mathbb{Q}$. Invoking Corollary 20.3, we see that $[\mathbb{Q}(\pi):\mathbb{Q}]$ cannot be finite.

Corollary 20.12

No method exists that allows the use of only straightedge and compass to construct a square equal in area to that of a circle whose radius is constructible.

In the exercises 20.3, you will find additional results concerning geometric constructions using straightedge and compass.

At this point you might ask why it took mathematicians over 2200 years to prove that none of the classical constructions can be performed. One answer is that the study of algebra and, in particular, the theory of field extensions, have taken all this time to develop to a level of maturity necessary to treat these problems. This is not to say no progress was made on these problems in the intervening years. Hippocrates of Chios is said to have connected the question of finding the geometric mean $\sqrt{ab}$ of two numbers a and b to the problem of duplicating the cube. Hippias of Ellis developed the first curve beyond the straight line and the circle. He used the "trisectrix of Hippias" to "solve" the trisecting problem. Dinostratus used this trisectrix curve to investigate the squaring of the circle.

While Pappus of Alexandria in 320 A.D. asserted that the three classical problems could not be done in euclidean geometry, no rigorous proof was given until the nineteenth century. During all this time, numerous modifications were made in both the statement of the problems and the approaches to solution. A great deal of mathematical insight was gained in the process. All the while, mathematicians took as much inspiration from early attempts at solving these problems as from the problems themselves. In many ways, current mathematics owes as much to the good problems studied by mathematicians as it owes to the mathematicians themselves.

Exercises 20.3

1. Verify that if $a_1x + b_1y = c_1$ and $a_2x + b_2y = c_2$ are not parallel lines, then $a_1b_2 - a_2b_1 \neq 0$.

2. Complete the proof of part ii of Theorem 20.10 in the case where $\alpha = 0$.

3. Prove part iii of Theorem 20.10.

4. Show how to construct angles of measure $30°$ and $45°$ using only straightedge and compass.

5. a. Use the trigonometric identity $\cos 3\theta = 4 \cos^3 \theta - 3 \cos \theta$ to verify that $\cos 20°$ is a root of $8x^3 - 6x - 1$.
   b. Prove that $p(x) = 8x^3 - 6x - 1$ is irreducible in $\mathbb{Q}[x]$.

6. Show how to construct angles of measure $120°$, $75°$, and $15°$ using a compass and straightedge.

7. a. Prove that if $\cos \alpha$ is constructible, then $\sin \alpha$ is also constructible.
   b. Use DeMoivre's Theorem to verify that $w = \cos \frac{2}{5}\pi + i \sin \frac{2}{5}\pi$ is a root of $x^4 + x^3 + x^2 + x + 1$.
   c. Compute $w + (1/w)$.
   d. Show that $w = \cos \frac{2}{5}\pi$ is a solution to $4x^2 + 2x - 1 = 0$.
   e. Conclude that $\frac{2}{5}\pi$ is constructible and, therefore, a regular pentagon can be constructed with straightedge and compass alone.

8. A regular 7-sided figure can be constructed if and only if the angle $\frac{2}{7}\pi$ can be constructed. Use the fact that $w = \cos \frac{2}{7}\pi + i \sin \frac{2}{7}\pi$ is a root of

$$x^7 - 1 = (x - 1)(x^6 + x^5 + x^4 + x^3 + x^2 + x + 1)$$

to conclude that no regular 7-sided figure can be constructed using only straightedge and compass.

9. Prove that it is impossible to trisect a $240°$ angle.

10. In geometry, we learned that $1° = \pi/180$. Show that $1°$ is not a constructible number.

11. Verify that a regular 18-sided polygon cannot be constructed using only straightedge and compass.

12. Show that although the angle $\pi/8$ can be constructed, the length $\pi/8$ cannot be constructed.

21

Finite Fields and

Polynomial Codes

Surprisingly, the theory of finite fields arises in many areas of mathematics. For example, in projective geometry, the theory of finite fields is utilized to determine the number of points in a finite desarguesian projective plane. Finite fields are used in combinatorics to construct Latin squares.[1] In statistics, finite fields can be applied to "confound" the effects of data in experimental design. In cryptology, schemes for secret codes can be implemented via finite fields.

In this chapter, we wish to determine the orders of all finite fields and describe them in terms of the theory of field extensions. We will also prove a theorem on the existence of a primitive element for any finite field. Finally, we will extend our work on group and matrix codes by investigating how finite fields allow the implementation of polynomial codes.

Throughout this chapter, F and E denote fields with identity and unity, symbolized by 0 and 1, respectively.

---

[1] For an excellent discussion of the application of finite fields to the construction of orthogonal Latin squares, see R. Brualdi's *Introductory Combinatorics*, New York: North Holland, 1979, Sections 9.1–9.3 and H. Ryser's *Combinatorial Mathematics*, Washington, D.C.: Math. Assn. of America, 1963.

21.1

The Structure of Finite Fields

Up to this point, we have not seen many examples of finite fields. Except for $\mathbb{Z}_2(x)$, the field of rational functions of polynomials with coefficients in $(\mathbb{Z}_2, \oplus, \odot)$ encountered in Example 20.7, the only finite fields discussed have been the fields of the form $(\mathbb{Z}_p, \oplus, \odot)$, where p is a prime. These are of paramount importance in the theory of finite fields, and we will soon see that any finite field must contain one of these fields as a subfield.

Suppose E is a finite field. We saw in Theorem 14.9 that E has a subring isomorphic either to $\mathbb{Z}$ or to $\mathbb{Z}_n$ for some $n \in \mathbb{N}$. Now the first situation cannot occur, for any ring that contains $\mathbb{Z}$ must contain an infinite number of elements. Therefore, E must contain $\mathbb{Z}_n$ (at least isomorphically), and the proof of Theorem 14.9 shows that n is the smallest natural number such that $n \cdot 1 = 0$ in E. But then n must be the characteristic of E.

Theorem 21.1

Let E be a field of characteristic $n > 0$. Then n is a prime.

Proof

Suppose $n = pn_1$, where p is a prime and $1 \leqslant n_1 < n$. Then $0 = n1 = (pn_1)1 = (p1)(n_1 1)$. Since a field can have no zero-divisors, either $p1 = 0$ or $n_1 1 = 0$. The characteristic of E is the smallest integer n such that $n1 = 0$. The only possibility is when $n_1 = 1$ and $n = p$.   □

Recall that the prime subfield of E is the smallest subfield of E and was shown in Exercise 3 of Section 20.1 to be the intersection of all the subfields of E. If E has characteristic p, let F_p denote the prime subfield of E. Then F_p must also have characteristic p (see Exercise 1), so that F_p contains a subfield isomorphic to $\mathbb{Z}_p$. Since any subfield of F_p is a subfield of E and F_p is the smallest subfield of E, we obtain the following result:

Corollary 21.1

If p is the characteristic of E, then the prime subfield of E is isomorphic to $\mathbb{Z}_p$.

Neither Theorem 21.1 nor Corollary 21.1 require E to be finite. The field $\mathbb{Z}_2(x)$ that we considered in Chapter 20 is an infinite field of characteristic 2. Its prime subfield is isomorphic to $\mathbb{Z}_2$. If E is finite, then E can be considered to be a vector space over its prime subfield. In this case, the order of E is determined.

Theorem 21.2

Let E be a finite field of order n and characteristic p. Then $n = p^m$ for some $m \in \mathbb{N}$.

Proof

E is a vector space over F_p, its prime subfield. If $m = \dim_{F_p}(E)$, then E has a basis of m elements $\alpha_1, \alpha_2, \ldots, \alpha_m$. Any element of E can be written as

$$a_1\alpha_1 + a_2\alpha_2 + \cdots + a_m\alpha_m \qquad \text{for} \quad a_1, a_2, \ldots, a_m \in F_p$$

Since $|F_p| = p$, there are exactly p^m distinct elements. $\qquad\square$

A finite field of order p^m has subfields of order p^k for $k = 1, 2, \ldots, m$. To show this, we need to use a result known as the **Binomial Theorem**, which can be stated as follows: If $a, b \in \mathbb{R}$, $n \in \mathbb{N}$, then

$$(a+b)^n = \sum_{k=0}^{n} \binom{n}{k} a^{n-k}b^k \qquad \text{where} \qquad \binom{n}{k} = \frac{n!}{k!(n-k)!}$$

This theorem can be proved by combining induction on n with the result that

$$\binom{n}{k} + \binom{n}{k+1} = \binom{n+1}{k+1}$$

(See Exercise 2.)

Lemma 21.1

Let E be a field of characteristic $p > 0$ and $\alpha, \beta \in E$. Then

$$(\alpha + \beta)^{p^n} = \alpha^{p^n} + \beta^{p^n} \qquad \text{and} \qquad (\alpha\beta)^{p^n} = \alpha^{p^n}\beta^{p^n}$$

for all $n \in \mathbb{N}$.

Proof

We proceed by induction on n. For $n = 1$, the Binomial Theorem implies

$$(\alpha + \beta)^p = \alpha^p + p\alpha^{p-1}\beta + \frac{p(p-1)}{2}\alpha^{p-2}\beta^2 + \cdots + p\alpha\beta^{p-1} + \beta^p$$

$$= a^p + \beta^p$$

since E has characteristic p and $\binom{p}{k}$ is a multiple of p for $k = 1, 2, \ldots, p-1$. (See Exercise 3.)

If $(\alpha + \beta)^{p^k} = \alpha^{p^k} + \beta^{p^k}$, then

$$(\alpha + \beta)^{p^{k+1}} = ((\alpha + \beta)^{p^k})^p = (\alpha^{p^k} + \beta^{p^k})^p$$

which equals

$$(\alpha^{p^k})^p + (\beta^{p^k})^p = \alpha^{p^{k+1}} + \beta^{p^{k+1}}$$

by our proof for the case $n = 1$. Thus $(\alpha + \beta)^{p^n} = \alpha^{p^n} + \beta^{p^n}$ for all $n \in \mathbb{N}$. A similar argument shows that $(\alpha\beta)^{p^n} = \alpha^{p^n}\beta^{p^n}$. (See Exercise 4.) $\qquad\square$

Theorem 21.3

Let E be a field of characteristic $p > 0$. Then, for each $k \in \mathbb{N}$, the set F' of all elements $\alpha \in E$ such that $\alpha^{p^k} = \alpha$ forms a subfield of E.

Proof

Lemma 21.1 shows that F' is closed under sums and products. Clearly F' is nonempty since $1^{p^k} = 1$. We need only show that F' contains additive and multiplicative inverses. Suppose $\alpha^{p^k} = \alpha$. If $p = 2$, then $-\alpha = \alpha \in F'$. If $p > 2$, then p is odd and $(-\alpha)^{p^k} = (-1)^{p^k}\alpha^{p^k} = -\alpha$, implying $-\alpha \in F'$. Also $(\alpha^{-1})^{p^k} = (\alpha^{p^k})^{-1} = \alpha^{-1}$ and $\alpha^{-1} \in F'$. Thus F' is a subfield of E. □

By Theorem 21.2, a finite field must have order equal to a prime power. This does not guarantee that there exists any field of order p^2 or p^3. To show that there exist finite fields other than those of the form $\mathbb{Z}_p$, let us utilize our theory of polynomial rings to construct a field E of order $4 = 2^2$:

Example 21.1

Consider $f(x) = x^2 + x + 1 \in \mathbb{Z}_2[x]$. The polynomial $f(x)$ is irreducible over $\mathbb{Z}_2$ since $f(0) = f(1) = 1 \neq 0$. By Corollary 20.5, $\mathbb{Z}_2[x]/(f(x))$ is a field and the elements of this field are of the form $a + b\delta$, where $a, b \in \mathbb{Z}_2$ and $\delta^2 + \delta + 1 = 0$. (Here we set δ equal to the coset $x + (f(x))$.) Then

$$f(\delta + 1) = (\delta + 1)^2 + (\delta + 1) + 1$$
$$= \delta^2 + 2\delta + 1 + \delta + 1 + 1 = \delta^2 + \delta + 1 = 0$$

so that $\delta + 1$ is the other root of $f(x)$. The field $E = \mathbb{Z}_2[x]/(f(x))$ consists of the four elements $(0, 1, \delta, \delta + 1\}$. The Cayley tables for addition and multiplication are given in Figure 21.1. The verification of the entries of the tables is the content in Exercise 5.

Constructions similar to Example 21.1 can be given to find fields of order p^m for other values of $m \in \mathbb{N}$. In fact the converse of Theorem 21.2 is true:

Theorem 21.4

If n is a prime power, then there exists a field of order n.

Proof

Let $n = p^k$ where p is a prime, and consider $f(x) = x^n - x \in \mathbb{Z}_p[x]$. If α is a root of $f(x)$ of multiplicity r, then $f(x) = (x - \alpha)^r f_1(x)$. We know from elementary

$+$	0	1	δ	$\delta + 1$
0	0	1	δ	$\delta + 1$
1	1	0	$\delta + 1$	δ
δ	δ	$\delta + 1$	0	1
$\delta + 1$	$\delta + 1$	δ	1	0

$\cdot$	1	δ	$\delta + 1$
1	1	δ	$\delta + 1$
δ	δ	$\delta + 1$	1
$\delta + 1$	$\delta + 1$	1	δ

Figure 21.1

calculus that the derivative of f has the form

$$f'(x) = r(x - \alpha)^{r-1}f_1(x) + (x - \alpha)^r f'_1(x)$$
$$= (x - \alpha)^{r-1}(rf_1(x) + (x - \alpha)f'_1(x))$$

Therefore, $f'(x)$ is divisible by $(x - \alpha)^{r-1}$. If $r > 1$, then $f(\alpha) = 0$. But

$$f'(x) = nx^{n-1} - 1 \equiv -1 \pmod p$$

Hence, $f(\alpha) \equiv -1 \pmod p$. This forces $r = 1$; so $f(x)$ cannot have multiple roots.

Let K be a splitting field of $f(x)$ over $\mathbb{Z}_p$. Then K is the smallest extension of $\mathbb{Z}_p$ containing all the n roots of $f(x)$. But the set of all roots of $f(x)$ forms a field by Theorem 21.3. Thus K consists of exactly the n roots of $f(x)$ and $|K| = p^k = n$. □

This result has two immediate consequences. The first (Corollary 21.2) gives an alternate description of finite fields. The second (Corollary 21.3) follows from the fact that any two splitting fields of the same polynomial are isomorphic.

Corollary 21.2

If E is a field of order p^k where p is a prime, then E is a splitting field of $f(x) = x^{p^k} - x$ over the prime subfield of E.

Corollary 21.3

Any two finite fields of the same order are isomorphic.

In Example 21.1, the field E of order $4 = 2^2$ was the splitting field over $\mathbb{Z}_2$ of $f(x) = x^2 + x + 1$. Corollary 21.2 says E should be the splitting field of $x^4 - x$. There is no contradiction here, since $x^4 - x = x(x - 1)(x^2 + x + 1)$ has the same splitting field as its factor $x^2 + x + 1$.

We can make two more observations about E, our example of a field of order 4. First, $E = \mathbb{Z}_2(\delta)$ is a simple extension of $\mathbb{Z}_2$. Secondly, $E - \{0\} = \{1, \delta, \delta + 1\}$ is a cyclic group generated by δ. These two properties hold in general. To prove them, we need to recall some results on finite abelian groups.

We saw in Theorem 10.6 that any finite abelian group A is the direct sum of its Sylow subgroups. Suppose

$$A = S_{p_1} \oplus S_{p_2} \oplus \cdots \oplus S_{p_t}$$

where $\{p_1, p_2, \ldots, p_t\}$ is the set of distinct prime divisors of $|A|$ and $S_{p_i} \in \mathrm{Syl}_{p_i}(A)$. Each of these Sylow subgroups is an abelian p-group for some prime p. By Lemma 10.2, each S_{p_i} contains an element b_i of order $p_i^{k_i}$, where $p_i^{k_i}$ is the smallest positive integer such that $p_i^{k_i}a = 0$ for all $a \in S_{p_i}$. Set

$$b = b_1 + b_2 + \cdots + b_t$$

Lemma 10.1 can be extended to prove that $b = b_1 + b_2 + \cdots + b_t$ has order $p_1^{k_1} p_2^{k_2} \cdots p_t^{k_t}$. (See Exercise 16.) We have proved the following lemma.

Lemma 21.2

Let A be an abelian group of order m. Suppose k is the smallest positive integer such that $ka = 0$ for all $a \in A$. Then there exists an element $b \in A$ of order k.

With the aid of this lemma, we can now prove that the multiplicative group of a finite field must be cyclic.

Theorem 21.5

Let $(E, +, \cdot)$ be a finite field. Then $(E - \{0\}, \cdot)$ is a cyclic group.

Proof

Let $|E| = p^n$, where p is a prime integer. By Corollary 21.2, $E - \{0\}$ can be identified with the nonzero roots of $x^{p^n} - x$, or equivalently, with the roots of $x^{p^n - 1} - 1$.

Set $m = p^n - 1$. $(E - \{0\}, \cdot)$ is an abelian group of order m, and by Lemma 21.2, there exists $b \in E - \{0\}$ of order $k \leqslant m$ and $x^k = 1$ for all $x \in E - \{0\}$. (Here we are using Lemma 21.2 in its multiplicative form.) If $k < m$, then the equation $x^k - 1$ has m distinct roots, contradicting the fact that a polynomial of degree k can have at most k distinct roots. Therefore $k = m$ and $b^m = 1$. We conclude that $(E - 0, \cdot) = \langle b \rangle$ and $E - \{0\}$ is cyclic. $\square$

Theorem 21.5 enables us to state a result analogous to Corollary 20.9 concerning simple extensions.

Corollary 21.4

If E is a finite extension of a finite field F, then there exists $\gamma \in E$ such that $E = F(\gamma)$.

Proof

By Theorem 21.5, there exists $\gamma \in E$ such that $E - \{0\} = \langle \gamma \rangle$. But then $E = F(\gamma)$. $\square$

We have utilized our work on finite abelian groups, rings, and field extensions to determine the structure of finite fields. The theory of finite fields demonstrates that when we place restrictions on an algebraic structure, such as finiteness on a field, we often are able to obtain numerous results concerning the nature of the structure. In this case, we determined completely all finite fields. To do so, however, required a great deal of preliminary material developed in previous chapters.

1. Let E be a field of characteristic p. If F is a subfield of E, prove that F has characteristic p.

2. Prove that

$$\binom{n}{k} + \binom{n}{k+1} = \binom{n+1}{k+1} \qquad \text{for} \quad k \leqslant n$$

3. Suppose p is a prime. Show that if $1 \leqslant k < p$, then

$$\binom{p}{k} \equiv 0 \ (\text{modulo } p)$$

4. Prove that if $\alpha, \beta \in E$, a field of characteristic $p > 0$, then

$$(\alpha\beta)^{p^n} = \alpha^{p^n}\beta^{p^n} \qquad \text{for all} \quad n \in \mathbb{N}$$

5. Verify the Cayley tables for the field of order 4 given in Example 21.1.

6. a. Show that $x^3 + x - 1$ is irreducible in $\mathbb{Z}_2[x]$.
   b. Prove that $\mathbb{Z}_2[x]/(x^3 + x - 1)$ is a field of order 8.
   c. Find a generator of this field.
   d. By Corollary 21.2, if K is the splitting field of $x^8 - x$ over $\mathbb{Z}_2$, then K is a field of order 8. Prove that K is isomorphic to $\mathbb{Z}_2[x]/(x^3 + x - 1)$.

7. a. Show that $x^3 - x + 1$ is irreducible over $\mathbb{Z}_3$.
   b. Construct a field of order 27.

8. Construct a field of 9 elements.

9. Find an irreducible polynomial of degree 3 in $\mathbb{Z}_5[x]$.

10. Prove that if $f(x) \in E[x]$ where E is a field of characteristic p, then $f(x^p) = (f(x))^p$.

11. If F is a field and $|F| = n$, why is it true that $a^{n-1} = 1$ for all $a \in F - \{0\}$?

12. Use Fermat's Little Theorem (Corollary 7.4) to show that if F is a field of prime order p, then every $a \in F$ is a root of $x^p - x$.

13. Suppose E is a field of order p^n, where p is a prime, and $\alpha \in E$. Prove that $m_{\alpha, F_p}(x)$, the minimal polynomial of α over the prime subfield of E, divides $x^{p^n} - x$.

14. Let E be a field of order p^n where p is a prime.
    a. If K is a subfield of E of order p^m, prove that $m \mid n$.
    b. Suppose $m \mid n$. Prove that $p^m - 1$ divides $p^n - 1$, and use this result to prove that E contains a unique subfield of order p^m.

15. Suppose F is a finite field of characteristic p.
    a. Prove that the mapping $\phi: F \to F$ defined by $\phi(a) = a^{p^k}$ is an automorphism for $k \geqslant 1$.
    b. Show that every element of F has a unique pth root.

16. Let A be an abelian group with elements $b_1, b_2, \ldots, b_t$ of order $p_1^{k_1}, p_2^{k_2}, \ldots, p_t^{k_t}$, respectively, where $p_1, p_2, \ldots, p_t$ are distinct primes. Verify that

$$b = b_1 + b_2 + \ldots + b_t$$

is an element of order $p_1^{k_1} p_2^{k_2} \cdots p_t^{k_t}$.

17. Do the nonzero elements of the field of rational numbers form a cyclic group under multiplication? Why?

18. If $(F, +, \cdot)$ is a finite field and S is a nonempty subset of $F - \{0\}$ closed under the operation $\cdot$, prove that $(S, \cdot)$ is a cyclic group.

*19. In Exercise 7, you were asked to construct a field K of order 27. Write a computer program that calculates the product and sum of any two elements in K.

21.2

Polynomial Codes

In Chapter 19, we saw that a block code can detect and correct all single errors in transmission if the columns of the parity-check matrix are nonzero and distinct. If H has 3 rows, then there are 7 possible distinct nonzero columns that could be used to form H. These columns correspond to the $2^3 - 1 = 7$ nonzero binary numbers from 1 to 7. Therefore, if H contained each and every other one of these columns, H would contain all possible syndromes. H would be a 3×7 matrix of the form $(B \vdots I_3)$, where B is a 3×4 binary matrix. The corresponding generator matrix G would then have $7 - 3 = 4$ rows and be of the form $(I_4 \vdots A)$ where A is a 4×3 binary matrix. (See Exercise 10 of Section 19.4.) The code generated by G is called a **Hamming (7, 4) code** after R.W. Hamming.[2] He first investigated these types of codes in 1950 and showed that codes of this type are **perfect error-correcting codes** because they correct all single errors and no larger errors.

Note that the code just described takes 4-bit binary words and adds 3 check-digits to produce the various codewords. In this section we will derive this code in another way through the use of some of our results on extensions of finite fields. We will then extend our new method to define similar polynomial codes that can detect and correct more than one error. We begin with an example:

Example 21.2

We wish to encode 4-bit binary numbers by adding three check-digits. We do this in the following way: Consider the polynomial $m(x) = x^3 + x^2 + 1$ in $\mathbb{Z}_2[x]$. This polynomial has no roots in $\mathbb{Z}_2$ and, therefore, no linear factors in $\mathbb{Z}_2[x]$. Since $\deg(m(x)) = 3$, $m(x)$ is irreducible in $\mathbb{Z}_2[x]$. By Corollary 20.5,

$$\mathbb{Z}_2[x]/(x^3 + x^2 + 1) \cong \mathbb{Z}_2(\alpha)$$

is a field, and this field of eight elements is generated by an element α such that $\alpha^3 + \alpha^2 + 1 = 0$. Suppose the binary word we wish to encode is

---

[2] R.W. Hamming, "Error detecting and error correcting codes", *Bell System Tech. J.*, Vol. 29 (1950), pp. 147–160.

$a_1a_2a_3a_4 = (a_1, a_2, a_3, a_4)$. Take the polynomial

$$f(x) = a_1x^6 + a_2x^5 + a_3x^4 + a_4x^3$$

$$
\begin{array}{r}
1100 \\
1101\overline{\smash{)}\,1011000} \\
\underline{1101} \\
1100 \\
\underline{1101} \\
1100 \\
\underline{1101} \\
100
\end{array}
$$

Figure 21.2

and divide $f(x)$ by $m(x)$. Then $f(x) = m(x)q(x) + r(x)$, where $r(x) = a_5x^2 + a_6x + a_7$, since $\deg(r(x)) < \deg(m(x)) = 3$. We take as our codeword the word $a_1a_2a_3a_4a_5a_6a_7$.

To illustrate, if we start with the word 1011 corresponding to $f(x) = x^6 + x^4 + x^3$ and divide by $m(x) = x^3 + x^2 + 1$, we can depict the division symbolically in Figure 21.2. Thus $r(x) = 1x^2 + 0x + 0 = x^2$, and we encode 1011 as 1011100.

Note that the procedure in Example 21.2 produces a **systematic code**, that is, one in which the original binary word appears at the front of the encoded word.

You are probably asking why we bother to do all this, since the methods we used to encode words in Chapters 12 and 19 were much simpler to implement. The answer is that this method has certain advantages when it comes to error correction. To see this, suppose that $f(x) = m(x)q(x) + r(x)$ where $m(\alpha) = 0$. Then

$$f(\alpha) = m(\alpha)q(\alpha) + r(\alpha) = 0q(\alpha) + r(\alpha) = r(\alpha)$$

If we set $F(x) = f(x) + r(x)$, then

$$F(\alpha) = f(\alpha) + r(\alpha) = r(\alpha) + r(\alpha) = 0$$

since $r(\alpha) \in \mathbb{Z}_2(\alpha)$, a field of characteristic 2. Suppose our codeword corresponding to the polynomial $F(x)$ is transmitted and received as a word corresponding to the polynomial $G(x)$. If no error occurred in transmission, then $G(\alpha) = F(\alpha) = 0$. If a single error occurred in transmitting our codeword, then $G(x) = F(x) + E(x)$, where $E(x) = x^k$ is an **error polynomial** related to an incorrect transmission of the kth digit. Therefore

$$G(\alpha) = F(\alpha) + E(\alpha) = 0 + \alpha^k = \alpha^k$$

Evaluating the polynomial $G(x)$ at $x = \alpha$ will enable us to detect the location of an error and correct the proper digit.

Let us illustrate this procedure by considering the polynomials in Example 21.2.

Example 21.3

In Example 21.2,

$$F(x) = f(x) + r(x) = a_1x^6 + a_2x^5 + a_3x^4 + \cdots + a_6x + a_7$$

Suppose 1011100 is transmitted and received as 1001100. Then $G(x) = x^6 + x^3 + x^2$ and we need to calculate $G(\alpha) = \alpha^6 + \alpha^3 + \alpha^2$. To do this, Figure 21.3 is helpful. It is obtained by using the fact that $\alpha^3 + \alpha^2 + 1 = 0$; in

$\mathbb{Z}_2(\alpha)$, we have

$$\alpha^3 = \alpha^2 + 1,$$

$$\alpha^4 = \alpha(\alpha^3) = \alpha(\alpha^2 + 1)$$

$$= \alpha^3 + \alpha$$

$$= (\alpha^2 + 1) + \alpha$$

$$= \alpha^2 + \alpha + 1, \quad \text{etc.}$$

$$
\begin{array}{l}
\alpha^0 = 1 \\
\alpha^1 = \alpha \\
\alpha^2 = \alpha^2 \\
\alpha^3 = \alpha^2 + 1 \\
\alpha^4 = \alpha^2 + \alpha + 1 \\
\alpha^5 = \alpha + 1 \\
\alpha^6 = \alpha^2 + \alpha \\
\alpha^7 = 1
\end{array}
$$

Figure 21.3

Figure 21.3 demonstrates that $\mathbb{Z}_2(\alpha)$ is a field of order 8, this field is cyclic, and $\{1, \alpha, \alpha^2\}$ is a basis of this field over $\mathbb{Z}_2$. With the aid of the figure, we see that

$$G(\alpha) = \alpha^6 + \alpha^3 + \alpha^2 = (\alpha^2 + \alpha) + (\alpha^2 + 1) + \alpha^2 = \alpha^2 + \alpha + 1 = \alpha^4$$

This says that the error occurs in the coefficient of x^4 and that the error message is 0010000. The corrected word is then $1001100 + 0010000 = 1011100$, which we decode as 1011.

This error-correcting scheme always works with single errors. It fails, however, to correct double errors:

Example 21.4

Consider again the code of Example 21.2. Suppose the word 1011100 is received as 1011111. Then

$$G(\alpha) = \alpha^6 + \alpha^4 + \alpha^3 + \alpha^2 + \alpha + 1$$

$$= (\alpha^2 + \alpha) + (\alpha^2 + \alpha + 1) + (\alpha^2 + 1) + \alpha^2 + \alpha + 1$$

$$= \alpha + 1 = \alpha^5$$

Our correction method in this case will lead us to conclude erroneously that the error message is 0100000 rather than 0000011.

To correct errors of size greater than one, we will need more check-digits. The number of check-digits in polynomial codes is related to the degree of the polynomial $m(x)$. In our first example, $m(x)$ was the minimal polynomial of a generator of the field $\mathbb{Z}_2(\alpha)$. To obtain more check-digits, we will increase the size of our finite field and the degree of $m(x)$ as well.

Since we wish to restrict our analysis to binary words, we turn now to an extension field of $\mathbb{Z}_2$ of degree 4. To do this, we need an irreducible polynomial of degree 4 in $\mathbb{Z}_2[x]$.

Example 21.5

Consider $m_1(x) = x^4 + x^3 + 1$. Neither 0 nor 1 is a root. While $m_1(x)$ has no linear factors in $\mathbb{Z}_2[x]$, $m_1(x)$ might still be a product of two quadratic factors. We leave it as Exercise 5 to show this is not the case. Therefore $m_1(x)$ is

$$
\begin{array}{lll}
\alpha^1 = \alpha & \alpha^6 = \alpha^3 + \alpha^2 + \alpha + 1 & \alpha^{11} = \alpha^3 + \alpha^2 + 1 \\
\alpha^2 = \alpha^2 & \alpha^7 = \alpha^2 + \alpha + 1 & \alpha^{12} = \alpha + 1 \\
\alpha^3 = \alpha^3 & \alpha^8 = \alpha^3 + \alpha^2 + \alpha & \alpha^{13} = \alpha^2 + \alpha \\
\alpha^4 = \alpha^3 + 1 & \alpha^9 = \alpha^2 + 1 & \alpha^{14} = \alpha^3 + \alpha^2 \\
\alpha^5 = \alpha^3 + \alpha + 1 & \alpha^{10} = \alpha^3 + \alpha & \alpha^{15} = 1
\end{array}
$$

Figure 21.4

irreducible and

$$\mathbb{Z}_2[x]/(x^4 + x^3 + 1) \cong \mathbb{Z}(\alpha)$$

is a field of order 16 in which $\alpha^4 + \alpha^3 + 1 = 0$. Figure 21.4 shows the powers of α.

In the code found in Example 21.2, we added 3 check-digits to 4-bit binary words to form 7-bit codewords, where $7 = 2^3 - 1$. Since there are $15 = 2^4 - 1$ possible nonzero columns in a 4-row parity-check matrix, let us add check-digits to our binary words to obtain 15-bit codewords. How long should our original words be, and how many checkdigits must we add? We were able to detect single errors in the code of Example 21.2 by evaluating at $x = \alpha$ the polynomial $G(x)$ corresponding to the codeword **received**. Perfect transmission yielded $G(\alpha) = 0$, while a single error in the kth digit gave the value $G(\alpha) = \alpha^k$. We were unable, however, to correct any double-errors by evaluating $G(x)$ at $x = \alpha$ only. In order to locate a second error, we desire another element β, such that $G(\beta) = 0$ when there is no error. In addition, we need a method by which we can detect and correct double errors by examining $G(\alpha)$ and $G(\beta)$.

Example 21.6

Let us utilize the polynomial $m_1(x)$ of Example 21.5. Suppose $a \in \mathbb{Z}$, then $a^2 \equiv a \pmod 2$ by Corollary 7.4, and $G(\alpha^2) = G(\alpha)$ in $\mathbb{Z}_2(\alpha)$. (See Exercise 10 of Section 21.1.) Therefore, α^2 is not a good choice for the value of β. Try $\beta = \alpha^3$. In Example 21.3, $F(\alpha) = 0$ since $m(x)$ divided $F(x)$ and $m(\alpha) = 0$. But

$$
\begin{aligned}
m_1(\beta) = m_1(\alpha^3) &= (\alpha^3)^4 + (\alpha^3)^3 + 1 \\
&= \alpha^{12} + \alpha^9 + 1 \\
&= (\alpha + 1) + (\alpha^2 + 1) + 1 \\
&= \alpha^2 + \alpha + 1 = \alpha^7 \neq 0
\end{aligned}
$$

In our new code, we will require the minimal polynomials of both α and α^3 to divide $F(x)$.

To find the minimal polynomial of α^3, we note that $\alpha^{15} = 1$. Therefore $(\alpha^3)^5 = 1$ and α^3 is root of

$$x^5 - 1 = (x - 1)(x^4 + x^3 + x^2 + x + 1)$$

Since $a^3 \neq 1$, α^3 is a root of $x^4 + x^3 + x^2 + x + 1$, which is irreducible by Example 17.7. If

$$m(x) = m_1(x)(x^4 + x^3 + x^2 + x + 1) = x^8 + x^4 + x^2 + x + 1$$

then $m(\alpha) = m(\alpha^3) = 0$. To implement our code, we add 8 check-digits to words of length 7 in the following way: Take a word $a_1 a_2 \cdots a_7$ and form the polynomial

$$F(x) = a_1 x^{14} + a_2 x^{13} + \cdots + a_7 x^8 + a_8 x^7 + a_9 x^6 + \cdots$$
$$+ a_{13} x^2 + a_{14} x + a_{15}$$

where $a_8 x^7 + a_9 x^6 + \cdots + a_{14} x + a_{15}$ is the remainder obtained by dividing

$$f(x) = a_1 x^{14} + a_2 x^{13} + \cdots + a_6 x^9 + a_7 x^8$$

by $m(x)$. Our codeword is then

$$a_1 a_2 \cdots a_7 a_8 a_9 \cdots a_{15}.$$

For example, dividing $x^{13} + x^{10} + x^8$ by $m(x)$ yields remainder $x^7 + x^2 + 1$, so that 0100101 is encoded as 010010110000101. Since $f(x) = m(x)q(x) + r(x) \in \mathbb{Z}_2[x]$, we have

$$F(x) = f(x) + r(x) = m(x)q(x) + 2r(x) = m(x)q(x)$$

so that $F(\alpha) = m(\alpha)q(\alpha) = 0$ and $F(\alpha^3) = m(\alpha^3)q(\alpha^3) = 0$.

To see how to use this scheme to correct errors, suppose that 010010110000101 were transmitted as 010010010000111. The error message is then 000000100000010, corresponding to the polynomial $E(x) = x^8 + x$. Now

$$E(\alpha) = \alpha^8 + \alpha = (\alpha^3 + \alpha^2 + \alpha) + \alpha = \alpha^3 + \alpha^2 = \alpha^{14}$$

while

$$E(\alpha^3) = (\alpha^3)^8 + \alpha^3 = \alpha^{24} + \alpha^3 = \alpha^{15}\alpha^9 + \alpha^3$$
$$= \alpha^9 + \alpha^3 = (\alpha^2 + 1) + \alpha^3$$
$$= \alpha^3 + \alpha^2 + 1 = \alpha^{11}$$

The receiver, however, does not know the polynomial $F(x)$ and must determine $E(\alpha)$ and $E(\alpha^3)$ by evaluating $G(x)$, which is

$$G(x) = x^{13} + x^{10} + x^7 + x^2 + x + 1$$

Again $G(\alpha) = F(\alpha) + E(\alpha) = 0 + E(\alpha) = E(\alpha)$, and

$$G(\alpha) = \alpha^{13} + \alpha^{10} + \alpha^7 + \alpha^2 + \alpha + 1$$
$$= (\alpha^2 + \alpha) + (\alpha^3 + \alpha) + (\alpha^2 + \alpha + 1) + \alpha^2 + \alpha + 1$$
$$= \alpha^3 + \alpha^2 = \alpha^{14}$$

Similarly, $G(\alpha^3) = E(\alpha^3)$ can be shown to equal α^{11}. (See Exercise 12.)

How can we correct the transmission with only the knowledge of the values of $E(\alpha)$ and $E(\alpha^3)$? In Example 21.3, the equation $G(\alpha) = E(\alpha) = \alpha^i$ indicated that there was an error in the ith term. Another way to look at this is to say that the polynomial $P(x) = x - E(\alpha)$ has as its root the power of α corresponding to the location of the single error in transmission. We now wish to find a polynomial $P(x)$ that has as its roots powers of α corresponding to locations of double errors. In other words, if $E(x) = x^j + x^k$, then we want $P(\alpha^j) = P(\alpha^k) = 0$ and the polynomial $P(x)$ should be determined by the values of $E(\alpha)$ and $E(\alpha^3)$. So suppose $P(x)$ is a polynomial of degree 2, say, $P(x) = x^2 + ax + b$. If $P(\alpha^j) = P(\alpha^k) = 0$, we have

$$0 = P(\alpha^j) = \alpha^{2j} + a\alpha^j + b \qquad \text{and} \qquad 0 = P(\alpha^k) = \alpha^{2k} + a\alpha^k + b$$

Subtracting, we obtain

$$(a^{2j} - a^{2k}) + a(\alpha^j - \alpha^j) = 0$$

Then

$$a = -\frac{\alpha^{2j} - \alpha^{2k}}{\alpha^j - \alpha^k} = -(\alpha^j + \alpha^k) = -E(\alpha) = E(\alpha)$$

since

$$E(x) = x^j + x^k = -x^j - x^k \in \mathbb{Z}_2[x]$$

Also

$$b = -\alpha^{2j} - a\alpha^j = -\alpha^{2j} + (\alpha^j + \alpha^k)\alpha^j = \alpha^{j+k}$$

Now $E(\alpha^2) = \alpha^{2j} + \alpha^{2k}$, while $E(\alpha^3) = \alpha^{3j} + \alpha^{3k}$. Therefore, in $\mathbb{Z}_2(\alpha)$,

$$\begin{aligned} E(\alpha)E(\alpha^2) &= (\alpha^j + \alpha^k)(\alpha^{2j} + \alpha^{2k}) \\ &= \alpha^{3j} + \alpha^{2j}\alpha^k + \alpha^j\alpha^{2k} + \alpha^{3k} \\ &= E(\alpha^3) + \alpha^j\alpha^k(\alpha^j + \alpha^k) \\ &= E(\alpha^3) + \alpha^{j+k}E(\alpha) \end{aligned}$$

We conclude that

$$b = \alpha^{j+k} = \frac{E(\alpha)E(\alpha^2) + E(\alpha^3)}{E(\alpha)}$$

and

$$P(x) = x^2 + E(\alpha)x + \frac{E(\alpha)E(\alpha^2) + E(\alpha^3)}{E(\alpha)}$$

| Example 21.7 |

For the code of Example 21.6, we have

$$P(x) = x^2 + \alpha^{14}x + \frac{\alpha^{14}\alpha^{28} + \alpha^{11}}{\alpha^{14}}$$

$$= x^2 + (\alpha^3 + \alpha^2)x + \alpha^{-14}(\alpha^{12} + \alpha^{11})$$

$$= x^2 + (\alpha^3 + \alpha^2)x + \alpha(\alpha^{12} + \alpha^{11}) \qquad \text{since } 1 = \alpha^{15} = \alpha^{14}\alpha$$

$$= x^2 + (\alpha^3 + \alpha^2)x + (\alpha^{13} + \alpha^{12})$$

$$= x^2 + (\alpha^3 + \alpha^2)x + ((\alpha^2 + \alpha) + (\alpha + 1))$$

$$= x^2 + (\alpha^3 + \alpha^2)x + (\alpha^2 + 1)$$

If we substitute powers of α, one-by-one, in the equation for x, we find

$$P(\alpha) = \alpha^2 + (\alpha^3 + \alpha^2)\alpha + (\alpha^2 + 1)$$

$$= \alpha^2 + \alpha^4 + \alpha^3 + \alpha^2 + 1$$

$$= \alpha^4 + \alpha^3 + 1 = 0$$

By further substitution or long division, we find $P(\alpha^8) = 0$, so that $P(x) = (x - \alpha^8)(x - \alpha)$. We conclude correctly that $E(x) = x^8 + x$.

The scheme we have just developed works for double errors, but what happens if there are less than two errors in transmission? If $E(\alpha) = 0$, we proceed as before and conclude that no error occurred. Suppose, however, there is a single error, say $E(x) = x^k$. Then $E(\alpha) = \alpha^k$, $E(\alpha^2) = \alpha^{2k}$, and $E(\alpha^3) = \alpha^{3k}$. The polynomial $P(x)$ takes the form

$$P(x) = x^2 + ax + b = x + \alpha^k x + \frac{(\alpha^k \alpha^{2k} + \alpha^{3k})}{\alpha^k}$$

$$= x^2 + \alpha^k x + \frac{2\alpha^{3k}}{\alpha^k}$$

$$= x^2 + \alpha^k x = x(x + \alpha^k)$$

since every element in $\mathbb{Z}_2[x]$ is its own inverse. Since $\alpha^n \neq 0$ for any n, we conclude that the only root of $P(x)$ among the powers of α is α^k and $E(x) = x^k$. We then correct and decode accordingly.

What we have done in developing a double error-correcting polynomial code can be extended to create codes that correct three or more errors. At the heart of this process is the theory of extensions of finite fields. Additional sources can be found in the references for those of you interested in pursuing these ideas.

1. Let

$$1 = \begin{bmatrix} 1 \\ 0 \\ 0 \end{bmatrix}, \qquad \alpha = \begin{bmatrix} 0 \\ 1 \\ 0 \end{bmatrix}, \qquad \alpha^2 = \begin{bmatrix} 0 \\ 0 \\ 1 \end{bmatrix}.$$

Express the elements in Figure 21.3 as linear combinations of these basis elements of the vector space $\mathbb{Z}_2[\alpha]$. Let G be the 3×7 matrix whose columns are $1, \alpha, \alpha^2, \alpha^3, \alpha^4, \alpha^5$, and α^6. Verify that G is a generator matrix for a block code, and determine the associated parity-check matrix H. Why is this code a perfect error-correcting mode?

2. Prove that $m(x) = x^3 + x + 1$ is irreducible in $\mathbb{Z}_2[x]$. Let β be a root of this polynomial in the field $\mathbb{Z}_2[x]/(m(x))$. Compute a table of powers of β as in Figure 21.3.

3. Encode the following message according to Example 21.3:

    1001   1101   1001

4. Suppose the following message is received:

    0111100   1011011   1000110

Decode this message according to the procedure outlined in Example 21.3.

5. Verify that $x^4 + x^3 + 1$ is irreducible in $\mathbb{Z}_2[x]$.

6. Encode the following message according to Example 21.6:

    1100111   0011001   1100000

7. Suppose the following message is received:

    110010110000100   110100010100001   011110010100001

Decode this message according to Example 21.6.

8. Find a block code determined by Example 21.6. (See Exercise 1.)

9. What would happen in Example 21.6 if we took $\beta = \alpha^5$, or $\beta = \alpha^6$?

10. Prove that $m(x) = x^4 + x + 1$ is irreducible in $\mathbb{Z}_2[x]$. Let β be a root of this polynomial in the field $\mathbb{Z}_2[x]/(m(x))$. Determine a table of powers of β as in Example 21.6.

11. Investigate the polynomial code obtained in the following way: Given the word $a_1 a_2 a_3$, divide $a_1 x^4 + a_2 x^3 + a_3 x^2$ by $x^2 + x + 1$. If the remainder is $a_4 x + a_5$, encode $a_1 a_2 a_3$ as $a_1 a_2 a_3 a_4 a_5$.

12. In Example 21.6, verify that $G(\alpha^3) = \alpha^{11}$.

13. Describe a polynomial binary code where the encoded words have a length of $31 = 2^5 - 1$ digits.

*14. Write a computer program to encode 4-bit words according to the scheme shown in Example 21.2 utilizing polynomial division by $x^3 + x^2 + 1$.

*15. Write a program to decode words transmitted according to the scheme developed in Example 21.3.

22

Galois Theory

In Chapter 17, we investigated methods whereby the roots of an arbitrary third- or fourth-degree polynomial with real coefficients can be found algebraically. What about polynomials of degree greater than four?

Evariste Galois and Neils Henrik Abel are generally credited with the solution to the classical problem of determining whether there exists a formula for finding the roots of a polynomial of degree greater than or equal to five. Abel proved that no such formula exists, while Galois treated the companion problem of determining just what types of polynomials possess formulas for finding their roots. We saw, for example, that one can always find the roots of an arbitrary polynomial of the form $x^n = a$, where $a \in \mathbb{C}$, by using DeMoivre's Theorem.

In this chapter, we will investigate the method Galois devised to describe how the roots of a polynomial are related to one another. This method greatly simplified the theory of equations. We will also apply this method to prove Abel's result on the insolvability of the quintic.

Three Basic Examples

The Galois Theory involves the construction of a bijection between the set of subgroups of a certain group defined by the roots of a given polynomial and the set of subfields of a splitting field for that polynomial. In this section, we wish to present the definitions underlying this construction and discuss three examples to illustrate this beautiful theory. These examples will motivate our analysis of the theory in the next section.

Recall that an automorphism of a field E is a bijection $\sigma: E \to E$ such that

$$\sigma(a + b) = \sigma(a) + \sigma(b) \qquad \text{and} \qquad \sigma(ab) = \sigma(a)\sigma(b)$$

for all $a, b \in E$. We showed in Theorem 17.13 that complex conjugation is an automorphism of $\mathbb{C}$.

Definition 22.1

> Let E be an extension field of F. An automorphism of E is said to **fix** F if $\sigma(a) = a$ for all $a \in F$. The set of all such automorphisms of E fixing F is denoted $G_{E/F}$ and is termed the **Galois group of E over F**. If $f(x) \in F[x]$, then the **Galois group of $f(x)$ over** F is $G_{E/F}$, where E is a splitting field for $f(x)$ over F.

We leave it to you to verify in Exercises 1 and 2 that the set of all automorphisms of E forms a group and that $G_{E/F}$ is a subgroup of that group. We speak of "the" Galois group of a polynomial since Theorem 20.7 guarantees that any two splitting fields for $f(x) \in F[x]$ are isomorphic by a mapping that fixes F. Therefore, any two splitting fields for a polynomial give rise to isomorphic Galois groups.

If E is a field of characteristic 0, then the prime subfield of E is isomorphic to $\mathbb{Q}$ by Exercise 5 of Section 20.1. An automorphism σ of a field must always take the unity element of E to itself. Since σ preserves addition and multiplication, σ fixes all terms of the form $n1$, for $n \in \mathbb{Z}$, and, therefore, all terms of the form $(n1)(m1)^{-1}$. We conclude that any automorphism of E must fix at least the prime subfield of E. This result is also true in the case where $\text{char}(E) = p$. (See Exercise 3.)

For $f(x) = x^2 + 1 \in \mathbb{R}[x]$, the splitting field of $f(x)$ over $\mathbb{R}$ is the field $\mathbb{R}(i) = \mathbb{C}$. In this case, there are exactly two automorphisms of $\mathbb{C}$ fixing $\mathbb{R}$. To see this, suppose that $\sigma: \mathbb{C} \to \mathbb{C}$ is an automorphism of f over $\mathbb{R}$. Since $0 = f(i) = i^2 + 1$, we have

$$0 = \sigma(0) = \sigma(i^2 + 1) = \sigma(i)^2 + 1$$

This says that $\sigma(i)$ must be a root of $x^2 + 1$ lying in $\mathbb{C}$. Now the only two roots of this polynomial are i and $-i$. If $\sigma(i) = i$, then

$$\sigma(a + bi) = \sigma(a) + \sigma(b)\sigma(i) = a + bi$$

and σ is the identity automorphism $1_{\mathbb{C}}(a + bi) = a + bi$. If $\sigma(i) = -i$, then σ is the automorphism

$$\sigma(a + bi) = \sigma(a) + \sigma(b)\sigma(i) = a + b(-i) = a - bi$$

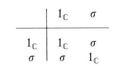

Figure 22.1

corresponding to complex conjugation. In this case, the set of all elements fixed by complex conjugation is much larger than the prime subfield $\mathbb{Q}$; it is $\mathbb{R}$, the field of real numbers. The Galois group G of the polynomial $f(x) = x^2 + 1$ is the group consisting of the two mappings $1_{\mathbb{C}}$ and σ whose group table is given in Figure 22.1. Observe that

$$(\sigma \circ \sigma)(a + bi) = \sigma(\sigma(a + bi)) = \sigma(a - bi) = a - (-b)i = a + bi$$

$$= 1_{\mathbb{C}}(a + bi)$$

This group is isomorphic to $(\mathbb{Z}_2, \oplus)$. Note that $|G| = 2 = [\mathbb{R}(i) : \mathbb{R}]$.

In this first example, we have utilized the fact that if σ is an automorphism of $f(x) = x^2 + 1 \in \mathbb{R}[x]$, then $\sigma(i)$ is another root of f. This result holds in general.

Lemma 22.1

Let E be a splitting field over F for $f(x) \in F[x]$. Suppose $f(\alpha) = 0$. If $\sigma \in G_{E/F}$ and $\sigma(\alpha) = \beta$, then $f(\beta) = 0$.

Proof

Let $f(x) = a_n x^n + a_{n-1} x^{n-1} + \cdots + a_1 x + a_0$. Then

$$0 = f(\alpha) = a_n \alpha^n + a_{n-1} \alpha^{n-1} + \cdots + a_1 \alpha + a_0$$

Since σ is an automorphism fixing all elements of F,

$$0 = \sigma(0)$$
$$= \sigma(a_n \alpha^n + a_{n-1} \alpha^{n-1} + \cdots + a_1 \alpha + a_0)$$
$$= a_n \sigma(\alpha)^n + a_{n-1} \sigma(\alpha)^{n-1} + \cdots + a_1 \sigma(\alpha) + a_0$$
$$= f(\sigma(\alpha))$$

But then, $f(\beta) = f(\sigma(\alpha)) = 0$. □

As a second example, we analyze the Galois group of the polynomial $f(x) = x^5 - 1 \in \mathbb{Q}[x]$. Now

$$f(x) = x^5 - 1 = (x - 1)(x^4 + x^3 + x^2 + x + 1)$$

and the polynomial

$$q(x) = x^4 + x^3 + x^2 + x + 1$$

was shown in Example 17.7 to be irreducible in $\mathbb{Q}[x]$ by Eisenstein's Criteria.

By DeMoivre's Theorem 17.16, we know the roots of $f(x) = x^5 - 1$ are $1, \omega, \omega^2, \omega^3$, and ω^4, where $\omega = \cos \frac{2}{5}\pi + i \sin \frac{2}{5}\pi$. Therefore, the splitting field of f is $E = \mathbb{Q}(\omega)$. If $\sigma \in G_{E/\mathbb{Q}}$, then

$$0 = \sigma(0) = \sigma(\omega^5 - 1) = \sigma(\omega)^5 - 1$$

is another root of $x^5 - 1$. The elements of $G_{E/\mathbb{Q}}$ are thus determined by their effect on ω since ω is a primitive element of the splitting field of $x^5 - 1$. Since 1 is an element fixed by all elements of $G_{E/\mathbb{Q}}$, we have $\sigma(\omega) \neq 1$ for all $\sigma \in G_{E/\mathbb{Q}}$. We conclude that

$$G_{E/\mathbb{Q}} = \{\sigma_i \mid \sigma_i(\omega) = \omega^i, \ i = 1, 2, 3, 4\}$$

	σ_1	σ_2	σ_3	σ_4
σ_1	σ_1	σ_2	σ_3	σ_4
σ_2	σ_2	σ_4	σ_1	σ_3
σ_3	σ_3	σ_1	σ_4	σ_2
σ_4	σ_4	σ_3	σ_2	σ_1

Figure 22.2

Now

$$(\sigma_2 \circ \sigma_3)(\omega) = \sigma_2(\sigma_3(\omega)) = \sigma_2(\omega^3) = \sigma_2(\omega)^3 = (\omega^2)^3 = \omega^6 = \omega = \sigma_1(\omega)$$

The complete Cayley table for $G_{E/\mathbb{Q}}$ is in Figure 22.2; σ_1 is the identity of this group, and we leave it to you to show that this group is cyclic. (See Exercise 7.) Its order is 4, which is also the degree of $\mathbb{Q}(\omega)$ over $\mathbb{Q}$.

Definition 22.1 describes a mapping from the set of all extensions of a field F to the set of all groups. There is another mapping we wish to investigate. Suppose E is an extension of F, and $H = \{\sigma_1, \sigma_2, \ldots, \sigma_n\}$ is a subgroup of the group of all automorphisms of E. Let

$$E_H = \{\alpha \in E \mid \sigma_i(\alpha) = \alpha \text{ for all } \sigma_i \in H\}$$

Theorem 22.1

E_H is a subfield of E.

Proof

To show that E_H is a subfield of E, we must first show $E_H \neq \varnothing$. Clearly $\sigma(0) = 0$ for all $\sigma \in H$; so $0 \in E_H$. Now suppose $\alpha, \beta \in E_H$. If $\sigma \in H$, then

$$\sigma(\alpha - \beta) = \sigma(\alpha) - \sigma(\beta) = \alpha - \beta$$

hence, E_H is closed under subtraction. Moreover, $\sigma(\alpha\beta) = \sigma(\alpha)\sigma(\beta) = \alpha\beta$ implies that E_H is a subring of E.

To complete our proof, we need only verify that E_H is closed under the taking of multiplicative inverses. Now $\sigma(\alpha^{-1}) = \sigma(\alpha)^{-1} = \alpha^{-1}$ for any $\sigma \in E_H$ by Theorem 14.4 since σ is a field automorphism. Thus E_H is a subfield of E. $\square$

E_H is known as the **fixed field of E by H**. The fixed field of the Galois group G of $f(x) = x^2 + 1 \in \mathbb{R}[x]$ is $\mathbb{R}$. To see this, suppose $a + bi \in \mathbb{C}_G$. Let σ be the automorphism in G corresponding to complex conjugation. Then $\sigma(a + bi) = a - bi$. Since $a + bi \in \mathbb{C}_G$, we also have $\sigma(a + bi) = a + bi$. Equating $a + bi = a - bi$, we have $b = 0$ and $a + bi = a \in \mathbb{R}$.

In our second example where $f(x) = x^5 - 1 \in \mathbb{Q}[x]$, the splitting field of

f over $\mathbb{Q}$ is $E = \mathbb{Q}(\omega)$. To calculate the fixed field of this group, we will find it convenient to represent the elements of E in the form

$$E = \{a_1\omega + a_2\omega^2 + a_3\omega^3 + a_4\omega^4 | a_i \in \mathbb{Q}\}$$

If $\beta = a_1\omega + a_2\omega^2 + a_3\omega^3 + a_4\omega^4$ is fixed by the whole Galois group G, then

$$\beta = \sigma_2(\beta) = \sigma_2(a_1\omega + a_2\omega^2 + a_3\omega^3 + a_4\omega^4)$$
$$= a_1\omega^2 + a_2\omega^4 + a_3\omega^6 + a_4\omega^8$$
$$= a_1\omega^2 + a_2\omega^4 + a_3\omega + a_4\omega^3 \qquad \text{since } \omega^5 = 1$$

Now $\{\omega, \omega^2, \omega^3, \omega^4\}$ is a basis of $E = \mathbb{Q}(\omega)$ over $F = \mathbb{Q}$; this implies

$$a_1 = a_2 = a_3 = a_4 \qquad \text{and} \qquad \beta = a_1(\omega + \omega^2 + \omega^3 + \omega^4)$$

Since $1 + \omega + \omega^2 + \omega^3 + \omega^4 = 0$, we conclude $\beta = a_1(-1) = -a_1 \in \mathbb{Q}$. Therefore, the fixed field of G is $\mathbb{Q}$.

Let us next consider the subgroup $H = \langle\sigma_4\rangle = \{\sigma_1, \sigma_4\}$ of this Galois group. What is E_H? If

$$\beta = a_1\omega + a_2\omega^2 + a_3\omega^3 + a_4\omega^4 \in E_H$$

then

$$\beta = \sigma_4(\beta) = \sigma_4(a_1\omega + a_2\omega^2 + a_3\omega^3 + a_4\omega^4)$$
$$= a_1\omega^4 + a_2\omega^3 + a_3\omega^2 + a_4\omega$$

Again the fact that $\{\omega, \omega^2, \omega^3, \omega^4\}$ constitutes a basis forces us to conclude that $a_1 = a_4$ and $a_2 = a_3$. Therefore,

$$\beta = a_1(\omega + \omega^4) + a_2(\omega^2 + \omega^3)$$

But then

$$\beta = a_2 + (a_1 - a_2)(\omega + \omega^4)$$

(See Exercise 5.) This shows that $E_H \subseteq \{b_0 + b_1(\omega + \omega^4) | b_0, b_1 \in \mathbb{Q}\}$. Moreover

$$\sigma_4(b_0 + b_1(\omega + \omega^4)) = b_0 + b_1(\omega^4 + \omega) = b_0 + b_1(\omega + \omega^4)$$

so that $E_H = \{b_0 + b_1(\omega + \omega^4) | b_0, b_1 \in \mathbb{Q}\}$. We leave to you to show that E_H, so defined, is a field.

Our goal is to exploit the ideas of Galois groups and fixed fields to develop a bijection between the set of subfields of an extension E of F and the subgroups of $G_{E/F}$. In order to do this, we place some restrictions on E.

Definition 22.2

A finite extension E of F is a **normal extension of F** if F is the fixed field of $G_{E/F}$.

We have encountered two examples of normal extensions. We saw first that the fixed field of the automorphism group of $f(x) = x^2 + 1 \in \mathbb{R}[x]$ is $\mathbb{R}$ itself. Then for our second example, we calculated the fixed field of the automorphism group of $\mathbb{Q}(\omega)$ and found it to be $\mathbb{Q}$. To find an extension that is not normal, we turn to our third example in this section.

Let $E = \mathbb{Q}(\sqrt[3]{5}) = \{a + b\sqrt[3]{5} + c\sqrt[3]{25} \mid a, b, c \in \mathbb{Q}\}$. If $G = G_{E/\mathbb{Q}}$, what is E_G? Let $\sigma \in G$. Then $\sigma(\sqrt[3]{5})^3 = \sigma((\sqrt[3]{5})^3) = \sigma(5) = 5$, so $\sigma(\sqrt[3]{5})$ must be a solution of $x^3 - 5 = 0$ lying in $\mathbb{Q}(\sqrt[3]{5})$. Because $\sqrt[3]{5}$ is the only real solution of $x^3 - 5 = 0$, we have

$$\sigma(\sqrt[3]{5}) = \sqrt[3]{5} \quad \text{and} \quad \sigma(\sqrt[3]{25}) = \sigma(\sqrt[3]{5}\sqrt[3]{5}) = \sigma(\sqrt[3]{5})\sigma(\sqrt[3]{5}) = \sqrt[3]{5}\sqrt[3]{5} = \sqrt[3]{25}$$

Thus σ fixes not only $\mathbb{Q}$ but **every** element of E, so that $E_G = E$ and E is *not* a normal extension of $\mathbb{Q}$. In addition, G is the trivial group consisting of just the identity mapping.

This last example provides us with an illustration to show that normal extensions are indeed a nontrivial classification of field extensions. In many ways, these are the "nicest" types of extensions and will be the type of extension field involved in the statement of the Fundamental Theorem of Galois Theory. If you happen to read some other texts on Galois Theory, however, you might find different definitions for *normal extension*. The reason is that there are a number of alternate versions of our definition. For fields of characteristic 0, all are equivalent. Since Galois and Abel worked with the fields $\mathbb{Q}$, $\mathbb{R}$, and $\mathbb{C}$, their original results were given for fields of characteristic 0. In the next section, we will present their results. While Galois Theory has been extended to fields of finite characteristic, we feel that an introduction to Galois Theory is best presented in the case of a field of characteristic 0. We therefore refer you to the references for generalizations to fields of prime characteristic.

Exercises 22.1

1. Verify that the set of all automorphisms of a field is a group.

2. If F is a subfield of E, prove that the set of all automorphisms of E fixing every element of F forms a subgroup of the group of all automorphisms of E.

3. Let σ be an automorphism of E, a field of characteristic p. Prove that σ fixes the prime subfield of E.

4. Let $f(x) = x^5 - x \in \mathbb{Z}_5[x]$. Why is the Galois group of $f(x)$ trivial?

5. Suppose $\omega^5 = 1$, $\omega \neq 1$. Prove that if $\sigma_4(\omega) = \omega^4$ and $H = \langle \sigma_4 \rangle$, then $E_H = \mathbb{Q}(\omega + \omega^4)$.

6. Verify that if σ is an automorphism of $\mathbb{R}$, then σ sends positive reals to positive reals.

7. Verify that the Galois group of $\mathbb{Q}(\omega)$ over $\mathbb{Q}$, where ω is a primitive nth root of unity, is an abelian group.

8. Prove that $\mathbb{Q}(\sqrt[3]{2})$ is not the splitting field of $f(x) = x^3 - 2 \in \mathbb{Q}[x]$. How is $\mathbb{Q}(\sqrt[3]{2})$ related to the splitting field?

9. Calculate the Galois group, its subgroups, and the corresponding fixed fields for the splitting field over $\mathbb{Q}$ of each of the following:
   a. $x^5 + 1$ 　　　　　　　　　　　　 b. $x^5 - 32$
   c. $x^4 - 1$ 　　　　　　　　　　　　 d. $x^7 - 1$
   e. $x^4 + x^2 + 1$ 　　　　　　　　 f. $x^6 - 1$
   g. $x^3 + x^2 + x + 1$ 　　　　　　 h. $x^4 + 1$

10. Find the monic polynomial f of least degree in $\mathbb{Q}[x]$ having $\sqrt{3}$ and $2+i$ as roots. Determine the splitting field of f. How is it related to $\mathbb{Q}(\sqrt{3})$ and $\mathbb{Q}(2+i)$? Find the Galois groups of these three fields. How are they related?

11. Find a primitive element γ for the splitting field of $f(x) = x^3 - 5$ over $\mathbb{Q}$. Is $\mathbb{Q}(\gamma)$ a normal extension of $\mathbb{Q}$?

12. Let ω be a primitive root of $x^3 - 1 = 0$, and set $F = \mathbb{Q}(\omega)$. Show that $f(x) = x^3 - 5$ is irreducible in F. If E is the splitting field of $f(x)$ over F and G the Galois group of E over F, prove that $|G| = [E : F]$ and that E is a normal extension of F.

13. Let p be a prime and γ a primitive root of $f(x) = x^p - 1 \in \mathbb{Q}[x]$. If $\sigma \in G$, the Galois group of $f(x)$ over $\mathbb{Q}$, prove $\sigma(\gamma) = \gamma^k$ for some $k = 1, 2, \ldots, p-1$, and G is a cyclic group of order p.

14. Prove that if $f(x) = x^2 - a \in \mathbb{Q}[x]$ and E is the splitting field of $f(x)$ over $\mathbb{Q}$, then $|G_{E/F}| = 1$ or 2.

15. An extension E of a field F is **abelian** if the Galois group of E over F is abelian. Suppose E is an abelian extension of F and K is a subfield of E containing F. Verify that E is an abelian extension of K.

16. Prove that if E is an extension of $\mathbb{Q}$ of degree 2, then E is a normal extension.

The Fundamental Theorem

Throughout this text, we have presented a good number of applications of abstract algebra. Many of these were directed toward systems of linear equations and congruences, while others concerned computer science. Galois Theory, however, is **the** classical application of algebra. It is from this application that the whole subject area of abstract algebra has sprung. It marked the point in history where arithmetic and algebra became distinct. Here we focus on this theory for fields of characteristic 0. In the next section, we apply these results to obtain Abel's application on the insolvability in terms of radicals of the quintic equation in $\mathbb{R}[x]$.

Let E be a finite algebraic extension of F, a field of characteristic 0. By Theorem 20.9, $E = F(\gamma)$ for some $\gamma \in E$. If $p(x)$ is the minimal polynomial of γ over F, then $p(\gamma) = 0$. If $\sigma \in G_{E/F}$, then σ must take γ to another root of $p(x)$ by Lemma 22.1. Let $n = \deg(p(x))$ and the set of (distinct) roots of $p(x)$ be $\Omega = \{\gamma = \gamma_1, \gamma_2, \ldots, \gamma_n\}$. Any automorphism of E must therefore be a permutation of Ω. Therefore, $G_{E/F}$ must be a subgroup of S_Ω, the symmetric group on Ω. Since S_Ω is isomorphic to S_n and $|S_n| = n!$, we can summarize our remarks in the following theorem.

Theorem 22.2

Suppose E is an extension of degree n over F, a field of characteristic 0. Then G is isomorphic to a subgroup of S_n and $|G_{E/F}|$ is a divisor of $n!$.

Example 22.1

Suppose E is the splitting field of $f(x) = (x^2 + 2)(x^2 + 1)$ over $\mathbb{Q}$. Then

$$E = \mathbb{Q}\left(\frac{1+i}{\sqrt{2}}\right) = \mathbb{Q}(i, \sqrt{2})$$

	$\sqrt{2}$	i
σ_1	$\sqrt{2}$	i
σ_2	$-\sqrt{2}$	i
σ_3	$\sqrt{2}$	$-i$
σ_4	$-\sqrt{2}$	$-i$

Figure 22.3

(see Exercise 2) and $[E : \mathbb{Q}] = 4$. If $G = G_{E/\mathbb{Q}}$, then Theorem 22.2 states that $|G|$ divides $4! = 24$. Let us find the actual order of G by determining the elements of G. Any element $\beta \in E$ can be written

$$\beta = a_0 + a_1 i + a_2\sqrt{2} + a_3 i\sqrt{2} \qquad \text{for} \quad a_i \in \mathbb{Q}$$

If $\sigma \in G$, then σ must fix $\mathbb{Q}$ and $\sigma(i)$ is another root of $x^2 + 1 \in \mathbb{Q}[x]$, while $\sigma(\sqrt{2})$ is another root of $x^2 + 2 \in \mathbb{Q}[x]$. The elements of G can then be given by Figure 22.3; for example, σ_1 is the identity mapping, while σ_2 acts on $\beta \in E$ according to the equation

$$\sigma_2(a_0 + a_1 i + a_2\sqrt{2} + a_3 i\sqrt{2}) = a_0 + a_1 i + a_2(-\sqrt{2}) + a_3 i(-\sqrt{2})$$

$$= a_0 + a_1 i - a_2\sqrt{2} - a_3 i\sqrt{2}$$

Clearly $|G| = 4$. We conclude that although $f(x)$ has 4 roots and there exist 24 permutations of the roots, only 4 of these permutations lie in the Galois group.

Under what conditions is $|G_{E/F}| = n!$? The next result provides the answer as well as giving two alternate definitions of **normal extensions**.

Theorem 22.3

Let E be a finite extension of F, where $\text{char}(F) = 0$. Then the following are equivalent:

    i. E is a normal extension of F.
    ii. $|G_{E/F}| = [E : F]$.
    iii. Any irreducible polynomial in $F[x]$ with one root in E splits in E.

Theorem 22.3 allows us to adapt previous investigations of splitting fields with our current analysis of Galois groups and extension fields. To ease the understanding of this somewhat involved result, we take a piecemeal approach to the theorem and divide the proof into a sequence of lemmas.

Lemma 22.2

Let E be an extension of degree n over a field F where $\text{char}(F) = 0$. If $G = G_{G/F}$, then $|G| \leqslant n$. Equality holds if the minimal polynomial of a primitive element of E over F splits in E.

Proof

By Theorem 20.9, there exists $\gamma \in E$ such that $E = F(\gamma)$. Let $p(x)$ be the minimal polynomial of γ over F. We know that an automorphism of E fixing F must take a root of $p(x)$ to another root of $p(x)$ **lying in** E. Moreover, since the powers of γ form a basis of E over F, any automorphism of E fixing F is completely determined by its action on γ. Since there are exactly $n = \deg(p(x))$ distinct powers of γ, some of which may not correspond to automorphisms, we have $|G_{E/F}| \leqslant [E:F] = n$.

Now suppose that E contains all the roots of $p(x) \in F[x]$. If γ' is another root of $p(x)$, then there exists an automorphism $\sigma: E \to E$ such that $\sigma(a) = a$ for all $a \in F$ and $\sigma(\gamma) = \gamma'$. (See Exercise 3.) From this we have $|G_{E/F}| \geqslant \deg(p(x)) = [E:F]$. The result follows. $\qquad \square$

Recall the field $\mathbb{Q}(\sqrt[3]{5})$ from Section 22.1. This field is an extension of degree 3 over $\mathbb{Q}$, and the minimal polynomial of the primitive element $\sqrt[3]{5}$ over $\mathbb{Q}$ is $f(x) = x^3 - 5$. Since $f(x)$ does not split in $\mathbb{Q}(\sqrt[3]{5})$, we have an alternate proof of the fact that this extension is not a normal extension of $\mathbb{Q}$.

Corollary 22.1

Suppose E is a finite extension of F where $\text{char}(F) = 0$, and every irreducible polynomial in $E[x]$ with one root in E splits in E. Then $|G_{E/F}| = [E:F]$.

Proof

By hypothesis, the minimal polynomial of a primitive element γ of E over F splits in E. But then E is a splitting field for that polynomial since any field containing F and γ contains E. $\qquad \square$

Lemma 22.3

If E is a finite extension of F, where $\text{char}(F) = 0$, and $|G_{E/F}| = [E:F]$, then E is a normal extension of F.

Proof

Let γ be a primitive element of E over F, so that $E = F(\gamma)$. If $[E:F] = n$, then γ has degree n over F, and $1, \gamma, \gamma^2, \dots, \gamma^{n-1}$ form a basis of E over F. Suppose K is the fixed field of $G_{E/F}$. We want to show $K = F$.

Let $\delta \in K$. Since $K \subseteq E$ and E is a finite and, hence, an algebraic extension of F, there exists $a_0, a_1, \dots, a_{n-1} \in F$ such that

$$\delta = \sum_{i=0}^{n-1} a_i \gamma^i$$

If $\sigma \in G_{E/F}$, then $\delta \in K$ implies

$$\delta = \sigma(\delta) = \sum_{i=0}^{n-1} a_i \sigma(\gamma^i) = \sum_{i=0}^{n-1} a_i \sigma(\gamma)^i$$

The polynomial

$$p(x) = \sum_{i=0}^{n-1} a_i x^i - \delta$$

has degree $n-1$ in $K[x]$. Moreover $\sigma(\gamma)$ is a root for all $\sigma \in G_{E/F}$. Now $|G_{E/F}| = [E:F] = n$ implies that $p(x)$ has n distinct roots since distinct elements of $G_{E/F}$ cannot have the same action on γ. But the degree of $p(x)$ is $n-1$. By Theorem 17.7, $p(x) = 0$, and thus, $p(0) = a_0 - \delta = 0$. But then $\delta = a_0 \in F$ and $K \subseteq F$. Since $F \subseteq K$, we conclude that $K = F$ and that E is a normal extension of F.   □

In Corollary 22.1, we showed that condition iii of Theorem 22.3 implies condition ii. We have just seen that ii implies i. To complete the proof of the theorem, we need only prove that condition iii follows from i. This is the purpose of the following lemma.

Lemma 22.4

Let E be a normal extension of F where char$(F) = 0$. Then E has the property that any irreducible polynomial in $F[x]$ with one root in E splits in E.

Proof

Let $p(x) \in F[x]$ be any irreducible monic polynomial with $p(\alpha) = 0$ for $\alpha \in E$. We must prove that every other root of $p(x)$ lies in E. If $\alpha \in F$, then the irreducibility of $p(x)$ in $F[x]$ forces $p(x) = x - \alpha$, and thus $p(x)$ splits in E. Without loss of generality, $\alpha \notin F$. Since F is the fixed field of E over F, there must exist some $\sigma \in G_{E/F}$ such that $\alpha_2 = \sigma(\alpha) \neq \alpha$. But then

$$0 = \sigma(0) = \sigma(p(\alpha)) = p(\sigma(\alpha)) = p(\alpha_2)$$

implies that α_2 is another root of $p(x)$. Therefore, any element of $G_{E/F}$ either fixes α or takes α to another root of $p(x)$ lying in E.

Suppose $\{\alpha, \alpha_2, \ldots, \alpha_m\}$ is the set of all distinct roots of $p(x)$ that lie in E, and let $s(x) = (x - \alpha)(x - \alpha_2) \cdots (x - \alpha_m)$. Then $s(x) \in E[x]$, and there exists $t(x) \in E[x]$ such that $p(x) = s(x)t(x)$. If $\deg(t(x)) = 0$, then $t(x)$ is constant, and $p(x)$ splits in E. We can assume, therefore, that $\deg(t(x)) > 0$. The remainder of the proof consists of showing that $s(x) \in F[x]$. Assume for the moment that this is the case and $s(x) \in F[x]$. Since $p(x) \in F[x]$, we know that there exist **unique** $q(x), r(x) \in F[x]$ such that

$$p(x) = q(x)s(x) + r(x), \qquad \text{where } r(x) = 0 \text{ or } \deg(r(x)) < m$$

But $s(x)$ is a factor of $p(x)$, so that $r(x) = 0$ and $t(x) = q(x) \in F[x]$, contrary to the fact that $p(x)$ is irreducible in $F[x]$. This contradiction forces $t(x)$ to be a constant and $p(x)$ to split in E.

To show $s(x) \in F[x]$, first note that any $\sigma \in G_{E/F}$ can be extended to an automorphism of $E[x]$. If

$$g(x) = \sum_{i=0}^{n} a_i x^i \in E[x]$$

simply define

$$\sigma(g(x)) = \sigma\left(\sum_{i=0}^{n} a_i x^i\right) = \sum_{i=0}^{n} \sigma(a_i) x^i$$

We leave it to you to show that σ is an automorphism of $E[x]$ that fixes $F(x)$. (See Exercise 8.) Suppose

$$s(x) = (x - \alpha)(x - \alpha_2) \cdots (x - \alpha_m) = x^m + c_{m-1} x^{m-1} + \cdots + c_1 x + c_0$$

If $s(x) \notin F[x]$, then at least one value c_i is not an element of F. But then there exists $\sigma \in G_{E/F}$ such that $\sigma(c_i) \neq c_i$. On the other hand,

$$\sigma(s(x)) = \sigma((x - \alpha)(x - \alpha_2) \cdots (x - \alpha_m))$$
$$= (x - \sigma(\alpha))(x - \sigma(\alpha_2)) \cdots (x - \sigma(\alpha_m)) = s(x)$$

since σ is injective and the values $\{\alpha, \alpha_2, \ldots, \alpha_n\}$ are distinct. But then $\sigma(c_i) = c_i$ for all i and $s(x) \in F[x]$. The result follows. $\square$

Before stating the Fundamental Theorem of Galois Theory, we give one last preliminary result relating normal extensions and intermediate fields.

Lemma 22.5

Let E be a normal extension of F where $\mathrm{char}(F) = 0$. If K is an extension of F contained in E, then E is a normal extension of K. Consequently, $|G_{E/K}| = [E : K]$.

Proof

The fact that E is a finite extension of F implies E is a finite extension of K. K is also a field of characteristic 0 since K contains F, a field of characteristic 0.

Let $K_1 = E_{G_1}$ where $G_1 = G_{E/K}$. We must show that $K_1 = K$. Clearly $K \subseteq K_1$. Now K_1, like K, is a field of characteristic 0. Theorem 20.9 states there exists $\omega \in E$ such that $E = K_1(\omega)$. Let $G_1 = \{1, \tau_2, \ldots, \tau_k\}$. If $q_1(x) \in K_1[x]$ is the minimal polynomial of ω over K_1, then $\deg(q_1(x)) = [E : K] = k_1$. As in the proof of the preceding lemma, extend the elements of G_1 to automorphisms of $K_1[x]$. Then

$$q(x) = (x - \omega)(x - \tau_2(\omega)) \cdots (x - \tau_k(\omega)) \in E[x]$$

and $\tau_i(q(x)) = q(x)$ for all $\tau_i \in G_1$, so that every coefficient of $q(x)$ must be fixed by the elements of G_1. Hence, $q(x) \in K_1[x]$. Since ω is a root of $q(x)$, while $q_1(x)$ is the minimal polynomial of ω over K_1, we can conclude that $k = \deg(q(x)) \geq \deg(q_1(x)) = k_1$. By Lemma 22.2,

$$k = |G_{E/K}| \leq [E : K] = k_1$$

We conclude that $k = k_1$ and that $[E : K] = [E : K_1]$; so $K = K_1$. $\square$

With these preliminaries completed, we can now state the main result of this section:

Theorem 22.4

Fundamental Theorem of Galois Theory

Let E be a normal extension of F where $\text{char}(F) = 0$. If K is an extension of F contained in E, then the mapping defined by $\phi(K) = G_{E/K}$ is a bijection between the set of extension fields of F contained in E and the set of subgroups of $G_{E/F}$. Moreover:

    i. $[E : K] = |G_{E/K}|$.
    ii. $[K : F] = [G_{E/F} : G_{E/K}]$.
    iii. K_1 is a proper subfield of K_2 if and only if G_{E/K_2} is a proper subgroup of G_{E/K_1}.
    iv. K is a normal extension of F if and only if $G_{E/K} \trianglelefteq G_{E/F}$.
    v. If K is a normal extension of F, then $G_{K/F} \cong G_{E/F}/G_{E/K}$.

Proof

Throughout this proof, we will apply Lemma 22.5 and assume E is a normal extension of every extension of F contained in E.

Suppose that K_1 and K_2 are extensions of F contained in E and $\phi(K_1) = \phi(K_2)$. Then $G_{E/K_1} = G_{E/K_2}$. Since E is a normal extension of K_1 and of K_2, we see that K_1 is the fixed field of G_{E/K_1} and K_2 is the fixed field of G_{E/K_2}. But now the uniqueness of a fixed field of $G_{E/K_1} = G_{E/K_2}$ forces $K_1 = K_2$, so that ϕ is an injection.

Suppose $H \leqslant G_{E/F}$, and $K = E_H = \{\alpha \in E | \tau(\alpha) = \alpha$ for all $\tau \in H\}$. Then K is an extension of F contained in E, so that E is a normal extension of K. Thus the fixed field of $G_{E/K}$ is K and $\phi(K) = G_{E/K} = H$. ϕ is therefore a bijection.

Statement i follows immediately from Lemma 22.5. If K is a field and $F \subseteq K \subseteq E$, then

$$[K : F] = [E : F]/[E : K]$$

by Theorem 20.1. Since E is a normal extension of both K and F, we have $[E : K] = |G_{E/K}|$ and $[E : F] = |G_{E/F}|$. By Lagrange's Theorem,

$$[K : F] = |G_{E/F}|/|G_{E/K}| = [G_{E/F} : G_{E/K}]$$

This proves part ii.

Now $K_1 \subseteq K_2$ if and only if $G_{E/K_2} \leqslant E_{E/K_1}$. (See Exercise 9.) Suppose that $K_1 \subset K_2$. If $G_{E/K_2} = G_{E/K_1}$, the fact that ϕ is injective implies that $K_1 = K_2$, contrary to the assumption. Thus G_{E/K_2} is a proper subgroup of G_{E/K_1}. Conversely, if G_{E/K_2} is a proper subgroup of G_{E/K_1}, we have that K_2, the fixed field of G_{E/K_2}, is not equal to K_1, the fixed field of G_{E/K_1}. Statement iii is therefore established.

To prove statement iv, suppose K is a normal extension of F, and $\tau \in G_{E/K}$ and $\sigma \in G_{E/F}$. What about $\sigma^{-1}\tau\sigma$? Let $\alpha \in K$. Now $\sigma(\alpha)$ is another root of the minimal polynomial $p(x)$ of α over F. Since $\alpha \in K$ is a root of $p(x)$ and K is a normal extension of F, Theorem 22.3 implies $p(x)$ splits in K and $\sigma(\alpha) \in K$ for every $\sigma \in G_{E/F}$. But then $\tau(\sigma(\alpha)) = \sigma(\alpha)$ since $\tau \in G_{E/K}$ implies τ fixes all elements of K. We conclude that $(\tau\sigma)(\alpha) = \sigma(\alpha)$ or $\sigma^{-1}\tau\sigma(\alpha) = \alpha$ for all $\alpha \in K$. So $\sigma^{-1}\tau\sigma \in G_{E/K}$ for all $\sigma \in G_{E/F}$ and $\tau \in G_{E/K}$. Therefore, $G_{E/K} \trianglelefteq G_{E/F}$.

Conversely, if $G_{E/K} \trianglelefteq G_{E/F}$, then let τ be an element of $G_{E/K}$. $G_{E/K} \trianglelefteq G_{E/F}$ implies that $\tau\sigma = \sigma\tau'$ for $\sigma \in G_{E/F}$, $\tau' \in G_{E/K}$. If $\alpha \in K$, then $(\sigma\tau')(\alpha) = \sigma(\tau'(\alpha)) = \sigma(\alpha)$. Since $\tau\sigma = \sigma\tau'$, we have $\sigma(\alpha) = \tau(\sigma(\alpha))$ for any $\tau \in G_{E/K}$. Thus $\sigma(\alpha)$ is in the fixed field of $G_{E/K}$; that is, $\sigma(\alpha) \in K$ for all $\sigma \in G_{E/F}$, $\alpha \in K$. For $\sigma \in G_{E/F}$, let $\sigma' = \sigma_K$, the restriction of σ to the field K. Since $\sigma(\alpha) \in K$ for all $\alpha \in K$, σ' is an automorphism of K fixing F, and thus an element of $G_{E/F}$. Let β be an element of the fixed field of $G_{K/F}$. Then $\sigma'(\beta) = \beta$, so that $\sigma(\beta) = \beta$ for any $\sigma \in G_{E/F}$. Since E is normal over F, $\beta \in F$; therefore, the fixed field of $G_{K/F}$ is F, and K is a normal extension of F. This completes the proof of statement iv.

Finally, define $\psi: G_{E/F} \to G_{K/F}$ by $\psi(\sigma) = \sigma_K$. If K is a normal extension of F, then a part of the proof in the preceding paragraph shows that ψ takes an automorphism of E to an automorphism of K. Moreover,

$$\psi(\sigma_1 \circ \sigma_2) = (\sigma_1 \circ \sigma_2)_K = (\sigma_1)_K(\sigma_2)_K = \psi(\sigma_1) \circ \psi(\sigma_2)$$

so that ψ is a group homomorphism. If $\psi(\sigma) = 1_K$, then $\sigma_K = 1_K$ implies $\sigma \in G_{E/K}$. Conversely, if $\sigma \in G_{E/K}$, then $\psi(\sigma) = \sigma_K = 1_K$, so ker $\psi = G_{E/K}$. We conclude that ψ is a group homomorphism. If $\psi(\sigma) = 1_K$, then $\sigma_K = 1_K$ implies $\sigma \in G_{E/K}$. Conversely, if $\sigma \in G_{E/K}$, then $\psi(\sigma) = \sigma_K = 1_K$, so that ker $\psi = G_{E/K}$. We conclude that

$$G_{E/F}/G_{E/K} = G_{E/K}/\text{ker } \psi \cong \text{Im } \psi$$

Now Im ψ is a subgroup of $G_{K/F}$ of order $|G_{E/F}|/|G_{E/K}| = [K : F] = |G_{K/F}|$ by part ii. Thus Im $\psi = G_{K/F}$.   □

Statements iv and v will prove exceedingly useful in Section 22.3, for they will allow us to solve polynomial equations in a step-by-step fashion by obtaining, where possible, a sequence of normal extensions.

Let us now turn to some examples of the Galois correspondence:

| Example 22.2 |

Recall $E = \mathbb{Q}(i)$. We saw in the last section that E is a finite normal extension of $\mathbb{Q}$ and that the Galois G of E over $\mathbb{Q}$ consists of the identity 1 and complex conjugation. The only subgroups of G are trivial. The fixed field of $\{1\}$ is E, while the fixed field of G is $\mathbb{Q}$. In Figure 22.4, we diagram these relationships. The field in Figure 22.4a is the fixed field of the group in part b. Note that the bijection is order-reversing.

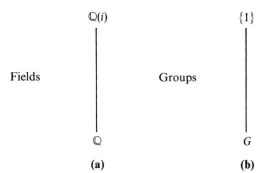

Figure 22.4   **(a)**   **(b)**

Example 22.3

Let $E = \mathbb{Q}(\omega)$ where ω is a primitive fifth root of unity. We saw in the previous section that the Galois group of E over $\mathbb{Q}$ is a cyclic group of order 4. The only subgroup of $G = \langle \sigma_1 \rangle$, other than G and the identity $\{\sigma_1\}$, is the group $H = \langle \sigma_4 \rangle$, whose fixed field is $\mathbb{Q}(\omega + \omega^4)$. The relationships guaranteed by Theorem 22.4 are given in Figure 22.5.

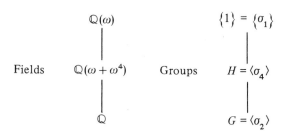

Figure 22.5

In Example 22.1, we considered $\mathbb{Q}(i, \sqrt{2})$, the splitting field of $f(x) = (x^2 + 2)(x^2 + 1) \in \mathbb{Q}[x]$. The Galois group of G is a noncyclic group of order 4. (See Exercise 6.) The proper subgroups of G are $H_1 = \langle \sigma_2 \rangle$, $H_2 = \langle \sigma_3 \rangle$, and $H_4 = \langle \sigma_4 \rangle$, and the corresponding fixed fields are $\mathbb{Q}(i)$, $\mathbb{Q}(\sqrt{2})$, and $\mathbb{Q}(i\sqrt{2})$, respectively. (See Exercise 10.)

We diagram these subfields and subgroups in Figure 22.6. Since every subgroup is normal in G, every extension is normal.

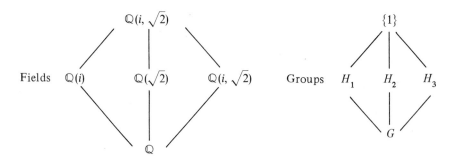

Figure 22.6

	e	ρ	ρ^2	τ_1	τ_2	τ_3
$\sqrt[3]{5}$	$\sqrt[3]{5}$	$\omega\sqrt[3]{5}$	$\omega^2\sqrt[3]{5}$	$\sqrt[3]{5}$	$\omega^2\sqrt[3]{5}$	$\omega\sqrt[3]{5}$
$\omega\sqrt[3]{5}$	$\omega\sqrt[3]{5}$	$\omega^2\sqrt[3]{5}$	$\sqrt[3]{5}$	$\omega^2\sqrt[3]{5}$	$\omega\sqrt[3]{5}$	$\sqrt[3]{5}$
$\omega^2\sqrt[3]{5}$	$\omega^2\sqrt[3]{5}$	$\sqrt[3]{5}$	$\omega\sqrt[3]{5}$	$\omega\sqrt[3]{5}$	$\sqrt[3]{5}$	$\omega^2\sqrt[3]{5}$

Figure 22.7

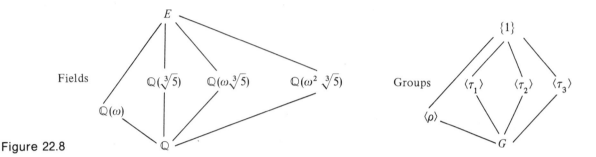

Figure 22.8

Example 22.4

Let $E = \mathbb{Q}(\sqrt[3]{5}, \omega)$ where $\omega^3 = 1$, $\omega \neq 1$. Then E is a normal extension of $\mathbb{Q}$ since it is the splitting field of $x^3 - 5$ over $\mathbb{Q}$. (See Exercise 15). We can write

$$E = \{a_1 + a_2\sqrt[3]{5} + a_3\sqrt[3]{25} + a_4\omega + a_5\omega\sqrt[3]{5}$$
$$+ a_6\omega\sqrt[3]{25} \mid a_1, a_2, a_3, a_4, a_5, a_6 \in \mathbb{Q}\}$$

The Galois group of E over $\mathbb{Q}$ is of order 6 and is determined by its effect on the roots of $x^3 - 5$. The group elements are described by Figure 22.7.

We leave it to you to determine a basis for each subfield and verify the correspondence given in Figure 22.8. (See Exercise 10.) Note that $\langle \tau_1 \rangle$ is not a normal subgroup of G. This corresponds to our earlier result that $\mathbb{Q}(\sqrt[3]{5})$ is not a normal extension of $\mathbb{Q}$.

Exercises 22.2

1. The polynomial $f(x) = x^3 + 2x + 2$ is irreducible in $\mathbb{Q}[x]$ by Eisenstein's Criteria (Theorem 17.12). If F is the splitting field of $f(x)$ over $\mathbb{Q}$ and $G = G_{E/\mathbb{Q}}$, list the possible orders for G.

2. Show that $\mathbb{Q}(i, \sqrt{2})$ is the splitting field of $f(x) = (x^2 + 2)(x^2 + 1)$ over $\mathbb{Q}$.

3. Let $p(x)$ be an irreducible polynomial in $F[x]$. Let γ and γ' be roots of $p(x)$ lying in some extension E of F. Prove that there exists an automorphism σ of E fixing F with $\sigma(\gamma) = \gamma'$.

4. Use part ii of Theorem 22.3 to prove that $\mathbb{Q}(\sqrt[3]{5})$ is not a normal extension of $\mathbb{Q}$.

5. Is the result of Lemma 22.2 true for a finite simple extension E of a field F of characteristic p, a prime?

6. Verify that the Galois group in Example 22.1 is isomorphic to the group of motions of the rectangle.

7. In Example 22.1, why is the mapping $\tau\colon E \to E$ defined by

$$\tau(a_0 + a_1 i + a_2\sqrt{2} + a_3 i\sqrt{2}) = a_0 + a_1\sqrt{2} + a_2 i + a_3 i\sqrt{2}$$

not an automorphism in G.

8. Prove that the mapping $\sigma\colon E(x) \to E(x)$ in the proof of Lemma 22.4 is an automorphism of $E[x]$ that fixes $F[x]$.

9. Let E be a normal extension of a field F of characteristic 0. If K_1 and K_2 are fields such that $F \subseteq K_1 \subseteq E$ and $F \subseteq K_2 \subseteq E$, prove $K_1 \subseteq K_2$ if and only if $G_{E/K_2} \leqslant G_{E/K_1}$.

10. Verify that the fixed fields of H_1, H_2, and H_3 of Example 22.4 are $\mathbb{Q}(i)$, $\mathbb{Q}(\sqrt{2})$, and $\mathbb{Q}(i\sqrt{2})$, respectively.

11. Let E be an extension of F and $p(x) \in F[x]$. Suppose there exist $f(x), g(x) \in E[x]$ such that $p(x) = f(x)g(x)$. Prove that if $f(x) \in F[x]$, then $g(x) \in F[x]$.

12. Suppose E is a normal extension of F, K is a normal extension of F and $F \subseteq K \subseteq E$. Prove that every element of $G_{K/F}$ can be extended to an element of $G_{E/F}$.

13. Prove that if E is a normal extension of F where $\mathrm{char}(F) = 0$, then there exists but a finite number of subfields of E containing F.

14. Let E be a normal extension of $\mathbb{Q}$, and suppose that $[E : \mathbb{Q}]$ is prime. Describe the group $G_{E/\mathbb{Q}}$.

15. Verify that $E = \mathbb{Q}(\sqrt[3]{5}, \omega)$, where $\omega^3 = 1$, $\omega \neq 1$, is the splitting field of $x^3 - 2 \in \mathbb{Q}[x]$. Let G denote the Galois group of E over $\mathbb{Q}$. Prove $G \cong S_3$.

16. Find a polynomial $f(x) \in \mathbb{Q}[x]$ such that the Galois group of f over $\mathbb{Q}$ is isomorphic to $(\mathbb{Z}_6, \oplus)$.

17. Find the Galois group G over $\mathbb{Q}$ of $f(x) = (x^2 - 2)(x^2 - 3)$. Determine all subgroups of G and all subfields of the splitting field.

18. Let $f(x) = x^4 - 5$. Show that $\mathbb{Q}(\sqrt[4]{5}, i)$ is the splitting field of f over $\mathbb{Q}$. Determine the corresponding Galois group and all its subgroups. Identify the group.

19. Let $f(x) = (x^2 - 2)(x^2 - 3)(x^2 - 5)$. Prove that the Galois group of f over $\mathbb{Q}$ is isomorphic to $\mathbb{Z}_2 \times \mathbb{Z}_2 \times \mathbb{Z}_2$.

22.3

The Insolvability of the Quintic

In this section we focus on Abel's result concerning the insolvability of a quintic, or fifth-degree polynomial, in $\mathbb{R}[x]$ in terms of radicals. As we did in Section 22.2, we require the fields in this section to be of characteristic 0. This amounts to studying extension fields of $\mathbb{Q}$.

Recall that in Chapter 17 we examined some classical methods of determining the roots of an arbitrary real or complex polynomial equation of degree three or four. These methods involved the computation of certain square roots and cube roots; in other words, we solved these equations in terms of radicals. How can we extend this concept?

Definition 22.3

Let E be an extension field of F. E is an **extension of F by radicals** if there exists a chain of subfields

$$F = F_0 \subseteq F_1 \subseteq \cdots \subseteq F_n = E$$

such that each subfield $F_i = F_{i-1}(\gamma_i)$, where $\gamma_i \in F_i$, has the property that $\gamma_i^{n_i} \in F_{i-1}$ for some $n_i \in \mathbb{N}$. A polynomial $f(x) \in F[x]$ is **solvable by radicals** if its splitting field over F is contained in an extension of F by radicals.

Clearly, each γ_i is a root of a polynomial of the form $x^{n_i} - a_i$ where $a_i \in F_{i-1}$. Both $\mathbb{Q}(i)$ and $\mathbb{Q}(\sqrt{2}, \sqrt[3]{3 - \sqrt{5}})$ are extensions of $\mathbb{Q}$ by radicals. In Chapter 20, we considered constructions by straightedge and compass. Constructible numbers were shown to be contained in extension fields obtained by successively adjoining square roots of elements. These quadratic extensions gave rise to extensions of $\mathbb{Q}$ by radicals. We leave it to you to verify that if E is an extension of K by radicals and K is an extension of F by radicals, then E is an extension of F by radicals. (See Exercise 1.)

Example 22.5

Consider the splitting field of the polynomial $x^n - 1 \in \mathbb{Q}[x]$. By DeMoivre's Theorem, the roots of the equation are

$$1, \omega, \omega^2, \ldots, \omega^{n-1} \qquad \text{where} \quad \omega = \cos\frac{2}{n}\pi + i\sin\frac{2}{n}\pi$$

Then $E = \mathbb{Q}(\omega)$ and ω is termed a **primitive nth root of unity**. E is then an extension of $\mathbb{Q}$ by radicals. If G is the Galois group of E over $\mathbb{Q}$ and $\sigma \in G$, then $\sigma(\omega)$ is another root of $x^n - 1$, so $\sigma(\omega) = \omega^{k_\sigma}$ for $1 < k_\sigma \leqslant n-1$. By analyzing automorphisms of this type, you should be able to show that G is an abelian group. (See Exercise 2.)

In our third example of Section 22.1, we showed that $E = \mathbb{Q}(\sqrt[3]{5})$ is not a normal extension of $\mathbb{Q}$. The minimal polynomial of $\sqrt[3]{5}$ over $\mathbb{Q}$ is $x^3 - 5$, whose roots are $\sqrt[3]{5}$, $\omega\sqrt[3]{5}$, and $\omega^2\sqrt[3]{5}$, where ω is a primitive third root of unity. If $K = E(\omega)$, then K is an extension of $\mathbb{Q}$ by radicals. We showed in Example 22.5 that K is, in fact, a normal extension of $\mathbb{Q}$ by radicals since K is the splitting field of $x^3 - 2$ over $\mathbb{Q}$. We thus extended a given extension by radicals to a normal extension by radicals by simply adjoining a primitive root of unity.

In general, if γ is a root of the polynomial $x^n - a \in F[x]$, then γ is one of the elements $\sqrt[n]{a}, \omega\sqrt[n]{a}, \ldots, \omega^{n-1}\sqrt[n]{a}$ where ω is a primitive nth root of unity. By adjoining ω to $F(\gamma)$, we obtain a new extension of F by radicals, that not only contains $F(\gamma)$ but also every root of $x^n - a$.

If E is an extension of F by radicals, then we can extend each element in the chain

$$F = F_0 \subseteq F_1 \subseteq \cdots \subseteq F_m = E$$

in the manner described so as to obtain a new chain

$$F = F_0 \subseteq F_1' \subseteq \cdots \subseteq F_n'$$

where $E \subseteq F_n$, and F_n' is a normal extension of F. We state our conclusion:

Theorem 22.5

If E is an extension of F by radicals, then E is contained in a normal extension of F by radicals.

For simplicity, we will *assume* throughout the remainder of this chapter that F is a field of characteristic 0 that *contains all* nth *roots* of unity for whatever values of n are necessary. This will certainly be the case if $F = \mathbb{C}$.

Under this assumption, every extension $F(\gamma)$, where γ is a root of the polynomial $x^n - a \in F[x]$, is a normal extension of F. To see this, note that $F(\gamma)$ must then contain the terms $\gamma, \omega\gamma, \ldots, \omega^{n-1}\gamma$, where ω is a primitive nth root of unity, so that $F(\gamma)$ is a splitting field for $x^n - a$ over F. If E is an extension of F by radicals, then each subfield of E in the chain is a normal extension of its predecessor.

We will shortly give a condition under which a given polynomial in $F[x]$ is solvable by radicals. The next lemma will be needed in the proof of that characterization:

Lemma 22.6

If $E = F(\gamma)$, where γ is a root of $p(x) = x^n - a \in F[x]$, then the Galois group of E over F is abelian.

Proof

The distinct roots of $p(x)$ are $\gamma, \omega\gamma, \omega^2\gamma, \ldots, \omega^{n-1}\gamma$, where γ is a primitive nth root of unity. The elements of $G_{E/F}$ permute the roots of $p(x)$ and are determined by their action on these roots. Let $\sigma, \tau \in G_{E/F}$. If $\sigma(\gamma) = \omega^i\gamma$ and $\tau(\gamma) = \omega^j\gamma$, then

$$\sigma\tau(\gamma) = \sigma(\tau(\gamma)) = \sigma(\omega^j\gamma) = \omega^j\sigma(\gamma) = \omega^j\omega^i\gamma = \omega^{i+j}\gamma$$

since F contains the nth roots of unity. But

$$\omega^{i+j}\gamma = \omega^i(\omega^j\gamma) = \omega^i(\tau(\gamma)) = \tau(\omega^i\gamma) = \tau\sigma(\gamma)$$

so that $\sigma\tau = \tau\sigma$. Thus $G_{E/F}$ is abelian.   □

We can now connect the theories of field extensions and group extensions. Recall that a group G is solvable if G contains subgroups G_i for $i =$

0, 1, 2, ..., n, such that

$$\{1\} = G_0 \trianglelefteq G_1 \trianglelefteq G_2 \cdots \trianglelefteq G_n = G$$

and G_i/G_{i-1} is an abelian group for each i. Lemma 22.6 allows us to develop the appropriate chain of subgroups of the Galois group of a polynomial in F that is solvable by radicals.

Theorem 22.6

Galois' Theorem

If $p(x) \in F[x]$ is solvable by radicals, then the Galois group of $p(x)$ over F is a solvable group.

Proof

Let K be a splitting field of $p(x)$ over F. By hypothesis, there exists an extension E of radicals $F = F_0 \subseteq F_1 \subseteq \cdots \subseteq F_n = E$ normal over F. Moreover F_i is normal over F_{i-1} for $i = 1, 2, ..., n$ by our assumption on the nature of F. By statement iv of Theorem 22.4, G_{E/F_i} is a normal subgroup of $G_{E/F_{i-1}}$, $i = 1, 2, ..., n$. We therefore obtain the chain

$$\{1\} = G_{E/E} \trianglelefteq G_{E/F_{n-1}} \trianglelefteq \cdots \trianglelefteq G_{E/F_1} \trianglelefteq G_{E/F_0} = G_{E/F}$$

For $i = 1, 2, ..., n$, we have by statement v of Theorem 22.4 that $G_{E/F_{i-1}}/G_{E/F_i} \cong G_{F_i/F_{i-1}}$, since E is a normal extension of each F_{i-1} by Lemma 22.5. But by Lemma 22.6, $G_{F_i/F_{i-1}}$ is abelian for $i = 1, 2, ..., n$. Thus $G_{E/F}$ is a solvable group. $\square$

Example 22.6

Let $f(x) = x^5 - 10x - 5 \in \mathbb{Q}[x]$. Then $f(x)$ is irreducible by Eisenstein's Criteria. (See Exercise 3.) If γ is a root of f, then the field $\mathbb{Q}(\gamma)$ is contained in the splitting field E of F. If G is the Galois group of $f(x)$ over $\mathbb{Q}$, then $|G|$ is divisible by 5, since $[\mathbb{Q}(\gamma) : \mathbb{Q}] = 5$ by Theorem 20.3 and $[\mathbb{Q}(\gamma) : \mathbb{Q}]$ divides $[E : F] = |G|$ by Lemma 22.3. By Cauchy's Theorem, G contains an element σ of order 5. Since G is isomorphic to a subgroup of S_5 by Theorem 22.2, we know that σ must be a 5-cycle.

From introductory calculus, we see $f'(x) = 5x^4 - 10 = 5(x^4 - 2)$ and $f''(x) = 20x^3$. Then $f(x)$ has a relative minimum at $x = \sqrt[4]{2}$ and a relative maximum at $x = -\sqrt[4]{2}$. From the graph of f given in Figure 22.9, we see f must have three real roots. But then f has two complex roots, which must be conjugates by Exercise 11 of Section 17.3. This says that complex conjugation is an automorphism of E fixing $\mathbb{Q}(\gamma)$. We conclude that G also contains an element τ that interchanges the two complex roots and fixes the remaining three real roots; hence, τ is a transposition.

It is not too difficult to show that any subgroup H of S_5 containing a 5-cycle and a transposition contains every transposition in S_5. (See Exercise 4.) But S_5 is generated by its transpositions according to Corollary 5.3.

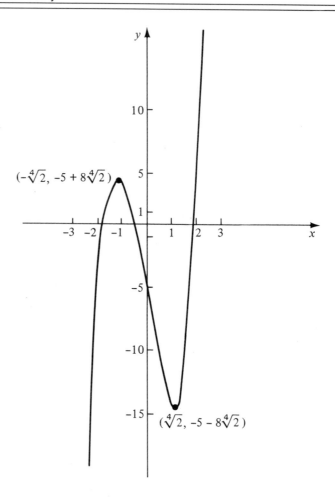

Figure 22.9

Therefore $G \cong S_5$, a nonsolvable group by Corollary 11.3. We conclude that $f(x)$ is not solvable by radicals.

Corollary 22.2

There exists no formula for solving a general real fifth-degree polynomial in terms of radicals.

To show there is no general formula for finding the roots of $f(x) \in \mathbb{C}[x]$ when $\deg(f(x)) = n \geqslant 5$, we need only find a polynomial of degree n whose Galois group is isomorphic to S_n or A_n, since we have proven in Chapter 11 that S_n and A_n are not solvable for $n \geqslant 5$.

We wish to establish that there exists a polynomial of degree $n \in \mathbb{C}[x]$ whose Galois group is isomorphic to S_n. To understand the definition we will introduce, consider the following examples:

$$p(x) = (x - 2)(x - 3)(x - 4)$$
$$= x^3 - 9x^2 + 26x - 24$$
$$= x^3 - (2 + 3 + 4)x^2 + (2 \cdot 3 + 2 \cdot 4 + 3 \cdot 4)x - (2 \cdot 3 \cdot 4)$$

$$q(x) = (x - 1)(x - 2)(x - 3)(x - 4)$$
$$= x^4 - 10x^3 + 35x^2 - 50x + 24$$
$$= x^4 - (1 + 2 + 3 + 4)x^3 + (1 \cdot 2 + 1 \cdot 3 + 1 \cdot 4 + 2 \cdot 3 + 2 \cdot 4 + 3 \cdot 4)x^2$$
$$- (1 \cdot 2 \cdot 3 + 1 \cdot 2 \cdot 4 + 1 \cdot 3 \cdot 4 + 2 \cdot 3 \cdot 4)x + (1 \cdot 2 \cdot 3 \cdot 4)$$

These examples illustrate how the roots of a polynomial determine the coefficients of the polynomial.

In general, if

$$g(x) = \prod_{i=1}^{n} (x - a_i)$$

then

$$g(x) = x^n - (a_1 + a_2 + \cdots + a_n)x^{n-1}$$
$$+ (a_1 a_2 + a_1 a_3 + \cdots + a_1 a_n + a_2 a_3$$
$$+ \cdots + a_2 a_n + \cdots + a_{n-1} a_n)x^{n-2}$$
$$+ \cdots + (-1)^{n-1}(a_1 a_2 \cdots a_{n-1} + \cdots + a_2 a_3 \cdots a_n)x$$
$$+ (-1)^n (a_1 a_2 \cdots a_n)$$

Now let us consider $F[s_1, s_2, \ldots, s_n]$, the set of all polynomials in the variables $s_1, s_2, \ldots, s_n$ with coefficients in F. Theorem 17.2 can be used to show that $F_0 = F[s_1, s_2, \ldots, s_n]$ is an integral domain. (See Exercise 6.) Let $K = F(s_1, s_2, \ldots, s_n)$ be its field of quotients. K consists of all quotients of polynomials in F_0 with nonzero denominator.

The polynomial

$$f(x) = x^n - s_1 x^{n-1} + s_2 x^{n-2} + \cdots + (-1)^{n-1}s_{n-1} + (-1)^n s_n \in K[x]$$

is termed the **general polynomial of degree n over** F. The motivation for this name comes from the observation that as $s_1, s_2, \ldots, s_n$ vary over F, $f(x)$ varies over $F[x]$. If there exists a formula for solving f in terms of radicals, then the roots of f should lie in some radical extension of K. We will outline a proof, however, that shows the Galois group of $f(x)$ over K is S_n, a nonsolvable group.

Let E be the splitting field of $g(x)$ over K and $\{y_1, y_2, \ldots, y_n\}$ the set of roots of $f(x)$ in K. Then

$$f(x) = \prod_{i=1}^{n} (x - y_i) = x^n - s_1 x^{n-1} + \cdots + (-1)^n s_n$$

implies that

$$s_1 = y_1 + y_2 + \cdots + y_n$$

$$s_2 = y_1 y_2 + y_1 y_3 + \cdots + y_1 y_n + y_2 y_3 + \cdots + y_2 y_n + \cdots + y_{n-1} y_n$$

$$\vdots$$

$$s_{n-1} = y_1 y_2 \cdots y_{n-1} + y_1 y_2 \cdots y_{n-2} y_n + \cdots + y_2 y_3 \cdots y_n$$

$$s_n = y_1 y_2 \cdots y_n$$

The elements $s_1, s_2, \ldots, s_n$ are called the **elementary symmetric functions** of $y_1, y_2, \ldots, y_n$. They have the property that any permutation of the elements $\{y_1, y_2, \ldots, y_n\}$ leaves these functions invariant.

Note that

$$E = K(y_1, y_2, \ldots, y_n)$$

$$= (F(s_1, s_2, \ldots, s_n))(y_1, y_2, \ldots, y_n)$$

$$= F(s_1, s_2, \ldots, s_n, y_1, y_2, \ldots, y_n)$$

$$= F(y_1, y_2, \ldots, y_n)$$

To determine the Galois group of $f(x)$ over K, we will look at this problem in a different way. Instead of working from the "bottom up," we will work from the "top down."

Let $t_1, t_2, \ldots, t_n$ be distinct indeterminates over F, and let $\bar{E}$ be the field $F(t_1, t_2, \ldots, t_n)$, the set of all rational functions in the t_i's.

If $\bar{f}(x) = (x - t_1)(x - t_2) \cdots (x - t_n)$, we have

$$\bar{f}(x) = x^n - z_1 x^{n-1} + z_2 s^{n-2} + \cdots + (-1)^{n-1} z_{n-1} x + (-1)^n z_n$$

where

$$z_1 = \sum_{i=1}^{n} t_i, \qquad z_2 = \sum_{1 \le i < j \le n} t_i t_j, \qquad \ldots, \qquad z_n = t_1 t_2 \cdots t_n$$

Then $\bar{K} = F(z_1, z_2, \ldots, z_n)$ is a subfield of $\bar{E}$.

First, we show that $\bar{E}$ is the splitting field of $\bar{f}(x)$ over $\bar{K}$ and that $G_{E/K} \cong S_n$. We will later prove that $G_{E/K} \cong G_{\bar{E}/\bar{K}}$.

Let $\sigma \in S_n$. Define $\bar{\sigma} \colon F[t_1, \ldots, t_n] \to F[t_1, \ldots, t_n]$ by

$$\bar{\sigma}(p(t_1, \ldots, t_n)) = p(t_{\sigma(1)}, t_{\sigma(2)}, \ldots, t_{\sigma(n)})$$

We leave it to you to verify that $\bar{\sigma}$ is an automorphism of the polynomial ring $F[t_1, t_2, \ldots, t_n]$. (See Exercise 7.) Extend $\bar{\sigma}$ to a mapping

$$\hat{\sigma} \colon F(t_1, t_2, \ldots, t_n) \to F(t_1, t_2, \ldots, t_n)$$

by defining

$$\hat{\sigma}\left(\frac{p(t_1, t_2, \ldots, t_n)}{q(t_1, t_2, \ldots, t_n)} \right) = \frac{\bar{\sigma}(p(t_1, t_2, \ldots, t_n))}{\bar{\sigma}(q(t_1, t_2, \ldots, t_n))}$$

Since $F(t_1, t_2, \ldots, t_n)$ consists of equivalence classes of quotients of polynomials, we must take care to assure that $\hat{\sigma}$ is well defined.

Suppose $p_1/q_1 = p_2/q_2$ for $p_1, p_2, q_1, q_2 \in F[t_1, t_2, \ldots, t_n]$. Then $p_1 q_2 - p_2 q_1 = 0$ and

$$0 = \bar{\sigma}(0) = \bar{\sigma}(p_1 q_2 - p_2 q_1) = \bar{\sigma}(p_1)\bar{\sigma}(q_2) - \bar{\sigma}(p_2)\bar{\sigma}(q_1)$$

Thus

$$\hat{\sigma}\left(\frac{p_1}{q_1}\right) = \frac{\bar{\sigma}(p_1)}{\bar{\sigma}(q_1)} = \frac{\bar{\sigma}(p_2)}{\bar{\sigma}(q_2)} = \hat{\sigma}\left(\frac{p_1}{q_1}\right)$$

and $\hat{\sigma}$ is well defined. It is a straightforward exercise to show that $\hat{\sigma}$, so defined, is an automorphism of $F(t_1, t_2, \ldots, t_n)$ fixing K. (See Exercise 8.) By defining $\hat{\sigma}(x) = x$, we can extend $\sigma \in S_n$ to an automorphism $\hat{\sigma} \in G_{\bar{E}/\bar{K}}$. Moreover, if $\sigma_1, \sigma_2 \in S_n$ with $\sigma_1 \neq \sigma_2$, then clearly $\hat{\sigma}_1 \neq \hat{\sigma}_2$, since $\hat{\sigma}_1(t_i) = \hat{\sigma}_2(t_i)$ for $i = 1, 2, \ldots, n$ implies $\sigma_1(i) = \sigma_2(i)$ for all i, contrary to choice of σ_1 and σ_2. Finally, if η is any other element of $G_{\bar{E}/\bar{K}}$, then η must permute the roots of $f(x)$. Thus $\eta(t_i) = t_{\tau(i)}$ for some $\tau \in S_n$ and $\eta = \hat{\tau}$. We conclude that $S_n \cong G_{\bar{E}/\bar{K}}$ under the isomorphism $\psi(\sigma) = \hat{\sigma}$.

To show that $G_{\bar{E}/\bar{K}} \cong G_{E/K}$, define the mapping

$$\phi: F[s_1, s_2, \ldots, s_n] \to F[z_1, z_2, \ldots, z_n] \qquad \text{by} \quad \phi(s_i) = z_i$$

ϕ is a linear transformation since the variables $s_1, s_2, \ldots, s_n$ form a basis of the F-vector space $F[s_1, s_2, \ldots, s_n]$. Moreover, ϕ is a ring homomorphism. (See Exercise 9.) The mapping

$$\theta: F[t_1, t_2, \ldots, t_n] \to F[y_1, y_2, \ldots, y_n]$$

defined by $\theta(t_i) = y_i$, is clearly a ring homomorphism for the same reason. Note that

$$\theta(z_1) = \theta(t_1 + t_2 + \cdots + t_n) = y_1 + y_2 + \cdots + y_n = s_1$$

$$\theta(z_2) = \theta\left(\sum_{1 \leq i < j \leq n} t_i t_j\right) = \sum_{1 \leq i < j \leq n} y_i y_j = s_2$$

$$\vdots$$

$$\theta(z_n) = \theta(t_1 t_2 \cdots t_n) = y_1 y_2 \cdots y_n = s_n$$

Consider the mapping $\theta\phi: F[s_1, s_2, \ldots, s_n] \to F[y_1, y_2, \ldots, y_n]$. Since $(\theta\phi)(s_i) = \theta(z_i) = s_i$ for $i = 1, 2, \ldots, n$, we can consider $\theta\phi$ as a homomorphism of $F[s_1, s_2, \ldots, s_n]$ with the property that $(\theta\phi)(g) = g$ for all $g \in F[s_1, s_2, \ldots, s_n]$. But then ϕ must be injective by Exercise 5 of Section 5.1. Since ϕ is surjective, this implies that ϕ is an isomorphism from the ring $F[s_1, \ldots, s_n]$ to the ring $F[z_1, \ldots, z_n]$.

As before, extend ϕ to an isomorphism $\bar{\phi}$ from the field of quotients $K = F(s_1, s_2, \ldots, s_n)$ to the field of quotients $\bar{K} = F(z_1, z_2, \ldots, z_n)$, and then to an isomorphism $\hat{\phi}$ from $K[x]$ to $\bar{K}[x]$ by setting $\hat{\phi}(x) = x$. Then

$$\hat{\phi}(f(x)) = \phi(x^n - s_1 x^{n-1} + \cdots + (-1)^n s_n)$$

$$= x - z_1 x^{n-1} + \cdots + (-1)^n z_n = \bar{f}(x)$$

Now E is a splitting field for $f(x)$ over K, and $\bar{E}$ is a splitting field for $\bar{f}(x)$ over $\bar{K}$. Since $K[x]$ and $\bar{K}[x]$ are isomorphic, Corollary 20.7 implies that E and $\bar{E}$ are isomorphic by a mapping that coincides with $\hat{\phi}$ on K. The Galois group $G_{E/K}$ is then isomorphic to $G_{\bar{E}/\bar{K}}$. We have proved the following theorem.

Theorem 22.7

The Galois group of the general polynomial of degree n is isomorphic to S_n.

We can now combine this theorem with Corollary 11.22 and Theorem 22.6 to conclude the following.

Theorem 22.8

Abel's Theorem

The general equation of degree n is not solvable by radicals for $n > 4$.

We have now proven there is no general formula for computing the roots of arbitrary polynomial of degree five or greater. There are, however, types of equations of degree five or greater that **are** solvable by radicals. For example, if $a \in \mathbb{C}$, DeMoivre's Theorem yields a way of determining all roots of $x^n - a$ for all $n \in \mathbb{N}$.

Then the key to understanding which types of equations are solvable by some formula lies in investigating whether the Galois groups of these types of equations are solvable groups.

We have only skimmed the surface of the exceedingly deep but interesting topic in algebra known as Galois Theory. It is one of the many areas of mathematics where concepts of abstract and linear algebra converge. You may wish to consult some of the references in regard to this subject.

Exercises 22.3

1. Prove that the phrase *is an extension by radicals of* is a transitive relation on the set of fields.

2. Verify that the Galois group of $x^n - 1 \in \mathbb{Q}[x]$ is abelian.

3. Show that $f(x) = x^5 - 10x - 5 \in \mathbb{Q}[x]$ is irreducible over $\mathbb{Q}$.

4. Suppose $\alpha = (1 \ \ 2 \ \ 3 \ \ 4 \ \ 5)$ and $\beta = (1 \ \ 2)$.
   a. Verify that $\alpha\beta\alpha^{-1} = (2 \ \ 3)$, $\alpha^2\beta\alpha^{-2} = (3 \ \ 4)$, $\alpha^3\beta\alpha^{-3} = (4 \ \ 5)$, and $\alpha^4\beta\alpha^{-4} = (1 \ \ 5)$.
   b. Show that $\{(1 \ \ 2), (2 \ \ 3), (3 \ \ 4), (4 \ \ 5), (1 \ \ 5)\}$ generates all transpositions in S_5.
   c. Prove that there exists no proper subgroup of S_5 containing a 5-cycle and a transposition.

5. Find the Galois group of the following fifth-degree polynomials in $\mathbb{Q}[x]$:

   a. $x^5 - 5x - 5$
   b. $x^5 + 4x^3 + 4x$
   c. $x^5 + 32$
   d. $x^5 - 4x^4 + 2x + 2$
   e. $x^5 - 19x^4 + 2x^3 - 24x^2 + 2$

6. If $s_1, s_2, \ldots, s_n$ are indeterminates over F, prove that $F[s_1, s_2, \ldots, s_n]$ is an integral domain.

7. If $\sigma \in S_n$ and $\bar{\sigma}(p(t_1, t_2, \ldots, t_n)) = p(t_{\sigma(1)}, t_{\sigma(2)}, \ldots, t_{\sigma(n)})$, prove $\bar{\sigma}$ is an automorphism of $F[t_1, t_2, \ldots, t_n]$.

8. If τ is an automorphism of $F[t_1, t_2, \ldots, t_n]$, show that

$$\bar{\tau}\left(\frac{p}{q}\right) = \frac{\tau(p)}{\tau(q)}, \qquad p, q \in F[t_1, t_2, \ldots, t_n]$$

   defines an automorphism of $F(t_1, t_2, \ldots, t_n)$.

9. Prove that the mapping $\phi\colon F[s_1, s_2, \ldots, s_n] \to F[z_1, z_2, \ldots, z_n]$ defined by $\phi(s_i) = z_i$ and $\phi(a) = a$ for all $a \in F$ is a ring homomorphism.

10. A polynomial function $f \in F[y_1, y_2, \ldots, y_n]$ is said to be a **symmetric function** if f is invariant under all permutations of $\{y_1, y_2, \ldots, y_n\}$.

    a. Show that

    $$g(y_1, y_2, y_3) = y_1^2 + y_2^2 + y_3^2$$

    and

    $$h(y_1, y_2, y_3) = y_1 y_2 y_3^2 + y_1 y_2^2 y_3 + y_1^2 y_2 y_3$$

    are symmetric functions in $F[y_1, y_2, y_3]$.

    b. If $s_1 = y_1 + y_2 + y_3, s_2 = y_1 y_2 + y_1 y_3 + y_2 y_3$, and $s_3 = y_1 y_2 y_3$ are the elementary symmetric functions in y_1, y_2, y_3, verify that $g = s_1^2 - s_2$ and $h = s_1 s_3$.

    c. If f is any symmetric function in $F[y_1, y_2, \ldots, y_n]$, prove that there exists $g \in F[y_1, y_2, \ldots, y_n]$ such that

    $$f(y_1, y_2, \ldots, y_n) = g(s_1, s_2, \ldots, s_n)$$

    where $s_1, s_2, \ldots, s_n$ are the elementary symmetric functions.

11. Let $\operatorname{char}(F) = 0$ and $f(x) \in F[x]$ have degree n and possess n distinct roots $u_1, u_2, \ldots, u_n$ in some splitting field E. Set

$$\Delta = \Pi(u_i - u_j) = (u_1 - u_2)(u_1 - u_3) \cdots (u_{n-1} - u_n)$$

The **discriminant** of f is the element Δ^2.

   a. If $\sigma \in G$, the Galois group of f over F, show that $\sigma(\Delta^2) = \Delta^2$.
   b. Show that $\Delta^2 \in F$.
   c. If $\tau \in G$ interchanges two roots of f and fixes all the others, show that $\tau(\Delta) = -\Delta$.
   d. Consider G as a subgroup of S_n. Show that $\sigma \in G$ is an even permutation if and only if $\sigma(\Delta) = \Delta$.
   e. Prove that G is a subgroup of A_n if and only if $\Delta \in F$.
   f. If $f(x) = x^2 + ax + b$, verify that $\Delta^2 = a^2 - 4b$, and if $g(x) = x^3 + px + q$, show that $\Delta^2 = -4p^3 - 27q^2$.
   g. Determine the Galois groups over $\mathbb{Q}$ of $g(x) = x^3 - 3x + 1$ and $g(x) = x^3 - 5x + 2$.

23 | Lattices

Every algebraic system that we have considered in this text involved an underlying ground set. For each set X, there is another set $\mathscr{P}(X)$, the power set of X, which is itself an algebraic structure under the set operations of union $\cup$, intersection $\cap$, and complement $-$. In this chapter, we will explore the connections between algebraic structures on a set X and those on $\mathscr{P}(X)$.

We first investigate special types of relations called partial orders. A **poset**, or **partially ordered set** (X, R), is a set X together with a partial order R on X. Posets can be used to aid in describing how subgroups are related to groups and how subrings are related to rings. We will define what we mean by *least upper bound* and *greatest lower bound* of two elements in a poset (X, R). A **lattice** is a poset in which every two elements have both a least upper bound and greatest lower bound. Lattices are widely used in algebra, computer science, geometry, and combinatorics.

We will discuss various kinds of lattices, including one particularly applicable type known as a Boolean algebra. A Boolean algebra is not only a lattice but also a commutative ring. In addition to providing a number of examples and discussing important properties of Boolean algebras, we will describe how Boolean algebras can be utilized to study switching circuits.

23.1
Partial Orders

Let R be a relation on a set X. In Section 7.1 we discussed special types of relations and saw that R is an equivalence relation on X when R is reflexive, symmetric, and transitive. Here we wish to treat a similar situation. We need the following definition:

Definition 23.1

> A relation R on a set X is **antisymmetric** if $a R b$ and $b R a$ imply that $a = b$.

The standard example for an antisymmetric relation is the relation **less than or equal to** on the real number system. There are a number of other important examples. The relation **is a subset of** is an antisymmetric relation on the collection of subsets of a set X. The relation **divides** is antisymmetric on the set of positive integers, but it is not antisymmetric on $\mathbb{Z}$. (See Exercise 1.)

Definition 23.2

> A relation R on a set X is a **partial order** if R is reflexive, antisymmetric, and transitive. If R is a partial order on X, then (X, R) is termed a **poset (partially ordered set)**.

Example 23.1

$(\mathbb{R}, \leqslant)$ is a poset. Here $\leqslant$ represents the standard *less than or equal to* ordering on the real number system. Note that if $a, b \in \mathbb{R}$, then either $a \leqslant b$ or $b \leqslant a$. This property does not have to hold in every partial order, as the next example illustrates.

Example 23.2

$(\mathscr{P}(X), \subseteq)$ is a poset since *is a subset of* is reflexive and transitive. That $\subseteq$ is antisymmetric follows from the basic definition of set equality. If $|X| \geqslant 2$, then it is not true that when $A, B \in \mathscr{P}(X)$, either $A \subseteq B$ or $B \subseteq A$. (See Exercise 2.)

Example 23.3

Let G be a group and define $\leqslant$ on the collection of subgroups of G by $H \leqslant K$ if H is a subgroup of K. Then $\leqslant$ is clearly reflexive and antisymmetric; the relation $\leqslant$ is transitive by virtue of Exercise 8 of Section 4.2. The relation *is a normal subgroup of* is *not*, however, a partial order on the set of subgroups of G. (See Exercise 3.)

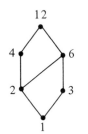

Figure 23.1

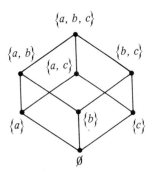

{a, b, c}

{a, b}   {b, c}
      {a, c}

{a}   {b}
                    {c}

∅

Figure 23.2

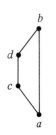

Figure 23.3

Let | denote *divides* on ℕ. Then $m \mid m$ for all $m \in \mathbb{N}$ and $m \mid n$ and $n \mid t$ imply $m \mid t$. If $m \mid n$ and $n \mid m$, then $m = n$ since we are working with positive integers only. Therefore $(\mathbb{N}, \mid)$ is a poset.

In discussing posets, it is common usage to denote an arbitrary partial order by the symbol $\leqslant$ even if the relation is not *less than or equal to* or *is a subgroup of*. We will adopt this convention.

Let $(X, \leqslant)$ be a poset. Define a new relation $<$ on X in the following way: for $a, b \in X$, write $a < b$ if $a \leqslant b$ but $a \neq b$. In Examples 23.1–4, this corresponds to *is strictly less than, is a proper subset of, is a proper subgroup of*, and *is a proper divisor of*. Now define a third relation *covers* on X by saying b **covers** a if $a < b$ and $a < c < b$ is not true for any $c \in X$. The **covering relation** can be used to give graphical representations of posets (X, R) when X is a finite set.

Suppose $X = \{x_1, x_2, \ldots, x_n\}$. Denote the elements of X by dots and draw a vertical line from x_i up to x_j if x_j covers x_i. For example, the positive divisors of 12 are $\{1, 2, 3, 4, 6, 12\}$. Figure 23.1 pictures the **covering diagram** of the relation $\mid$ on this set. In Figure 23.2, we give the covering diagram of Example 23.2 for $X = \{a, b, c\}$.

The covering diagram for a finite poset allows you to view the poset geometrically. Not every diagram of this form, however, comes from a poset. For example, the diagram given in Figure 23.3 does not represent a poset, for it says that b covers a, but $a < c < d < b$.

In a poset $(X, \leqslant)$, we can often find elements that have special properties. An element $x \in X$ is termed **minimal** if there does not exist a $y \in X$ such that $x > y$. Similarly $x \in X$ is a **maximal** element if there does not exist a $y \in X$ such that $x < y$. The poset of *proper* subsets of $\{a, b, c\}$ contains three maximal elements but just one minimal element. The poset $(\mathbb{Z}, \leqslant)$ contains no maximal or minimal elements. An element $z \in (X, \leqslant)$ is a **zero element** if $z \leqslant x$ for all $x \in X$, while z is a **unit element** if $x \leqslant z$ for all $x \in X$. The empty set is the zero element of the poset $(\mathcal{P}(X), \subseteq)$, while X is the unit element of this poset. For the poset of divisors of 12, the zero element is 1 and the unit element is 12. There can be at most one zero and one unit in a poset. (See Exercise 4.) It might happen that a poset contains no zero or no unit. Any zero element is a minimal element, but the converse is not true. (See Exercise 5.)

One relationship between zero elements and minimal elements is given in the next result.

If there is a unique minimal element in a finite poset $(X, \leqslant)$, then that minimal element is a zero element. Similarly, a unique maximal element in a finite poset must be a unit element.

Proof

Suppose x is the unique minimal element of X, and let $y \in X$. We want to show $x \leqslant y$. If y is not equal to x, then y is not minimal. Thus, there exists

$y_2 \in X$ such that $y > y_2$. If y_2 is not minimal, then there exists $y_3 \in X$ such that $y_2 > y_3$. Continuing this process, we arrive at a set of elements such that $y > y_2 > y_3 > \cdots$. Since X is a finite set, this chain must end with a minimal element. But x is the only possibility; so we have $x \leqslant y$.

The proof of the second assertion is similar to the first and is left to you as Exercise 6. □

Zero and unit elements are customarily denoted by the symbols 0 and 1. If a poset $(X, \leqslant)$ has a zero element 0, then the elements of X that cover 0 are termed **atoms**. For the poset of subsets of X, the atoms are the singleton subsets. For the poset of the divisors of a natural number n under the relation *divides*, the atoms are the prime divisors of n.

If a poset $(X, \leqslant)$ has unit 1, then the elements covered by 1 are called the **coatoms** of X. For the posets just described, the coatoms are of the form $X - \{x_i\}$ for $x_i \in X$ and n/p, where p is a prime divisor of n.

Suppose $(X, \leqslant)$ is a poset. Define a relation $\geqslant$ on X by $y \geqslant x$ if $x \leqslant y$. Then $\geqslant$ is a partial order on X (see Exercise 8) called the **converse** of $\leqslant$ and $(X, \geqslant)$ is called the **dual** of the poset $(X, \leqslant)$. Then the atoms of $(X, \geqslant)$ are the coatoms of $(X, \geqslant)$ and conversely. This relationship between a poset and its dual can be extended. If we prove any result about a poset $(X, \leqslant)$, then we can replace the relation $\leqslant$ by its converse $\geqslant$ to obtain another result about the dual poset $(X, \geqslant)$. This idea is called the **Duality Principle** for posets.

Whenever we considered the algebraic structures, whether they were groups, rings, fields, or vector spaces, we discussed structure-preserving mappings. We do this for posets also.

Definition 23.3

> Let $(X, \leqslant)$ and $(Y, \leqslant')$ be posets. A bijection $f\colon X \to Y$ is an **isomorphism** of posets when $x_1 \leqslant x_2$ if and only if $f(x_1) \leqslant' f(x_2)$ for all $x_1, x_2 \in X$. If $f\colon X \to Y$ is an isomorphism, then $(X, \leqslant)$ and $(Y, \leqslant')$ are termed **order-isomorphic**.

If we prove a theorem on the structure of a poset, then the theorem will be true for any other poset isomorphic to the original. The idea of poset isomorphism also allows us to rephrase part of the Fundamental Theorem of Galois Theory.

Example 23.5

Consider the poset of subfields of $E = \mathbb{Q}(\sqrt[3]{5}, \omega)$ and the poset of subgroups of $G = G_{E/\mathbb{Q}}$ pictured in Figure 22.8. Note that we have drawn the covering diagram for the dual of the poset $(G, \leqslant)$. The Fundamental Theorem of Galois asserts, in part, that there is a bijection between the set of extension fields of $\mathbb{Q}$ contained in E and the set of subgroups of $G_{E/\mathbb{Q}}$. This bijection is order-reversing. Therefore, there exists a poset isomorphism between the poset of subfields of E and the dual of the poset $(G, \leqslant)$.

Among the many concepts and special elements in a poset, the next is most important.

Definition 23.4

Let $(X, \leqslant)$ be a poset. If $x, y \in X$, then an element $u \in X$ is the **greatest lower bound (glb)** of x and y if

i. $u \leqslant x$ and $u \leqslant y$; and
ii. whenever there exists $w \in X$ such that $w \leqslant x$ and $w \leqslant y$, then $w \leqslant u$.

Similarly, an element $v \in X$ is the **least upper bound (lub)** of x and y if

i. $x \leqslant v$ and $y \leqslant v$, and
ii. whenever there exists $w \in X$ such that $x \leqslant w$ and $y \leqslant w$, then $v \leqslant w$.

The abbreviation glb is often pronounced "glub." The glb of x and y, if it exists, is symbolized by $x \wedge y$, while the lub (pronounced "lub") of x and y is denoted $x \vee y$. If $(X, \leqslant)$ is a poset and $\geqslant$ is the converse of $\leqslant$, then the glb of x and x in $(X, \leqslant)$ is the lub of x and y in the dual poset $(X, \geqslant)$.

Example 23.6

In the poset $(\mathscr{P}(X), \leqslant)$, we have $A \wedge B = A \cap B$ and $A \vee B = A \cup B$. While intersections and unions are probably the easiest way to envision glb's and lub's, care must be taken so as not to assume that all properties of these operations on $\mathscr{P}(X)$ hold in all posets.

Example 23.7

Let $n \in \mathbb{N}$, and consider the divisors of n under the relation *divides*. If $a \mid n$ and $b \mid n$, then $a \wedge b$ is the greatest common divisor of a and b, while $a \vee b$ is the least common multiple defined in Exercise 14 of Section 16.2. (See Exercise 9.)

Example 23.8

Let G be a group, and let $(X, \leqslant)$ be the poset of subgroups of G under the relation $\leqslant$. If $A \leqslant G$, $B \leqslant G$, then $A \wedge B = A \cap B$ and $A \vee B = \langle A, B \rangle$, the smallest subgroup of G containing both A and B.

Suppose $(X, \leqslant)$ is a poset with $x, y \in X$. Can more than one element of X satisfy the requirements for being the glb of x and y? Suppose u_1 and u_2 are both glb's of x and y. Then $u_1 \leqslant x$ and $u_1 \leqslant y$. Since u_2 is a glb of x and y, $u_1 \leqslant u_2$. Interchanging the roles of u_1 and u_2, we similarly obtain $u_2 \leqslant u_1$.

By antisymmetry, $u_1 = u_2$. By this argument, and the Duality Principle, we have the following:

If $(X, \leqslant)$ is a poset and $x, y \in X$ have a glb or lub in X, then that glb or lub is unique.

You may encounter a poset in which there are elements having no lub or glb. For example, the set $\{2, 3, 5, 7\}$ forms a poset under *divides*, but no two elements have a lub or a glb. This situation is disconcerting to many mathematicians and computer scientists. Most would prefer to investigate relationships among the special elements of a poset. To insure that such elements exist, we need only restrict our attention to the appropriate type of poset.

A **lattice** $(L, \leqslant)$ is a poset in which every two elements have a lub and a glb.

We have already seen examples of lattices. For example, $(\mathscr{P}(X), \subseteq)$ is a lattice as are the divisors of a natural number n under *divides*. There are others:

Example 23.9

Consider $(\mathbb{R}, \leqslant)$. If $x \vee y = \max\{x, y\}$ and $x \wedge y = \min\{x, y\}$, then $(\mathbb{R}, \leqslant)$ is an infinite lattice.

Example 23.10

Suppose $B = \{0, 1\}$ and $a_1 a_2 \cdots a_n, b_1 b_2 \cdots b_n \in B^n$. Define $a_1 a_2 \cdots a_n \leqslant b_1 b_2 \cdots b_n$ if $a_i \leqslant b_i$ for all i. Thus $10100 \leqslant 11101$ and $01100 \leqslant 01111$. Set

$$a_1 a_2 \cdots a_n \wedge b_1 b_2 \cdots b_n = c_1 c_2 \cdots c_n$$

where $c_i = \min\{a_i, b_i\}$ and

$$a_1 a_i \cdots a_n \vee b_1 b_2 \cdots b_n = d_1 d_2 \cdots d_n$$

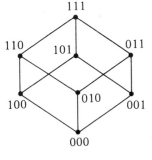

where $d_i = \max\{a_i, b_i\}$. Then $(B^n, \leqslant)$ is a lattice. (See Exercise 11.) The diagram of the lattice B^3 is contained in Figure 23.4. To find the glb of x and $y \in B^3$, say 110 and 011, simply locate the element, in this case 010, that is assigned to the place where lines descending from x and y first come together. Similarly, one finds the lub of 110 and 011 by finding the point 111 where lines ascending from 110 and 011 intersect.

Figure 23.4

In the next section, we investigate lattices in more detail. We will show

how to transform the definition of lattice into an algebraic form. In addition, we will explore some of the various types of lattices that can occur.

1. Verify that *divides* is antisymmetric on $\mathbb{N}$ but not on $\mathbb{Z}$.

2. A relation R on a set X is **complete** if $x, y \in X$ implies either $x\, R\, y$ or $y\, R\, x$. Prove that *is a subset of* is a partial order on $\mathscr{P}(X)$ that is not complete on $\mathscr{P}(X)$.

3. Let G be the collection of subgroups of a group G. Prove that $(G, \trianglelefteq)$ is not a poset.

4. Show that a zero element or a unit element in a poset must be unique.

5. Let (A, R_1) and (B, R_2) be posets. Define $\leq$ on $A \times B$ by $(a, b) \leq (c, d)$ if $a\, R_1\, c$ and $b\, R_2\, d$.
   a. Prove that $(A \times B, \leq)$ is a poset. (This poset is termed the **direct product of A and B**.)
   b. Let (A, R_1) be the poset $(\mathbb{Z}, \leq)$ and (B, R_2) be the poset $(\mathscr{P}(X), \subseteq)$ where $X = \{a, b, c\}$. Show that $\mathbb{Z} \times \mathscr{P}(X)$ has maximal and minimal elements but no zero or unit elements.

6. Prove that a unique maximal element in a finite poset must be a unit element.

7. Draw a covering diagram for each of the following posets:
   a. $(\mathscr{P}(X), \subseteq)$ where $X = \{1, 2, 3, 4\}$.
   b. $(X, |)$ where X is the set of positive divisors of 125.
   c. $(X, |)$ where X is the set of positive divisors of 72.
   d. $(X, \leq)$ where X is the set of subgroups of A_4, the alternating group on 4 letters.
   e. $(X, \leq)$ where X is the set of subgroups of $\mathbb{Z}_8 \times \mathbb{Z}_9$.
   f. $(X, \leq)$ where X is the set of subgroups of S_5, the symmetric group on 5 letters.

8. Let $(X, \leq)$ be a poset. Prove that $\geq$, the converse of $\leq$, is a partial order on X.

9. Let $n \in \mathbb{N}$ and X be the set of positive divisors of n. In the poset $(X, |)$, prove that

$$a \wedge b = \gcd(a, b) \qquad \text{and} \qquad a \vee b = \operatorname{lcm}(a, b) = \frac{ab}{a \wedge b}$$

10. For the lattice of subgroups of S_4, the symmetric group on 4 letters, under the relation $\leq$, compute
    a. $A_4 \wedge S_3$
    b. $A_4 \vee S_3$
    c. $\langle (1\ \ 2\ \ 3\ \ 4) \rangle \vee S_3$
    d. $\langle (1\ \ 2\ \ 3\ \ 4) \rangle \wedge S_3$
    e. $\langle (1\ \ 2\ \ 3\ \ 4) \rangle \vee \langle (1\ \ 2)(3\ \ 4) \rangle$
    f. $\langle (1\ \ 2)(3\ \ 4) \rangle \vee \langle (1\ \ 4)(2\ \ 3) \rangle \vee \langle (1\ \ 3)(2\ \ 4) \rangle$

11. Prove that $(B, \leq)$, defined in Example 23.10, is a lattice.

12. Let $(V, \oplus)$ be a vector space and $\mathscr{S}$ the set of subspaces of V.
    a. Verify that $(\mathscr{S}, \subseteq)$ is a poset.
    b. If $A, B \in \mathscr{S}$, show $A \vee B = A \oplus B$ and $A \wedge B = A \cap B$.
    c. Construct the lattice of subspaces for $V = \mathbb{Z}_2 \oplus \mathbb{Z}_2 \oplus \mathbb{Z}_2$.

13. Let R be a ring and $\mathscr{I}$ the set of ideals of R.
    a. Prove that $(\mathscr{I}, \subseteq)$ is a lattice.
    b. Determine the covering diagram for this lattice in the case $R = (\mathbb{Z}_{48}, \oplus, \odot)$.

14. A lattice $(L, \leqslant)$ is termed **self-dual** if it is isomorphic to $(L, \geqslant)$ where $\geqslant$ is the converse of $\leqslant$.
    a. Verify that the subgroups of the Klein Four-Group form a self-dual lattice under the relation $\leqslant$.
    b. Determine all self-dual lattices L for $|L| \leqslant 5$.

15. Suppose G is a cyclic group of order p^n where p is a prime. Determine the lattice diagram for the lattice of subgroups of G.

16. Consider the set of all partitions of the set $X = \{a, b, c, d\}$. We will say that the partition $\{\{a, b\}, \{c\}, \{d, e\}\}$ **is finer than** $\{\{a, b\}, \{c, d, e\}\}$ because every subset in the first is contained in a subset in the second.
    a. Find all 15 partitions of X.
    b. Order the partitions by *is finer than* and draw the covering diagram.
    c. Verify that this set of partitions is a lattice under the relation *is finer than*.

17. Define $f: \mathscr{P}(X) \to \mathscr{P}(X)$ by $f(A) = X - A$. Prove that f is an isomorphism between $(\mathscr{P}(X), \subseteq)$ and its dual.

Types of Lattices

In the last section we introduced the concept of lattice as a special type of poset. In this section we propose to investigate several different kinds of lattices. We will prove a number of properties that all lattices share; in addition, we will give an alternate definition of lattice that is similar in form to definitions given for groups, rings, fields, and vector spaces.

The first question you might ask about lattices is, "Given a poset, is there an easy way I can tell whether or not the poset is a lattice?" In certain situations, the answer is yes.

Example 23.11

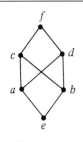

Figure 23.5

Let R be the relation on the set $X = \{a, b, c, d, e, f\}$ given by

$$R = \{(e, a), (e, b), (e, c), (e, d), (e, e), (e, f), (a, a), (a, c), (a, d), (a, f),$$

$$(b, b), (b, c), (b, d), (b, f), (c, c), (c, f), (d, d), (d, f), (f, f)\}$$

Then (X, R) is a poset whose covering diagram is given in Figure 23.5. Note that $a \vee b = c$ and $a \vee b = d$. By Theorem 23.2, the lub of two elements in a lattice is unique. Therefore (X, R) is not a lattice.

In order to develop useful techniques for dealing with lattices, we must investigate some basic properties:

| Lemma 23.1 |

If $(L, \leqslant)$ is a lattice, then the following are equivalent:

 i. $a \leqslant b$,
 ii. $a \vee b = b$, and
 iii. $a \wedge b = a$

Proof

We will show how i and ii are equivalent and leave the verifications of the remainder of this lemma as Exercise 2. Suppose $a \leqslant b$. Then $a \leqslant b$ and $b \leqslant b$ imply $a \vee b \leqslant b$. On the other hand, $b \leqslant a \vee b$ by definition. Therefore $a \vee b = b$ by antisymmetry. Conversely, if $a \vee b = b$, then $a \leqslant a \vee b = b$ implies $a \leqslant b$.   □

In a lattice $(L, \leqslant)$, not only does L contain the lub and glb for any two elements, but also any finite subset of L. Let $\{a_1, a_2, \ldots, a_n\} \subseteq L$. Set

$$a_1 \vee a_2 \vee a_3 = \text{lub}\ \{a_1 \vee a_2, a_3\} \quad \text{and} \quad a_1 \wedge a_2 \wedge a_3 = \text{lub}\ \{a_1 \wedge a_2, a_3\}$$

Similarly, define

$$\bigvee_{i=1}^{n} a_i = \text{lub}\ \left\{\bigvee_{k=1}^{n-1} a_i, a_n\right\} \quad \text{and} \quad \bigwedge_{i=1}^{n} a_i = \text{glb}\ \left\{\bigwedge_{k=1}^{n-1} a_i, a_n\right\}$$

If

$$b = \bigvee_{i=1}^{n} a_i$$

then $a_i \leqslant b$ for all i; moreover, if $a_i \leqslant c$ for all i, then $b \leqslant c$. Therefore our inductive definition satisfies the basic requirements of a lub. Analogous results hold for our definition of a glb of $\{a_1, a_2, \ldots, a_n\}$.

| Theorem 23.3 |

If $(L, \leqslant)$ is a lattice and $a, b, c \in L$, then the following properties hold:

i. $a \vee a = a$,	$a \wedge a = a$	(Idempotency Laws)
ii. $a \vee b = b \vee a$,	$a \wedge b = b \wedge a$	(Commutative Laws)
iii. $a \vee (b \vee c) = (a \vee b) \vee c$,	$a \wedge (b \wedge c) = (a \wedge b) \wedge c$	
		(Associative Laws)
iv. $a \vee (a \wedge b) = a$,	$a \wedge (a \vee b) = a$	(Absorption Laws)

Proof

Assertions i and ii follow immediately from the definition of lub and glb. To prove iii, note that

$$a \vee (b \vee c) = \text{lub}\ \{a, b \vee c\}$$

$$= \text{lub}\ \{b \vee c, a\}$$

$$= \text{lub}\ \{b, c, a\}$$

$$= \text{lub}\ \{a, b, c\} = (a \vee b) \vee c$$

By the Duality Principle, $a \wedge (b \wedge c) = (a \wedge b) \wedge c$.

Finally, consider $a \vee (a \wedge b)$. Since $a \leqslant a$ and $a \wedge b \leqslant a$, we have $a \vee (a \wedge b) \leqslant a$ by the definition of lub. But $a \leqslant a \vee (a \wedge b)$. Thus $a = a \vee (a \wedge b)$. Similarly $a = a \wedge (a \vee b)$.   □

The four properties of Theorem 23.3 are similar to those encountered in definitions of rings and vector spaces. To give an alternate axiomatic definition for lattice, we need a way of recovering a partial ordering from these rules.

Lemma 23.2

Suppose L is a set and $\vee$, $\wedge$ are two operations on L that satisfy the absorption laws of Theorem 23.3. Then $a \wedge b = a$ if and only if $a \vee b = b$.

Proof
Suppose $a \wedge b = a$. Then $a \vee b = (a \wedge b) \vee b = b$ by one of the absorption laws. Conversely, if $a \vee b = b$, then $a \wedge b = a \wedge (a \vee b) = a$ by the other absorption law.   □

Theorem 23.4

Let L be a set on which two operations, $\vee$ and $\wedge$, are defined that satisfy the four laws of Theorem 23.3. Define $\leqslant$ on L by $a \leqslant b$ if and only if $a \vee b = b$. Then $(L, \leqslant)$ is a lattice in which lub $(a, b) = a \vee b$ and glb $(a, b) = a \wedge b$.

Proof
We first show $\leqslant$ is a partial order on L. The idempotency laws says $a \vee a = a$, so that $a \leqslant a$ and $\leqslant$ is reflexive. If $a \leqslant b$ and $b \leqslant a$, then $a \vee b = b$ and $b \vee a = a$. By the commutativity rule,

$$a = b \vee a = a \vee b = b$$

and $\leqslant$ is antisymmetric. Next assume $a \leqslant b$ and $b \leqslant c$ for $a, b, c \in L$. Then $a \vee b = b$ and $b \vee c = c$, so that

$$a \vee c = a \vee (b \vee c) = (a \vee b) \vee c = b \vee c = c$$

by associativity. Hence, $a \leqslant c$ and $(L, \leqslant)$ is a poset.

We still must show that $\vee$ and $\wedge$ yield the lub and glb for our definition of $\leqslant$. Since

$$a \vee (a \vee b) = (a \vee a) \vee b = a \vee b$$

our definition of $\leqslant$ forces $a \leqslant a \vee b$. In a similar manner, $b \leqslant a \vee b$. Suppose $a \leqslant c$ and $b \leqslant c$, so that $a \vee c = c$ and $b \vee c = c$. Then

$$(a \vee b) \vee c = a \vee (b \vee c) = a \vee c = c$$

and $a \vee b \leqslant c$. Therefore $a \vee b$ is, indeed, the lub of a and b in the poset $(L, \leqslant)$.

It remains to show that $a \wedge b$ is the glb of a and b in $(L, \leqslant)$. By absorption, $a \vee (a \wedge b) = a$. But then $a \wedge b \leqslant a$. Similarly $a \wedge b \leqslant b$. If $c \leqslant a$

and $c \leqslant b$, we have $c \vee a = a$ and $c \vee b = b$. By Lemma 23.2, $c \wedge a = c$ and $c \wedge b = c$. Thus

$$c \wedge (a \wedge b) = (c \wedge a) \wedge b = c \wedge b = c$$

and $c \vee (a \wedge b) = a \wedge b$ by Lemma 23.2. This says that $c \leqslant a \wedge b$ and $a \wedge b$ is the glb of a and b in $(L, \leqslant)$.   ☐

The alternate definition for lattice contained in this last theorem is not too different from that given for rings; in fact, it is not uncommon to see the definition for lattice given in terms of two operations symbolized by $+$ and $\cdot$. But the definition for a ring also included statements about identities, inverses, and distributive laws. By adding such requirements to our new definition, we will produce a number of interesting types of lattices.

Definition 23.6

A lattice $(L, \leqslant)$ is **bounded** if L contains a zero and a unit element.

The lattice of integers $\mathbb{Z}$ under $\leqslant$ is an example of a lattice that is not bounded. Most of our other examples of lattices, however, are bounded. This is due to the fact that these lattices are finite.

Theorem 23.5

Every finite lattice is bounded.

Proof

Let $(L, \leqslant)$ be a lattice with $L = \{a_1, a_2, \dots, a_n\}$. Then

$$u = \bigwedge_{i=1}^{n} a_i$$

has the property that $u \leqslant a_i$ for all i, while

$$v = \bigvee_{i=1}^{n} a_i$$

satisfies the inequality $a_i \leqslant v$ for all i.   ☐

If every finite lattice is bounded, is every infinite lattice unbounded? The answer is no. Consider the real numbers in the interval $[0, 1]$ under the standard relation $\leqslant$ for $\mathbb{R}$. $([0, 1], \leqslant)$ is an infinite lattice with zero 0 and unit 1. Note that this lattice is contained in the lattice $(\mathbb{R}, \leqslant)$. The relationship between these two lattices is given in the next definition:

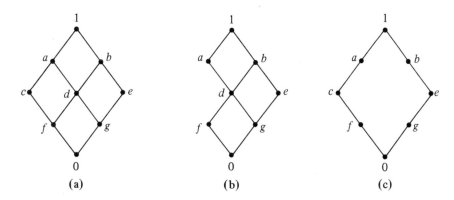

Figure 23.6   (a)   (b)   (c)

Definition 23.7

Let $(L, \vee, \wedge)$ be a lattice. A subset L_1 of L is a **sublattice** of L if L_1 is a lattice under the restrictions of $\vee$ and $\wedge$ to L_1.

Note the similarity between this definition and those for subgroup, subring, and subspace. Here we are saying that L_1 is a sublattice of L if the lub and glb (in L) of every pair of elements in L_1 is also in L_1. Consider the covering diagrams of Figure 23.6. The covering diagram in Figure 23.6a is that of a lattice. The diagram in Figure 23.6b does represent a sublattice; but that in Figure 23.6c does not. In Figure 23.6c, note that $a \wedge b = 0$, while $a \wedge b = d$ in the original lattice.

We will make use of the idea of sublattice in analyzing the next type of lattice.

Definition 23.8

Let $(L, \leqslant)$ be a lattice. L is a **modular lattice** if $a \leqslant c$ implies $(a \vee b) \wedge c = a \vee (b \wedge c)$.

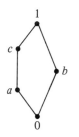

Figure 23.7

At first glance, it might appear as if a modular lattice satisfies an associative law. This is, however, a pseudo-associative law in that different operations are involved. The set of subsets of a set X forms a modular lattice under $\subseteq$, because Theorem 1.3 and $A \subseteq C$ imply

$$(A \cup B) \cap C = (A \cap C) \cup (B \cap C) = A \cup (B \cap C)$$

Not every lattice is modular, however. Consider the **pentagonal lattice** given in Figure 23.7. On one hand, $(a \vee b) \wedge c = 1 \wedge c = c$. On the other we have, $a \vee (b \wedge c) = a \vee 0 = a$. Therefore, this lattice is not modular.

If a given lattice contains a pentagonal sublattice, that is, a sublattice whose covering diagram is the pentagonal lattice, then the lattice is clearly not modular. It is somewhat surprising that the converse is true, also.

Let $(L, \leqslant)$ be a lattice. L is modular if and only if L does not contain a pentagonal sublattice.

Proof

Since the pentagonal sublattice is not modular, the elements in L involved in this sublattice fail to satisfy the definition. Therefore $(L, \leqslant)$ is not modular.

Conversely, if $(L, \leqslant)$ is not modular, there exist a, b, $c \in L$ such that $a \leqslant c$ and $(a \vee b) \wedge c \neq a \vee (b \wedge c)$. If $a = c$, then

$$(a \vee b) \wedge c = (a \vee b) \wedge a = a = a \vee (b \wedge a) = a \vee (b \wedge c)$$

Thus, $a < c$; moreover, we can also assume $(a \vee b) \wedge c < a \vee (b \wedge c)$, since the relation $a \vee (b \wedge c) \leqslant (a \vee b) \wedge c$ holds in any lattice when $a \leqslant c$. (See Exercise 14.) Set $x = (a \vee b) \wedge c$, $z = a \vee (b \wedge c)$. Then $x < z$ and

$$x \wedge b = [(a \vee b) \wedge c] \wedge b = (a \vee b) \wedge (b \wedge c)$$
$$= [(a \vee b) \wedge b] \wedge c$$
$$= [b \wedge (a \vee b)] \wedge c = b \wedge c$$

by the absorption law. In addition,

$$z \wedge b = [a \vee (b \wedge c)] \wedge b \leqslant [c \vee (b \wedge c)] \wedge b = c \wedge b = b \wedge c$$

Since $b \wedge c = x \wedge b \leqslant z \wedge b \leqslant b \wedge c$, we have $z \wedge b = b \wedge c = x \wedge b$.

On the other hand,

$$z \vee b = [a \vee (b \wedge c)] \vee b = a \vee [(b \wedge c) \vee b] = a \vee b$$

while

$$a \vee b = z \vee b \leqslant x \vee b = [(a \vee b) \wedge c] \vee b \leqslant [(a \vee b) \wedge a] \vee b = a \vee b$$

so $z \vee b = x \vee b = a \vee b$.

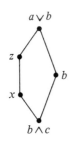

Figure 23.8

To complete our proof and show that x, z, b, $a \vee b$, and $b \vee c$ form the sublattice in Figure 23.8, we must verify that $b < a \vee b$, $b \wedge c < b$, $x > b \wedge c$, and $z < a \vee b$.

If $b = a \vee b$, then $a \leqslant b$ and

$$z = a \vee (b \wedge c) \leqslant b \vee (b \wedge c) = b$$

so that

$$z = z \wedge b \leqslant b \wedge c = (a \vee b) \wedge c = x$$

contradicting $x < z$. Similarly $b \wedge c = b$ leads to the same contradiction. If

$x = b \wedge c$, then

$$b = (b \wedge c) \vee b = x \vee b = a \vee b$$

which we just showed is impossible. Thus $x < b \wedge c$ and a similar argument (see Exercise 15) demonstrates $z < a \vee b$. Therefore, $(L, \leqslant)$ contains the pentagonal sublattice of Figure 23.8.   □

A most important type of lattice is described by our next definition.

Definition 23.9

Let $(L, \leqslant)$ be a lattice. L is a **distributive lattice** if

$$(a \vee b) \wedge c = (a \wedge c) \vee (b \wedge c) \qquad \text{for all} \quad a, b, c \in L$$

A second distributive law $a \vee (b \wedge c) = (a \vee b) \wedge (a \vee c)$ follows from the first by setting $x = a \vee b$, $y = a$, and $z = c$. (See Exercise 17.)

The fact that set intersection $\cap$ distributes over set union $\cup$ shows that $(\mathscr{P}(X), \subseteq)$ is a distributive lattice for all sets X. In our discussion of modular lattices, we used this same fact to prove that $(\mathscr{P}(X), \subseteq)$ is a modular lattice. This relationship holds in general:

Theorem 23.7

Every distributive lattice is modular.

Proof

If $(L, \leqslant)$ is a lattice and $a, b, c \in L$ with $a \leqslant c$, then

$$(a \vee b) \wedge c = (a \wedge c) \vee (b \wedge c) = a \wedge (b \wedge c)$$ □

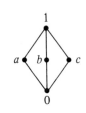

Figure 23.9

Since the pentagonal lattice is not modular, we know that it is not distributive. Are there other lattices that are not distributive? Consider the lattice in Figure 23.9. It is isomorphic to the lattice of subgroups of Klein Four-Group K. We will refer to this as the K-lattice. Note that

$$(a \vee b) \wedge c = 1 \wedge c = c$$

while

$$(a \wedge c) \vee (b \wedge c) = 0 \vee 0 = 0$$

Therefore, the K-lattice is not distributive. It *is* modular, however, by Theorem 23.6.

There exists a result analogous to Theorem 23.6 that relates the pentagonal and K-lattices to nondistributive lattices. We state this result in Theorem 23.8, but we refer you to Rutherford's text in the references for the proof.

Theorem 23.8

A lattice $(L, \leqslant)$ is distributive if and only if it does not contain a pentagonal or K-sublattice.

In studying sets and their relationships, we encountered the idea of set complement. This concept is also involved in the study of lattices.

Definition 23.10

A lattice $(L, \leqslant)$ is **complemented** if it is bounded and for each $a \in L$, there exists $\bar{a} \in L$ such that $a \vee \bar{a} = 1$ and $a \wedge \bar{a} = 0$.

If $(L, \leqslant)$ is a **complemented lattice**, then $\bar{a}$ is called a **complement of** a. While the complement of a set $A \subseteq X$ is $X - A$ in the lattice $(\mathscr{P}(X), \subseteq)$, the Examples 23.12–14 show that the idea of *complemented lattice* allows some interesting variations:

Example 23.12

Consider the lattice K given in Figure 23.9. Both b and c are complements of a, since $a \vee b = a \vee c = 1$ and $a \wedge b = a \wedge c = 0$. Therefore complements need not be unique.

Example 23.13

Let us consider the lattice of positive divisors of 30 under the relation *divides*. If $a \mid 30$, then $\bar{a} = 30/a$ since gcd $(30, 30/a) = 1$, while lcm $(30, 30/a) = 30$. This is an example of a **uniquely complemented lattice**.

Example 23.14

The lattice of positive divisors of 12 given in Figure 23.1 is not a complemented lattice. Although 4 is the complement of 3, 2 does not have a complement.

We have seen that lattices exist that contain more than one complement for some element. A bounded, distributive lattice, however, can have at most one complement for any given element.

Theorem 23.9

In a bounded, distributive lattice $(L, \leqslant)$, complements are unique.

Proof

Suppose $a \in L$ and both $\hat{a}$ and $\bar{a}$ are complements of a. Then

$$\hat{a} = \hat{a} \vee 0 = \hat{a} \vee (a \wedge \bar{a}) = (\hat{a} \vee a) \wedge (\hat{a} \vee \bar{a}) = 1 \wedge (\hat{a} \vee \bar{a})$$

$$= \hat{a} \vee \bar{a} = \bar{a} \vee \hat{a} = 1 \wedge (\bar{a} \vee \hat{a})$$

$$= (\bar{a} \vee a) \wedge (\bar{a} \vee \hat{a}) = \bar{a} \vee (a \wedge \hat{a}) = \bar{a} \vee 0 = \bar{a} \qquad \square$$

In the next section, we will focus on lattices that are both complemented and distributive. By Theorem 23.9, these lattices have a unique complement for each element.

Exercises 23.2

1. Which of the following poset-covering diagrams are lattices?

a.   b.   c.

d.   e.   f.

2. If $(L, \leqslant)$ is a lattice, prove that $a \leqslant b$ if and only if $a \wedge b = a$.

3. For the lattice of divisors of 180 under the relation *divides* find

   a. $2 \vee 3 \vee 5$ b. $(2 \vee 3) \wedge 5$

   c. $(4 \vee 9) \wedge 15$ d. $6 \wedge 9 \wedge 15$

   e. $5 \wedge 9 \wedge 4$

4. Suppose $(L, \leqslant)$ is a lattice and $a, b \in L$. Prove $a \wedge (a \vee b) = a$.

5. Let $(L, \leqslant)$ be a lattice and $a, b \in L$ with $a \leqslant b$. Set

   $$[a, b] = \{c \in L \mid a \leqslant c \leqslant b\}$$

   $[a, b]$ is called an **interval** in L. Prove that $[a, b]$ is a bounded sublattice of $(L, \leqslant)$.

6. Find all sublattices for each of the following lattices:

   a. Figure 23.1   b. Figure 23.2

   c. Exercise 7c of Section 23.1   d. Exercise 7d of Section 23.1

7. Give an example of a lattice $(L, \leqslant)$ with a subset L_1 such that $(L_1, \leqslant)$ is a lattice but not a sublattice.

8. Prove that if $(L, \leqslant)$ is a lattice and $a, b, c, d \in L$, then

   a. $a \leqslant b$ implies $(a \vee c) \wedge b \leqslant (a \wedge c) \vee b$,

   b. $a \vee (b \wedge c) \leqslant (a \vee b) \wedge (a \vee c)$, and

   c. $(a \wedge b) \vee (a \wedge c) \leqslant a \wedge (b \vee c)$.

9. For $a + bi, c + di \in \mathbb{C}$, define $a + bi \trianglelefteq c + di$ if $a \leqslant c$ and $b \leqslant d$. Is $(\mathbb{C}, \trianglelefteq)$ a poset? Is $(\mathbb{C}, \trianglelefteq)$ a lattice?

10. Prove that a lattice $(L, \leqslant)$ is modular if and only if the relation $a \geqslant c$ implies $(a \wedge b) \vee c = a \wedge (b \vee c)$.

11. Verify that the lattice whose covering diagram follows is a modular lattice.

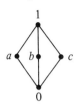

12. Let A, B, C be normal subgroups of a group G. If $A \leqslant C$, verify that $AB \cap C = A(B \cap C)$. This proves that the lattice of normal subgroups of a group is modular.

13. Show that the lattice of subgroups of A_4 (see Exercise 7d of Section 23.1) contains a pentagonal sublattice.

14. Verify that if $(L, \leqslant)$ is a lattice and a, b, $c \in L$ with $a \leqslant c$, then

$$a \vee (b \wedge c) \leqslant (a \vee b) \wedge c$$

15. Suppose $(L, \leqslant)$ is a lattice, a, b, $c \in L$, with $a < c$ and

$$(a \vee b) \wedge c < a \vee (b \wedge c)$$

Prove $a \vee (b \wedge c) \neq a \vee b$.

16. If $(L, \leqslant)$ is a lattice, prove that L is modular if and only if no interval $[a, b]$ for a, $b \in L$ contains an element having two comparable complements in $[a, b]$. (See Exercise 5.)

17. Let $(L, \leqslant)$ be a distributive lattice with a, b, $c \in L$. Prove that

$$a \vee (b \wedge c) = (a \vee b) \wedge (a \vee c)$$

18. Let $(L, \leqslant)$ be a lattice. Prove that $(L, \leqslant)$ is distributive if and only if

$$(a \vee b) \wedge (b \vee c) \wedge (c \vee a) = (a \wedge b) \vee (b \wedge c) \vee (c \wedge a)$$

for all a, b, $c \in L$.

19. If $(L, \leqslant)$ is a lattice, prove that L is distributive if no interval $[a, b]$ for a, $b \in L$ contains an element having two distinct complements.

20. Prove that the positive divisors of the positive integer n form a complemented lattice if and only if n is square-free.

21. Which of the lattices in Exercise 1 are modular? distributive? complemented? uniquely complemented?

22. Is the lattice of subgroups of S_4 a complemented lattice? Do the normal subgroups of S_4 form a complemented lattice?

23. Prove that the subspaces of a real vector space V form a complemented lattice with $0 = \{0\}$ and $1 = V$. Is this lattice modular?

*24. Let $L = \{a_1, a_2, \ldots, a_n\}$ be a lattice. Suppose the relation $\leqslant$ on L is given by a set of ordered pairs of the form (a_i, a_j). Write a program that accepts this relation for $n \leqslant 20$ points and returns the $n \times n$ relation matrix $R = (r_{ij})$, where $r_{ij} = 1$ if $a_i \leqslant a_j$ and $r_{ij} = 0$ if $a_i > a_j$.

*25. Extend the program of Exercise 24 to find the $n \times n$ covering matrix $C = (c_{ij})$, where $c_{ij} = 1$ if a_j covers a_i and $c_{ij} = 0$ if a_j does not cover a_i.

23.3

Boolean Algebras

In 1854, George Boole's *An Investigation into the Laws of Thought* was published. In it, he applied the theory of complemented, distributed lattices to analyze the fundamental structures of logic. He was the first to apply mathematics to the study of logic although a number of distinguished mathematicians, including Augustus DeMorgan and Gottfried Leibnitz, had previously indicated that such an application would be fruitful. In this section, we analyze the lattice structure that carries his name.

Definition 23.11

> A **Boolean algebra** $(B, \vee, \wedge)$ is a complemented, distributive lattice.

It is common practice to use $+$ and $\cdot$, or simply juxtaposition, to indicate the lub symbol $\vee$ and the glb symbol $\wedge$ in a Boolean algebra. For completeness, we list the axioms that govern a Boolean algebra $(B, +, \cdot)$.

B is a set together with two binary operations $+$ and $\cdot$ such that, for all $a, b, c \in B$, the following properties hold:

    i. $a + b = b + a, \quad ab = ba$.

    ii. $a + (b + c) = (a + b) + c, \quad a(bc) = (ab)c$.

    iii. There exist $0, 1 \in B$ such that $a + 0 = a$, $a0 = 0$, $a + 1 = 1$, and $a1 = a$.

    iv. For each $a \in B$, there exists $\bar{a} \in B$ such that $a + \bar{a} = 1$ and $a\bar{a} = 0$.

    v. $a(b + c) = (ab) + (ac), \quad a + (bc) = (a + b)(a + c)$.

    vi. $a + a = a, \quad aa = a$.

    vii. $a + ab = a, \quad a(a + b) = a$.

Except for the existence of inverses for $+$ and $\cdot$, a Boolean algebra satisfies the axioms of a commutative ring. Unlike most rings, however, each operation distributes over the other; in addition, the absorption and idempotency axioms do not hold in general in most rings. However, there are many properties of Boolean algebras that are quite similar to some familiar results from our study of groups and rings. We will see some of these in the following discussion.

Remember, too, that a Boolean algebra is also a poset with a corresponding partial order defined by $a \leqslant b$ if and only if $a + b = b$. By this definition, we have $0 \leqslant a \leqslant 1$ for all $a \in B$. By Theorem 23.9, a Boolean algebra is a uniquely complemented lattice; that is, for each $a \in B$, $\bar{a}$ is unique. Finally, recall that the Duality Principle holds in a Boolean algebra. By the symmetry of the definition, any property derived from the axioms of a Boolean algebra gives rise to a second property obtained by interchanging the operations of $+$ and $\cdot$ and the elements 0 and 1. This new property is also true and can be proved by making the corresponding changes in the proof of the first.

Example 23.15

Let X be a set, and consider $(\mathscr{P}(X), \cup, \cap)$. Then $\mathscr{P}(X)$ is a Boolean algebra; in fact, it is the classic example of a complemented, distributive lattice. For $X = \{a\}$, we obtain the trivial Boolean algebra $B = \{\varnothing, \{a\}\}$ whose covering diagram is given in Figure 23.10. For $X = \{a, b\}$, we obtain a Boolean algebra whose covering diagram is drawn in Figure 23.11. We saw the diagram for $X = \{a, b, c\}$ in Figure 23.2.

$$1 = \{a\}$$
$$0 = \varnothing$$

Figure 23.10

Figure 23.11

Example 23.16

Let B be the set of true-false statements with the operations of $\vee$ and $\wedge$ defined in Section 1.1. Let F represent the statement that is always false and T be the statement that is always true. $(B, \vee, \wedge)$ is a Boolean algebra with zero F and unit T. Many of the axioms for this example were proved in Chapter 1 through the use of truth tables. The verification of the remaining axioms is left to you as Exercise 7.

The next result shows that a type of cancellation holds in a Boolean algebra:

Theorem 23.10

In a Boolean algebra $(B, +, \cdot)$, $a + b = a + c$ and $ab = ac$ imply $b = c$.

Proof

$$b = b + ba = b + ab = b + ac = (b + a)(b + c)$$
$$= (a + b)(b + c) = (a + c)(b + c) = ab + c = ac + c = c \qquad \square$$

This last result did not make use of the existence of complements. It holds in any distributive lattice. Note that the identities $1 + b = 1 + c = 1$ and $0b = 0c = 0$ for any b, $c \in B$ show that neither condition alone implies $b = c$. The cancellation property can be used to prove the uniqueness of complements. (See Exercise 8.)

Duality is utilized in the proof of the next result, which gets its name from one of the founding fathers of set theory.

Theorem 23.11

DeMorgan's Laws

In Boolean algebra $(B, +, \cdot)$, the properties $\overline{(a + b)} = \bar{a}\,\bar{b}$ and $\overline{(ab)} = \bar{a} + \bar{b}$ hold for all $a, b \in B$.

Proof

Since

$$(a + b) + \bar{a}\,\bar{b} = a + (b + \bar{a}\,\bar{b}) = a + (b + \bar{a})(b + \bar{b})$$

$$= a + (b + \bar{a})(1)$$

$$= a + (\bar{a} + b)$$

$$= (a + \bar{a}) + b = 1 + b = 1$$

and

$$(a + b)(\bar{a}\,\bar{b}) = a(\bar{a}\,\bar{b}) + b(\bar{a}\,\bar{b}) = (a\bar{a})\bar{b} + b(\bar{b}\,\bar{a})$$

$$= (a\bar{a})\bar{b} + (b\bar{b})\bar{a}$$

$$= 0\bar{b} + 0\bar{a} = 0 + 0 = 0$$

we see that $\bar{a}\,\bar{b}$ acts as a complement of $a + b$. But complements are unique in a Boolean algebra by Theorem 23.9. Therefore, $\overline{(a + b)} = \bar{a}\,\bar{b}$.

The second law follows by the Duality Principle.   □

DeMorgan's Laws are quite useful in verifying identities in a Boolean algebra:

Example 23.17

If $(B, +, \cdot)$ is a Boolean algebra and $a, b \in B$, then $a = ab$ if and only if $a\bar{b} = 0$. To see this, suppose first that $a = ab$, then

$$a\bar{b} = (ab)\bar{b} = a(b\bar{b}) = a0 = 0$$

Conversely if $a\bar{b} = 0$, then

$$a = a(1) = a(b + \bar{b}) = (ab) + (a\bar{b}) = (ab) + 0 = ab$$

In Section 23.1, we defined what is meant by *poset isomorphism*. We now extend that definition to lattices.

Definition 23.12

Let $(L_1, \vee, \wedge)$ and $(L_2, \cup, \cap)$ be two lattices. A bijection $\phi: L_1 \rightarrow L_2$ is a **lattice isomorphism** if

$$\phi(a \vee b) = \phi(a) \cup \phi(b) \qquad \text{and} \qquad \phi(a \wedge b) = \phi(a) \cap \phi(b)$$

for all $a, b \in L_1$.

In other words, a bijection between lattices is an isomorphism if it preserves lub's and glb's. Every lattice isomorphism is an isomorphism of posets; in fact, if two lattices are order-isomorphic as posets, then they are isomorphic as lattices. (See Exercise 16.) When working with lattices, however, you will probably find it easier to utilize Definition 23.12.

We now give an interesting application of a lattice isomorphism to provide another example of a Boolean algebra:

Example 23.18

Let X be a set and $B = \{f: X \to \{0, 1\}\}$. For $f, g \in B$, define

$$(f + g)(x) = \max\{f(x), g(x)\} \quad \text{and} \quad (fg)(x) = \min\{f(x), g(x)\}$$

Then $(B, +, \cdot)$ is a Boolean algebra. To see this, let $f \in B$ and $S_f = \{x \in X \mid f(x) = 1\}$. S_f is termed the **support** of f. It is a straightforward exercise to prove that

$$S_{f+g} = S_f \cup S_g \quad \text{and} \quad S_{fg} = S_f \cap S_g$$

(See Exercise 12.) Then the mapping $\phi: B \to \mathscr{P}(X)$ defined by $\phi(f) = S_f$ is a bijection that preserves the Boolean algebra operations. Since $(B, +, \cdot)$ is isomorphic to the Boolean algebra $(\mathscr{P}(X), \cup, \cap)$, B is a Boolean algebra.

Recall that if $(L, \leqslant)$ is a bounded lattice, then the elements of L that cover 0 are termed the atoms of L. Elements in a finite Boolean algebra can be expressed uniquely in terms of atoms. To prove this requires a lemma.

Lemma 23.3

If $(B, +, \cdot)$ is a Boolean algebra and a_1, a_2 are distinct atoms of B, then $a_1 a_2 = 0$.

Proof

By definition of glb, $a_1 a_2 \leqslant a_2$. If $a_1 a_2 = a_2$, then $a_2 \leqslant a_1$. Since a_1 is an atom and $a_2 > 0$, we conclude that $a_1 = a_2$ contrary to the choice of a_2. Thus, $a_1 a_2 < a_2$. Again the fact that a_2 is an atom forces $a_1 a_2 = 0$. □

Theorem 23.12

Let $(B, +, \cdot)$ be a finite Boolean algebra. Every $b \in B$ can be written as the sum of atoms. The atoms that appear in such a sum are unique up to rearrangement.

Proof

We first prove that such a decomposition of $b \in B$ into a sum of atoms exists. If b is an atom, there is nothing to prove. If b is not an atom, then there exists an atom $a_1 < b$ such that $b = a_1 + (\bar{a}_1 b)$, since $a_1 < b$ implies

$$a_1 + \bar{a}_1 b = (a_1 + \bar{a}_1)(a_1 + b) = 1(a_1 + b) = b$$

If $\bar{a}_1 b$ is an atom, we are done. Otherwise, there exists an atom $a_2 \in B$ such that $a_2 \neq a_1$ and $\bar{a}_1 b = a_2 + (\bar{a}_1 b)\bar{a}_2$. Then

$$b = a_1 + a_2 + \bar{a}_1 \bar{a}_2 b$$

If $\bar{a}_1 \bar{a}_2 b$ is an atom, our proof of existence is finished; otherwise, we can repeat our argument again and again. Since L is finite, there exists but a finite number of atoms. Therefore, this procedure must terminate, and b can be written as a sum of atoms.

To prove uniqueness, assume

$$a = a_1 + a_2 + \cdots + a_s = b_1 + b_2 + \cdots + b_t$$

where the a_i's are distinct atoms of B, as are the b_j's. We must show that $s = t$ and that there exists $\sigma \in S_t$ such that $a_{\sigma(i)} = b_i$. We proceed by induction on t. If $t = 1$, then a is an atom and $a = a_1$. Assume that the result is true for such decompositions into a sum of less than t atoms. Since

$$b_1 \leqslant b_1 + b_2 + \cdots + b_t = a = a_1 + a_2 + \cdots + a_s$$

we have

$$b_1 = b_1 a = b_1(a_1 + a_2 + \cdots + a_s) = b_1 a_1 + b_1 a_2 + \cdots + b_1 a_s$$

Then $b_1 a_i \leqslant b_1$ for all $i = 1, 2, \ldots, s$. By Lemma 23.3, there exists a_k such that $b_1 a_k = b_1$ and $b_1 a_i = 0$ for all $i \neq k$. But $b_1 a_k = b_1$ implies $b_1 \leqslant a_k$, so that $b_1 = a_k$ since a_k is an atom. This says

$$a_k + (a_1 + \cdots + a_{k-1} + a_{k+1} + \cdots + a_s) = b_1 + (b_2 + b_3 + \cdots + b_k)$$

Since a_k and b_1 are atoms,

$$a_k(a_1 + \cdots + a_{k-1} + a_{k+1} + \cdots + a_s) = b_1(b_2 + b_3 + \cdots + b_k)$$

By cancellation (Theorem 23.10), we have

$$a_1 + \cdots + a_{k-1} + a_{k+1} + \cdots + a_s = b_2 + b_3 + \cdots + b_t$$

The result now follows by induction.   □

Note the similarity between the statement of this theorem and that of the Fundamental Theorem of Arithmetic (Corollary 16.3). If the Boolean algebra B in Theorem 23.12 is $\mathscr{P}(X)$ for some finite set X, then the atoms of B are simply the singleton subsets of X. This theorem then states that every subset of X can be written uniquely as a union of singleton sets. The next result shows how this example essentially determines all finite Boolean algebras.

| Theorem 23.13 |

If $(B, +, \cdot)$ is a finite Boolean algebra, there exists a set X such that $(B, +, \cdot)$ is lattice isomorphic to $(\mathscr{P}(X), \cup, \cap)$.

Proof

Let $X = \{a_1, a_2, \ldots, a_n\}$ be the set of distinct atoms of B. For each $b \in B$, we know by Theorem 23.12 that there is a unique set of atoms $\{a_{i_1}, a_{i_2}, \ldots, a_{i_m}\}$ such that

$$b = a_{i_1} + a_{i_2} + \cdots + a_{i_m}$$

Define $\phi: B \to \mathscr{P}(X)$ by $\phi(b) = \{a_{i_1}, a_{i_2}, \ldots, a_{i_m}\}$. Since the representation of an element of B as such a sum is unique, ϕ is clearly surjective and injective. It remains to show that if $b_1, b_2 \in B$, then

$$\phi(b_1 + b_2) = \phi(b_1) \cup \phi(b_2) \qquad \text{and} \qquad \phi(b_1 b_2) = \phi(b_1) \cap \phi(b_2)$$

Suppose $\phi(b_1) \cap \phi(b_2) = \{c_1, c_2, \ldots, c_t\}$, where $c_1, c_2, \ldots, c_t$ are atoms of B. Then

$$b_1 = c_1 + c_2 + \cdots + c_t + d_1 + d_2 + \cdots + d_r$$

and

$$b_2 = c_1 + c_2 + \cdots + c_t + e_1 + e_2 + \cdots + e_s$$

where the d_i's and e_j's are atoms in X. But by the idempotency rule,

$$
\begin{aligned}
b_1 + b_2 &= (c_1 + \cdots + c_t) + (d_1 + \cdots + d_r) \\
&\quad + (c_1 + \cdots + c_t) + (e_1 + \cdots + e_s) \\
&= c_1 + \cdots + c_t + d_1 + \cdots + d_r + e_1 + \cdots + e_s
\end{aligned}
$$

so that

$$
\begin{aligned}
\phi(b_1 + b_2) &= \{c_1, c_2, \ldots, c_t, d_1, d_2, \ldots, d_r, e_1, e_2, \ldots, e_s\} \\
&= \phi(b_1) \cup \phi(b_2)
\end{aligned}
$$

By repeated use of Lemma 23.1 and other lattice axioms, we have

$$\phi(b_1 b_2) = c_1 + c_2 + \cdots + c_t = \phi(b_1) \cap \phi(b_2)$$

Thus ϕ is a lattice isomorphism. $\square$

If X contains n elements, we know that X has 2^n subsets. This enables us to determine the size of all finite Boolean algebras:

Corollary 23.1   If $(B, +, \cdot)$ is a finite Boolean algebra, then $|B| = 2^n$ for some $n > 0$.

In our next section, we show how the atoms in a Boolean algebra have a special form, and we develop a procedure to determine exactly what atoms are contained in the representation guaranteed in Theorem 23.12.

1. Find all lattices on four or five elements. Which ones are Boolean algebras?

2. A ring $(B, +, \cdot)$ is **Boolean** if every $a \in B$ satisfies the identity $a^2 = a$.
   a. Prove that any Boolean ring $(B, +, \cdot)$ is a Boolean algebra.
   b. Prove that any Boolean algebra $(B, \vee, \wedge)$ is a Boolean ring.

3. Is the lattice of normal subgroups of a group a Boolean algebra?

4. Do the ideals in a principal ideal domain form a Boolean algebra under the relation $\subseteq$?

5. If $(B, +, \cdot)$ is a Boolean algebra with $a, b \in B$, show that:
   a. $(\bar{a}) = a$ b. $a + \bar{a}b = b + \bar{b}a$
   c. $a + b = b$ if and only if $\bar{a} + b = 1$ d. $(a + b)(\overline{ab}) = (a\bar{b}) + (\bar{a}b)$

6. The set of true-false statements forms a Boolean algebra with unit T and zero F under $\vee$, $\wedge$, and $\sim$. Express $p \to q$ in terms of these operations.

7. Verify that if p, q, and r represent true-false statements, then
   a. $(p \wedge \sim p) \leftrightarrow F$, $(p \vee \sim p) \leftrightarrow T$
   b. $[p \vee (p \wedge q)] \leftrightarrow p$, $[p \wedge (p \vee q)] \leftrightarrow p$
   c. $[p \wedge (q \vee r)] \leftrightarrow [(p \wedge q) \vee (p \wedge r)]$, $[p \vee (q \wedge r)] \leftrightarrow [(p \vee q) \wedge (p \vee r)]$
   d. $[p \wedge (q \wedge r)] \leftrightarrow [(p \wedge q) \wedge r]$, $[p \vee (q \vee r)] \leftrightarrow [(p \vee q) \vee r]$
   e. $(p \wedge q) \leftrightarrow (q \wedge p)$, $(p \vee q) \leftrightarrow (q \vee p)$

8. Use Theorem 23.9 to prove that complements are unique in a Boolean algebra.

9. Prove the generalized DeMorgan's Laws: if $(B, +, \cdot)$ is a Boolean algebra and $\{a_1, a_2, \dots, a_n\} \subseteq B$, then

$$\overline{\left(\sum_{i=1}^{n} a_i \right)} = \prod_{i=1}^{n} \bar{a}_i \quad \text{and} \quad \overline{\left(\prod_{i=1}^{n} a_i \right)} = \sum_{i=1}^{n} \bar{a}_i$$

10. Show that each of DeMorgan's Laws implies the other.

11. Prove that the lattice B^n of Example 23.10 is a Boolean algebra.

12. In Example 23.18, verify

$$S_{f+g} = S_f \cup S_g \quad \text{and} \quad S_{fg} = S_f \cap S_g$$

13. Draw covering diagrams for a Boolean algebra with 8 elements and with 16 elements.

14. Prove that any element in a finite Boolean algebra can be written uniquely as a product of coatoms.

15. Suppose B_1 and B_2 are Boolean algebras with the same finite number of elements. Prove that B_1 is isomorphic to B_2.

16. Suppose L_1 and L_2 are lattices that are isomorphic as posets. Prove that L_1 and L_2 are isomorphic as lattices.

17. Give an example of a lattice L in which $\bar{\bar{a}} = a$ does not hold for all $a \in L$.

18. Suppose B is set together with two commutative binary operations $+$ and $\cdot$ such that, for all $a, b, c \in B$,
   a. $a + (b \cdot c) = (a + b) \cdot (a + c)$, $a \cdot (b + c) = (a \cdot b) + (a \cdot c)$
   b. There exist $0, 1 \in B$ such that $a + 0 = a$, $a \cdot 1 = a$.
   c. For each $a \in B$, there exists $\bar{a} \in B$ such that $a + \bar{a} = 1$, and $a \cdot \bar{a} = 0$.
   Prove that $(B, +, \cdot)$ is a Boolean algebra.

23.4

Normal Form

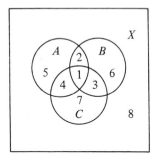

Figure 23.12

Consider the sets A, B, C pictured in Figure 23.12. Certain regions or **cells** are labelled in this figure. Let us use $\bar{A}$, $\bar{B}$, and $\bar{C}$ to indicate $X - A$, $X - B$, and $X - C$, respectively. The cells are:

1. $A \cap B \cap C$	2. $A \cap B \cap \bar{C}$
3. $\bar{A} \cap B \cap C$	4. $A \cap \bar{B} \cap C$
5. $A \cap \bar{B} \cap \bar{C}$	6. $\bar{A} \cap B \cap \bar{C}$
7. $\bar{A} \cap \bar{B} \cap C$	8. $\bar{A} \cap \bar{B} \cap \bar{C}$

Any expression in A, B, or C can be written uniquely as the union of one or more of these cells. For example, the set

$$A \cap (B \cup (A - C)) = \{1, 2, 5\}$$

$$= (A \cap B \cap C) \cup (A \cap B \cap \bar{C}) \cup (A \cap \bar{B} \cap \bar{C})$$

If A, B, C represent arbitrary sets in the Boolean algebra $(\mathscr{P}(X), \cup, \cap)$, then the set of all expressions involving A, B, and C forms a Boolean algebra whose atoms are the cells just listed. Therefore, describing our set as a union of cells is equivalent to writing an element in this Boolean algebra as a sum of atoms.

In this section, we will give a detailed analysis of expressions in a Boolean algebra. We will investigate a natural way to write arbitrary expressions in a systematic way to simplify the identification of equivalent expressions. The discussion accompanying Figure 23.12 will serve as a guide to the type of *canonical* or *normal form* we seek for a Boolean expression. We will then apply our methods to the algebra of logical expressions and the theory of switching circuits that we first encountered in Chapter 1.

We begin with the following definition:

Definition 23.13

$B(n)$, the set of **Boolean expressions** or **Boolean forms** in the variables $x_1, x_2, \ldots, x_n$, is defined by

i. $x_i \in B(n)$ for $i = 1, 2, \ldots, n$;
ii. $0, 1 \in B(n)$; and
iii. if $f, g \in B(n)$, then $f + g$, fg, and $\bar{f} \in B(n)$.

Boolean expressions are those that can be formed by applying a finite sequence of operations corresponding to addition, multiplication, and complementation on the set of elements $\{0, 1, x_1, x_2, \ldots, x_n\}$. For example,

$$\overline{((x_1 + x_2)(x_2 + x_3) + x_2)}$$

is a Boolean expression in $B(3)$. Boolean expressions can take different forms depending upon the type of Boolean algebra under consideration. For the algebra of logical expressions,

$$\sim (p \wedge (q \vee (\sim r)) \vee (q \wedge (\sim p)))$$

is a Boolean expression. In the case where we are working with the Boolean algebra of subsets of X, then our original example $A \cap (B \cup (A - C))$ is a Boolean expression. We want to simplify arbitrary Boolean expressions in the same way that we did with our first example. This will enable us to determine when two Boolean expressions are **equal**, that is, when one can be obtained from the other by a finite number of applications of the axioms of a Boolean algebra.

In $B(n)$, the expressions of the form x_i or $\overline{x_i}$ are known as **literals**. A literal or a product of two or more literals in which no two literals contain the same variable is termed a **fundamental product**. For example $x_1 \overline{x_2}$ and $\overline{x_1} x_2 x_3$ are fundamental products, but $x_1 x_2 \overline{x_1} x_3$ is not. A fundamental product in $B(n)$, in which all the variables are involved either as x_i or as its complement $\overline{x_i}$, is called a **complete fundamental product** or a **minterm**. The product $x_1 \overline{x_2} x_3$ is a minterm in $B(3)$, but $x_1 \bar{x}_3$ is not.

Definition 23.14

> A Boolean expression $f \in B(n)$ is in **disjunctive normal form (dnf)** if f is a fundamental product or a sum of two or more fundamental products in which no summand is contained in or can be absorbed into any other. If f is in dnf and each summand is a minterm, then f is said to be in **full disjunctive normal form (full dnf)**.

The expressions $x_1 x_2 \bar{x}_3$ and $x_1 + x_2 x_3$ are in dnf, but the expression $x_1 \bar{x}_2 x_3 + x_1 x_3 + x_1 x_2$ is not, since

$$x_1 \bar{x}_2 x_3 + x_1 x_3 + x_1 x_2 = x_1 (\bar{x}_2 x_3 + x_3) + x_1 x_2 = x_1 x_3 + x_1 x_2$$

If a Boolean expression f is not in dnf, then the axioms of a Boolean algebra can be used to find an equivalent expression of f that is in dnf.

Example 23.19

We put $f = \overline{(x_1 \bar{x}_3)} + \bar{x}_1 x_2$ in dnf. By the DeMorgan laws and the fact that $\bar{\bar{x}} = x$ in a Boolean algebra, we obtain

$$f = \overline{(x_1 \bar{x}_3)} + \bar{x}_1 \bar{x}_2 = \bar{x}_1 + \bar{\bar{x}}_3 + \bar{x}_1 \bar{x}_2 = \bar{x}_1 + x_3 + \bar{x}_1 \bar{x}_2$$

Then

$$f = (\bar{x}_1 + \overline{x}_1 \overline{x}_2) + x_3 = \overline{x}_1 + x_3$$

follows from the commutative, associative, and absorption laws.

While the final expression in Example 23.19 is in dnf, it is not in full dnf, since neither summand involves all three variables. An expression in dnf can be put into full dnf by using the fact that $1 = x_i + \bar{x}_i$.

Example 23.20

Put $f = \bar{x}_i + x_3$ in full dnf.

$$f = \bar{x}_1 + x_3$$

$$= \bar{x}_1(x_2 + \bar{x}_2)(x_3 + \bar{x}_3) + (x_1 + \bar{x}_1)(x_2 + \bar{x}_2)x_3$$

$$= \bar{x}_1 x_2 x_3 + \bar{x}_1 \bar{x}_2 x_3 + \bar{x}_1 x_2 \bar{x}_3 + \bar{x}_1 \bar{x}_2 \bar{x}_3$$

$$+ x_1 x_2 x_3 + x_1 \bar{x}_2 x_3 + \bar{x}_1 x_2 x_3 + \bar{x}_1 \bar{x}_2 x_3$$

$$= \bar{x}_1 x_2 x_3 + \bar{x}_1 \bar{x}_2 x_3 + \bar{x}_1 x_2 \bar{x}_3 + \bar{x}_1 \bar{x}_2 \bar{x}_3 + x_1 x_2 x_3 + x_1 \bar{x}_2 x_3$$

Note the similarity between the final form of f in this example and the decomposition of the subset of X written in terms of A, B, and C in our introductory example. What we have done in these examples holds in general:

Theorem 23.14

Every nonzero Boolean expression can be written in full disjunctive normal form. The minterms that appear in such a representation are unique.

Proof

We first show that if $f \in B(n)$, then f can be written as a sum of minterms. We then prove that the representation is unique.

To prove existence, we follow the procedure outlined in Example 23.20. Use DeMorgan's Laws and involution (the fact that $\bar{\bar{x}} = x$) to transform f so that complements apply to literals. Next use the distributive laws to write f as a sum of products. The commutative, associative, idempotency, and absorption laws can then be applied to modify f so that no product in f is part of any other. At this point, f is in dnf. If any fundamental product in f is not a minterm, use the identities $1 = x_1 + \bar{x}_1 = x_2 + \bar{x}_2$, etc., and the distributive laws again to see that each fundamental product in f is a minterm. Finally, combine together any duplicate minterms through the use of the idempotency, commutative, and associative laws. The expression f is now in full dnf.

To show uniqueness, suppose that $f = p_1 + p_2 + \cdots + p_r$ and $f = q_1 + q_2 + \cdots + q_s$ are two distinct representations of f as sums of minterms. Then some p_i, say p_1, is not equal to any q_j, so that $p_1 q_j = 0$ for $j = 1, 2, \ldots, s$. (See Exercise 5.) But then $p_1 = p_1 f$ by Lemma 23.2 and

$$p_1 = p_1 f = p_1(q_1 + q_2 + \cdots + q_s)$$

$$= p_1 q_1 + p_1 q_2 + \cdots + p_1 q_s$$

$$= 0 + 0 + \cdots + 0 = 0$$

contrary to the choice of p_1. Uniqueness follows.   □

The atoms of $B(n)$ are its minterms. (See Exercise 6.) If we know that there are but a finite number of expressions in $B(n)$, we could obtain the proof of Theorem 23.14 by appealing to Theorem 23.12. But there are an infinite number of (unsimplified) Boolean expressions in $B(n)$ (see Exercise 7), and Theorem 23.12 does not immediately apply. Only after we have proved Theorem 23.14 can we learn that there exist but a finite number of Boolean expressions in full dnf.

Corollary 23.2

Every Boolean expression in $B(n)$ is equal to one of a finite number of expressions in full dnf.

We now show how Theorem 23.14 can be applied to some of our more familiar examples of Boolean algebras. In set theory, it is enough to show that two subsets of a set have the same disjunctive normal form to conclude that they are equal.

Example 23.21

Verify $(A \cup B) \cap (\bar{A} \cup C) = (A \cap C) \cup (\bar{A} \cap B)$.

$$(A \cup B) \cap (\bar{A} \cup C) = [(A \cup B) \cap \bar{A}] \cup [(A \cup B) \cap C]$$
$$= [(A \cap \bar{A}) \cup (B \cap \bar{A})] \cup [(A \cap C) \cup (B \cap C)]$$
$$= (\varnothing \cup (\bar{A} \cap B)) \cup (A \cap C) \cup (B \cap C)$$
$$= (\bar{A} \cap B) \cup (A \cap C) \cup (B \cap C)$$

This expression is in dnf, as is the term on the right side of the statement we are attempting to verify. We can complete the proof if we can show both expressions have the same sum of minterm representation. We leave it to you to show in Exercise 12 that both sides of the original statement can be expressed as

$$(A \cap B \cap C) \cup (A \cap \bar{B} \cap C) \cup (\bar{A} \cap B \cap C) \cup (\bar{A} \cap B \cap \bar{C})$$

If we are working with the Boolean algebra of TF-statements, then showing that both sides of a biconditional statement have the same full dnf is equivalent to showing that the statement is a tautology. When we put a logical expression in full dnf, then, by examining the minterms present, we can determine what truth values of the variables cause the statement to be true. In particular, an expression involving n statements will be a tautology when the full dnf representation of the expression contains each and every one of the 2^n possible minterms. (See Exercise 13.)

Example 23.22

Determine under what conditions the statement $[(\sim p) \wedge (q \vee p)] \vee (\sim q)$ is true.

$$[(\sim p) \wedge (q \vee p)] \vee (\sim q) = ([(\sim p) \wedge q] \vee [(\sim p) \wedge p]) \vee (\sim q)$$
$$= [(\sim p) \wedge q] \vee \mathrm{F} \vee (\sim q)$$
$$= [(\sim p) \wedge q] \vee (\sim q)$$
$$= [(\sim p) \wedge q] \vee [(\sim q) \wedge \mathrm{T}]$$
$$= [(\sim p) \wedge q] \vee [(\sim q) \wedge (p \vee (\sim p))]$$
$$= [(\sim p) \wedge q] \vee ([(\sim q) \wedge p] \vee [(\sim q) \wedge (\sim p)])$$
$$= [(\sim p) \wedge q] \vee [(\sim q) \wedge p] \vee [(\sim q) \wedge (\sim p)]$$

From this representation, we see that the statement is false only when both p and q are true.

In Chapter 1, we introduced switching functions. We said then that a switching function $f: (\mathbb{Z}_2)^n \to \mathbb{Z}_2$ can always be constructed from AND-, OR-, and NOT-gates. Why is this true? Well, first note that the set of all switching functions of n-variables forms a Boolean algebra isomorphic to $(\mathscr{P}(X), \cup, \cap)$ for $X = \{1, 2, \ldots, n\}$. Secondly, the fact that every Boolean expression has a full dnf representation says that any switching function can be constructed from n-input AND-gates and a multiple-input OR-gate, with NOT-gates used to obtain complements of the variables. Finally, any multiple-input AND-gate or OR-gate can be implemented from two-input gates of the type discussed in Chapter 1. (See Exercise 15.)

In practice, switching functions are seldom constructed from their full dnf representation. Instead, switching functions are described as Boolean expressions and the properties of Boolean algebra are used to simplify the construction.

Example 23.23

Let us design circuitry to implement the switching circuit $f(x, y, z) = (x(y + z) + \bar{x}\,\bar{y})\bar{z}$. Simplifying, we see

$$f = (x(y + z) + \bar{x}\,\bar{y})\bar{z} = (xy + xz + \bar{x}\,\bar{y})\bar{z}$$
$$= xy\bar{z} + xz\bar{z} + \bar{x}\,\bar{y}\,\bar{z} = xy\bar{z} + x(0) + \bar{x}\,\bar{y}\,\bar{z} = xy\bar{z} + \bar{x}\,\bar{y}\,\bar{z}$$

Therefore, f can be implemented by two 3-input AND-gates, four NOT-gates, and one 2-input OR-gate. This circuit is given in Figure 23.13.

A number of excellent texts have been written detailing several methods to design, implement, and simplify circuitry. We list some in the references.

We have shown how Boolean algebras, and, in fact, the logical procedures developed throughout this text have far-reaching applications, especially in the area of computer science. It is here that the greatest number of new applications will probably be developed, but these advances will require a solid understanding of the fundamental structures of algebra. Our

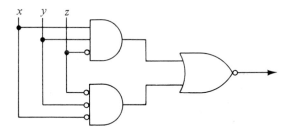

Figure 23.13

hope is that this text has given you the basics necessary for additional study in the theory and application of abstract algebra.

1. Write an expression involving A, B, C that represents the region in Figure 23.10 corresponding to
   a. $\{1, 7\}$ b. X c. $\{2, 3, 4\}$ d. $\{1, 2, 4, 5\}$

2. Give the full disjunctive normal form representation for each of the following elements in $B(3)$:
   a. $\overline{x_1 \bar{x}_3 + x_2}$
   b. $(x_1 + \bar{x}_3)(x_1 \bar{x}_2 + x_1 x_3)$
   c. $(x_1 x_2 + x_2 x_3) + x_1 \bar{x}_2 x_3$
   d. $(x_1 + \bar{x}_2 + x_3)(x_1 x_2 + x_1 x_3)$

3. In the Boolean algebra of logical expressions, put each of the following in full dnf:
   a. $p \to q$
   b. $p \leftrightarrow q$
   c. $\sim(p \wedge q)$
   d. $[(\sim p) \to r] \wedge (q \leftrightarrow p)$

4. Show that the following are tautologies in the algebra of logical expressions:
   a. $(p \to q) \leftrightarrow [(\sim q) \to (\sim p)]$ b. $q \vee [p \wedge (\sim q)] \vee [(\sim p) \wedge (\sim q)]$
   c. $[p \wedge (p \to q)] \to q$ d. $(\sim p) \to (p \to q)$

5. Suppose p and q are two distinct minterms in $B(n)$. Prove $pq = 0$.

6. Verify that the minterms in $B(n)$ are its atoms.

7. Show that $B(n)$ contains an infinite number of unsimplified Boolean expressions.

8. Prove that $B(n)$ has 2^n minterms and $2^{(2^n)}$ distinct elements.

9. $B(n)$, the lattice of Example 23.10, was shown to be a Boolean algebra in Exercise 11 of Section 23.3. Prove that the minterms of $B(n)$ are simply the words of weight one.

10. Determine the validity of the following statement: If f, $g \in B(n)$ and f and g possess distinct disjunctive normal forms, then f and g are not equal.

11. Verify that the following identities hold in any Boolean algebra:
    a. $ac + \bar{a}b + \bar{a}\,\bar{b} + a\bar{c} = 1$ b. $a + (ab)(ac) = 1$
    c. $a + bc = ab + ac + bc$ d. $ac + \bar{a}b = (a + b)(\bar{a} + c)$

12. Verify that both

    $$(A \cap C) \cup (\bar{A} \cap B) \qquad \text{and} \qquad (\bar{A} \cap B) \cup (A \cap C) \cup (B \cap C)$$

    have the same sum of minterms representation.

13. Prove that a logical expression in n variables is a tautology if and only if its full dnf representation involves all 2^n minterms.

14. An expression m in $B(n)$ is a **maxterm** if m is a sum of literals in which each variable or its complement appear once. For example, both $x_1 + \bar{x}_2 + x_3$ and $\bar{x}_1 + \bar{x}_2 + x_3$ are maxterms in $B(3)$.
    a. Prove that the maxterms of $B(n)$ are the complements of the minterms of $B(n)$.
    b. Show that if $f \in B(n)$, then f can be written uniquely as a product of maxterms. (This is known as **conjunctive normal form**.)
    c. Write each expression in Exercise 2 as a product of maxterms.

15. Show how an m-input AND-gate can be constructed from 2-input AND-gates.

16. Design circuitry to implement the following switching circuits:
    a. $f = \overline{(\bar{x} + \bar{y}z)}$ b. $g = \overline{xy\bar{z} + \bar{x}(y+z)}$
    c. $h = xy + \overline{x(y+z)}$

17. The switching circuit pictured in the following figure is known as a NOR-gate.

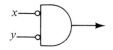

    a. Show that 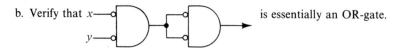 is equivalent to a NOT-gate.

    b. Verify that is essentially an OR-gate.

    c. Construct an AND-gate from three NOR-gates.
    d. Prove that any switching circuit can be constructed from NOR-gates alone.

*18. Write a program to enumerate all 2^8 distinct Boolean expressions in the set of variables x, y, and z.

Selected Answers

Chapter 1

Section 1.1, page 3

1. $T(F \rightarrow F)$
4. $T(F \leftrightarrow F)$
9. F, use $(T \rightarrow F)$ since if r were rational, then r^2 would be rational.

Section 1.2, page 7

1. b. Converse: If mathematics is fun, then $1 + 1 = 2$.
   Contrapositive: If mathematics is not fun, then $1 + 1 \neq 2$.
   Inverse: If $1 + 1 \neq 2$, then mathematics is not fun.

3. e. $(p \rightarrow q) \leftrightarrow [(\sim p) \vee q]$:

p	q	$p \rightarrow q$	$\sim p$	$\sim p \vee q$	$(p \rightarrow q) \leftrightarrow [(\sim p) \vee q]$
T	T	T	F	T	T
T	F	F	F	F	T
F	T	T	T	T	T
F	F	T	T	T	T

i. $[\sim(p \rightarrow q)] \leftrightarrow [p \wedge (\sim q)]$:

p	q	$p \rightarrow q$	$\sim q$	$p \wedge (\sim q)$	$\sim(p \rightarrow q)$	$[\sim(p \rightarrow q)] \leftrightarrow [p \wedge (\sim q)]$
T	T	T	F	F	F	T
T	F	F	T	T	T	T
F	T	T	F	F	F	T
F	F	T	T	F	F	T

Section 1.3, page 13

1. b. $(D \cup C) \cup A = \{1, 2, 3, 4, 5, 6, 7, 9, 10\}$ g. $A - C = \varnothing$
   h. $B - C = \{2, 7, 8\}$

4. Theorem 1.2: To prove that for any set A, $\emptyset \subseteq A$, prove the contrapositive; that is, prove that if $x \notin A$ then $x \notin \emptyset$.
Corollary 1.1: For any set A, $\emptyset \subseteq A$. If $A = \emptyset$, then $\emptyset = \emptyset$ so the empty set is not a proper subset of itself. If $A \neq \emptyset$, then there is $x \in A - \emptyset$ and $x \notin \emptyset$ so $\emptyset$ is a proper subset of A.

5. Suppose $\emptyset$ and $\emptyset'$ are distinct empty sets and show $\emptyset = \emptyset'$.

6. To prove equality, take an element on one side and show that it is also an element of the other side; then reverse the process.
Proof of Theorem 1.3c: $A \cap (B \cup C) = (A \cap B) \cup (A \cap C)$. If $x \in A \cap (B \cup C)$,

then $x \in A$ and $x \in B \cup C$;

then $x \in A$ and $(x \in B$ or $x \in C)$;

then $(x \in A$ and $x \in B)$ or $(x \in A$ and $x \in C)$;

then $x \in A \cap B$ or $x \in A \cap C$;

then $x \in (A \cap B) \cup (A \cap C)$

Thus $A \cap (B \cup C) \subseteq (A \cap B) \cup (A \cap C)$. Now you need to show that if $x \in (A \cap B) \cup (A \cap C)$, then $x \in A \cap (B \cup C)$.

8. a. Use the tautology $[\sim(p \wedge q)] \leftrightarrow [(\sim p) \vee (\sim q)]$.

9. a. Let $A = \{1, 2, 3\}$. The five partitions are

$$\{\{1\},\{2, 3\}\}, \qquad \{\{1\}, \{2\}, \{3\}\}, \qquad \{\{2\}, \{1, 3\}\},$$
$$\{\{3\}, \{1, 2\}\}, \qquad \{1, 2, 3\}$$

10. Yes.

11. a. You need to show $x \in A \triangle B$ if and only if $x \in B \triangle A$.

$x \in A \triangle B$ $\qquad$ if and only if $x \in (A - B) \cup (B - A)$

$\qquad\qquad\qquad$ if and only if $x \in (B - A) \cup (A - B)$

$\qquad\qquad\qquad$ if and only if $x \in B \triangle A$

Section 1.4, page 20

1. c. $861 = 3 \cdot 7 \cdot 41$

2. a. $(84, 16) = 4$ since $84 = 4 \cdot 21$, and $16 = 4 \cdot 4$ and $(21, 4) = 1$.

4. a. When $n = 1$,

$$1 = \frac{1(1 + 1)(2 \cdot 1 + 1)}{6} = \frac{1 \cdot 2 \cdot 3}{6}$$

is true. Assume $P(k)$ is true; that is, we suppose that for some integer k,

$$1^2 + 2^2 + \cdots + k^2 = \frac{k(k + 1)(2k + 1)}{6}$$

Then

$$1^2 + \cdots + k^2 + (k + 1)^2 = [1^2 + \cdots + k^2] + (k + 1)^2$$

$$= \frac{k(k+1)(2k+1)}{6} + (k+1)^2$$

$$= \frac{k(k+1)(2k+1) + 6(k+1)^2}{6}$$

$$= \frac{(k+1)[k(2k+1) + 6(k+1)]}{6}$$

$$= \frac{(k+1)(2k^2 + 7k + 6)}{6}$$

$$= \frac{(k+1)(k+2)(2k+3)}{6}$$

$$= \frac{(k+1)[(k+1)+1][2(k+1)+1]}{6}$$

But then $P(k+1)$ is true.

7. If $U \subseteq \mathbb{N}$ and U does not have a least element, let $P(n)$ be the statement "$n \notin U$." Then show $U = \varnothing$.

8. No, $\mathbb{Z}$ doesn't have a smallest integer. Yes, consider the set of absolute values. It has a smallest number so the negative of it is the largest number.

10. a. Set $U = \{n | P(n) \text{ is false}\}$.

Section 1.5, page 25

1. d. State Diagram: $\overline{\overline{xy} + x\bar{y}}$
    Truth Table:

x	y	$\bar{y}$	xy	$\overline{xy}$	$x\bar{y}$	$\overline{xy} + x\bar{y}$	$\overline{\overline{xy} + x\bar{y}}$
1	1	0	1	0	0	0	1
1	0	1	0	1	1	1	0
0	1	0	0	1	0	1	0
0	0	1	0	1	0	1	0

2. c. Left Side State Diagram: $(x + y)(\bar{x} + z)(y + z)$
    Left Side Truth Table:

x	y	z	$\bar{x}$	$x + y$	$\bar{x} + z$	$(x + y)(\bar{x} + z)$	$y + z$	$(x + y)(x + z)(y + z)$
1	1	1	0	1	1	1	1	1
1	1	0	0	1	0	0	1	0
1	0	1	0	1	1	1	1	1
1	0	0	0	1	0	0	0	0
0	1	1	1	1	1	1	1	1
0	1	0	1	1	1	1	1	1
0	0	1	1	0	1	0	1	0
0	0	0	1	0	1	0	0	0

Right Side State Diagram: $(x + y)(\bar{x} + z)$
Right Side Truth Table:

x	y	z	$x + y$	$\bar{x}$	$\bar{x} + z$	$(x + y)(\bar{x} + z)$
1	1	1	1	0	1	1
1	1	0	1	0	0	0
1	0	1	1	0	1	1
1	0	0	1	0	0	0
0	1	1	1	1	1	1
0	1	0	1	1	1	1
0	0	1	0	1	1	0
0	0	0	0	1	1	0

3. b.

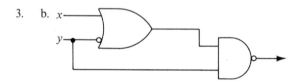

Chapter 2

Section 2.1, page 32

2. b. $\{x | x \geq 0\}$
7. b. $y \in f^{\rightarrow}(A_1 \cap A_2)$ implies there is $x \in A_1 \cap A_2$ such that $f(x) = y$. But then $x \in A_1$ and $x \in A_2$ and $f(x) = y$, so that $y \in f^{\rightarrow}(A_1)$ and $y \in f^{\rightarrow}(A_2)$. Therefore $y \in f^{\rightarrow}(A_1) \cap f^{\rightarrow}(A_2)$.

Section 2.2, page 37

1. a. not an operation on $\mathbb{N}$, since $2 * 3 = -1$
    c. not an operation on $\mathbb{N}$, $\mathbb{Z}$, or $\mathbb{Q}$

4. b. $\begin{bmatrix} 2 & 3 \\ 3 & 4 \end{bmatrix} + \begin{bmatrix} 1 & 2 \\ 3 & -4 \end{bmatrix} = \begin{bmatrix} 3 & 5 \\ 6 & 0 \end{bmatrix}$, $\begin{bmatrix} 2 & 3 \\ 3 & 4 \end{bmatrix} \times \begin{bmatrix} 1 & 2 \\ 3 & -4 \end{bmatrix} = \begin{bmatrix} 11 & -8 \\ 15 & -10 \end{bmatrix}$

7. $*$ is not associative. $\#$ is commutative and associative. (*Note*: Later you will realize that this is addition modulo 4 with $d = 0$, $a = 1$, $b = 2$, $c = 3$. See 10c, which follows.) $\triangle$ is neither commutative nor associative.

10. c.

$\oplus$	0	1	2	3
0	0	1	2	3
1	1	2	3	0
2	2	3	0	1
3	3	0	1	2

11. c.

$\odot$	0	1	2	3
0	0	0	0	0
1	0	1	2	3
2	0	2	0	2
3	0	3	2	1

13. If $a, b \in A$, then $a * b = b * a$ since $a, b \in X$ and $*$ is commutative on X.

Section 2.3, page 43

2. Isosceles: there are only two motions; leave it alone, and flip it on the altitude to the unequal side.

3. Isosceles: The only motion is to leave it alone.

8. There are 10 basic motions: leave it alone, rotate it through 1, 2, 3, or 4 vertices (clockwise); and flip it on the axis, which is the bisector of any one of the point angles, and then rotate through 1, 2, 3, or 4 vertices.

9. $2n$

12. b.

	e	ρ	ρ^2	τ_A	τ_B	τ_C
e	e	ρ	ρ^2	τ_A	τ_B	τ_C
ρ	ρ	ρ^2	e	τ_C	τ_A	τ_B
ρ^2	ρ^2	e	ρ	τ_B	τ_C	τ_A
τ_A	τ_A	τ_B	τ_C	e	ρ	ρ^2
τ_B	τ_B	τ_C	τ_A	ρ^2	e	ρ
τ_C	τ_C	τ_A	τ_B	ρ	ρ^2	e

Section 3.1, page 49

1. d. No identity for composition exists. To see this, suppose $e(x) = c$ where e is the identity. If $f(x) = a$, then $ef = f$ implies $c = a$ for any $a \in \mathbb{Z}$.

3. $(\mathbb{Z}, *)$ is a group. 1 is the identity for $*$ since $a * 1 = a + 1 - 1 = a$ and $1 * a = 1 + a - 1 = a$. The inverse of a is $2 - a$.

6.

$\triangle$	$\{\ \}$	$\{a\}$	$\{b\}$	$\{c\}$	$\{a, b\}$	$\{a, c\}$	$\{b, c\}$	X
$\{\ \}$	$\{\ \}$	$\{a\}$	$\{b\}$	$\{c\}$	$\{a, b\}$	$\{a, c\}$	$\{b, c\}$	X
$\{a\}$	$\{a\}$	$\{\ \}$	$\{a, b\}$	$\{a, c\}$	$\{b\}$	$\{c\}$	X	$\{b, c\}$
$\{b\}$	$\{b\}$	$\{a, b\}$	$\{\ \}$	$\{b, c\}$	$\{a\}$	X	$\{c\}$	$\{a, c\}$
$\{c\}$	$\{c\}$	$\{a, c\}$	$\{b, c\}$	$\{\ \}$	X	$\{a\}$	$\{b\}$	$\{a, b\}$
$\{a, b\}$	$\{a, b\}$	$\{b\}$	$\{a\}$	X	$\{\ \}$	$\{b, c\}$	$\{a, c\}$	$\{c\}$
$\{a, c\}$	$\{a, c\}$	$\{c\}$	X	$\{a\}$	$\{b, c\}$	$\{\ \}$	$\{a, b\}$	$\{b\}$
$\{b, c\}$	$\{b, c\}$	X	$\{c\}$	$\{b\}$	$\{a, c\}$	$\{a, b\}$	$\{\ \}$	$\{a\}$
X	X	$\{b, c\}$	$\{a, c\}$	$\{a, b\}$	$\{c\}$	$\{b\}$	$\{a\}$	$\{\ \}$

Remember that $\triangle$ is commutative. (See Exercise 11a of Section 1.3.)

11. If p is prime, then $k \in \mathbb{Z}_p^*$ implies $(k, p) = 1$; so there exist s, t such that $sk + tp = 1$. Thus $s \odot k = 1$ in $\mathbb{Z}_p^*$, so s (or the element in $\mathbb{Z}_p^*$ that corresponds to s) is the inverse of k. Since $\odot$ is commutative, associative, and 1 is the identity, $(\mathbb{Z}_p^*, \odot)$ is a group.

14. Only the elements relatively prime to 9 have inverses. These are 1, 2, 4, 5, 7, and 8. To show they form a group, you only need to show closure. Construct the Cayley table.

Section 3.2, page 55

5. Since $abab = (ab)(ab) = 1 = 1 \cdot 1 = (aa)(bb) = aabb$, we obtain $ba = ab$ by using the cancellation property on both the left and right side.

7. If $ac = bc$, then using c^{-1} to denote the inverse of c,

$(ac)c^{-1} = (bc)c^{-1}$ (closure)

$a(cc^{-1}) = b(cc^{-1})$ (associative Law)

$a1 = b1$ (inverse)

$a = b$ (identity)

9. There are only 7 Cayley tables to consider. (There are 4 possible tables each containing 4 elements.)

10. Do you have scuba gear?

13. Since $a = e * a = (a * a') * a$ and e is unique by ii), we have $a * a' = e$. Similarly $a' * a = e$. So e is the identity; a' is the inverse of a.

14. Use the Second Principle of Mathematical Induction given in Exercise 10 of Section 1.4.

Chapter 4

Section 4.1, page 62

1. a. $3\mathbb{Z} = \langle 3 \rangle$ f. $\mathbb{Z}_7^* = \langle 3 \rangle$ h. not cyclic

5. Since G is cyclic, $b = a^h$ for some $h \in \mathbb{N}$. Then, $b^n = (a^h)^n = a^{nh} = (a^n)^h = 1^h = 1$.

9. Assume $t > s$. Since $(a^t)^{-1} = a^{n-t}$ by Exercise 6, $1 = a^t a^{n-t} = a^s a^{n-t} = a^{n+s-t}$, which is a contradiction. (Why?)

10. See Exercise 8.

12. Use the fact that there exist $s, t \in \mathbb{Z}$ such that $1 = sk + tn$.

Section 4.2, page 65

2. a. not a subgroup   c. subgroup   d. subgroup

4. H and K are subgroups. L isn't a subgroup since

$$\begin{bmatrix} a & 0 \\ 0 & 1 \end{bmatrix} + \begin{bmatrix} a & 0 \\ 0 & 1 \end{bmatrix} = \begin{bmatrix} 2a & 0 \\ 0 & 2 \end{bmatrix}$$

7. $o(0) = 1$; $o(1) = 6$; $o(2) = 3$; $o(3) = 2$; $o(4) = 3$; $o(5) = 6$.

Section 4.3, page 70

4. Yes.

5. No, closure does not hold. There is also no identity element in H.

11. No, $H \leqslant G$ implies $H \neq \varnothing$. The two conditions are satisfied (vacuously) by $\varnothing$, but $\varnothing$ is not a subgroup of G. (Why?)

13. Set $m = qn + r$ where $0 \leqslant r < n$. Show that r must equal 0.

14. $H = \langle a^h \rangle$ where $n = hk$.

19. $N_G(H)$ is a subgroup by Theorem 4.4: $N_G(H) \neq \varnothing$ since $1h1^{-1} = h \in H$ implies $1 \in N_G(H)$. If $a, b \in N_G(H)$, then

$$(ab)h(ab)^{-1} = (ab)h(b^{-1}a^{-1}) = a(bhb^{-1})a^{-1} \in H$$

since $bhb^{-1} \in H$. Also, $aha^{-1} = h_1 \in H$ implies $a^{-1}h_1(a^{-1})^{-1} \in H$.

Chapter 5

Section 5.1, page 75

2.   b. surjective; injective   d. neither, if $B \neq \varnothing$ g. surjective

4.   For 2b, $f^{-1}: \mathbb{N} \cup \{0\} \to \mathbb{N}$, $f^{-1}(n) = n + 1$.

5.   Suppose $g \circ f: X \to Z$ is injective and $f(x_1) = f(x_2)$. Then $(g \circ f)(x_1) = g(f(x_1)) = g(f(x_2)) = (g \circ f)(x_2)$ implies $x_1 = x_2$. For example, let $X = \mathbb{R}^+$, $Y = Z = \mathbb{R}$, $f(x) = \sqrt{x}$, $g(x) = x^2$.

9.   a. What happens if we restrict the domain so that f is also injective? Cardinality is involved here!

Section 5.2, page 82

1.   See Figure 5.6 on page 92.

3.   a. $\sigma\tau = (1 \quad 2)$ b. $\sigma^{-1} = (1 \quad 4)(2 \quad 3) = \sigma$

4.   a. For α, $1 \to 2 \to 1$, $3 \to 5 \to 4 \to 6 \to 3$, $7 \to 8 \to 7$, so

$$\alpha = (1 \quad 2)(3 \quad 5 \quad 4 \quad 6)(7 \quad 8)$$

c. $\alpha^{-1} = (1 \quad 2)(3 \quad 6 \quad 4 \quad 5)(7 \quad 8)$.

8.   The orbits of α are $\{\{1, 2\}, \{3, 5, 4, 6\}, \{7, 8\}\}$.

16. Show this for S_3, and then use permutations with a similar cycle representation in S_n with $n > 3$.

Section 5.3, page 88

1.   c. $(1 \quad 2 \quad 3)(2 \quad 3 \quad 4) = (1 \quad 3)(1 \quad 2)(2 \quad 4)(2 \quad 3) = (1 \quad 2)(3 \quad 4)$, which is even.

2. $\alpha = (1 \quad 2)(3 \quad 6)(3 \quad 4)(3 \quad 5)(7 \quad 8)$, an odd permutation.

8.   No, the identity permutation is even.

9.   Consider $(1 \quad 3 \quad 2)(1 \quad 2)(1 \quad 2 \quad 3)$ and generalize.

14. Use Theorem 3.2, induction, and the fact that $\tau_i^{-1} = \tau_i$.

Section 5.4, page 93

2. $e \sim (1)$, $\rho \sim (1 \quad 2 \quad 3)$, $\rho^2 \sim (1 \quad 3 \quad 2)$, $\tau_A \sim (1 \quad 3)$, $\tau_B \sim (2 \quad 3)$, $\tau_C \sim (1 \quad 2)$.

5.

	(1)	(1 2 3)	(1 3 2)	(1 3)	(2 3)	(1 2)
(1)	(1)	(1 2 3)	(1 3 2)	(1 3)	(2 3)	(1 2)
(1 2 3)	(1 2 3)	(1 3 2)	(1)	(2 3)	(1 2)	(1 3)
(1 3 2)	(1 3 2)	(1)	(1 2 3)	(1 2)	(1 3)	(2 3)
(1 3)	(1 3)	(1 2)	(2 3)	(1)	(1 3 2)	(1 2 3)
(2 3)	(2 3)	(1 3)	(1 2)	(1 2 3)	(1)	(1 3 2)
(1 2)	(1 2)	(2 3)	(1 3)	(1 3 2)	(1 2 3)	(1)

12. Consider $(1\ 2)(3\ 4) = (1\ 4\ 2)(1\ 4\ 3)$ and $(1\ 2)(1\ 3) = (1\ 3\ 2)$.

18. Since $(1\ s)(1\ t)(1\ s) = (s\ t)$, any subgroup that contains all transpositions of the form $(1\ t)$ contains *all* transpositions. But any permutation can be written as a product of transpositions.

Chapter 6

Section 6.1, page 99

3. Since $\phi(a \oplus b) = 3(a \oplus b) = 3a \oplus 3b = \phi(a) \oplus \phi(b)$ and $\phi(-a) = 3(-a) = -3a = -\phi(a)$, ϕ is a homomorphism and $\phi^{\rightarrow}(\mathbb{Z}_{12}) \cong \mathbb{Z}_4$. No; No.

8. $\mathbb{Z}_6$ is abelian, but S_3 isn't. (See Exercise 16 of Section 5.2.)

9. If G is abelian, then $\phi(ab) = \phi(ba) = (ba)^{-1} = a^{-1}b^{-1} = \phi(a)\phi(b)$.

11. b. Let $G = \{\sigma \in S_4 | \sigma(3) = 3$ and $\sigma(4) = 4\}$. Show $G \cong S_2$.

Section 6.2, page 104

2. $\ker(\phi) = \{4, 8, 0\}$

7. Subgroups $\{0\}, \{0, 2\}, \mathbb{Z}_4$; all are kernels. Images are $\mathbb{Z}_4, \mathbb{Z}_2, \{0\}$ isomorphically.

8. See Exercise 9 of Section 5.1.

14. If $b = \phi(a^k)$, then $b^n = \phi(a^k)^n = \phi([a^k]^n) = \phi([a^n]^k) = \phi(1^k) = \phi(1) = 1$. Now apply Exercise 13 of Section 4.3 to the group $\langle b \rangle$.

Section 6.3, page 107

2. $\mathbb{Z}_4 \cong \langle (1\ 2\ 3\ 4) \rangle = \langle (1\ 4\ 3\ 2) \rangle$

     $\cong \langle (1\ 2\ 4\ 3) \rangle = \langle (1\ 3\ 4\ 2) \rangle$

     $\cong \langle (1\ 3\ 2\ 4) \rangle = \langle (1\ 4\ 2\ 3) \rangle$

3. In addition to the subgroups in Exercise 2, S_4 contains the subgroups

     $K = \{(1), (1\ 2)(3\ 4), (1\ 3)(2\ 4), (1\ 4)(2\ 3)\}$

     $\cong \{(1), (1\ 2), (3\ 4), (1\ 2)(3\ 4)\}$

     $\cong \{(1), (1\ 3), (2\ 4), (1\ 3)(2\ 4)\}$

     $\cong \{(1), (1\ 4), (2\ 3), (1\ 4)(2\ 3)\}$

11.   Suppose $G = \langle a \rangle$ has order p. Show that the mapping $\phi: G \to G$ defined by $\phi(a) = a^k$ is an automorphism for $k = 1, 2, \ldots, p - 1$.

Chapter 7

Section 7.1, page 112

2.   It is neither.

5.   Since equivalence classes are disjoint or identical, show that $\bar{x} = \bar{r}$.

7. $\{(1, 3), (1, 1), (1, 5), (3, 3), (3, 1), (5, 5), (5, 1), (3, 5), (5, 3), (2, 2), (2, 6), (6, 6),$ $(6, 2), (4, 4)\}$

8.   c. all

9.   b. neither reflexive nor transitive   f. not symmetric

13.   b. No, only if f is surjective.

14.   No, not if for some $a \in X$, a is not related to anything.

Section 7.2, page 117

2.   b. $\{(1), (1 \quad 2 \quad 3), (1 \quad 3 \quad 2)\}$ and $\{(1 \quad 2), (1 \quad 3), (2 \quad 3)\}$
       f. $5\mathbb{Z}, 5\mathbb{Z} + 1, 5\mathbb{Z} + 2, 5\mathbb{Z} + 3,$ and $5\mathbb{Z} + 4$

7.   b. 1, 2, 3, 4, 6, 12

10.   a. If $y \in x(H \cap K)$ then there is $h \in H \cap K$ such that $y = xh$. Then $xh \in xH$ since $h \in H$ and $xh \in xK$ since $h \in K$; so $y = xh \in xH \cap xK$. If $y \in xH \cap xK$, then there are $h \in H$ and $k \in K$ such that $y = xh$ and $y = xk$. Thus $xh = xk$, which implies $h = k$ by cancellation; so $h \in H \cap K$ and $y = xh \in x(H \cap G)$.

18.   The proper subgroups have order 1, p, or q by Lagrange's Theorem. Any group of prime order is cyclic.

19. $\mathbb{Z}_8, \mathbb{Z}_4 \times \mathbb{Z}_2, \mathbb{Z}_2 \times \mathbb{Z}_2 \times \mathbb{Z}_2$

Section 7.3, page 124

2.   No subgroup of order 2 is normal in S_3; yes.

6. $gN = Ng$ means there is $n' \in N$ such that $gn = n'g$. Thus $gng^{-1} = n' \in N$.

14.   Show $Z(G) \leqslant G$. If $z \in Z(G)$, $g \in G$, then $gzg^{-1} = zgg^{-1} = z1 = z \in Z(G)$. Apply Theorem 7.6.

18.   Since 97 is a prime, $\{1\}$ and G are the only subgroups. Both are normal. If $\ker \phi = \{1\}$, ϕ is a monomorphism. If $\ker \phi = G$, then $\phi(g) = 1_{G'}$, for all $g \in G$.

19.   See Exercise 8 of Section 6.2.

21.   See Exercise 3 of Section 6.3.

Chapter 8

Section 8.1, page 130

1. $*$ is not well defined since $(1)N * (1 \quad 3)N = (1 \quad 3)N$ but $(1 \quad 2)N * (1 \quad 3)N = (1 \quad 3 \quad 2)N \neq (1 \quad 3)N$.

3.   If $n \in N$ and $x \in G$, then $N = x^{-1}xN = (x^{-1}N)(xN) = (x^{-1}N)(nN)(xN) = (x^{-1}nx)N$ implies $x^{-1}nx \in N$. Thus $x^{-1}Nx \subseteq N$. Use Theorem 7.6.

8.   b. Since $G = \langle a \rangle$, $G/N = \langle aN \rangle$.

12.   97 is prime; the only factor groups have orders 1 and 97.

15.   Since $g^{-1}h^{-1}ghN = N$, $ghN = hgN$ and $gNhN = hNgN$ for all gN, $hN \in G/N$.

16.   a. Consider $x(aba^{-1}b^{-1})x^{-1}$, and use $1 = x^{-1}x$.

Section 8.2, page 137

3.   Consider $nkn^{-1}k^{-1}$.

4. A_n is normal in S_n. S_n/A_n is a group of order 2. So is S_2. Use Theorem 6.4.

5. $AB = \{(1), (1 \;\; 2), (2 \;\; 3), (1 \;\; 2 \;\; 3), (1 \;\; 3), (1 \;\; 3 \;\; 2)\}$

10. $|A \cap B| = 2$, $A \cap B = \{0, 12\}$, $AB = \mathbb{Z}_{24}$

13.   Define $\phi: G \to \mathrm{Im}(G)$ by $\phi(g) = \phi_g$ where $\phi_g(x) = gxg^{-1}$. Show that $\ker \phi = Z(G)$.

Chapter 9

Section 9.1, page 145

1. $\mathrm{cl}\{(1)\} = \{(1)\}$, $\mathrm{cl}\{(1 \;\; 2 \;\; 3 \;\; 4)\} = \{(1 \;\; 2 \;\; 3 \;\; 4), (1 \;\; 4 \;\; 3 \;\; 2)\}$,
     $\mathrm{cl}\{(1 \;\; 3)(2 \;\; 4)\} = \{(1 \;\; 3)(2 \;\; 4)\}$,
     $\mathrm{cl}\{(1 \;\; 2)(3 \;\; 4)\} = \{(1 \;\; 2)(3 \;\; 4), (1 \;\; 4)(2 \;\; 3)\}$,
     $\mathrm{cl}\{(2 \;\; 4)\} = \{(2 \;\; 4), (1 \;\; 3)\}$

4.   Yes, $(2 \;\; 4 \;\; 3)H_1(2 \;\; 4 \;\; 3)^{-1} = H_2$.

7. $1a = a1$ implies $1 \in C_G(a)$; so $C_G(a) \neq \varnothing$. If b, $c \in C_G(a)$, then $(bc)a = b(ca) = b(ac) = (ba)c = a(bc)$ and $bc \in C_G(a)$. Also $ba = ab$ implies $ab^{-1} = b^{-1}a$; therefore $b^{-1} \in C_G(a)$. The result now follows from Theorem 4.4.

9.   a. $C_{A_4}\{(1 \;\; 2)(3 \;\; 4)\} = \{(1), (1 \;\; 2)(3 \;\; 4), (1 \;\; 3)(2 \;\; 4), (1 \;\; 4)(2 \;\; 3)\}$
     d. $C_{A_4}\{(1 \;\; 2 \;\; 3)\} = \{(1), (1 \;\; 2 \;\; 3), (1 \;\; 3 \;\; 2)\}$
     g. $C_{A_5}\{(1 \;\; 2 \;\; 3 \;\; 4 \;\; 5)\} = \{(1), (1 \;\; 2 \;\; 3 \;\; 4 \;\; 5), (1 \;\; 3 \;\; 5 \;\; 2 \;\; 4),$
                       $(1 \;\; 4 \;\; 2 \;\; 5 \;\; 3), (1 \;\; 5 \;\; 4 \;\; 3 \;\; 2)\}$

11.   By Exercise 1, $Z(D) = \{(1), (1 \;\; 3)(2 \;\; 4)\}$,

$$8 = |D| = |Z(D)| + |\mathrm{cl}\{(1 \;\; 2 \;\; 3 \;\; 4)\}| + |\mathrm{cl}\{(1 \;\; 2)(3 \;\; 4)\}| + |\mathrm{cl}\{(2 \;\; 4)\}|$$

$$= 2 + 2 + 2 + 2$$

13.   Reflexive: $1H1^{-1} = 1H1 = H$ implies $H \approx H$.
     Symmetric: if $xHx^{-1} = K$, then $x^{-1}Kx = (x^{-1})H(x^{-1})^{-1} = H$.
     Transitive: If $xHx^{-1} = K$ and $yKy^{-1} = L$, then

$$(yx)H(yx)^{-1} = (yx)H(x^{-1}y^{-1}) = y(xHx^{-1})y^{-1} = yKy^{-1} = L$$

15.   a. $N_{S_3}(\langle (1 \;\; 2 \;\; 3) \rangle) = S_3$
     c. $N_{A_4}(\langle (1 \;\; 2)(3 \;\; 4) \rangle) = \{(1), (1 \;\; 2)(3 \;\; 4), (1 \;\; 3)(2 \;\; 4), (1 \;\; 4)(2 \;\; 3)\}$
     e. $N(\langle (1) \rangle) = S_4$

17.   Since $N \trianglelefteq G$, $xHx^{-1} \subseteq xNx^{-1} = N$. Therefore $\bigcup_{x \in G} xHx^{-1} \subseteq \bigcup_{x \in G} N = N$.

19.   a. Since $\psi_n(gh) = n(gh)n^{-1} = ng(n^{-1}n)hn^{-1} = (ngn^{-1})(nhn^{-1}) = \psi_n(g)\psi_n(h)$, ψ_n is a homomorphism. If $\psi_n(h_1) = \psi_n(h_2)$, then $nh_1n^{-1} = nh_2n^{-1}$ and $h_1 = h_2$; therefore ψ_n is injective. If $h \in H$, set $g = n^{-1}hn$; then $\psi_n(g) = ngn^{-1} = n(n^{-1}hn)n^{-1} = h$; thus ψ_n is surjective.

     b. If $\psi_n(h) = h$ for all $h \in H$, then $nhn^{-1} = h$ and $nh = hn$ for all $h \in H$. Therefore, $n \in C_G(H)$. Conversely, if $n \in C_G(H)$, then $\psi_n(h) = nhn^{-1} = hnn^{-1} = h$ for all $h \in H$, and ψ_n is the identity mapping.

Section 9.2, page 151

2. $Z(D) = \{(1), (1 \quad 3)(2 \quad 4)\}$

4. $Z(S_3) = \{(1)\}$

6.   By Lagrange's Theorem, if $|H| = p^n$ and $|G/H| = p^m$, then $|G| = [G:H] \cdot |H| = |G/H| \cdot |H| = p^m p^n = p^{m+n}$.

8.   No; consider the group of symmetries of a rectangle.

11.   Let $K \trianglelefteq P$, $|K| = p^{n-1}$, $|P| = p^n$. Choose $x \in Z(P)$ where $o(x) = p$. If $N = \langle x \rangle$, then $|N| = p$ and $N \trianglelefteq P$. If $N_P(K) = P$, we are done. If $N_P(K) \neq P$, then $N_P(K) = K$ since $K \leqslant N_P(K)$. Since $x \in Z(P)$, $x \in N_P(K) = K$ and $K/N \leqslant P/N$, a group of order p^{n-1}. By induction, $K/N \trianglelefteq P/N$. By Exercise 20 of Section 8.2, $K \trianglelefteq P$.

13.   By Theorem 9.9, $|Z(P)| = p$, p^2, or p^3. P nonabelian implies $|Z(P)| \neq p^3$. If $|Z(p)| = p^2$, then let $x \in P - Z(P)$. Then $Z(P)$ is a proper subgroup of $C_P(x)$. Since $p^2 = |Z(P)| < |C_P(x)| \leqslant |P| = p^3$, $C_P(x) = P$ and $x \in Z(P)$, contrary to choice of x. Thus $|Z(P)| = p$.

Section 9.3, page 159

1. $P \trianglelefteq G$ if and only if $gPg^{-1} = P$ for all $g \in G$ if and only if $\text{Syl}_p(G) = \{P\}$ (by Sylow's Third Theorem).

4.   Reflexive: $1_\Omega(\alpha) = \alpha$ implies $\alpha \approx a$.
     Symmetric: $g(\alpha) = \beta$ implies $\alpha = g^{-1}(\beta)$.
     Transitive: $g(\alpha) = \beta$, $h(\beta) = \gamma$ implies $(hg)(\alpha) = h(g(\alpha)) = h(\beta) = \gamma$. Thus $\alpha \approx \gamma$.
     Clearly $1_\Omega \in G_\alpha$. If g, $h \in G_\alpha$, then $gh(\alpha) = g(h(\alpha)) = g(\alpha) = \alpha$, so $gh \in G_\alpha$. Moreover $g(\alpha) = \alpha$ implies $\alpha = g^{-1}(\alpha)$, so $g^{-1} \in G_\alpha$.

6.   If $\phi: N_G(P) \to N_G(P)/P$ is the natural map, consider $\phi^{\to}(P')$.

7.   Let $g \in P_1$. Then $xgx^{-1} \in xP_1x^{-1} = xP_2x^{-1}$ implies $xgx^{-1} = xhx^{-1}$ for $h \in P_2$. But then $g = h \in P_2$ and $P_1 \subseteq P_2$. Similarly $P_2 \subseteq P_1$.

9.   The stabilizer of P in

$$G = \{x \in G | P^x = P\} = \{x \in G | xPx^{-1} = P\} = N_G(P)$$

     The stabilizer of P in $P' = \{x \in P' | P^x = P\} = N_G(P) \cap P'$.

11.   Use Theorem 9.10.

12.   If $|G| = p^n m$ where $(p, m) = 1$, then $|PN/N| = |P/P \cap N|$ is a power of p. Since $[G/N : PN/N] = [G : PN]$ is a divisor of $[G : P] = m$, then p does not divide $[G/N : PN/N]$.

15. Since $P \leqslant N_G(P) \leqslant H$, then $P \in \mathrm{Syl}_p(H)$ and $[H : N_G P)] \equiv 1$ (modulo p) by Sylow's Third Theorem. But $N_G(P) \leqslant H$ implies $N_H(P) = N_G(P) \cap H = N_G(P)$.

17. Let $g \in N_G(H)$. Then $P^g \leqslant H^g = H$, so $P^g \in \mathrm{Syl}_p(H)$. Therefore, there exists $h \in H$ such that $P^{gh} = (P^g)^h = P$. But then $gh \in N_G(P) \leqslant H$ and $g \in H$. Thus $N_G(H) \leqslant H$.

Chapter 10

Section 10.1, page 168

2. If $(h_1, k_1), (h_2, k_2) \in H \times K$, then $(h_1, k_1)(h_2, k_2) = (h_1 h_2, k_1 k_2) = (h_2 h_1, k_2 k_1) = (h_2, k_2)(h_1, k_1)$.

5. Since $\mathbb{Z}_2 \times \mathbb{Z}_2 \times \mathbb{Z}_2$ contains no element of order 4, it cannot be isomorphic to the group of symmetries of the square.

7. Let $H = \langle 5 \rangle$, $K = \langle 3 \rangle$. Then $|H| = 3$, $|K| = 5$; so $|H \cap K| = 1$ and $H \cong \mathbb{Z}_3$, $K \cong \mathbb{Z}_5$. Since $\mathbb{Z}_{15}$ is abelian, $H \trianglelefteq \mathbb{Z}_{15}$, $K \trianglelefteq \mathbb{Z}_{15}$. The result follows from Corollary 10.1.

10. As the semi-direct product of N by K, $N \trianglelefteq G$, $G = NK$, and $N \cap K = \{1\}$. If $K \trianglelefteq G$, then $G = N \times K$ by Corollary 10.1.

12. Take $N = A_3$, $G = S_3$, $K = \langle (1 \quad 2) \rangle$.

15. Note that $(1 \quad 3 \quad 4 \quad 2) = (1 \quad 3 \quad 2)(3 \quad 4) \in S_3 B$, but

$$(1 \quad 4)(2 \quad 3) = (1 \quad 3 \quad 4 \quad 2)(1 \quad 3 \quad 4 \quad 2) \notin S_3 B$$

18. $|AB| = |A||B|/|A \cap B| = |B||A|/|B \cap A| = |BA|$

Section 10.2, page 174

3. Let $x, y, z \in G$ have orders 3, 5, 7, respectively. If $a = xyz$, then $a^2 = x^2 y^2 z^2$, $a^3 = x^3 y^3 z^3 = y^3 z^3$, ..., $a^{104} = x^{104} y^{104} z^{104} = x^2 y^4 z^6$, $a^{105} = 1$. The order of a is the least common multiple of 3, 5, 7, that is, 105.

6. a. Since $(p, q) = 1$, there exists $s, t \in \mathbb{Z}$ such that $sq + tq = 1$. But then $sm + tqm_1 = spm_1 + tqm_1 = m_1$. The left side of this equation is divisible by q.
   b. Let $m = pm_1$. Since $q \mid m$, we know by part a that $q \mid m_1$ and $m_1 = qm_2$. But then $m = pm_1 = p(qm_2) = (pq)m_2$ and $pq \mid m$.

7. Use the Basis Theorem and Lemma 10.2.

9. Let $a, b \in G$, $o(a) = m$, $o(b) = n$. Then $(ab)^{m+n} = a^{m+n} b^{m+n} = 1^n 1^m = 1$ and $(a^{-1})^m = (a^m)^{-1} = 1^{-1} = 1$. Since $o(1) = 1$, the set of elements of G of finite order form a subgroup by Theorem 4.4.

11. a.

Elementary Divisors	Group
$\{5^2\}$	$\mathbb{Z}_{25}$

c.

Elementary Divisors	Group
$\{2^4, 3^2\}$	$\mathbb{Z}_{16} \oplus \mathbb{Z}_9$
$\{2^3, 2, 3^2\}$	$\mathbb{Z}_8 \oplus \mathbb{Z}_2 \oplus \mathbb{Z}_9$
$\{2^2, 2, 2, 3^2\}$	$\mathbb{Z}_4 \oplus \mathbb{Z}_2 \oplus \mathbb{Z}_2 \oplus \mathbb{Z}_9$
$\{2^2, 2^2, 3^2\}$	$\mathbb{Z}_4 \oplus \mathbb{Z}_4 \oplus \mathbb{Z}_9$
$\{2, 2, 2, 2, 3^2\}$	$\mathbb{Z}_2 \oplus \mathbb{Z}_2 \oplus \mathbb{Z}_2 \oplus \mathbb{Z}_2 \oplus \mathbb{Z}_9$

$$\{2^4, 3, 3\} \qquad\qquad \mathbb{Z}_{16} \oplus \mathbb{Z}_3 \oplus \mathbb{Z}_3$$
$$\{2^3, 2, 3, 3\} \qquad\quad \mathbb{Z}_8 \oplus \mathbb{Z}_2 \oplus \mathbb{Z}_3 \oplus \mathbb{Z}_3$$
$$\{2^2, 2^2, 3, 3\} \qquad\; \mathbb{Z}_4 \oplus \mathbb{Z}_4 \oplus \mathbb{Z}_3 \oplus \mathbb{Z}_3$$
$$\{2^2, 2, 2, 3, 3\} \qquad \mathbb{Z}_4 \oplus \mathbb{Z}_2 \oplus \mathbb{Z}_2 \oplus \mathbb{Z}_3 \oplus \mathbb{Z}_3$$
$$\{2, 2, 2, 2, 3, 3\} \qquad \mathbb{Z}_2 \oplus \mathbb{Z}_2 \oplus \mathbb{Z}_2 \oplus \mathbb{Z}_2 \oplus \mathbb{Z}_3 \oplus \mathbb{Z}_3.$$

12. a. | Invariant Factors | Group |
             |---|---|
             | $\{5^2\}$ | $\mathbb{Z}_{25}$ |

     b. | Invariant Factors | Group |
             |---|---|
             | $\{144)$ | $\mathbb{Z}_{144}$ |
             | $\{72, 2\}$ | $\mathbb{Z}_{72} \oplus \mathbb{Z}_2$ |
             | $\{36, 2, 2\}$ | $\mathbb{Z}_{36} \oplus \mathbb{Z}_2 \oplus \mathbb{Z}_2$ |
             | $\{36, 4\}$ | $\mathbb{Z}_{36} \oplus \mathbb{Z}_4$ |
             | $\{18, 2, 2, 2\}$ | $\mathbb{Z}_{18} \oplus \mathbb{Z}_2 \oplus \mathbb{Z}_2 \oplus \mathbb{Z}_2$ |
             | $\{48, 3\}$ | $\mathbb{Z}_{48} \oplus \mathbb{Z}_3$ |
             | $\{24, 6\}$ | $\mathbb{Z}_{24} \oplus \mathbb{Z}_6$ |
             | $\{12, 12\}$ | $\mathbb{Z}_{12} \oplus \mathbb{Z}_{12}$ |
             | $\{6, 6, 2, 2\}$ | $\mathbb{Z}_6 \oplus \mathbb{Z}_6 \oplus \mathbb{Z}_2 \oplus \mathbb{Z}_2$ |
             | $\{12, 6, 2\}$ | $\mathbb{Z}_{12} \oplus \mathbb{Z}_6 \oplus \mathbb{Z}_2$ |

13. 3, 5, 7, and 11

16. $\mathbb{Z}_4 \oplus \mathbb{Z}_2$ contains an element (1, 0) of order 4. If $\mathbb{Z}_4 \oplus \mathbb{Z}_2$ were isomorphic to $\mathbb{Z}_2 \oplus \mathbb{Z}_2 \oplus \mathbb{Z}_2$, then $\mathbb{Z}_2 \oplus \mathbb{Z}_2 \oplus \mathbb{Z}_2$ would contain an element of order 4, but $\mathbb{Z}_2 \oplus \mathbb{Z}_2 \oplus \mathbb{Z}_2$ does not.

Section 11.1, page 180

1.   Since G is abelian, every subgroup in a normal series is normal, not only in its successor but also in G.

3.   b. $\{0\} \trianglelefteq \langle 4 \rangle \trianglelefteq \langle 2 \rangle \trianglelefteq \mathbb{Z}_8$.

     f. $\langle ((1), (1)) \rangle \trianglelefteq K \times \langle (1) \rangle \trianglelefteq A_4 \times \langle (1) \rangle \trianglelefteq A_4 \times A_3$
         $\langle ((1), (1)) \rangle \leqslant \langle (1) \rangle \times A_3 \leqslant K \times A_3 \leqslant A_4 \times A_3$
         $\langle ((1), (1)) \rangle \leqslant K \times \langle (1) \rangle \leqslant K \times A_3 \leqslant A_4 \times A_3$

6.   The permutation (1   3) works since the first series has factors isomorphic to $\mathbb{Z}_2$, $\mathbb{Z}_2$, and $\mathbb{Z}_3$, while the second has factors $\mathbb{Z}_3$, $\mathbb{Z}_2$, and $\mathbb{Z}_2$.

8.   Since $(H_{k+j}/N) \trianglelefteq (H_{k+j+1}/N)$ by Theorem 8.7 and

$$\frac{H_{k+j+1}/N}{H_{k+j}/N} \cong H_{k+j+1}/H_{k+j}$$

is a nontrivial simple group, $\{H_k/N, \ldots, H_n/N\}$ is a composition series.

10. Let $x \in H_i$, $y \in H_{i+1}$. Then $x, y \in H$, so that $yxy^{-1} \in H$. Also $x \in G_i$, $y \in G_{i+1}$, so that $yxy^{-1} \in G_i$. But then $yxy^{-1} \in H \cap G_i = H_i$ and $H_i \trianglelefteq H_{i+1}$.

14. Since the series $\{0\} \trianglelefteq \cdots \trianglelefteq 2^n\mathbb{Z} \trianglelefteq \cdots \trianglelefteq 2^5\mathbb{Z} \trianglelefteq 16\mathbb{Z}$ is infinite in length, the series could never be refined to a composition series.

17. Let $x \in \mathbb{Z}_p$, $o(x) = p$, $N = \langle x \rangle$. If $P' = P/N$, then P' is a group of order p^{n-1}. By induction, the composition factors of P' are isomorphic to $\mathbb{Z}_p$. Any composition

series of P gives rise to a composition series between N and P under the action of the natural mapping from P to P/N. The remaining composition factor is $N/\{1\} \cong \mathbb{Z}_p$.

Section 11.2, page 189

2. No, A_3 is not normal in A_4.

5. Since $G_1 \cong G_1 \times \{1\} \trianglelefteq G_1 \times G_2$ and $(G_1 \times G_2)/(G_1 \times \{1\}) \cong G_2$ are both solvable, $G_1 \times G_2$ is solvable by Theorem 11.8.

8. Let $x \in NG_1$, $y \in NG_2$. Then $x = n_1 g_1$, $y = n_2 g_2$ and

$$yxy^{-1} = (n_2 g_2)(n_1 g_1)(n_2 g_2)^{-1} = n_2 g_2 n_1 g_1 g_2^{-1} n_2^{-1}$$

$$= n_2 n_3 g_2 g_1 g_2^{-1} n_2^{-1} = n_2 n_3 g_3 n_2^{-1} = n_2 n_3 n_4 g_3 \in NG_1$$

for $n_3, n_4 \in N$, $g_3 \in G_1$, since $N \trianglelefteq G$ and $G_1 \trianglelefteq G_2$. Therefore $NG_1 \trianglelefteq NG_2$.

10. If $N \trianglelefteq G$, $|N| = p^\alpha$, then $G' = G/N$ has order q^β and is solvable by Theorem 11.9. By Theorem 11.8, G is solvable.

13. No, the class of the identity must be one of the conjugate classes.

15. Consider $\alpha(1 \quad 2 \quad 5)\alpha(1 \quad 5 \quad 2)$.

16. If G is nilpotent, then $G = P_1 \times P_2 \times \cdots \times P_k$ where $P_k \in \mathrm{Syl}_{p_k}(G)$. Each P_i is a p-group and hence solvable by Theorem 11.9. The result follows from Exercise 5 and induction.

Section 11.3, page 194

1. Both have composition factors isomorphic to $\mathbb{Z}_2$ and $\mathbb{Z}_3$.

4. a. If G is simple and $|G| = 12$, then G must contain four 3-Sylow subgroups. Then G has 4×2 elements of order 3 and just four other elements to make up the 2-Sylow subgroup(s) of G. Therefore, there is but one 2-Sylow subgroup. It must be normal.

   f. If G is simple and $|G| = 392 = 2^3 \cdot 7^2$ and $P \in \mathrm{Syl}_7(G)$, then $[G : N_G(P)] = 1$ or 8. If 1, then $P \trianglelefteq G$. If 8, then G simple implies $|G|$ divides 8! by Theorem 11.12. But $392 \nmid 8!$. Thus G is not simple.

6. If $\phi_x(g_i H) = \phi_x(g_j H)$, then $xg_i H = xg_j H$; so $g_i H = g_j H$ by cancellation, and ϕ_x is injective. If $g_j H \in \Omega$, then $x^{-1} g_j \in g_k H$ for some k since Ω is a partition of G. Then $\phi_x(g_k H) = \phi_x(x^{-1} g_j H) = x(x^{-1} g_j H) = g_j H$, implying ϕ_x is surjective. Finally $\phi(xy) = \phi_{xy}$. But $\phi_{xy}(g_i H) = xy(g_i H) = x(yg_i H) = \phi_x(yg_i H) = \phi_x \phi_y(g_i H)$. Therefore $\phi(xy) = \phi_{xy} = \phi_x \phi_y = \phi(x)\phi(y)$, and ϕ is a homomorphism.

9. $\{(1 \quad 4)(2 \quad 3), (1 \quad 4 \quad 2 \quad 3)\}$

11. $$\begin{bmatrix} p-1 & 0 \\ 0 & p-1 \end{bmatrix} \begin{bmatrix} p-1 & 0 \\ 0 & p-1 \end{bmatrix} = \begin{bmatrix} p^2 - 2p + 1 & 0 \\ 0 & p^2 - 2p + 1 \end{bmatrix} = \begin{bmatrix} 1 & 0 \\ 0 & 1 \end{bmatrix}$$

    in $SL(2, p)$ since $p^2 - 2p + 1 \equiv 1$ (modulo p).

15. See Exercise 20 of Section 9.1.

Section 12.1, page 201

1. a. 2.43×10^{-7}　　b. 3.76×10^{-3}　　c. 4.00×10^{-6}
4. a. $.1374$　　b. $.9709$　　c. $.0291$
7. a. If $E = W + R$, then $W + E = W + (W + R) = (W + W) + R = 0 + R = R$ since every word in $B(n)$ is its own inverse.
10. 111010

Section 12.2, page 207

1. No. Yes. No.　　b. 2.14×10^{-3}　　e. $.7657$
4. The distance between code words is 4. The code can detect 3 errors and correct 1.
7. The code can detect 2 and correct 1 error.
9. Suppose W_1 is sent, W_2 is received, and that $H(W_1, W_2) \leqslant k$. Use Exercise 8d of Section 12.2 to show that if W_2 is decoded as W_1, then $H(W_3, W_2) \geqslant k + 1$ for every other code word W_3.
10. Let W_3 be the word that disagrees with the word W_1 in the first k entries and agrees with W_1 in the remaining entries. Then $H(W_1, W_3) = k$ and because $H(W_1, W_2) < 2k + 1$, W_3 can differ from W_2 in less than $(2k + 1) - k = k + 1$ bit locations. Therefore $H(W_3, W_2) < k$. If W_3 is received, then W_3 could be properly decoded as both W_1 and W_2, contradicting the fact that the code corrects k or fewer errors.

Section 12.3, page 211

1. Since $011010 + 000100 = 011110 \in C$, decode as 011.
4. Any group homomorphism $f: B(n) \to B(t)$ must take the identity of $B(n)$ to the identity of $B(t)$, namely, $000 \cdots 0$.
9. a.

$$f(0010) = f(0011 + 0001)$$
$$= f(0011) + f(0001)$$
$$= 0011000 + 0001101$$
$$= 0010101$$

$$f(0100) = f(0111 + 0011)$$
$$= f(0111) + f(0011)$$
$$= 0111011 + 0011000$$
$$= 0100011$$

$$f(1000) = f(1111 + 0111)$$
$$= f(1111) + f(0111)$$
$$= 1111101 + 0111011$$
$$= 1000110, \quad \text{etc.}$$

Chapter 13

Section 13.1, page 218

1. b. $\overline{1}, \overline{5}$　　c. $1, -1$

4. $\begin{bmatrix} 1 & 2 \\ 2 & 4 \end{bmatrix}$ and $\begin{bmatrix} 2 & 0 \\ 0 & 3 \end{bmatrix}$ do not have inverses in $M_2(\mathbb{Z})$.

6. b. $\bar{2} + \bar{2}x + \bar{3}x^2 + x^4$

10. b. $\bar{+}$: (0, 0); $\bar{\cdot}$: (1, 1)

13. Consider $(a + bi) \times (a - bi)$.

15. 0 is the unity.

18. $*$ is left distributive but not right distributive.

Section 13.2, page 226

3. No, the second operation must be commutative.

9. b. $\bar{2}, \bar{4}, \bar{6}$

12. If $n = mk$ where $1 < m < n, 1 < k < n$, then $\bar{m}\bar{k} = \bar{0}$, contradicting the assumption that $\mathbb{Z}_n$ is an integral domain. Thus n is prime.

15. $(a, 0)(0, b) = (0, 0)$

20. Consider bab.

22. a. Consider $1 \oplus 0 \odot 1$.

Chapter 14

Section 14.1, page 233

4. No, $(\sqrt[3]{2})(\sqrt[3]{2}) = \sqrt[3]{4} \notin \mathbb{Z}[\sqrt[3]{2}]$.

9. There are three possibilities: $c = 0$, or $b = 0$, or $b = c = 0$.

13. $R = \{a + b\sqrt[3]{5} + c\sqrt[3]{25} | a, b, c \in \mathbb{Q}\}$

15. If $a^n = 0$ and $b^m = 0$ for $a, b \in R$, then $(ab)^n = a^n b^n = 0b^n = 0$ and $(-a)^n = -a^n = -0 = 0$. Also $(a - b)^{m+n} = 0$ by the Binomial Theorem. Thus the nilpotent elements of R form a subring of R.

20. By Lagrange's Theorem, the order of the subgroup $(S, +)$ divides the order of the group $(R, +)$.

Section 14.2, page 239

2. $\phi(ab) = 2(ab) \neq (2a)(2b) = \phi(a)\phi(b)$

5. Assume there are two inverses of a, say a' and a''. Consider $a'aa''$.

9. Cancellation is handy here.

12. The kernel of ϕ must be an ideal, but the only ideals in a field are 0 and F. If ker $\phi = \{0\}$, then ϕ is a monomorphism. If ker $\phi = F$, then $\phi(a) = 0$ for all $a \in F$.

17. By Exercise 10 of Section 14.1 and induction, the intersection of subrings is another subring. If $r \in R$ and a is an element of R lying in the intersection of all the ideals, then ar and ra are elements in each and every idea. Hence ar and ra are contained in the intersection.

24. If $a^n = 0$, then $\phi(a)^n = \phi(a^n) = \phi(0) = 0$.

Section 14.3, page 245

5. x^2 acts like -1.

6.   b. $(-2, 0) \cdot (2, -2) = (0, 0)$

8.   Recall that if $(R, +, \times)$ is a field, then $(R, +)$ and $(R - \{0\}, \cdot)$ are abelian groups.

12. a. $\begin{bmatrix} 1 & 2 \\ 3 & 4 \end{bmatrix}\begin{bmatrix} 1 & 1 \\ 0 & 0 \end{bmatrix} = \begin{bmatrix} 1 & 1 \\ 3 & 3 \end{bmatrix} \notin S$ b. $\begin{bmatrix} 0 & 1 \\ 0 & 0 \end{bmatrix}\begin{bmatrix} 1 & 2 \\ 3 & 4 \end{bmatrix} = \begin{bmatrix} 3 & 4 \\ 0 & 0 \end{bmatrix} \notin T$

16. If a is a unit of I, then there exists $b \in R$ such that $ab = 1 \in I$. But then $r = r \cdot 1 \in I$ for all $r \in R$.

Section 15.1, page 255

4.   b. Consider $(a - b\sqrt{d})(a + b\sqrt{d})$.

    d. If $(a + b\sqrt{d})(x + y\sqrt{d}) = 0$ with $x \neq 0$, $y \neq 0$, and $(a, b) = 1$, then $ax + dby = 0$ and $ay + bx = 0$. Show $a^2 = db^2$ and contradict the assumption that d is square-free.

8.   Yes, it's generated by x.

9.   c. What about $12 = 6 \cdot 2$?

14. If $a, b \in R - I$ implies $ab \in R - I$, then $a \notin I$, $b \notin I$ implies $ab \notin I$. Then $ab \in I$ implies $a \in I$ or $b \in I$, and I is prime.

20. $P(A)$ forms a subring under the operations $\triangle$ and $\cap$. If $A_1 \subseteq A$ and $B \subseteq X$, then $A_1 \cap B = B \cap A_1 \subseteq A_1 \subseteq A$.

Section 15.2, page 260

3.   The elements of $\mathbb{Q}$ are equivalence classes and $0/b = 0/1$.

5.   If $ab = 0$, then $(a/1)(b/1) = 0/1$.

8.   The field of quotients of $\mathbb{Z}[i]$ is $\mathbb{Q}[i] = \{a + bi \mid a, b \in \mathbb{Q}\}$.

13. $\mathbb{Q}\sqrt{2} = \{a + b\sqrt{2} \mid a, b \in \mathbb{Q}\}$

16. R is isomorphic to $F = \{a/1 \mid a \in R\}$, a subring of the field of quotients of R.

Section 15.3, page 267

3.   b. $sa + tn = 1$ implies $sa \equiv sa + 0 \equiv sa + tn \equiv 1 \pmod{n}$. If $ax \equiv b \pmod{n}$, then $x \equiv 1x \equiv sax \equiv sb \pmod{n}$.

6.   See Exercise 3b of Section 15.3.

9.   a. 218 (mod 280)   b. 31 (mod 44)   e. no solution

12. Assume $(pr, q) \neq 1$; let d be a prime divisior of (pr, q). Then $d \mid pr$.

Section 15.4, page 276

2.   4

4. $147 \to (147, 147, 147) = (3, 2, 0)$; $251 \to (251, 251, 251) = (8, 1, 6)$

5. Use the Fundamental Theorem of Arithmetic to show that multiplication of numbers in factored form is done by addition of exponents.

6. If $\phi: \mathbb{Z}_{p^m q^n} \to \mathbb{Z}_{p^m} \oplus \mathbb{Z}_{q^n}$ is the isomorphism, then

$$\phi^{-1}((a, b)) = \phi^{-1}(a(1, 0) + b(0, 1))$$
$$= a\phi^{-1}((1, 0)) + b\phi^{-1}((0, 1))$$
$$= ax + by$$

For $\phi: \mathbb{Z}_{5^2 2^4} \to \mathbb{Z}_{5^2} \oplus \mathbb{Z}_{2^4}$, $\phi(176) = (176, 176) = (1, 0)$ and $\phi(225) = (225, 225) = (0, 1)$. (21, 15) has preimage $21(176) + 15(225) = 7071$, which is congruent modulo 400 to 271. Verify that $\phi(271) = (21, 15)$.

9. a. $G = \{1, a, a^2, a^3, b, ab, a^2 b, a^3 b\}$. Note that, since $bab = a^3$ and $b^2 = 1$, we have $ba = a^3 b$ and $ab = ba^3$.

b. $\langle a \rangle \cap \langle b \rangle = \{1, a, a^2, a^3\} \cap \{1, b\} = \{1\}$ shows that no nonidentity element lies in all nontrivial subgroups.

c. G is not isomorphic to $\mathbb{Z}_8$.

12. Define $f: \mathbb{Z}_{16} \to B(4)$ by setting $f(x)$ equal to the binary representation of x.

13. Use the proof of Theorem 15.13 to show that if $\mathbb{Z}_{p^n} = \langle a \rangle$ and $H = \langle a^{p^{n-1}} \rangle$, then H lies in every subgroup of $\mathbb{Z}_{p^n}$.

15. Since $(\mathbb{Z}_{p^m}, \oplus) = \langle 1 \rangle$ in additive notation, $a^{p^{m-1}}$ corresponds to $p^{m-1}(a) = p^{m-1}(1) = p^{m-1}$.

Chapter 16

Section 16.1, page 281

2. b. If $(a + b\sqrt{5})(c + d\sqrt{5}) = 1$, then

$$(a - b\sqrt{5})(a + b\sqrt{5})(c + d\sqrt{5}) = a - b\sqrt{5}$$

so that $(a - b\sqrt{5})(a + b\sqrt{5}) = a^2 - 5b^2$ divides $a - b\sqrt{5}$. If $d = (a, b)$, then $d^2 | d$, forcing d to be a unit. Therefore $a^2 - 5b^2$ divides $a - b\sqrt{5}$ implies $a^2 - 5b^2$ divides $d = (a, b)$, and $a^2 - 5b^2 = \pm 1$.

3. Now $c | a$ and $c | b$ since $a = c(c^{-1}a)$ and $b = c(c^{-1}b)$. If $d | a$ and $d | b$, then $d = c(c^{-1}d)$ shows $c | d$.

5. Let $p \in D$ be prime and suppose $p = ab$ for $a, b \in D$. Then $p | ab$. Since p is prime, $p | a$ or $p | b$. If $p | a$, then $a = pa_1$ and $p = ab = (pa_1)b = p(a_1 b)$ implies $1 = a_1 b$; so b is a unit. If $p | b$, a similar argument shows that a is a unit. Thus p is irreducible.

9. Reflexive: $a = a \cdot 1$ implies $a \# a$ for all $a \in D$.
Symmetric: $a \# b$ implies $a = bv$, where $v | 1$; that is, $vu = 1$ for some $u \in D$. But then $au = (bv)u = b(vu) = b \cdot 1 = b$ and $b \# a$.
Transitive: $a \# b$ and $b \# c$ imply $a = bv_1$, $b = cv_2$, where $v_1 | 1$ and $v_2 | 1$. Since $a = bv_1 = (cv_2)v_1 = c(v_1 v_2)$ and $(v_2 v_1) | 1$ (why?), $a \# c$.

16. In $\mathbb{Z}[i]$, $2 + 3i$ has associates $-2 - 3i$, $3 - 2i$, and $-3 + 2i$. In $\mathbb{C}$, every nonzero multiple of $2 + 3i$ is an associate of $2 + 3i$.

Section 16.2, page 287

2. Since every ideal in $2\mathbb{Z}$ is also an ideal in $\mathbb{Z}$, every ideal in $2\mathbb{Z}$ is principal. However, $2\mathbb{Z}$ is not a PID since $2\mathbb{Z}$ has no unit.

5. Suppose $d_1 = (a_1, b_1)$. Then $a_1 = d_1 a_2$, $b_1 = d_1 b_2$ and $a = da_1 = (dd_1)a_2$, $b = db_1 = (dd_1)b_2$. But then (dd_1) is a common divisor of a and b, forcing $(dd_1) \mid d$. Since $d \mid dd_1$, $d = dd_1 v$ where v is a unit, and $1 = d_1 v$ shows that $d_1 = (a_1, b_1)$ is a unit.

7. a. $\overline{33}$　　c. $\overline{15}$

10. By Theorem 16.3, $d = (a, b) = sa + tb \in U + V$. If $U + V = (d_1)$, then $d \in (d_1)$ implies $d_1 \mid d$. Since $a = 1a + 0b \in U + V$, $b = 0a + 1b \notin U + V$, we have $d_1 \mid a$ and $d_1 \mid b$. But then $d \mid d_1$ forcing d and d_1 to be associates. Thus $(d_1) = (d)$.

12. If $a, b \in I$, then $a, b \in I_n$ for some n, so that $a - b$ and $ab \in I_n \subseteq I$. Therefore I is a subring. Furthermore if $r \in R$, the $ra \in I_n \subseteq I$. Thus I is an ideal.

15. a. $(a) \subseteq (b)$ implies $a \in (b)$ implies $a = bd$ implies $b \mid a$. Conversely, if $b \mid a$, then $a = bd$. If $x \in (a)$, then $x = ac = (bd)c = b(dc) \in (b)$; hence $(a) \subseteq (b)$.
   b. No. $\mathbb{Z} \supset 2\mathbb{Z} \supset 4\mathbb{Z} \supset \cdots \supset 2^n\mathbb{Z} \supset \cdots$ is such a chain.

Section 16.3, page 291

1. a. Suppose $a^2 + b^2 = p$, a prime, and $(a + bi) = (u + vi)(x + yi)$. Then $a - bi = (u - vi)(x - yi)$. (Check.) Therefore

$$p = (a + bi)(a - bi)$$
$$= (v + vi)(u - vi)(x + yi)(x - yi)$$
$$= (u^2 + v^2)(x^2 + y^2)$$

Therefore $u^2 + v^2 = 1$ or $x^2 + y^2 = 1$, and $u + vi$ or $x + yi$ is a unit.
   b. $25 = 5 \cdot 5$ in $\mathbb{Z}$. In $\mathbb{Z}[i]$, $5 = (2 + i)(2 - i)$; therefore $25 = (2 + i)^2(2 - i)^2$.

4. a. Let $p \mid ab$. Then $ab = p_1 p_2 \cdots p_n$ where the p_i's are irreducible. Since $p \mid ab$, the uniqueness of the factorizations of a and b forces $p = p_i$ for some i. But the product of irreducible factorizations of a and b is an irreducible factorization of ab. Therefore p is an irreducible factor of either a or b.
   b. Induction on n. If $p \mid a^{n+1} = a^n a$, then $p \mid a^n$ or $p \mid a$ by part a. By induction $p \mid a^n$ implies $p \mid a$.

6. If $a = p_1^{e_1} p_2^{e_2} \cdots p_k^{e_k}$ and $b = p_1^{f_1} p_2^{f_2} \cdots p_k^{f_k}$, where $e_i \geqslant 0$, $f_i \geqslant 0$, then

$$(a) \cap (b) = (p_1^{g_1} p_2^{g_2} \cdots p_k^{g_k})$$

where g_i is the larger of e_i and f_i. Since 3 and $2 + \sqrt{-5}$ are both irreducible in D, (check) if $\mathbb{Z}[2 + \sqrt{-5}]$ is a UFD, then

$$(3) \cap (2 + \sqrt{-5}) = (3(2 + \sqrt{-5})) = (6 + 3\sqrt{-5})$$

But $9 \in (3) \cap (2 + \sqrt{-5})$ and $9 \notin (6 + 3\sqrt{-5})$.

9. If $a = p_1^{e_1} p_2^{e_2} \cdots p_k^{e_k}$ and $b = p_1^{f_1} p_2^{f_2} \cdots p_k^{f_k}$, then

$$\text{lcm}(a, b) = p_1^{g_1} p_2^{g_2} \cdots p_k^{g_k}$$

where $p_i = \max\{e_i, f_i\}$.

12. Since every PID is a UFD, Exercise 9 of Section 16.2 shows $\mathbb{Z}_n$ is a UFD. $\mathbb{Z}_5$ is a field; every nonzero element is a unit. The only nonzero nonunits in $\mathbb{Z}_9$ are 3 and 6; both are prime.

Chapter 17

Section 17.1, page 299

1. Since R is closed under ring addition and multiplication, if $f, g \in R[x]$, then $f \oplus g$ and $f \odot g$ are in $R[x]$.

4. Let $f = (a_0, a_1, \ldots)$, $g = (b_0, b_1, \ldots)$, $h = (c_0, c_1, \ldots)$. Then

$$f \odot (g \oplus h) = f \odot (b_0 + c_0, b_0 + c_1, \ldots) = (d_0, d_1, \ldots)$$

where

$$d_n = \sum_{k+m=n} a_k(b_m + c_m) = \sum_{k+m=n} (a_k b_m + a_k c_m)$$

Since

$$f \odot g = (r_0, r_1, \ldots) \qquad \text{where} \quad r_n = \sum_{k+m=n} a_k b_m$$

and

$$f \odot h = (s_0, s_1, \ldots) \qquad \text{where} \quad s_n = \sum_{k+m=n} a_k c_m$$

we have $(f \odot g) \oplus (f \odot h) = (t_0, t_1, \ldots)$ where

$$t_n = r_n + s_n = \sum_{k+m=n} a_k b_m + \sum_{k+m=n} a_k c_m = d_n$$

Therefore $f \odot (g \oplus h) = (f \odot g) \oplus (f \odot h)$.

7. $ax^i = (b_0, b_1, \ldots)$ where $b_k = \begin{cases} a & k = i \\ 0 & k \neq i \end{cases}$

$bx^j = (c_0, c_1, \ldots)$ where $c_m = \begin{cases} b & m = j \\ 0 & m \neq j \end{cases}$

Then $ax^i \odot bx^j = (d_0, d_1, \ldots)$ where

$$d_n = \sum_{k+m=n} b_k c_m = \begin{cases} 0, & k \neq i \text{ or } m \neq j \\ ab, & k = i \text{ and } m = j \end{cases}$$

Thus

$$d_n = \begin{cases} 0 & n \neq i+j \\ ab & n = i+j \end{cases} \qquad \text{and} \qquad ax^i \odot bx^j = abx^{i+j}$$

10. $f(x) + g(x) = \bar{5}x + \bar{4}$; $f(x)h(x) = \bar{3}x^3 + \bar{2}x^2 + \bar{1}$

Section 17.2, page 308

1. If $f(x) = a$ where $a \in F - \{0\}$, then set $g(x) = a^{-1}$. Clearly $f(x)g(x) = aa^{-1} = 1$, and $f(x)$ is a unit. If $f(x)$ is not a constant polynomial, then $\deg(f(x)) \geqslant 1$ and

$\deg(f(x)g(x)) \geqslant 1$ for any nonzero polynomial $g(x) \in F[x]$. Thus $f(x)g(x) \neq 1$ since $\deg(1) = 0$.

4. $f(x) = (2x - 1)(x + 2)$ is not irreducible, but $f(1) = 3$.

7. $f(3) = 22$, $f(1) = 0$, $f(0) = 1$, $f(-2) = -3$, $f(2) = 5$. Thus 1 is a root.

10. $f(0) = f(1) = f(3) = f(4) = 0$

12. a. $x - 1$ c. $x - 2$

14. Since $x^2 - 2$ has no roots in $\mathbb{Q}$, $(x^2 - 2, x^2 - 3x + 2) = 1$. By Theorem 16.3, the answer is yes.

Section 17.3, page 317

1. If $(r, s^n) = d \neq 1$, then let p be a prime divisor of d. Then $p \mid r$ and $p \mid s^n$. By Exercise 4 of Section 16.3, $p \mid s$, so that $p \mid (r, s)$, contradicting the fact that $(r, s) = 1$.

3. By Theorem 17.10, if r/s is a rational root of $f(x)$, then s divides the leading coefficient 1. If $s \mid 1$, then $r/s = \pm r \in \mathbb{Z}$.

7. a. $(x + 1)(x^2 - x + 1)$ d. $(x - 2)(x + 2)(x + 1)(x + 5)$
    g. $x^8 + 1$ h. $(2x^2 + 5)(x^2 + 4)$

10. Let $z = r(\cos\theta + i\sin\theta)$. Then

$$z^n z^{-n} = r^n r^{-n}(\cos n\theta + i\sin n\theta)(\cos(-n)\theta + i\sin(-n)\theta)$$

$$= 1(\cos n\theta + i\sin n\theta)(\cos n\theta - i\sin n\theta)$$

$$= (\cos n\theta)^2 + (\sin n\theta)^2 = 1$$

Thus $z^{-n} = (z^n)^{-1}$.

13. Each complex root z of $f(x)$ can be paired with its complex conjugate z to obtain a quadratic real factor of $f(x)$. The quotient of $f(x)$ by all such possible quadratic factors will be a real polynomial of odd degree all of whose roots are real.

15. a. $\{\sqrt[3]{2}, \sqrt[3]{2}(-\frac{1}{2} + i\frac{\sqrt{3}}{2}), \sqrt[3]{2}(-\frac{1}{2} - i\frac{\sqrt{3}}{2})\}$
    c. $x^4 = 2/i = -2i = 2(\cos(\pi + 2\pi k) + i\sin(\pi + 2\pi k))$ implies

$$x = \sqrt[4]{2}\left(\cos\left(\frac{\pi}{4} + \frac{\pi}{2}k\right) + i\sin\left(\frac{\pi}{4} + \frac{\pi}{2}k\right)\right) \text{ for } k = 0, 1, 2, 3$$

j. $x = \cos\left(\dfrac{\pi}{12} + \dfrac{\pi}{6}k\right) + i\sin\left(\dfrac{\pi}{12} + \dfrac{\pi}{6}k\right)$ for $k = 0, 1, 2, \ldots, 11$

18. Use induction on n. If $n = 1$, then $f(x) = a_1 x + a_0 = a_1(x - (-a_0 a_1^{-1}))$. If $n > 1$, then $f(x)$ has a root z_1 by the Fundamental Theorem of Algebra. Therefore $f(x) = (x - z_1)g(x)$, where $g(x) \in \mathbb{C}[x]$ has degree $n - 1$ and leading coefficient a_n. Any root of $g(x)$ is a root of $f(x)$. Thus, by our induction hypothesis,

$$f(x) = (x - z_1)(a_n(x - z_2) \cdots (x - z_n))$$

$$= a_n(x - z_1) \cdots (x - z_n)$$

Section 17.4, page 323

2. a. Irreducible by Eisenstein's Criteria.
    d. $(x-1)^2(x+2)(x^2+x+1)$
    f. $(x-1)(x^2+10x+28)$

5. a. $x^3+6x^2-3x-148 = (x-4)(x^2+10x+37)$

$$= (x-4)(x+5-2i\sqrt{3})(x+5+2i\sqrt{3})$$

f. $x^5+32 = (x+2)(x^4-2x^3+4x^2-8x+16)$

$$= (x+2)\left(x^2-4\left(\cos\frac{\pi}{5}\right)x+4\right)\left(x^2-4\left(\cos\frac{3\pi}{5}\right)x+4\right)$$

$$= \prod_{k=0}^{4}\left(x-2\left(\cos\left(\frac{\pi}{5}+\frac{2\pi k}{5}\right)+i\sin\left(\frac{\pi}{5}+\frac{2\pi k}{5}\right)\right)\right)$$

g. $x^4-10x^2+32x-7$

$$= (x^2-4x+7)(x^2-4x-1)$$

$$= (x^2-4x+7)(x+2-\sqrt{5})(x+2+\sqrt{5})$$

$$= (x-2-i\sqrt{3})(x-2+i\sqrt{3})(x+2-\sqrt{5})(x+2+\sqrt{5})$$

Chapter 18

Section 18.1, page 329

7. b. $[3, 0, 20, 27]$
10. Show that zero divisors exist.
14. b. The set V fails to contain the zero vector.
15. $a\mathbf{u} = a\mathbf{v}$ implies $a(\mathbf{u} \ominus \mathbf{v}) = \mathbf{0}$. Apply Theorem 18.1.

Section 18.2, page 337

4. If $V = \{ap(x)|a \in F\}$, then $ap(x)+bp(x) = (a+b)p(x) \in V$ and $c(ap(x)) = (ca)p(x) \in V$.
7. Exercise 4 is a special case of Exercise 7, the case where $\mathbf{v} = p(x)$ and $V = F[x]$.
11. Careful, $B \subseteq A$ does not mean it is a subspace of A.
14. a. $[2, -1, 3] = (-5/3)[1, 0, -1] + 5[0, 2, 1] + (-11/3)[-1, 3, 1]$
19. Any linear combination of the vectors $\mathbf{v}_1, \mathbf{v}_2, \ldots, \mathbf{v}_n$ is a linear combination of vectors in S.

Section 18.3, page 347

2. c. not independent   d. independent
8. dim $W = 4$
10. The vector $[1, 0, 0]$ is one possible answer.

13.　b. If $\{\mathbf{w}_1, \mathbf{w}_2, \ldots, \mathbf{w}_k\}$ is independent, then it can be extended to a basis of more than n vectors, an obvious contradiction.

15.　$\dim_F(A + B) = 3$

Section 18.4, page 356

1.　Parts a and e are transformations.

3.　b. $4\tau_S([x, y, z, w] = [4x - 4w + 4y + 20z, 4y - 8x + 24w, 4z, 4w]$

6.　If $\tau(V) = V$, then $\dim(N_\tau) = \dim(V) - \dim(\tau(V)) = 0$.

8.　a. $\{\mathbf{0}\}$　　b. 2

13.　N_D consists of all constant functions.

16.　e. $[1, 2, 3]$

19.　a. $N_{\tau_1} = \{\mathbf{0}\}$; rank is 2.

Section 19.1, page 370

2.　Show both have the same images.

5.　$M(\tau) = \begin{bmatrix} 1 & 1 & 0 \\ 0 & 0 & 1 \end{bmatrix}$

7.　$(\rho\sigma)\left(\begin{bmatrix} x_1 \\ x_2 \end{bmatrix}\right) = \begin{bmatrix} x_1 \\ x_2 \end{bmatrix}$

8.　a. $AB = \begin{bmatrix} 2 & 2 \\ -2 & -1 \\ 4 & 5 \end{bmatrix}$　　b. BA isn't defined.

14.　$\begin{bmatrix} 1 & -1 & 0 \\ 1 & 1 & 2 \end{bmatrix}^T = \begin{bmatrix} 1 & 1 \\ -1 & 1 \\ 0 & 2 \end{bmatrix}$

Section 19.2, page 382

4.　The transition matrix is $\begin{bmatrix} x_1 & x_2 \\ x_3 & x_4 \end{bmatrix}$ where

$$x_1\begin{bmatrix} 1 \\ 1 \end{bmatrix} + x_3\begin{bmatrix} -1 \\ 1 \end{bmatrix} = \begin{bmatrix} 2 \\ 1 \end{bmatrix}, \qquad x_2\begin{bmatrix} 1 \\ 1 \end{bmatrix} + x_4\begin{bmatrix} -1 \\ 1 \end{bmatrix} = \begin{bmatrix} 0 \\ 2 \end{bmatrix}$$

7.　$P = \begin{bmatrix} 1 & 0 & -1 \\ 0 & 1 & 0 \\ 0 & 1 & 1 \end{bmatrix}$

11.　See Exercises 11 and 20 of Section 18.4.

Section 19.3, page 401

2. $\tau\left(\sum_{i=1}^{n} a_i\mathbf{v}_i + \sum_{i=1}^{n} b_i\mathbf{v}_i\right) = \tau\left(\sum_{i=1}^{n} (a_i + b_i)\mathbf{v}_i\right) = \sum_{i=1}^{n} (a_i + b_i)\mathbf{w}_i$

$$= \sum_{i=1}^{n} a_i\mathbf{w}_i + \sum_{i=1}^{n} b_i\mathbf{w}_i$$

$$= \tau\left(\sum_{i=1}^{n} a_i\mathbf{v}_i\right) + \tau\left(\sum_{i=1}^{n} b_i\mathbf{v}_i\right)$$

Similarly, $\tau\left(c \sum_{i=1}^{n} a_i\mathbf{v}_i\right) = \tau\left(\sum_{i=1}^{n} (ca_i)\mathbf{v}_i\right) = \sum_{i=1}^{n} (ca_i)\mathbf{w}_i$

$$= c \sum_{i=1}^{n} a_i\mathbf{w}_i = c\left(\tau\left(\sum_{i=1}^{n} a_i\mathbf{v}_i\right)\right)$$

8.   b. $\begin{bmatrix} 1 & 0 & 2 \\ 0 & 1 & -1 \\ 0 & 0 & 0 \end{bmatrix}$ d. $\begin{bmatrix} 1 & 0 & 0 \\ 0 & 1 & 0 \\ 0 & 0 & 1 \\ 0 & 0 & 0 \end{bmatrix}$

11.   a. independent   b. independent
12.   a. $x = y = z = 0$ d. no solution
15.   For symmetry, use the fact that the inverse of any elementary operation is another elementary operation.

Section 19.4, page 410

1.   00000, 10101, 01111, 11010   3. $\begin{bmatrix} 1 & 0 & 0 & 1 & 0 \\ 0 & 1 & 0 & 1 & 1 \\ 0 & 0 & 1 & 0 & 1 \end{bmatrix}$

7.   The message is: 0010   0011   1111   1010.
12. $H(Y_1 + Y_2)^T = H(Y_1^T + Y_2^T) = HY_1^T + HY_2^T = \mathbf{0} + \mathbf{0} = \mathbf{0}$ and $H(cY)^T = c(HY^T)$
     $= c\mathbf{0} = \mathbf{0}$.

Chapter 20

Section 20.1, page 421

1.   Since $|F| \geqslant 2$, there exists $a \in F - \{0\}$. Therefore $0 = a - a$ and $1 = aa^{-1} \in F$. By Theorem 4.6, both $(F, +)$ and $(F - \{0\}, \cdot)$ are groups. Then $b^{-1} = 1 \cdot b^{-1} \in F$ for all $b \in F - \{0\}$, $a - b$ and $ab \in F$ for all $ab \in F$. Thus, F is a subring of the commutative ring E. Because $(F - \{0\}, \cdot)$ is a group, F is a subfield.
4.   Let F be a subfield of $\mathbb{Q}$. Then $1 \in F$; furthermore, the set $\{n \cdot 1 | n \in \mathbb{Z}\}$ is a subring of F isomorphic to $\mathbb{Z}$. Since F is a field, it must contain the field of quotients of $\mathbb{Z}$. But then $\mathbb{Q} \subseteq F$.
7.   By induction and Exercise 2, the intersection of all subfields of E containing

both F and S is a subfield F_1 of E. This subfield F_1 clearly contains both F and S. The smallest subfield containing F and S must therefore contain F_1, forcing F_1 to be this smallest subfield.

10. The set $\{1, x, 1/x, x^2, 1/x^2, x^3, 1/x^3, \ldots\}$ is an infinite subset of $F(x)$. This set is linearly independent in the F-vector space $F(x)$. Thus, $\dim_F F(x)$ is infinite.

13. Since $\sqrt{336} = 4\sqrt{21} = 4\sqrt{3}\sqrt{7} \in \mathbb{Q}(\sqrt{3}, \sqrt{7})$, $\mathbb{Q}(\sqrt{3}, \sqrt{7}, \sqrt{336}) = \mathbb{Q}(\sqrt{3}, \sqrt{7})$. But $[\mathbb{Q}(\sqrt{3}, \sqrt{7}):\mathbb{Q}] = 4$.

16. If $m(x) = m_{\alpha, F}(x) = p(x)q(x)$, $\deg(p(x)) < \deg(m(x))$ and $\deg(q(x)) < \deg(m(x))$, then $0 = m(\alpha) = p(\alpha)q(\alpha)$. Since $F[x]$ is an integral domain, $p(\alpha) = 0$ or $q(\alpha) = 0$, contradicting the nature of $m(x)$ as the minimal polynomial of α.

18. $m_{\sqrt[3]{2}, \mathbb{Q}}(x) = x^3 - 2$. $m_{\sqrt[3]{2}, \mathbb{R}}(x) = m_{\sqrt[3]{2}, \mathbb{C}}(x) = x - \sqrt[3]{2}$.

21. By Exercise 7, $F(\alpha, \beta)$ is the smallest subfield of the extension containing F, α, and β. But this is also $F(\beta, \alpha)$.

Section 20.2, page 430

1. If $a_0(1 + I) + a_1(x + I) + \cdots + a_{n-1}(x^{n-1} + I) = 0 + I = I$, then

$$(a_0 + a_1 x + \cdots + a_{n-1} x^{n-1}) + I = I$$

and

$$a_0 1 + a_1 x + \cdots + a_{n-1} x^{n-1} \in (p(x))$$

Since $\deg(p(x)) = n$, we must have $a_0 = a_1 = \cdots = a_{n-1} = 0$, and the set

$$S = \{1 + I, x + I, \ldots, x^{n-1} + I\}$$

is independent. Furthermore, if $f(x) + I \in F[x]/I$, then by the Division Algorithm for $F[x]$, there exist $q(x), r(x) \in F[x]$ such that $f(x) = q(x)p(x) + r(x)$, where $r(x) = 0$ or $\deg(r(x)) < \deg(p(x)) = n$. Thus $f(x) + I = r(x) + I$. If $r(x) = a_0 + a_1 x + \cdots + a_{n-1} x^{n-1}$, then

$$r(x) + I = a_0(1 + I) + a_1(x + I) + \cdots + a_{n-1}(x^{n-1} + I)$$

This shows that S is a spanning set.

3. If $p(x) = x^3 - 5$, then $p(\sqrt[3]{5}) = 0$, and $p(x)$ is irreducible over $\mathbb{Q}$ by Eisenstein's Criteria. Therefore $p(x) = m_{\sqrt[3]{5}, \mathbb{Q}}(x)$ and $[\mathbb{Q}(\sqrt[3]{5} : \mathbb{Q}] = 3$ by Theorem 20.3.

5. a. $\mathbb{Q}(i, \sqrt{3})$ d. $\mathbb{Q}(i)$ g. $\mathbb{Q}(\sqrt{3/2}, i)$ i. $\mathbb{C}$

8. Define $\bar{\phi}: F[x] \to F_1[x]$ by

$$\bar{\phi}(a_0 + a_1 x + \cdots + a_n x^n) = \bar{a}_0 + \bar{a}_1 x + \cdots + \bar{a}_n x^n, \text{ where } \bar{a}_i = \phi(a_i)$$

Then $\bar{\phi}$ is an isomorphism and $\bar{\phi}(a_0) = \phi(a_0)$. Let $p(x)$ be an irreducible factor of $f(x)$ and $p_1(x) = \bar{\phi}(p(x))$. Then $p_1(x)$ is an irreducible factor of $f_1(x)$. If u is a root of $p(x)$, then $F(u) \cong F[x]/(p(x))$. If v is a root of $p_1(x)$, then $F_1(v) \cong F_1[x]/(p_1(x))$. Define $\bar{\phi}: F(u) \to F(v)$ by

$$\bar{\phi}(g(x) + (p(x))) = \bar{\phi}(g(x)) + (p_1(x))$$

Show that $\bar{\phi}$ is an isomorphism and $\bar{\phi}(a_0) = \bar{\phi}(a_0) = \phi(a_0)$. If K is a splitting field of f over F and L is a splitting field of f_1 over F_1, we can now prove the result by induction on $[K : F]$. If $[K : F] = 1$, then $K = F$, and all roots of f are in F. Then $L = F_1$ and ϕ is itself the desired isomorphism. If $[K : F] > 1$, then we have seen that $F(u)$ is isomorphic to $F(v)$ and $[K : F(u)] < [K : F]$. Therefore, there exists an isomorphism from K to L that agrees with $\bar{\phi}$, and hence ϕ, on F.

10.   Let $K = F(\alpha_1, \alpha_2, \ldots, \alpha_n)$ be an extension of characteristic 0. If $n = 1$, then $K = F(\alpha_1)$ is simple. Proceed by induction on n, and assume that any extension generated by $n - 1$ elements is simple. But

$$K = (F(\alpha_1, \ldots, \alpha_{n-1}))(\alpha_n) = F(\gamma, \alpha_n)$$

by induction. The result follows from Theorem 20.9.

13.   a. $i + \sqrt{2}$ d. 1   f. $\sqrt{3} \equiv 1$ (modulo 2); therefore, i is primitive.   h. $i + \sqrt{2}$

14.   a. $x^4 - 2x^2 + 9$ d. $x^4 - 10x^2 + 1 = 0$

16.   If $f(x) = (x - a)^m g(x)$, then $f'(x) = (x - a)^{m-1}[mg(x) + (x - a)g'(x)]$. Then $(x - a)^{m-1}$ divides $(f(x), f'(x))$.

Section 20.3, page 437

1.   If $b_1 = 0$ and $b_2 = 0$, then both lines are parallel. If $b_1 = 0$ and $b_2 \neq 0$, then $a_1 \neq 0$ so that $a_1 b_2 - a_2 b_1 = a_1 b_2 \neq 0$. If $b_1 - 0$, subtract b_2 times the first equation from b_1 times the second equation to obtain

$$(a_2 b_1 - a_1 b_2)x = c_2 b_1 - b_2 c_1$$

If the lines intersect at (x, y) then

$$x = \frac{c_2 b_1 - b_2 c_1}{a_2 b_1 - a_1 b_2} \qquad \text{so} \qquad a_2 b_1 - a_1 b_2 \neq 0$$

4.   Construct an equilateral triangle and bisect any vertex to obtain a 60° angle. Bisect any right angle to obtain a 45° angle.

5.   b. The only possible rational roots are $\pm 1, \pm 1/2, \pm 1/4; \pm 1/8$. $p(a) \neq 0$ for any of these values of a.

7.   d. Consider $\dfrac{w^4 + w^3 + w^2 + w + 1}{w^2} = 0$.

8. $f(x) = x^6 + x^5 + x^4 + x^3 + x^2 + x + 1$ is irreducible over $\mathbb{Q}$ by Example 17.7. But $m_{w, \mathbb{Q}}(x) = f(x)$ has degree 6, not a power of 2. Therefore the angle $2\pi/7$ cannot be constructed.

10.   If 1° is a constructible number, then $\pi = 180(1°)$ is a constructible number.

Chapter 21

Section 21.1, page 444

1.   Since $p \cdot 1 = 0$ in E, $p \cdot 1 = 0$ also in F.

4.   If $\alpha, \beta \in F - \{0\}$, then $(\alpha\beta)^{p^n} = \alpha^{p^n}\beta^{p^n}$ since $(F - \{0\}, \cdot)$ is an abelian group. If either α or β is 0, then both $(\alpha\beta)^{p^n}$ and $\alpha^{p^n}\beta^{p^n}$ are 0.

8. $x^2 - 2$ is irreducible in $\mathbb{Z}_3[x]$. Let β be a root of this polynomial in some

splitting field. Then $\beta^2 = 2$. Then set

$$\{0, 1, 2, \beta, 2\beta, 1 + \beta, 1 + 2\beta, 2 + \beta, 2 + 2\beta\}$$

forms a field with multiplication given by

$$(a + b\beta)(c + d\beta) = ac + (ad + bc)\beta + bd\beta^2$$
$$= (ac + 2bd) + (ad + bc)\beta$$

12. If $|F| = p$, then F is isomorphic to $(\mathbb{Z}_p, +, \cdot)$. By Corollary 7.4, $a^p \equiv a$ (modulo p) for all $a \in \mathbb{Z}_p$. Therefore $a^p - a \equiv 0$ (modulo p) for all $a \in F \cong \mathbb{Z}_p$.

15. $\phi(a + b) = (a + b)^{p^k} = a^{p^k} + b^{p^k} = \phi(a) + \phi(b)$ by Lemma 21.1. Similarly $\phi(ab) = \phi(a)\phi(b)$. If $\phi(a) = \phi(b)$, then $(a - b)^{p^k} = a^{p^k} - b^{p^k} = 0$. Since $(F - \{0\}, \cdot)$ is a group, $a - b$ is not an element of $F - \{0\}$ and $a = b$. Therefore ϕ is injective. Since $|F|$ is finite, ϕ is an automorphism.

18. $(F - \{0\}, \cdot)$ is a finite cyclic group. By Theorem 4.5, $(S, \cdot)$ is a subgroup and by Theorem 4.7, $(S, \cdot)$ is cyclic.

Section 21.2, page 451

1. $\alpha^3 = \begin{bmatrix} 1 \\ 0 \\ 1 \end{bmatrix}$, $\alpha^4 = \begin{bmatrix} 1 \\ 1 \\ 1 \end{bmatrix}$, $\alpha^5 = \begin{bmatrix} 1 \\ 1 \\ 0 \end{bmatrix}$, $\alpha^6 = \begin{bmatrix} 0 \\ 1 \\ 1 \end{bmatrix}$

4. 0111100 corresponds to $\alpha^5 + \alpha^4 + \alpha^3 + \alpha^2 = \alpha^2 + 1 = \alpha^3$. Therefore correct as $0111100 + 0001000 = 0110100$ and decode as 0110.

7. a. $G(x) = x^{14} + x^{13} + x^{10} + x^8 + x^7 + x^2$ implies

$$G(\alpha) = \alpha^{14} + \alpha^{13} + \alpha^{10} + \alpha^8 + \alpha^7 + \alpha^2 = \alpha^{11}$$

$$G(\alpha^3) = \alpha^{42} + \alpha^{39} + \alpha^{30} + \alpha^{24} + \alpha^{21} + \alpha^6$$
$$= \alpha^{12} + \alpha^9 + 1 + \alpha^9 + \alpha^6 + \alpha^6 = \alpha$$

$$P(x) = x^2 + \alpha^{11}x + \frac{\alpha^{11}\alpha^{22} + \alpha}{\alpha^{11}}$$
$$= x^2 + \alpha^{11}x + (\alpha^7 + \alpha^5)$$
$$= x^2 + \alpha^{11}x + \alpha^{14} = (x + \alpha^{14})(x + 1)$$

Therefore $E(x) = x^{14} + 1$. Message should have been 010010110000101. Decode as 0100101.

10. $m(0) = m(1) = 1$. If $m(x) = (x^2 + ax + 1)(x^2 + bx + 1) = x^4 + x + 1$, then $a + b = 0$, $ab = 0$ and $a + b = 1$, which cannot occur. Thus $m(x)$ is irreducible.

12. $G(\alpha^3) = \alpha^{39} + \alpha^{30} + \alpha^{21} + \alpha^6 + \alpha^3 + 1$
$$= \alpha^9 + 1 + \alpha^6 + \alpha^6 + \alpha^3 + 1$$
$$= \alpha^9 + \alpha^3 = (\alpha^2 + 1) + \alpha^3 = \alpha^3 + \alpha^2 + 1 = \alpha^{11}$$

Chapter 22

Section 22.1, page 458

2. Let G be the group of automorphisms of E and

$$G_{E/F} = \{\sigma \in G \mid \sigma(\alpha) = \alpha \text{ for all } \alpha \in F\}$$

Clearly $1_G \in G_{E/F}$. If σ, $\tau \in G_{E/F}$ and $\sigma \in F$, then $(\sigma\tau)(a) = \sigma(\tau(a)) = \sigma(a) = a$, so that $\sigma\tau \in G_{E/F}$. Furthermore, $\sigma(a) = a$ implies $a = \sigma^{-1}(a)$, so $\sigma^{-1} \in G_{E/F}$. Thus, $G_{E/F} \leqslant G$ by Theorem 4.4.

4. $f(x)$ splits over $\mathbb{Z}_5$.

6. Consider $\sigma(a^2)$.

8. The splitting field of $x^3 - 2$ over $\mathbb{Q}$ is $\mathbb{Q}(\sqrt[3]{2}, \omega)$, where $\omega = (-1 + i\sqrt{3})/2$. $x^3 - 2$ has two imaginary roots, so $\mathbb{Q}(\sqrt[3]{2})$ is not the splitting field, since $\mathbb{Q}(\sqrt[3]{2}) \subseteq \mathbb{R}$. $\mathbb{Q}(\sqrt[3]{2})$ is a subfield of $\mathbb{Q}(\sqrt[3]{2}, \omega)$.

9. a. The splitting field of $x^5 + 1$ over $\mathbb{Q}$ is $\mathbb{Q}(\omega)$, where $\omega = \cos(\pi/5) + i \sin(\pi/5)$. The Galois group

$$G = \{\sigma_i \mid \sigma_i(\omega) = \omega^i, \ i = 1, 2, 3, 4\}$$

is cyclic, and $|G| = 4$. The only nontrivial subgroup of G is $H = \langle \sigma_4 \rangle = \{\sigma_1, \sigma_4\}$. The corresponding fixed field is

$$E = \{b_0 + b_1(\omega + \omega^4) \mid b_0, b_1 \in \mathbb{Q}\}$$

   f. $x^6 - 1 = (x^3 - 1)(x^3 + 1) = (x + 1)(x - 1)(x^2 - x + 1)(x^2 + x + 1)$

The splitting field is $\mathbb{Q}(i\sqrt{3})$. The Galois group has order 2, generated by complex conjugation.

11. $\gamma = \sqrt[3]{5} + i\sqrt{3}$. $[\mathbb{Q}(\gamma) : \mathbb{Q}] = 6$. Yes, $\mathbb{Q}(\gamma)$ is a normal extension.

14. If $a = b^2$ for $b \in \mathbb{Q}$, then $f(x) = (x - b)(x + b)$, and $E = \mathbb{Q}$. If $a \neq b^2$ for $b \notin \mathbb{Q}$, then f is irreducible in $\mathbb{Q}[x]$. If $a > 0$, then $E = \mathbb{Q}(\sqrt{a})$; if $a < 0$, then $E = \mathbb{Q}(i\sqrt{a})$. In both cases, $[E : \mathbb{Q}] = 2$.

Section 22.2, page 467

1. Since $|G|$ divides $3! = 6$ and 3 divides $|G|$ by Theorem 20.5, $|G| = 3$ or 6.

4. $[\mathbb{Q}(\sqrt[3]{5}) : \mathbb{Q}] = 3$, but $|G| = 6$.

7. Since $x^2 + 1 \in \mathbb{Q}[x]$, $\tau(x^2 + 1) = x^2 + 1$. But

$$\tau(x^2 + 1) = \tau((x + i)(x - i)) = \tau(x + i)\tau(x - i) = (x + \sqrt{2})(x - \sqrt{2}) = x^2 - \sqrt{2}$$

10. $a + bi + c\sqrt{2} + di\sqrt{2} = \sigma(a + bi + c\sqrt{2} + di\sqrt{2}) = a + bi - c\sqrt{2} - di\sqrt{2}$ implies $c = -c$, $d = -d$ so $a + bi + c\sqrt{2} + di\sqrt{2} = a + bi$. Thus $F = \mathbb{Q}(i)$.

14. $G_{E/\mathbb{Q}}$ is a cyclic group of order p.

17. $E = \mathbb{Q}(\sqrt{2}, \sqrt{3})$. Subfields: $\mathbb{Q}(\sqrt{2})$, $\mathbb{Q}(\sqrt{3})$, $\mathbb{Q}(\sqrt{6})$.

$$G = \langle \sigma, \tau \mid \sigma(\sqrt{2} + \sqrt{3}) = -\sqrt{2} + \sqrt{3}, \ \tau(\sqrt{2} + \sqrt{3}) = \sqrt{2} - \sqrt{3} \rangle$$

$|G| = 4$.

Section 22.3, page 476

3. Use Eisenstein's Criteria.

4. b. $(1\quad 3) = (2\quad 3)(1\quad 2)(2\quad 3)$,
    $(1\quad 4) = (3\quad 4)(1\quad 3)(3\quad 4)$, $(2\quad 4) = (3\quad 4)(2\quad 3)(3\quad 4)$, etc.

5. b. $\mathbb{Z}_4$ d. S_5

9. Use Exercise 1 of Section 19.2.

10. c. Show that f is contained in the fixed field of $G_{E/K}$, where $K = F(s_1, s_2, \ldots, s_n)$ and E is its splitting field.

11. b. $\sigma(\Delta^2) = \Delta^2$ implies $\Delta^2 \in E_G = F$.
    d. Any even permutation is the product of an even number, say $2m$, of transpositions. By part c, $\tau(\Delta) = (-1)^{2m}\Delta = \Delta$.

Section 23.1, page 484

1. $a = a \cdot 1$ implies $a \mid a$. If $a \mid b$ and $b \mid a$, then $b = av$, $a = bu$, so that $b = buv$ and $uv = 1$. Since $u \in \mathbb{N}$, $u = 1$ and $a = b$. Since $2 \mid (-2)$ and $(-2) \mid 2$ in $\mathbb{Z}$, "divides" is not antisymmetric on $\mathbb{Z}$.

3. See Exercise 10 of Section 7.3.

4. If 0, $0'$ are zero elements, then $0 \leqslant 0'$ and $0' \leqslant 0$, so that $0 = 0'$.

7. a.   d.

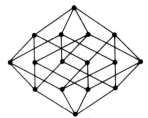
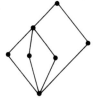

10. b. S_4 e. D of Example 8.6

15.

$$G = \langle a \rangle$$
$$\langle a^p \rangle$$
$$\langle a^{p^2} \rangle$$
$$\vdots$$
$$\langle a^{p^{n-1}} \rangle$$
$$\{1\}$$

Section 23.2, page 493

1. Parts a, c, d, e, and f

4. $a \wedge (a \vee b) \leqslant a$ by definition of $\wedge$. Conversely $a \leqslant a$ and $a \leqslant a \vee b$ imply $a = a \wedge a \leqslant a \wedge (a \vee b)$. By antisymmetry, $a = a \wedge (a \vee b)$.

6.   b. Equivalent to • ,  , or the original lattice.

9.   Set $(a + bi) \vee (c + di) = u + vi$, where $u = \max\{a, c\}$, $v = \max\{b, d\}$, and
$(a + bi) \wedge (c + di) = r + si$, where $r = \min\{a, c\}$, $s = \min\{b, d\}$.

12.   Let $x = ab \in AB \cap C$. Show that $b \in B \cap C$.

13.   contains three pentagonal sublattices.

17.   Let $x = a \vee b$, $y = a$, $z = c$. Then

$$(a \vee b) \wedge (a \vee c) = x \wedge (y \vee z)$$

$$= (x \wedge y) \vee (x \wedge z)$$

$$= [(a \vee b) \wedge a] \vee [(a \vee b) \wedge c]$$

$$= [a] \vee [(a \wedge c) \vee (b \wedge c)]$$

$$= [a \vee (a \wedge c)] \vee (b \wedge c)$$

$$= a \vee (b \wedge c)$$

20.   Suppose n is square-free and $m \mid n$. Then $\bar{m} = n/m$ since $(m, n/m) = 1$. If p^2/n, then p
has no complement: if $\bar{p}$ is the complement of p, then $1 = p \wedge \bar{p} = (p, \bar{p})$ implies
$p \dagger \bar{p}$. But then $n = p \wedge \bar{p} = p\bar{p}$ is not divisible by p^2.

22.   The normal subgroups of S_4 are 1, K, A_4, and S_4. Their lattice is not
complemented.

Section 23.3, page 501

2.   a. Define $a \vee b = a + b - ab$ and $a \wedge b = ab$.

3.   No. The lattice of subgroups of K, the Klein Four-Group, is nonmodular.

5.   b. $a + \bar{a}b = a(b + \bar{b}) + \bar{a}b = ab + a\bar{b} + \bar{a}b = ab + \bar{a}b + \bar{b}a$

$$= (a + \bar{a})b + \bar{b}a = b + \bar{b}a$$

8. $1 = a + \bar{a} = a + a'$ and $0 = a\bar{a} = aa'$ imply $\bar{a} = a'$.

13.   a.

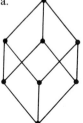

15. If $|B_1| = |B_2| = 2^n$, both are isomorphic to the Boolean algebra of subsets of $A = \{a_1, a_2, \ldots, a_n\}$.

17. Figure 23.9. Here b is a complement for a, c is a complement for b, and $c \neq a$.

18. You must show that the idempotent, absorption, and associative laws, as well as $a + 1 = 1$ and $a \cdot 0 = 0$, follow from the rules stated in this exercise.

Section 23.4, page 507

1. c. $(A \cap B \cap \bar{C}) \cup (\bar{A} \cap B \cap C) \cap (A \cap \bar{B} \cap C)$

2. a. $\overline{x_1 \bar{x}_3 + x_2} = (\overline{x_1 \bar{x}_3})(\bar{x}_2) = (\bar{x}_1 + \bar{\bar{x}}_3)\bar{x}_2$

$$= (\bar{x}_1 + x_3)\bar{x}_2 = \bar{x}_1\bar{x}_2 + \bar{x}_2 x_3$$

$$= \bar{x}_1\bar{x}_2(x_3 + \bar{x}_3) + (x_1 + \bar{x}_1)\bar{x}_2 x_3$$

$$= \bar{x}_1\bar{x}_2 x_3 + \bar{x}_1\bar{x}_2\bar{x}_3 + x_1\bar{x}_2 x_3$$

3. b. $p \leftrightarrow q$ is equivalent to $pq + (\sim p)(\sim q)$.

4. c. $[p \wedge (p \to q)]$ is equivalent to $p(pq + \bar{p}q + \bar{p}\bar{q}) = pq$ where $\bar{p}$ denotes $\sim p$. Thus

$$[p \wedge (p \to q)] \to q$$

is equivalent to $pq \to q$, which can be written as

$$(pq)q + \overline{pq}\,q + \overline{pq}\,\bar{q} = pq + (\bar{p} + \bar{q})q + (\bar{p} + \bar{q})\bar{q} = pq + \bar{p}q + \bar{p}q + \bar{q}$$

$$= pq + \bar{p}q + p\bar{q} + (p + \bar{p})\bar{q} = pq + \bar{p}q + p\bar{q} + \bar{p}\bar{q}$$

Since all minterms are present, we have a tautology.

6. Any element of $B(n)$ (including the atoms) can be written as a sum of minterms. Thus the minterms are contained in the atoms. Since the atoms are minimal, the atoms must be minterms.

8. The minterms of $B(n)$ are of the form $x_1^{e_1} x_2^{e_2} \cdots x_n^{e_n}$ where $x_i^{e_i} = x_i$ or $\bar{x}_i$. These correspond to the vectors in $(\mathbb{Z}_2)^n$, where $x_i^{e_i} = x_i$ yields a 1 and $x_i^{e_i} = \bar{x}_i$ yields a 0. There are 2^n such minterms. Every element of $B(n)$ can be written uniquely as a sum of (some of) these minterms. There are $2^{(2^n)}$ subsets of the set of minterms, and, therefore, $2^{(2^n)}$ such expressions.

11. a. $ac + \bar{a}b + \bar{a}\bar{b} + a\bar{c} = a(c + \bar{c}) + \bar{a}(b + \bar{b}) = a + \bar{a} = 1$

14. b. If $f = \bar{g}$ and $g = \sum x_1^{e_1} x_2^{e_2} \cdots x_n^{e_n}$, then

$$\overline{\sum x_1^{e_1} x_2^{e_2} \cdots x_n^{e_n}} = \prod (\bar{x}_1^{e_1} + \bar{x}_2^{e_2} + \cdots + \bar{x}_n^{e_n})$$

by Exercise 9 of Section 23.3.

16. a. $f = \overline{(\bar{x} + \bar{y}z)}$

$$= \overline{\bar{x}}\,(\overline{\bar{y}z})$$

$$= \bar{\bar{x}}(\bar{\bar{y}} + \bar{z})$$

$$= x(y + \bar{z})$$

$$= xy + x\bar{z}$$

Bibliography

Abstract Algebra (General)

Birkhoff, G., and S. MacLane. *A Survey of Modern Algebra* 4th ed. New York: Macmillan, 1977.

Burton, D.M. *Abstract and Linear Algebra*. Reading, Mass.: Addison-Wesley, 1972.

Dean, R.A. *Elements of Abstract Algebra*. New York: Wiley, 1966.

Durbin, J.R. *Modern Algebra*. New York: Wiley, 1979.

Fraleigh, J.B. *A First Course in Abstract Algebra*. 3rd ed. Reading, Mass.: Addison-Wesley, 1983.

Gilbert, J., and L. Gilbert. *Modern Algebra*. Boston: Prindle, Weber & Schmidt, 1984.

Herstein, I.N. *Topics in Algebra*. 2nd ed. New York: Wiley, 1975.

Hungerford, T.W. *Algebra*. New York: Holt; Rinehart and Winston, 1974.

Jacobson, N. *Basic Algebra I*. San Francisco: W.F. Freeman and Company, 1974.

Larney, V.C. *Abstract Algebra: A First Course*. Boston: Prindle, Weber & Schmidt, 1975.

McCoy, N.H., and T.R. Berger. *Algebra: Groups, Rings, and Other Topics*. Boston: Allyn and Bacon, 1977.

Pinter, C.C. *A Book of Abstract Algebra*. New York: McGraw Hill, 1982.

Shapiro, L.W. *Introduction to Abstract Algebra*. New York: McGraw Hill, 1975.

Computer Applications and Discrete Structures

Birkhoff, G., and T.C. Bartee. *Modern Applied Algebra*. New York: McGraw-Hill, 1970.

Brualdi, R. *Introductory Combinatorics*. New York: North Holland, 1979.

Hamming, R.W. *Coding and Information Theory*. Englewood Cliffs, N.J.: Prentice Hall, 1980.

Laufer, H.B. *Discrete Mathematics and Applied Modern Algebra*. Boston: Prindle, Weber & Schmidt, 1984.

Prather, R.E. *Discrete Mathematical Structures for Computer Science*. Boston: Houghton Mifflin, 1976.

Ryser, H. *Combinatorial Mathematics*. Washington D.C.: Math. Assn. of Amer., 1963.

Stone, H.S. *Discrete Mathematical Structures and Their Applications*. Chicago: SRA, 1973.

Field Extensions and Galois Theory

Goldstein, L.J. *Abstract Algebra*. Englewood Cliffs, N.J.: Prentice Hall, 1973.

Maxfield, J.E., and M.W. Maxfield. *Abstract Algebra and Solution by Radicals*. Philadelphia: Saunders, 1971.

Group Theory

Fuchs, L. *Abelian Groups*. Oxford: Pergamon Press, 1960.

Gorenstein, D. *Finite Simple Groups, and Introduction to Their Classification*. New York: Plenum Publishing, 1982.

Rose, J.S. *A Course in Group Theory*. Cambridge: Cambridge University Press, 1978.

Rotman, J.J. *The Theory of Groups*. 2nd ed. Boston: Allyn and Bacon, 1973.

Schenkman, E. *Group Theory*. Englewood Cliffs, N.J.: Prentice Hall, 1964.

History

Boyer, C.B. *A History of Mathematics*. New York: Wiley, 1968.

Kline, M. *Mathematical Thought from Ancient to Modern Times*. New York: Oxford, 1972.

Struik, D.J. *A Concise History of Mathematics*. New York: Dover, 1967.

Lattice Theory

Birkhoff, G. *Lattice Theory*. Providence, R.I.: Amer. Math. Society Colloquium Publications, 1967.

Donnellan, T. *Lattice Theory*. Oxford: Pergamon Press, 1968.

Rutherford, D.E. *Introduction to Lattice Theory*. New York: Hafner, 1965.

Linear Algebra

Anton, H. *Elementary Linear Algebra*. 4th ed. New York: Wiley, 1984.

Grossman, S.I. *Elementary Linear Algebra*. 2nd ed. Belmont, Ca.: Wadsworth, 1984.

Hoffman, K., and R. Kunze. *Linear Algebra*. 2nd ed. Englewood Cliffs, N.J.: Prentice Hall, 1971.

Williams, G. *Linear Algebra with Applications*. Boston: Allyn and Bacon, 1984.

Ring Theory

Burton, D.N. *A First Course in Rings and Ideals*. Reading, Mass.: Addison-Wesley, 1970.

Herstein, I.N. *Noncommutative Rings*. Buffalo, N.Y.: Math. Assn. of Amer., 1968.

McCoy, N.H. *Rings and Ideals*. Washington, D.C.: Math. Assn. of Amer., 1968.

McCoy, N.H. *The Theory of Rings*. New York: Chelsea, 1972.

Zariski, O., and P. Samuel. *Commutative Algebra I*. New York: Springer-Verlag, 1975.

Theory of Equations

Dickson, L.E. *Elementary Theory of Equations*. New York: Wiley, 1914.

Uspensky, J.V. *Theory of Equations*. New York, 1948.

Miscellaneous Articles and Publications

Burn, R.P. "Cayley tables and associativity." *Math. Gazette* Vol. 62 (1978): 278–81.

Dickson, L.E. "Definition of a group and a field by independent postulates." *Trans. of the Amer. Math. Soc.* Vol. 6 (1905): 198–204.

Feit, W., and J.G. Thompson, "Solvability of groups of odd order." *Pacific J. of Math.* Vol. 13, No. 3 (1963).

Gallian, J.A. "Another proof that A_5 is simple." *Amer. Math. Monthly* Vol. 91 (1984): 134–35.

Gallian, J.A. "Group theory and the design of a letter facing machine." *Amer. Math. Monthly* Vol. 84 (1977): 285–87.

Gallian, J.A. "The search for finite simple groups." *Mathematics Magazine* Vol. 49 (1976): 163–80.

Gardner, M. *Mathematical Games and Diversions*. New York: Simon and Schuster, 1959.

Gorenstein, D. "The Enormous Theorem." *Sci. Amer.* Vol. 6 (1985): 104–15.

Hamming, R. W. "Error detecting and error correcting codes." *Bell System Tech. J.* Vol. 29 (1950): 147–60.

Hofstadter, D.R. "Metamagical Themas." *Sci. Amer.* (March 1981): 20–39.

Huntington, E.V. "Note on the definition of abstract groups and fields by sets of independent postulates." *Trans. of the Amer. Math. Soc.* Vol. 6 (1905): 181–97.

Johnson, W., and M. Silver. "A model for permutations." *Amer. Math. Monthly* Vol. 81 (1974): 503–06.

Langdon, G.G., R.J. Ord-Smith, and H.F. Trotter. "Algorithms for generating permutations." *Communications of the ACM* Vol. 5 (1962): 434–35; Vol. 10 (1967): 298–99, 452; Vol. 11 (1968): 117.

Liebeck, H. "Even and odd permutations." *Amer. Math. Monthly* Vol. 85 (1978): 90–94.

Marsden, J.E. *Basic Complex Analysis*. San Francisco: W.H. Freeman and Company, 1973.

Pless, V. "Error-correcting codes: Practical origins and mathematical implications." *Amer. Math. Monthly* Vol. 85 (1978): 90–94.

Spitznagel, E.L., Jr. "Note on the alternating group." *Amer. Math. Monthly* Vol. 75 (1968): 68–69.

Winograd, S. "On the time required to perform addition." *J. of the Assn. for Computer Machinery* Vol. 12, No. 2 (1965): 277–85.

Index